Global Analysis

PAPERS IN HONOR OF K. KODAIRA

K. KODAIRA

Global Analysis

Papers in Honor of K. Kodaira

Edited by
D. C. Spencer and S. Iyanaga

UNIVERSITY OF TOKYO PRESS
PRINCETON UNIVERSITY PRESS
1969

Copublished by
UNIVERSITY of TOKYO PRESS
and
PRINCETON UNIVERSITY PRESS
Library of Congress Catalogue: 70-106393
SBN 691-08077-1

Foreword

Kunihiko Kodaira, to whom this volume is dedicated, lived in the United States from 1949, when he was thirty four years old, until 1967. Thus he spent there eighteen years of his scholarly life.

During the summer of 1967, an informal meeting of some twenty mathematicians took place at Stanford University. At this meeting it was agreed to bring out a collection of papers in honor of Kodaira. This volume is the outcome of that agreement. Most of the authors who have contributed papers to this volume were present at Stanford during at least part of the summer, 1967.

Kodaira was invited in 1949 by the late professor H. Weyl to the Institute for Advanced Study at Princeton, New Jersey. During his eighteen years in the United States, he was a member of the Institute for Advanced Study, professor at Princeton University until 1963, from 1963–65 professor at the Johns Hopkins University, and from 1965–67 professor at Stanford University; he was also a visiting professor at the Johns Hopkins University (1950–51) and at Harvard University (1962–63). He had the good fortune of being always surrounded by distinguished colleagues; he had such teachers as Professor H. Weyl, and such friends as Professor D. C. Spencer. His students during his stay in the United States include W. L. Baily (professor at the University of Chicago), A. Kas (assistant professor at the University of California, Berkeley), J. Wavrik (instructor at Columbia University), and James Morrow (assistant professor at the University of California, Berkeley). However, his mathematical influence had a much wider range and helped lay the foundation for modern complex analysis.

During his years in the United States, Kodaira made remarkable progress in his mathematical research. For his achievement in the theory of harmonic integrals and its applications to kählerian and, in particular, to algebraic varieties, he was co-recipient, with J-P. Serre, of the Fields Medal at the International Congress of Mathematicians, Amsterdam, 1954. Professor Weyl, as chairman of the Fields Medal Committee, handed Kodaira the medal and his report on Kodaira's work appears in the Proceedings of that Congress.

Kodaira's native country was also keen to recognize his scientific achievements. He received the Japan Academy Award in 1957. In the same year he was also awarded the Japan Culture Medal (the second mathematician to receive this honor, the first was Takagi in 1940), which is esteemed as the highest prize for cultural achievement in this country. Since 1965, he has been a member of the Japan Academy. In 1967, he returned to Japan and was appointed professor at the University of Tokyo, the university from which he graduated and in which he spent the early years of his academic life.

I had the good fortune of teaching him when he entered the Department of Mathematics of the Faculty of Science of the University of Tokyo in 1935, and of watching with pleasure how his study and research developed from that time. I am grateful to Kodaira's American colleagues whose friendship and collaboration contributed to his success. I know that Kodaira is also grateful to his many American friends, in particular to the authors who have contributed to this volume. It is our sincere hope that he will further develop in his native land the lines of research which he began with these people.

Our thanks are due to Mrs. Fanny H. Rosenblum, Technical Editor, Annals of Mathematics, who kindly helped us in editing all the manuscripts, and to Mr. S. Minowa, Director of the University of Tokyo Press, who assumed the responsibility for publishing this volume, and finally to Princeton University for providing funds to help defray costs in preparing the manuscripts for publication.

S. IYANAGA

TABLE OF CONTENTS

The second Lefschetz theorem on hyperplane sections*

By Aldo Andreotti and Theodore Frankel

0. Introduction

The Lefschetz theorems on hyperplane sections describe how the topology of a projective algebraic manifold X is related to the topology of a (generic) hyperplane section X_0. We have (letting n be the complex dimension of X)

$$H_i(X_0; \mathbf{Z}) \longrightarrow H_i(X; \mathbf{Z}) \qquad \text{is} \begin{cases} \text{bijective if} & i < n-1 \\ \text{surjective if} & i = n-1 \,. \end{cases}$$

We shall call this the first Lefschetz theorem. The second Lefschetz theorem describes in detail

$$\ker H_{n-1}(X_0; \mathbf{Z}) \longrightarrow H_{n-1}(X; \mathbf{Z})$$

in terms of a pencil of hyperplanes having X_0 as generic member.

In an unpublished lecture given at Princeton in 1957, René Thom described the geometric situation of the Lefschetz theorems in terms of Morse's theory of critical points as applied to the function associated with the pencil.

In [1] we gave a proof of the first Lefschetz theorem which did not involve the use of a pencil of hyperplanes, but which was inspired by Thom's remarks.

The second Lefschetz theorem explicitly involves a pencil. Here we use Thom's picture and methods. If there is any new contribution it can only be in overcoming the technical difficulty arising from the fact that the meromorphic function associated with a pencil of hyperplanes is not well defined on the whole of X. This is done by passing to a new space $\tilde{X}$ obtained from X by "blowing up" the axis of the pencil (see § 1), a very natural procedure in this situation. The bulk of the present paper is concerned with relating the results obtained for $\tilde{X}$ back to the original space X. The resulting picture is, we believe, a rigorous and intuitive

* Work supported in part by AF-EOAR 67-38 (Andreotti) and NSF-GP-5258 (Frankel).

description of some aspects of the topology of algebraic varieties. In [4] Alan Howard has shown how such methods can be successfully applied to the topology of affine algebraic varieties. Finally the Lefschetz theorems (and especially the "vanishing cycles," i.e., $\ker H_{n-1}(X_0; \mathbf{Z}) \to H_{n-1}(X; Z)$) seem to be of interest in questions of the residue calculus as applied in modern physics.

1. Blowing up the manifold X

Let $X = X^n \subset P = P^N(\mathrm{C})$ be an n dimensional non-singular algebraic subvariety of projective N-space, and let $[Z_0, Z_1, \cdots, Z_N]$ be homogeneous coordinates for P.

First assumption. The hyperplanes P_0: $Z_0 = 0$ and P_∞: $Z_1 = 0$ and the $(N-2)$ linear subspace $P_0 \cap P_\infty$: $Z_0 = Z_1 = 0$ are nowhere tangent to X.

The quotient Z_0/Z_1 defines a holomorphic map

$$\varphi: P - (P_0 \cap P_\infty) \longrightarrow P^1$$

which can be written in terms of homogeneous coordinates on the projective line P^1 as

$$\varphi[Z_0, Z_1, \cdots, Z_N] = [Z_0, Z_1] .$$

By restricting φ to X we get a holomorphic map

$$X - X \cap (P_0 \cap P_\infty) \longrightarrow P^1 .$$

We shall use the notation

$$X_0 = X \cap P_0$$
$$X_\infty = X \cap P_\infty$$
$$Y = X_0 \cap X_\infty$$

in terms of which the map becomes $X - Y \to P^1$.

Now by means of a monoidal transformation, we "blow up" P along the subvariety $P_0 \cap P_\infty$; i.e., we consider the submanifold

$$\tilde{P} = \tilde{P}^N \subset P^N \times P^1$$

defined by pairs $([Z_0, \cdots, Z_N], [t_0, t_1])$ subject to the single requirement

$$Z_0 t_1 = Z_1 t_0 .$$

We have the projection $\pi^N: P^N \times P^1 \to P^N$ which projects the submanifold $\tilde{P}^N$ holomorphically down onto P^N. Note that if $(Z_0, Z_1) \neq (0, 0)$, then for the restriction $\tilde{\pi} = \pi^N \mid \tilde{P}^N$ we have

$\tilde{\pi}^{-1}[Z_0, Z_1, \cdots, Z_N]$ is the unique point $([Z_0, Z_1, \cdots, Z_N], [Z_0, Z_1])$, and one sees that

$$\tilde{\pi}: \tilde{P}^N - \tilde{\pi}^{-1}(P_0 \cap P_\infty) \longrightarrow P^N - P_0 \cap P_\infty$$

is biholomorphic. On the other hand

$$\tilde{\pi}^{-1}[0, 0, Z_2, \cdots, Z_N] = \{([0, 0, Z_2, \cdots, Z_N], [t_0, t_1])\}$$

with no restriction on $[t_0, t_1]$ at all, i.e., $\tilde{\pi}^{-1}(P_0 \cap P_\infty)$ is biholomorphically $(P_0 \cap P_\infty) \times P^1 \subset P^N \times P^1$.

Now let $\pi^1: P^N \times P^1 \rightarrow P^1$ be projection onto the second factor. For the restriction $\pi^1 \mid \tilde{P}^N$ we see

$$\pi^1 \mid \tilde{P}^N([Z_0, \cdots, Z_N], [t_0, t_1]) = [t_0, t_1] = [Z_0, Z_1]$$

and thus $\pi^1 \mid \tilde{P}: \tilde{P} \rightarrow P^1$ *is a globally defined holomorphic map that extends the lifting of the map* $\varphi: P - (P_0 \cap P_\infty) \rightarrow P^1$ *to all of* $\tilde{P}$.

We note further that if $[t_0, t_1] \in P^1$, then $(\pi^1 \mid \tilde{P})^{-1}[t_0, t_1]$ projects down *via* $\tilde{\pi}: \tilde{P}^N \rightarrow P^N$ biholomorphically onto the hyperplane

$$t_0 Z_1 = t_1 Z_0$$

of P^N; thus the points of P^1 are in 1 to 1 correspondence with the hyperplanes of the "pencil" of hyperplanes in P^N having $P_0 \cap P_\infty$ as "axis."

By our *first assumption*, $X \subset P^N$ meets $P_0 \cap P_\infty$ transversally. Also $\tilde{\pi}: \tilde{\pi}^{-1}(P_0 \cap P_\infty) \rightarrow P_0 \cap P_\infty$ is the projection of $(P_0 \cap P_\infty) \times P^1$ onto its first factor. This implies that when we consider $\tilde{\pi}: \tilde{P}^N \rightarrow P^N$, we see that $\tilde{\pi}$ is "transverse regular" on the submanifold $X \subset P^N$ and hence by a standard theorem

$$\tilde{X} \equiv \tilde{\pi}^{-1}(X)$$

is a holomorphic submanifold of $\tilde{P}^N$ of complex dimension n. We note again that $\pi \equiv \tilde{\pi} \mid \tilde{X}$ is a holomorphic map $\tilde{X} \rightarrow X$ that maps $\pi^{-1}(X - Y)$ biholomorphically onto $X - Y$ and

$$\tilde{Y} \equiv \pi^{-1}(Y)$$

is biholomorphically $Y \times P^1$, i.e., $\tilde{Y}$ is that part of the product $(P_0 \cap P_\infty) \times P^1$ "over" the submanifold $Y \subset P_0 \cap P_\infty$.

We see that in the process of "blowing up" P^N along $P_0 \cap P_\infty$ we have "blown up" X^n along $Y^{n-2} = (P_0 \cap P_\infty) \cap X$ to obtain a new manifold $\tilde{X}^n$ with $\tilde{Y} = Y \times P^1$ "over" Y.

Finally we define $f = \pi^1 \mid \tilde{X}$. Thus $f: \tilde{X} \rightarrow P^1$ with

$$f([Z_0, \cdots, Z_N], [t_0, t_1]) = [t_0, t_1] ,$$

where again $Z_0 t_1 = Z_1 t_0$. Note that for $t = [t_0, t_1] \in P^1$, $f^{-1}(t)$ projects down *via* π onto the hyperplane section $t_0 Z_1 = t_1 Z_0$ of X.

Notation. If a roman capital letter, say A, is used for an analytic submanifold of X, then the same script capital $\mathcal{A}$ will be used for a subset of $\tilde{X}$ that projects *via* π biholomorphically onto A. For example

$$f^{-1}[0, 1] = \mathcal{X}_0$$

since π maps $\mathcal{X}_0$ biholomorphically onto X_0. Note that

$$\mathcal{X}_0 \neq \pi^{-1}(X_0) = \mathcal{X}_0 \cup \pi^{-1}(Y) .$$

2. Critical points of the holomorphic map $f\colon \tilde{X} \to P^1$

Recall that a point $\tilde{x} \in \tilde{X}$ is a critical *point* for f if and only if the induced differential map on tangent spaces

$$f_* : \tilde{X}_{\tilde{x}} \longrightarrow P^1_{f(\tilde{x})}$$

is not surjective. A point $t \in P^1$ is a critical *value* if $f^{-1}(t)$ contains at least one critical point. A point or value is regular if it is not critical.

Consider first a point $\tilde{x} \in \tilde{X} - \tilde{Y}$, where again $\tilde{Y} = \pi^{-1}(Y) = Y \times P^1$. Now $\tilde{X} - \tilde{Y}$ is biholomorphically $X - Y$ and f on $\tilde{X} - \tilde{Y}$ is merely the lift of the "function" $\varphi = Z_0/Z_1$ restricted to $X - Y$. Then $\tilde{x}$ is a critical point for f if and only if $x = \pi(\tilde{x})$ is a critical point for φ on $X - Y$, and x, in turn, is critical for φ on $X - Y$ if and only if the level set

$$\{p \in P^N - (P_0 \cap P_\infty)\colon \varphi(p) = \varphi(x)\}$$

does not meet X transversally at x. Now if we put $\varphi(x) = [t_0, t_1]$, the level set in question is merely the hyperplane $t_0 Z_1 = t_1 Z_0$, and thus x is critical for φ if and only if this hyperplane is tangent to X at x. Summarizing, the *critical points of f on $\tilde{X} - \tilde{Y}$ project down via π onto the points of $X - Y$ where the hyperplanes of the given pencil are tangent to $X - Y$.*

We next need to consider critical points of f that lie on $\tilde{Y} = \pi^{-1}(Y)$. Let $\tilde{y} \in \tilde{Y}$ and consider $\tilde{y}$ as a point on $P^N \times P^1$. f is $\pi^1 \mid \tilde{X}$ where $\pi^1\colon P^N \times P^1 \to P^1$ is projection onto the second factor. Now $\tilde{y} \in \tilde{Y}$ means $\tilde{y}$ is of the form

$$\tilde{y} = ([0, 0, Z_2, \cdots, Z_N], [t_0, t_1])$$

with $[0, 0, Z_2, \cdots, Z_N] \in X$. Note that for fixed $[0, 0, Z_2, \cdots, Z_N] \in X$

the entire projective line $[0, 0, Z_2, \cdots, Z_N] \times P^1$ lies in $\tilde{X}$ and f restricted to this line is essentially the identity map. In particular f_* at $\tilde{y} \in \tilde{Y}$ is surjective and we conclude that *there are no critical points of f on $\tilde{Y}$.* We have thus proved the following

LEMMA 1. *The critical points of $f: \tilde{X} \to P^1$ are all in $\tilde{X} - \tilde{Y}$ and project down via $\pi: \tilde{X} \to X$ onto the points of X where the hyperplanes of the pencil whose base is $P_0 \cap P_\infty$ are tangent to X.*

In particular, by our first assumption, $\mathfrak{X}_0 = f^{-1}[0, 1]$ and $\mathfrak{X}_\infty = f^{-1}[1, 0]$ consist entirely of regular points for f.

Second assumption. We assume that there are only a finite number (say μ) of points of X where the hyperplanes of the pencil with base $P_0 \cap P_\infty$ are tangent to X.

These μ tangency points, when lifted *via* $\pi: \tilde{X} \to X$, are precisely the critical points of $f: \tilde{X} \to P^1$. The nature of a critical point is described by means of the hessian matrix of second partial derivatives. Let $\tilde{x}$ be a critical point for f. Then $x = \pi(\tilde{x})$ is a contact point of X with a hyperplane $t_0 Z_1 = t_1 Z_0$ of the pencil. The "value" of φ at x is $Z_0/Z_1 = t_0/t_1 \equiv \alpha$. From Lemma 1 we know $\alpha \neq 0$ and $\alpha \neq \infty$, i.e., we know $x \in X - (X_0 \cup X_\infty)$. Near α we use non-homogeneous coordinates in P^N defined by

$$\zeta_0 = Z_0/Z_1,\ \zeta_1 = Z_2/Z_1,\ \cdots,\ \zeta_{N-1} = Z_N/Z_1$$

(this being valid since $Z_1 \neq 0$). Thus in the affine space $\mathbf{C}^N$, with coordinates $\zeta_0, \zeta_1, \cdots, \zeta_{N-1}$, we have that the affine algebraic manifold X is tangent to the hyperplane $\zeta_0 = \alpha$. By a linear change of the coordinates $\zeta_1, \cdots, \zeta_{N-1}$ we may assume that $\zeta_1, \cdots, \zeta_n$ can be used as local coordinates for X near x, and $\zeta_1(x) = \cdots = \zeta_n(x) = 0$. Thus X is locally described near x by

$$\begin{cases} \zeta_0 = \zeta_0(\zeta_1, \cdots, \zeta_n) \\ \zeta_{n+1} = \zeta_{n+1}(\zeta_1, \cdots, \zeta_n) \\ \vdots \\ \zeta_{N-1} = \zeta_{N-1}(\zeta_1, \cdots, \zeta_n) \end{cases}$$

and on this submanifold of $\mathbf{C}^N$ we must consider the function

$$\varphi(\zeta_1, \cdots, \zeta_n) = \zeta_0(\zeta_1, \cdots, \zeta_n)\ .$$

Since X is tangent to the hyperplane $\zeta_0 = \alpha$ at x, we have

$$\varphi(\zeta_1, \cdots, \zeta_n) = \alpha + \textstyle\sum_{i,j=1}^{n} a_{ij}\zeta_i\zeta_j + \text{higher order terms}\ .$$

Again by a linear change of coordinates $\zeta_1, \cdots, \zeta_n$ we may assume that φ is of the form

$$\varphi(\zeta_1, \cdots, \zeta_n) = \alpha + \sum_{i=1}^{r} \zeta_i^2 + \text{higher order terms},$$

where $r \leqq n$. We shall call the critical point x *non-degenerate* if $r = n$. This is equivalent to having

$$\det (a_{ij}) \neq 0 .$$

Third assumption. We assume that each critical point of f is non-degenerate.

3. Morse structure of $\tilde{X}$

We now define a real valued function on $\tilde{X} - \mathfrak{X}_\infty$ by

$$F(\tilde{x}) = |f(\tilde{x})|^2$$

(and we shall occasionally consider F as an extended real valued function on all of $\tilde{X}$ by assigning the value $+\infty$ to points of $\mathfrak{X}_\infty$).

To discuss the critical points of F we need only notice that $F = \nu \circ f$, where $\nu\colon P^1 - \infty \rightarrow \mathbf{R}$ is defined by $\nu(t) = |t|^2$. The only critical point for ν is the point 0 where ν attains its absolute minimum. Since $f\colon \tilde{X} \rightarrow P^1$ has as its only critical points the μ points of "tangency," we conclude that *the critical points of F consist of the absolute minimum manifold $\mathfrak{X}_0 = F^{-1}(0)$ together with the μ tangency points from the pencil.*

The nature of a critical point is now determined by the real hessian matrix of second partial derivatives. As we have seen, near a point of tangency f (or φ) has the form

$$f(\zeta_1, \cdots, \zeta_n) = \alpha \cdot (1 + \sum_{i=1}^{n} \zeta_i^2 + \text{higher order terms})$$

where we have made the substitution $\zeta_i \rightarrow \zeta_i \alpha^{1/2}$, $i = 1, \cdots, n$, which is admissible since $\alpha \neq 0$. If we introduce real coordinates by $\zeta_k = u_k + iv_k$, then we have

$$\begin{aligned} F(u_1, v_1, \cdots, u_n, v_n) &= |f(\zeta_1, \cdots, \zeta_n)|^2 \\ &= |\alpha|^2 \big(1 + 2\sum_{k=1}^{n} (u_k^2 - v_k^2) \\ &\qquad + \text{higher order terms}\big) . \end{aligned}$$

Recall that the Morse *index* of a critical point is the number of negative squares in the diagonal version of the quadratic part of the function. We see that in our case, *at each tangency point the index of F is precisely n* (i.e., half the real dimension of $\tilde{X}$). This observation by R. Thom is the crucial point in this Morse theoretic approach to the Lefschetz theorems.

Let $D_\varepsilon = \{t \in P^1: |t^2| \leqq \varepsilon\}$. Since $0 \in P^1$ is a regular value of $f: \tilde{X} \to P^1$, we know that if $\varepsilon > 0$ is sufficiently small

$$N_0 = \{\tilde{x} \in \tilde{X}: F(\tilde{x}) \leqq \varepsilon\} = f^{-1}(D_\varepsilon)$$

is a tubular 2 disc neighborhood of $\mathfrak{X}_0$ in $\tilde{X}$ that is topologically $\mathfrak{X}_0 \times D_\varepsilon$ and $\mathfrak{X}_0$ is a deformation retract of N_0. The fundamental theorem of Morse theory (as given, for example, in Milnor's book [5]) then gives the following (since each critical point of F on $\tilde{X} - (\mathfrak{X}_0 \cup \mathfrak{X}_\infty)$ is non-degenerate and of index n).

THEOREM 1. *$\tilde{X} - \mathfrak{X}_\infty$ is of the same homotopy type as $\mathfrak{X}_0$ with cells of dimension n attached.*

$$\tilde{X} - \mathfrak{X}_\infty \sim \mathfrak{X}_0 \cup e_n^1 \cup \cdots \cup e_n^\mu .$$

There is one n-cell e_n^k emanating from each tangency point and each is attached by means of its boundary $\dot{e}_n^k$ along the $(n-1)$ skeleton of $\mathfrak{X}_0$ in some triangulation of $\mathfrak{X}_0$.

Definition. The vanishing cycle group on $\mathfrak{X}_0$ is the subgroup of $H_{n-1}(\mathfrak{X}_0; \mathbf{Z})$ generated by the boundaries $\dot{e}_n^1, \cdots, \dot{e}_n^\mu$.

COROLLARY. $\ker H_{n-1}(\mathfrak{X}_0; \mathbf{Z}) \to H_{n-1}(\tilde{X} - \mathfrak{X}_\infty; \mathbf{Z})$ *consists precisely of the vanishing cycles on $\mathfrak{X}_0$.*

Definition. The vanishing cycle group on X_0 is the subgroup of $H_{n-1}(X_0; \mathbf{Z})$ generated by the cycles $(\pi \mid \mathfrak{X}_0)_* \dot{e}_n^1, \cdots, (\pi \mid \mathfrak{X}_0)_* \dot{e}_n^\mu$.

The vanishing cycle group on X_0 is isomorphic to that on $\mathfrak{X}_0$ since $\pi \mid \mathfrak{X}_0$ is a diffeomorphism. The second Lefschetz theorem asserts that the vanishing cycles on X_0 form precisely

$$\ker H_{n-1}(X_0; \mathbf{Z}) \longrightarrow H_{n-1}(X; \mathbf{Z}) .$$

The proof of this involves a digression on the relationship between the topologies of X and $\tilde{X}$.

4. Topology of $\pi: \tilde{X} \to X$

Recall that $\tilde{Y} = \pi^{-1}(Y)$ is topologically $Y \times P^1$, the product of Y with an oriented 2-sphere. We then have the natural maps

$$H_k(Y) \xrightarrow{\beta} H_{k+2}(\tilde{Y}) \xrightarrow{i_*} H_{k+2}(\tilde{X})$$

where $i: \tilde{Y} \to \tilde{X}$ is inclusion and β results from taking the "cross product" of a k-cycle of Y with the 2-cycle $P^1 = S$, i.e., $\beta(z_k) = z_k \times P^1$. Define $\sigma: H_k(Y) \to H_{k+2}(\tilde{X})$ by

$$\sigma = (-1)^k i_* \circ \beta .$$

THEOREM 2. *The sequence (with any coefficient group)*

$$0 \longrightarrow H_k(Y) \xrightarrow{\sigma} H_{k+2}(\tilde{X}) \xrightarrow{\pi_*} H_{k+2}(X) \longrightarrow 0$$

is exact for all k, and splits.

PROOF. π is clearly a map of compact oriented $2n$ dimensional manifolds of degree 1. By Hopf's theorem

$$\pi_*: H_j(\tilde{X}) \longrightarrow H_j(X)$$

admits a right inverse $h: H_j(X) \to H_j(\tilde{X})$, i.e., $\pi_* \circ h$ is the identity on $H_j(X)$ for all j (in fact h is given explicitly by $\lambda_{\tilde{X}} \circ \pi^* \circ \lambda_X^{-1}$ where λ_X and $\lambda_{\tilde{X}}$ are the Poincaré duality isomorphisms on X and $\tilde{X}$ respectively). In particular π_* is surjective.

Given $\pi: \tilde{X} \to X$ we consider the *mapping cylinder* $J(\tilde{X}, X)$ of π, i.e., the topological space obtained from the disjoint union $\tilde{X} \times [0, 1] \cup X$ by identifying $(\tilde{x}, 1)$ with $\pi(\tilde{x})$. In $J(\tilde{X}, X)$ we have the subspaces $J(\tilde{Y}, Y)$ and $A = \{J(\tilde{X}, X) - J(\tilde{Y}, Y)\} \cup \tilde{Y}$. We then have the exact commutative diagram

$$\begin{array}{ccccccc} \cdots \longrightarrow H_{k+3}(J(\tilde{X}, X), A) & \xrightarrow{\partial} & H_{k+2}(A) & \longrightarrow & H_{k+2}(J(\tilde{X}, X)) & \longrightarrow \cdots \\ & & \approx \downarrow r_* & & \approx \downarrow u_* & \\ & & H_{k+2}(\tilde{X}) & \xrightarrow{\pi_*} & H_{k+2}(X) & \end{array}$$

where $r: A \to \tilde{X}$ is retraction "upward" along the line segments and $u: J(\tilde{X}, X) \to X$ is retraction "downward" along the line segments. Both r and u induce homology isomorphisms. Since π_* is surjective, we get

$$0 \longrightarrow H_{k+3}(J(\tilde{X}, X), A) \xrightarrow{r_* \circ \partial} H_{k+2}(\tilde{X}) \xrightarrow{\pi_*} H_{k+2}(X) \longrightarrow 0 .$$

We now excise the subset $J(\tilde{X}, X) - J(\tilde{Y}, Y)$. This excision induces an isomorphism $H_{k+3}(J(\tilde{X}, X), A) \approx H_{k+3}(J(\tilde{Y}, Y), \tilde{Y})$ because $J(\tilde{Y}, Y)$ is a neighorhood retract in $J(\tilde{X}, X)$. Our sequence becomes

$$0 \longrightarrow H_{k+3}(J(\tilde{Y}, Y), \tilde{Y}) \xrightarrow{r_* \circ \partial} H_{k+2}(\tilde{X}) \xrightarrow{\pi_*} H_{k+2}(X) \longrightarrow 0 .$$

Note that in our situation $\tilde{Y} = Y \times P^1$ is a (product) 2-sphere bundle over Y and thus $J(\tilde{Y}, Y)$ is the (product) closed 3-disc bundle over Y obtained by "filling in" each 2-sphere fiber with a 3-cell. The Thom isomorphism

$$T\colon H_k(Y) \xrightarrow{\approx} H_{k+3}(J(\tilde{Y}, Y), \tilde{Y})$$

is obtained as follows. One takes an oriented 3-cell fiber e_3 and to each k-cycle z_k on Y one forms the relative $(k + 3)$ cycle $z_k \times e_3$ on $J(\tilde{Y}, Y) \bmod \tilde{Y}$. Then from $\partial(z_k \times e_3) = (-1)^k z_k \times \dot{e}_3 = (-1)^k z_k \times S$ ($S = \dot{e}_3$ is a fixed 2-sphere fiber of $\tilde{Y} \to Y$) we see that $r_* \circ \partial \circ T = \sigma$ and thus our short exact sequence becomes

$$0 \longrightarrow H_k(Y) \xrightarrow{\sigma} H_{k+2}(\tilde{X}) \xrightarrow{\pi_*} H_{k+2}(X) \longrightarrow 0$$

as desired. The sequence splits since π_* has a right inverse h.

5. A remark on Euler characteristics

Let again $\mathfrak{X}_\infty = f^{-1}(\infty)$ be the subspace of $\tilde{X}$ that projects down biholomorphically *via* $\pi\colon \tilde{X} \to X$ onto the hyperplane section X_∞ of X. As in our discussion of $\mathfrak{X}_0$ in § 3, we easily see that there is a closed tubular neighborhood N_∞ of $\mathfrak{X}_\infty$ that realizes the normal 2-disc bundle to $\mathfrak{X}_\infty$ in $\tilde{X}$, and, in fact, $N_\infty = \mathfrak{X}_\infty \times D$, where D is a closed 2-disc. Also, from Theorem 1 we know that $\tilde{X} - N_\infty$ is of the same homotopy type as $\mathfrak{X}_0$ with μ cells of dimension n attached, providing N_∞ is small enough. Letting $\dot{N}_\infty$ be the boundary circle bundle, we have for Euler characteristics

$$\begin{aligned}\chi(\tilde{X}) &= \chi(\text{closure of } \tilde{X} - N_\infty) + \chi(N_\infty) - \chi(\dot{N}_\infty)\\ &= \chi(\mathfrak{X}_0 \cup e_n^1 \cup \cdots \cup e_n^\mu) + \chi(N_\infty)\end{aligned}$$

because the Euler characteristic of a circle bundle vanishes. Since N_∞ has $\mathfrak{X}_\infty$ as deformation retract, we have

$$\chi(\tilde{X}) = \chi(\mathfrak{X}_0) + (-1)^n\mu + \chi(\mathfrak{X}_\infty)\ .$$

Note, however, that all non-singular "hyperplane sections" are diffeomorphic. Indeed $f^{-1}\,(P^1 - \text{critical values } f)$ is a locally trivial fiber space over the regular values of f on P^1. Then $\chi(\mathfrak{X}_\infty) = \chi(\mathfrak{X}_0) = \chi(X_0)$, and so $\chi(\tilde{X}) = 2\chi(X_0) + (-1)^n\mu$. But from the short exact sequence of Theorem 2 we have $\chi(\tilde{X}) = \chi(X) + \chi(Y)$. We then have the following relation between the number of tangency points μ and the Euler characteristics of X, the hyperplane section X_0, and the axis Y of the pencil.

COROLLARY. $\chi(X) = 2\chi(X_0) - \chi(Y) + (-1)^n\mu.$

The affine version of this corollary can be found in the paper of Fàry [2]. For projective algebraic surfaces, the formula we have proved is the so called Zeuthen-Segre formula.

6. The second Lefschetz theorem

Let $X^n \subset P^N$ be a connected projective algebraic manifold, and let us assume that X does not lie in any projective subspace of P^N. Let us denote by P^{N*} the dual space of P^N, i.e., P^{N*} is the space of all hyperplanes of P^N. A pencil Λ of hyperplane sections on X^n is thus determined by the choice of a projective line in P^{N*} and therefore by a point on the Grassmann manifold $\mathrm{Gr}(P^{N*})$ of projective lines in P^{N*}. We denote by B_Λ the set $\{x \in P^N : x \in$ all hyperplanes of $\Lambda\}$ and by f_Λ the map of $X - Y_\Lambda \to P^1$ determined (up to an automorphism of P^1) by the pencil Λ.

LEMMA 2. *The set of pencils* Λ *such that*

(1) B_Λ *is nowhere tangent to* X

(2) f_Λ *has only a finite number of critical points, all non-degenerate,*

fills up a non-emply Zariski open subset of $\mathrm{Gr}(P^{N*})$.

Once a pencil of this type is chosen, we can select on Λ two distinct elements P_0 and P_∞ which are not tangent to X and then assumptions 1, 2, and 3 (specified in the previous paragraphs) are all satisfied.

For the proof of this lemma we refer the reader to the appendix of this paper.

We now assume that such a pencil Λ has been chosen. As we have seen in § 3, we have the following:

(I) $\tilde{X} - \mathfrak{X}_\infty \sim \mathfrak{X}_0 \cup e_n^1 \cup \cdots \cup e_n^\mu$.

(II) $\ker H_{n-1}(\mathfrak{X}_0; \mathbf{Z}) \to H_{n-1}(\tilde{X} - \mathfrak{X}_\infty; \mathbf{Z})$ is generated by the boundaries $\dot{e}_n^1, \cdots, \dot{e}_n^\mu$, and forms the vanishing cycle group for $\mathfrak{X}_0$.

(III) The vanishing cycle group of X_0 is defined to be the bijective image of that of $\mathfrak{X}_0$ under the biholomorphic map $(\pi \mid \mathfrak{X}_0)$: $\mathfrak{X}_0 \to X_0$.

THEOREM 3. (The second Lefschetz theorem). *The kernel of the homomorphism* $H_{n-1}(X_0; \mathbf{Z}) \to H_{n-1}(X: \mathbf{Z})$ *is exactly the vanishing cycle group on* X_0.

PROOF. Put

$$\mathfrak{Y} = \tilde{Y} \cap \mathfrak{X}_\infty = \tilde{Y} \cap f^{-1}(\infty) .$$

This is a cross-section of the (product) bundle $\tilde{Y} \xrightarrow{\pi} Y$. We now identify Y with this cross section $\mathcal{Y}$. Then in Theorem 2 the map $\sigma\colon H_k(\mathcal{Y}) \to H_{k+2}(\tilde{X})$ is given by $z_k \to (-1)^k z_k \times S$ followed by inclusion in $\tilde{X}$. We then consider the exact diagram (integer coefficients throughout)

$$\begin{array}{ccccccccc}
 & & H_{n-3}(\mathcal{X}_\infty) & \xrightarrow[\approx]{\tilde{T}} & H_{n-1}(\tilde{X}, \tilde{X} - \mathcal{X}_\infty) & & & & \\
 & & \uparrow{\scriptstyle l_*} & & \uparrow{\scriptstyle k_*} & & & & \\
0 & \longrightarrow & H_{n-3}(\mathcal{Y}) & \xrightarrow{\sigma} & H_{n-1}(\tilde{X}) & \xrightarrow{\pi_*} & H_{n-1}(X) & \longrightarrow & 0 \\
 & & & & \uparrow{\scriptstyle j_*} & & \uparrow{\scriptstyle i_*} & & \\
 & & & & H_{n-1}(\tilde{X} - \mathcal{X}_\infty) & \xleftarrow{\tau_*} & H_{n-1}(X_0) & & \\
 & & & & \uparrow{\scriptstyle \partial} & & & & \\
 & & H_{n-2}(\mathcal{X}_\infty) & \xrightarrow[\approx]{\tilde{T}} & H_n(\tilde{X}, \tilde{X} - \mathcal{X}_\infty) & & & & \\
 & & \uparrow{\scriptstyle l_*} & & \uparrow{\scriptstyle k_*} & & & & \\
 & & H_{n-2}(\mathcal{Y}) & \xrightarrow{\sigma} & H_n(\tilde{X}) \, . & & & &
\end{array}$$

The long vertical sequence is that of the pair $(\tilde{X}, \tilde{X} - \mathcal{X}_\infty)$. $\tau_* : H_{n-1}(X_0) \to H_{n-1}(\tilde{X} - \mathcal{X}_\infty)$ is induced by the homeomorphism $X_0 = \mathcal{X}_0$ followed by inclusion $\mathcal{X}_0 \subset \tilde{X} - \mathcal{X}_\infty$. The homomorphism $l_* : H_k(\mathcal{Y}) \to H_k(\mathcal{X}_\infty)$ is induced by inclusion $\mathcal{Y} = \tilde{Y} \cap \mathcal{X}_\infty \subset \mathcal{X}_\infty$. $\tilde{T}$ is defined as follows. By excision

$$H_k(\tilde{X}, \tilde{X} - \mathcal{X}_\infty) \approx H_k(N_\infty, \dot{N}_\infty) \, ,$$

where again N_∞ is a closed normal 2-disc bundle to $\mathcal{X}_\infty$ in $\tilde{X}$, and $\dot{N}_\infty$ is the boundary circle bundle. The Thom isomorphism states that

$$H_k(\mathcal{X}_\infty) \xrightarrow{T} H_{k+2}(N_\infty, \dot{N}_\infty)$$

is an isomorphism, and $\tilde{T}$ is the composition of T with the excision isomorphism together with the multiplicative factor $(-1)^k$.

We first claim that the diagram is commutative. The only possible question arises in a square involving $\tilde{T}$ and σ, which we write in full as

$$\begin{array}{ccccc}
H_k(\mathcal{X}_\infty) & \xrightarrow[\approx]{(-1)^k T} & H_{k+2}(N_\infty, \dot{N}_\infty) & \xrightarrow[\approx]{} & H_{k+2}(\tilde{X}, \tilde{X} - \mathcal{X}_\infty) \\
\uparrow{\scriptstyle l_*} & & & & \uparrow{\scriptstyle k_*} \\
H_k(\mathcal{Y}) & & \xrightarrow{\qquad\qquad\sigma\qquad\qquad} & & H_{k+2}(\tilde{X}) \, .
\end{array}$$

Let z_k be a k-cycle on $\mathcal{Y}$. Then $\sigma(z_k) = (-1)^k z_k \times S$. Geometrically $k_*\sigma(Z_k)$ is again $(-1)^k z_k \times S$ but considered as a relative cycle modulo $\tilde{X} - \mathcal{X}_\infty$. Again $l_*(z_k)$ is geometrically the cycle z_k, considered as a cycle on $\mathcal{X}_\infty$. By Lemma 1, we know that S meets each of the "hyperplane sections" $f^{-1}(t)$ transversally. Thus S meets N_∞ in a closed 2-disc e_2' whose boundary $\dot{e}_2'$ is merely $S \cap \dot{N}_\infty$. Thus e_2' is a generator of $H_2(N_\infty, \dot{N}_\infty)$ and $Tl_*(z_k)$ is geometrically the relative cycle $z_k \times e_2'$ mod $\dot{N}_\infty$. Note that the chain $z_k \times S - z_k \times e_2'$ lies entirely on $\tilde{X} - N_\infty$, and hence

$$k_*\sigma(z_k) - (-1)^k Tl_*(z_k) \equiv 0 \qquad \text{mod } \tilde{X} - \mathcal{X}_\infty$$

which proves commutativity in the square. To prove the second Lefschetz theorem we need to show that

$$\ker i_*\colon H_{n-1}(X_0) \longrightarrow H_{n-1}(X)$$

is the same as

$$\ker \tau_*\colon H_{n-1}(X_0) \longrightarrow H_{n-1}(\tilde{X} - \mathcal{X}_\infty)$$

for this last kernel is, by the corollary to Theorem 1, the vanishing cycle group on X_0. First note that $\ker \tau_* \subset \ker i_*$ follows immediately from the square

$$\begin{array}{ccc} H_{n-1}(\tilde{X}) & \xrightarrow{\pi_*} & H_{n-1}(X) \\ \uparrow{\scriptstyle j_*} & & \uparrow{\scriptstyle i_*} \\ H_{n-1}(\tilde{X} - \mathcal{X}_\infty) & \xleftarrow{\tau_*} & H_{n-1}(X_0) \end{array}$$

We need only show that $\ker i_* \subset \ker \tau_*$.

LEMMA 3. *$l_*\colon H_{n-3}(\mathcal{Y}) \to H_{n-3}(\mathcal{X}_\infty)$ is bijective, and $l_*\colon H_{n-2}(\mathcal{Y}) \to H_{n-2}(\mathcal{X}_\infty)$ is surjective.*

PROOF. $(\mathcal{X}_\infty, \mathcal{Y})$ is diffeomorphic to the pair (X_∞, Y) under π, and $Y = X_\infty \cap \{z_0 = 0\}$ is a hyperplane section of X_∞. Our lemma follows directly from the first Lefschetz theorem.

The proof of $\ker i_* \subset \ker \tau_*$ is a simple diagram chase. Let $i_* z = 0 \in H_{n-1}(X)$. Then $\pi_* j_* \tau_*(z) = 0$ and so $j_*\tau_*(z) = \sigma(y)$ for $y \in H_{n-3}(\mathcal{Y})$. But then $\tilde{T}l_*(y) = k_*\sigma(y) = k_* j_* \tau_*(z) = 0$ since $k_* \circ j_* = 0$. But $\tilde{T}$ is an isomorphism, and so is l_* by Lemma 3, hence $y = 0$. We conclude that $j_*\tau_*(z) = 0$. Then $\tau_* z = \partial x$ for $x \in H_n(\tilde{X}, \tilde{X} - \mathcal{X}_\infty)$. Again $\tilde{T}$ is an isomorphism and l_* is surjective, hence $x = \tilde{T}l_* w$ for $w \in H_{n-2}(\mathcal{Y})$. Thus $x = k_*\sigma(w)$ and therefore

$\tau_* z = \partial k_* \sigma(w) = 0$ since $\partial \circ k_* = 0$. Hence $z \in \ker \tau_*$, as desired.

Remark. The vanishing cycle group on $\mathfrak{X}_0$ has been described as $\ker H_{n-1}(\mathfrak{X}_0) \to H_{n-1}(\tilde{X} - \mathfrak{X}_\infty)$. From our diagram we note $k_* : H_n(\tilde{X}) \to H_n(\tilde{X}, \tilde{X} - \mathfrak{X}_\infty)$ is surjective and hence

$$\partial : H_n(\tilde{X}, \tilde{X} - \mathfrak{X}_\infty) \longrightarrow H_{n-1}(\tilde{X} - \mathfrak{X}_\infty)$$

is 0 and $j_* : H_{n-1}(\tilde{X} - \mathfrak{X}_\infty) \to H_{n-1}(\tilde{X})$ is then injective. We conclude that *the vanishing cycle group on* $\mathfrak{X}_0 = \ker H_{n-1}(\mathfrak{X}_0) \to H_{n-1}(\tilde{X})$.

7. Invariant cycles and cocycles

(a) In this section we shall define the invariant cocycles. The "invariance" property involved will be discussed in the section (c) below.

For cohomology, the first Lefschetz theorem asserts that

$$H^i(X; \mathbf{Z}) \to H^i(X_0; \mathbf{Z}) \qquad \text{is} \begin{cases} \text{bijective for } i < n-1 \\ \text{injective for } i = n-1\,, \end{cases}$$

and our purpose here is to describe the image

$$\operatorname{im} H^{n-1}(X; \mathbf{Z}) \longrightarrow H^{n-1}(X_0; \mathbf{Z})\,.$$

Following the dual procedure of the previous paragraphs, one defines the *invariant cocycles on* $\mathfrak{X}_0$ as the elements of

$$\operatorname{im} H^{n-1}(\tilde{X} - \mathfrak{X}_\infty; \mathbf{Z}) \longrightarrow H^{n-1}(\mathfrak{X}_0; \mathbf{Z})\,.$$

From Theorem 1 and the exact cohomology sequence

$$\cdots \longrightarrow H^{n-1}(\tilde{X} - \mathfrak{X}_\infty; \mathbf{Z}) \longrightarrow H^{n-1}(\mathfrak{X}_0; \mathbf{Z}) \xrightarrow{\delta^*} H^n(\tilde{X} - \mathfrak{X}_\infty, \mathfrak{X}_0; \mathbf{Z}) \longrightarrow \cdots$$

one verifies the following

COROLLARY. *The invariant cocycle group on* $\mathfrak{X}_0$ *is the subgroup of* $H^{n-1}(\mathfrak{X}_0; \mathbf{Z})$ *consisting of those cohomology classes which vanish on the vanishing cycles, i.e.,*

$$\{\alpha \in H^{n-1}(\mathfrak{X}_0; \mathbf{Z}) : \alpha(\dot{e}_n^k) = 0,\ k = 1, \cdots, \mu\}\,.$$

The identification $X_0 = \mathfrak{X}_0$ under π allows one to define the *invariant cocycles on* X_0 by this last characterization, i.e., they are *the cohomology classes on* X_0 *that vanish on the vanishing cycles of*

X_0. We shall draw no sharp distinction between cycles on X_0 and cycles on $\mathfrak{X}_0$.

The sequence of Theorem 2 is now replaced by the split exact sequence (any coefficient group)

$$0 \longrightarrow H^k(X) \xrightarrow{\pi^*} H^k(\tilde{X}) \longrightarrow H^{k-2}(Y) \longrightarrow 0 .$$

We now copy the diagram of the proof of the second Lefschetz theorem, replacing homology by cohomology and reversing arrows. The same type of argument used there now gives

THEOREM 4. *The invariant cocycle group*, i.e.,

$$\{\alpha \in H^{n-1}(X_0; \mathbf{Z}) \colon \alpha(\dot{e}_n^k) = 0, k = 1, 2, \cdots, \mu\}$$

coincides with im $\{H^{n-1}(X; \mathbf{Z}) \to H^{n-1}(X_0; \mathbf{Z})\}$ *which in turn is the same as*

$$\text{im}\,\{H^{n-1}(\tilde{X}; \mathbf{Z}) \longrightarrow H^{n-1}(\mathfrak{X}_0; \mathbf{Z})\} .$$

The first equality gives the required description of the image under the restriction homomorphism in terms of the vanishing cycles and may be considered as the cohomological form of the second Lefschetz theorem.

In [1] the authors have shown that the quotient group of $H^{n-1}(X_0; \mathbf{Z})$ modulo the image of $H^{n-1}(X; Z)$ in $H^{n-1}(X_0; Z)$ is torsion free, hence

COROLLARY. *The torsion* $(n-1)$ *cocycles of* $H^{n-1}(X_0; \mathbf{Z})$ *are invariant cocycles.*

(b) Consider the following diagram

$$\begin{array}{ccc} H_{2n-i}(X; \mathbf{Z}) & \xrightarrow{\mu} & H_{2n-i-2}(X_0; \mathbf{Z}) \\ \uparrow \lambda_X & & \uparrow \lambda_{X_0} \\ H^i(X, \mathbf{Z}) & \xrightarrow{i^*} & H^i(X_0; \mathbf{Z}) \end{array}$$

where λ_X, λ_{X_0} are the Poincaré duality isomorphisms on X and X_0 respectively, and where μ is the composition of the maps

$$H_{2n-i}(X; \mathbf{Z}) \xrightarrow{j^*} H_{2n-i}(X, X - X_0; \mathbf{Z}) \xrightarrow{T} H_{2n-i-2}(X_0; \mathbf{Z}) ,$$

where T is obtained by the excision map

$$H_k(X, X - X_0) \longrightarrow H_k(N, \dot{N})$$

(where N is a closed normal 2-disc bundle of X_0 in X) followed by the Thom isomorphism $H_k(N, \dot{N}) \to H_{k-2}(X_0)$.

One verifies that this diagram is commutative and that μ is the homomorphism which to every cycle $z \in H_{2n-i}(X; \mathbf{Z})$ associates the cycle on X_0 "intersection of z with X_0."

We now define for $i = n - 1$ the space of *invariant cycles* on X_0 as the space $\lambda_{X_0} i^* H^{n-1}(X; \mathbf{Z})$, i.e., the space of Poincaré duals on X_0 of the invariant cocycles. The previous diagram shows the following

THEOREM 5. *Each invariant cycle in $H_{n-1}(X_0; \mathbf{Z})$ is characterized as being the intersection with X_0 of an $(n+1)$ cycle of $H_{n+1}(X; Z_0)$.*

On $H_{n-1}(X_0; \mathbf{Q})$ one can consider the bilinear form given by the intersection number $I(,)$. From our characterization of invariant cocycles given in (a) we see that *with respect to the intersection form I the vector space $H_{n-1}(X_0; \mathbf{Q})$ splits in two orthogonal subspaces, the space of vanishing cycles and the space of invariant cycles.*

(c) We now wish to explain the denomination "invariant" cocycles. By our definition, the space of invariant cocycles on $\mathcal{X}_0$ is the image $i^* H^{n-1}(\tilde{X} - \mathcal{X}_\infty; \mathbf{Z})$ in $H^{n-1}(\mathcal{X}_0; \mathbf{Z})$. Consider again the map

$$f\colon (\tilde{X} - \mathcal{X}_\infty) \longrightarrow (P^1 - \infty) = \mathbf{C}\ .$$

This map is not a fibration due to the critical fibers. For $q \geqq 0$, we have the sheaf of cohomology groups with stalks $\mathcal{H}^q(t) =$ direct limit $H^q(f^{-1}(U); \mathbf{Z})$ as U runs over subsets of $\mathbf{C}$ holding $t \in \mathbf{C}$. Since f is a proper map, we have in fact (see [3] for a discussion of the spectral sequence of a map)

$$\mathcal{H}^q(t) = H^q(f^{-1}(t); \mathbf{Z})\ .$$

We note that $H^q(f^{-1}(t); \mathbf{Z}) = H^q(X_t; \mathbf{Z})$ since $f^{-1}(t) = \mathcal{X}_t$ is homeomorphic under π to the hyperplane section X_t of X. But the first Lefschetz theorem states that $H^q(X_t; \mathbf{Z})$ is naturally isomorphic to $H^q(X; \mathbf{Z})$ for $0 \leqq q \leqq n-2$. Thus each sheaf $\mathcal{H}^q$ is naturally isomorphic to the constant sheaf $\mathbf{C} \times H^q(X; \mathbf{Z})$ for $0 \leqq q \leqq n-2$, i.e.,

$$\mathcal{H}^q = H^q(X_0; \mathbf{Z}) \qquad \text{for } 0 \leqq q \leqq n-2\ .$$

In the resulting spectral sequence for the map $f\colon \tilde{X} - \mathcal{X}_\infty \to \mathbf{C}$ we have $E_2^{p,q} = H^p(C; \mathcal{H}^q)$, and so $E_2^{p,q} = 0$ for $p > 0$ and $0 \leqq q \leqq n - 2$. It then follows that

$$H^{n-1}(\widetilde{X} - \mathfrak{X}_\infty; \mathbf{Z}) = E_\infty^{0,n-1} = E_2^{0,n-1} = H^0(C; \mathcal{H}^{n-1}) ,$$

where $H^0(C; \mathcal{H}^{n-1})$ is the group of global cross sections of the sheaf $\mathcal{H}^{n-1}$. By evaluating a section at the point $0 \in C$ we get an injective map $H^0(C; \mathcal{H}^{n-1}) \to H^{n-1}(\mathfrak{X}_0; \mathbf{Z})$ and the composition

$$H^{n-1}(X - \mathfrak{X}_\infty; \mathbf{Z}) \xrightarrow{\approx} H^0(C; \mathcal{H}^{n-1}) \longrightarrow H^{n-1}(\mathfrak{X}_0; \mathbf{Z})$$

is merely the restriction homomorphism. We thus conclude that the invariant cocycle group coincides with the image $H^0(C; \mathcal{H}^{n-1}) \to H^{n-1}(\mathfrak{X}_0; \mathbf{Z})$.

Since $\infty \in P^1$ is a regular value of f, we conclude that

$$H^0(C; \mathcal{H}^{n-1}) = H^0(P^1; \mathcal{H}^{n-1}) ,$$

because any section over $P^1 - \infty$ can be extended uniquely to a section over P^1. Thus the invariant cocycle group on $\mathfrak{X}_0$ coincides with the image of the injective map[1]

$$H^0(P^1; \mathcal{H}^{n-1}) \longrightarrow H^{n-1}(\mathfrak{X}_0; \mathbf{Z})$$

and this, in turn, can be viewed as follows.

Mark on P^1 the critical values $t_1, \cdots, t_\mu$ of f, and consider the fiber space with base $P^1 - \bigcup_i t_i$, total space $f^{-1}(P^1 - \bigcup_i t_i)$ and projection f. The fundamental group $\pi_1(P^1 - \bigcup_i t_i)$ operates by automorphisms on the cohomology groups $H^{n-1}(\mathfrak{X}_0; \mathbf{Z})$ of the typical fiber. From the above considerations we have the following

THEOREM 6. *The space of invariant cocycles is the subgroup of $H^{n-1}(\mathfrak{X}_0; \mathbf{Z})$ consisting of elements that are invariant under the action of $\pi_1(P^1 - \bigcup_i t_i)$ on that group.*

Indeed, Lefschetz has described this action explicitly in terms of homology groups as follows. Let l_j be a loop on $P^1 - \bigcup_i t_i$ going from the base point $0 \in P^1$ around the critical point t_j. The loops $l_1, \cdots, l_\mu$ can be taken as generators of $\pi_1(P^1 - \bigcup_i t_j)$. Denote by θ_j the automorphism of $H_{n-1}(\mathfrak{X}_0; \mathbf{Z})$ induced by l_j. Lefschetz proves

$$\begin{cases} \theta_i(\dot{e}_n^i) = (-1)^n \dot{e}_n^i \\ z - \theta_i(z) = I(z, \dot{e}_n^i)\dot{e}_n^i \end{cases}$$

[1] This can also be seen by considering the generalized Wang sequence for the map $f: \widetilde{X} \to P^1$. This is not a fibration but it behaves like one as far as low dimensional cohomology groups are concerned.

where again $I(z, \dot{e}_n^i)$ denotes the intersection number. A proof of this formula can also be found in Fáry's paper [2].

Appendix on the hyperplane pencil

(1) Let X^n be an algebraic submanifold of P^N and let P^{N*} be the dual projective space of hyperplanes in P^N. Let Φ be the graph in $X \times P^{N*}$ of the correspondence that associates to each $x \in X$ the system of hyperplanes that are tangent to X at x.

LEMMA A.1. *The graph Φ is an algebraic manifold of complex dimention $N-1$.*

PROOF. Let $x^0 = [Z_0^0, \cdots, Z_N^0] \in X$ and let, for example, $Z_0^0 \neq 0$. In a small neighborhood U of x^0 in P^N we take as non-homogeneous coordinates $y_i = Z_i/Z_0$, $1 \leqq i \leqq N$, and by linear change of these coordinates we may assume that X in U is given by equations

$$\begin{cases} y_{n+1} = f_{n+1}(y_1, \cdots, y_n) \\ \vdots \\ y_N = f_N(y_1, \cdots, y_n) \end{cases}$$

where $n = \dim_{\mathbf{C}} X$. The hyperplane $u = a_0 + \sum a_i y_i = 0$ will be tangent to X at a point $x = (y_1, \cdots, y_n)$ if and only if the following equations are satisfied

$$(1) \qquad \begin{cases} a_0 + \sum_{i=1}^n a_i y_i + \sum_{\alpha=n+1}^N a_\alpha f_\alpha(y_1, \cdots, y_n) = 0 \\ a_i + \sum_{\alpha=n+1}^N a_\alpha \dfrac{\partial f_\alpha}{\partial y_i}(y_1, \cdots, y_n) = 0 \end{cases}$$

for $i = 1, \cdots, n$. Therefore the part of Φ over U has the parametric equations

$$(2) \qquad \begin{cases} a_0 = -\sum_{\alpha=n+1}^N a_\alpha\left(f_\alpha - \sum_{i=1}^n y_i \dfrac{\partial f_\alpha}{\partial y_i}\right) \\ a_i = -\sum_{\alpha=n+1}^N a_\alpha \dfrac{\partial f_\alpha}{\partial y_i} \qquad\qquad i = 1, \cdots, n \end{cases}$$

in the parameters $(y_1, \cdots, y_n) \in U$ and $(a_{n+1}, \cdots, a_N) \in P^{N-n-1}$ (note that the a_α's cannot all vanish for $n + 1 \leqq \alpha \leqq N$). These equations show that Φ is a manifold of dimension $N - 1$ and Φ is a closed analytic subset of $X \times P^{N*}$, and Φ is thus an algebraic manifold of dimension $N - 1$.

COROLLARY A.1. *The set of tangent hyperplanes to X fill up in P^{N*} an algebraic variety V of dimension $\leqq N - 1$.*

PROOF. In fact $V = pr_{P^{N*}}(\Phi)$, and since the projection map is proper, we have by a theorem of Remmert [6, Th. 23] that V is an analytic subset of P^{N*} and thus an algebraic variety. Moreover, $\dim_{\mathbf{C}} V \leqq \dim_{\mathbf{C}} \Phi$.

Let us denote by π the projection map $pr_{P^{N*}}$ restricted to Φ.

COROLLARY A.2. *Consider the set*

$$H = \{p \in \Phi : (\operatorname{rank} \pi)_p < N - 1\} .$$

Then $W = \pi(H)$ is an algebraic subvariety of P^{N*} of complex dimension $\leqq N - 2$.

PROOF. Again this follows from the theory of Remmert. Indeed H is an analytic subset of Φ. On an irreducible component Φ_0 of Φ, $H_0 = H \cap \Phi_0$ is either all of Φ_0 or a proper analytic subset. In the first case, the rank of the map in the sense of Remmert and Stein is $< N - 1$ [6, Lemma p. 348 and Th. 15]. In the second case H_0 is of dimension $\leqq N - 2$. Thus by [6, Th. 23] we get in both cases that $\dim \pi(H_0) \leqq N - 2$.

(2) We now assume that X is connected and does not lie in a projective subspace of P^N. We use the same notation as in the proof of Lemma A.1. The function $u|_X = (a_0 + \sum_{k=1}^N a_k y_k)$ restricted to X will have a critical point at x when the hyperplane $u = 0$ is tangent to X at the point $x \in U$. The hessian of $u|_X$ at x is represented by the matrix

$$\left(\sum_{\alpha=n+1}^N a_\alpha \frac{\partial^2 f_\alpha}{\partial y_i \partial y_j}(x)\right) \qquad \begin{matrix} 1 \leqq i \leqq n \\ 1 \leqq j \leqq n . \end{matrix}$$

Thus the critical point x of $u|_X$ is degenerate if and only if

$$\det\left(\sum_{\alpha=n+1}^N a_\alpha \frac{\partial^2 f_\alpha}{\partial y_i \partial y_j}(x)\right) = 0 . \tag{3}$$

At the point $(x, u) \in \Phi$ we can take as local coordinates the functions $(y_1, \cdots, y_n)$ and the ratios of the coefficients $(a_{n+1}, \cdots, a_N)$. It follows then that if (3) is satisfied, $(\operatorname{rank} \pi)_{(x,u)} < N - 1$, and conversely. We thus deduce the following

LEMMA A.2. *Consider a pencil of hyperplanes*

$$\lambda\left(\sum_{i=0}^N a_i Z_i\right) + \mu\left(\sum_{i=0}^N b_i Z_i\right) = 0$$

in P^N, where $a = [a_0, \cdots, a_N]$, $b = [b_0, \cdots, b_N]$ are two distinct points of P^{N*}. Let

$$Y = \left\{\sum_{i=0}^N a_i Z_i = 0 = \sum_{i=0}^N b_i Z_i\right\} \cap X .$$

Then the map $f\colon (X - Y) \to P^1$ *defined by*

$$f = \frac{\sum a_i Z_i}{\sum b_j Z_j}$$

has only non-degenerate critical points if the projective line $\{u_i = \lambda a_i + \mu b_i, 0 \leqq i \leqq N\}$ *in* P^{N*} *does not meet the variety* W *of Corollary* A.2.

PROOF. Indeed by a proper choice of coordinates in P^N we may assume $\sum_{i=0}^N b_i Z_i \equiv Z_0$, and then $f = a_0 + \sum_{i=1}^N a_i y_i \equiv u$. Assume there exists an $x \in X - Y$ that is a degenerate critical point for f. We may assume that $u(x) = 0$, replacing a if necessary by another point on the line joining a and b and different from b. Then (1) and (3) are satisfied; thus $(x, u) \in H$ and $u \in W$.

(3) For the proof of Lemma 2 of the text, it is sufficient to show that the points of $\mathrm{Gr}(P^{N*})$, representing lines of P^{N*} for which

(i) B_Λ is nowhere tangent to X,

(ii) Λ does not meet W,

fill up an open Zariski subset of $\mathrm{Gr}(P^{N*})$. For this it is enough to show that

(α) $\{\Lambda \in \mathrm{Gr}(P^{N*})\colon \Lambda$ satisfies (i)$\}$ is a non-empty Zariski open set.

(β) $\{\Lambda \in \mathrm{Gr}(P^{N*})\colon \Lambda$ satisfies (ii)$\}$ is a non-empty Zariski open set.

PROOF OF (β). Consider in $W \times \mathrm{Gr}(P^{N*})$ the graph Ψ of the correspondence that associates to each $u \in W$ the projective lines of P^{N*} passing thru the point u. For a line $u_i = \lambda a_i + \mu b_i$, a, b in P^{N*}, $a \neq b$, introduce the Plücker coordinates $P_{ik} = a_i b_k - a_k b_i$. The graph Ψ of the correspondence is then given by the equations obtained from the condition

$$\left(\textstyle\sum_{i,k=0}^N P_{ik}\, dt_i \wedge dt_k\right) \wedge \left(\textstyle\sum_{j=0}^N u_j\, dt_j\right) = 0$$

identically satisfied in $dt_0, dt_1, \cdots, dt_N$. The graph Ψ is thus an algebraic variety. Then $A = pr_{\mathrm{Gr}(P^{N*})}(\Psi)$ is an algebraic subvariety of $\mathrm{Gr}(P^{N*})$. It is enough then to show that $A \neq \mathrm{Gr}(P^{N*})$. But if $A = \mathrm{Gr}(P^{N*})$, then every projective line of P^{N*} meets W. This is not the case since W is of codimension $\geqq 2$.

PROOF OF (α). Consider in $X \times \mathrm{Gr}(P^{N*})$ the graph Ξ of the correspondence that associates to each $x \in X$ the projective lines $\Lambda \in \mathrm{Gr}(P^{N*})$ such that $x \in B_\Lambda$ and B_Λ is not transversal to X at x.

Again using Plücker coordinates, one verifies that Ξ is given by the system of equations

$$\{\textstyle\sum_{i=0}^{N} P_{ik}Z_i = 0, 0 \leqq k \leqq N\}$$

expressing the fact that $x \in B_\Lambda$, and the system of equations given by

$$\operatorname{rank}\left(a_i, b_i, \frac{\partial f_\nu(x)}{\partial Z_i}\right) < N - n + 2$$

where $\{f_\nu(Z)\}$ is a basis of the ideal of all homogeneous polynomials in $Z_0, \cdots, Z_N$ vanishing on X, expressing the non-transversality of B_Λ and X at x. It is clear that this second system of equations is also expressible as a system of algebraic equations between the Plücker coordinates $P_{ik} = a_ib_k - a_kb_i$ and the coordinates of a variable point in X. Thus Ξ is again an algebraic variety and therefore its projection B on Gr (P^{N^*}) is also an algebraic variety. We have only to show that $B \neq$ Gr (P^{N^*}). By Corollary A.1, we can choose a hyperplane $u \equiv \sum a_iZ_i = 0$ in P^N that is nowhere tangent to X. Let $X_1 = X \cap \{u = 0\}$. This is again an algebraic manifold of pure dimension $n - 1$. Again by Corollary A.2, as applied to X_1 we can choose a second hyperplane $v \equiv \sum b_iz_i = 0$ in P^N that is nowhere tangent to X_1. Certainly u and v are independent, and at a point x where $u = 0$ and $v = 0$ the tangent lines to X which lie in $u = 0$ and $v = 0$ fill a space of dimension $n - 2$. This shows that for the pencil $\Lambda \equiv \lambda u + \mu v = 0$, B_Λ is transversal to X, i.e., $\Lambda \notin B$. This completes the proof.

UNIVERSITÀ DI PISA
UNIVERSITY OF CALIFORNIA, SAN DIEGO

REFERENCES

[1] A. ANDREOTTI and T. FRANKEL, *The Lefschetz theorem on hyperplane sections*, Ann. of Math., **69** (1959) 713–717.

[2] I. FÁRY, *Cohomologie des variétés algébriques*, Ann. of Math. **65** (1957), 21–73.

[3] R. GODEMENT, Topologie algébriques et théorie des faisceaux, Hermann, Paris, 1958.

[4] A. HOWARD, *On the homotopy groups of an affine algebraic hypersurface*, Ann. of Math. **84** (1966), 197–216.

[5] J. MILNOR, Morse Theory, Princeton Univ. Press, Princeton, 1963.

[6] R. REMMERT, *Holomorphe und meromorphe abbildungen komplexer Räume*, Math. Ann. **133** (1957), 328–370.

(Received September 18, 1968)

Algebraization of formal moduli: I*

By M. Artin

Introduction

Since the fundamental work of Kodaira and Spencer [7] on deformation theory appeared, their methods have become standard tools for the study of complex analytic spaces. Grothendieck (FGA, 195) and others have treated the algebro-geometric analogues, whose formal aspects are now well understood. In particular, readily applicable criteria for the existence of a formal universal deformation of a given structure are available. The criterion of Schlessinger [14] is the simplest of these. However an algebraic analogue of a convergence theorem was lacking, and it is the purpose of this paper to derive such a theorem from the approximation theory of [3]. Our result (1.6) is that in certain cases a formal universal deformation is represented by an algebraic family.

The most striking result of this type in the analytic case is the theorem of Kuranishi [9] on the existence of deformations of smooth compact analytic spaces. Now since analytic, and hence formal, deformations of an algebraic variety are not necessarily algebraic, it is clear from the start that Kuranishi's theorem will *not* admit a direct algebraic analogue. Thus we have to restrict our attention to more amenable deformation problems. What we do is to impose a condition of *effectivity* on a formal deformation (cf. § 1). The effectivity may be viewed as a half way step towards algebraization, and our Theorem (1.6) asserts that, under mild additional hypotheses, a formal universal deformation which is effective is

* Supported by the Sloan Foundation and the National Science Foundation.

actually algebraizable. Fortunately, the existence theorem for formal sheaves of Grothendieck (EGA III, 5) can be used to conclude effectivity for many functors.

Since the approximation results [3] are available only in those cases, we have to assume that the base scheme S is of finite type over a field or over an excellent Dedekind domain.

In §§ 3 and 5, we use the algebraization theorem to derive criteria, based on effective pro-representability, for representability of an abstract functor as an algebraic space [1], [8]. It seems clear to us that for representability as a *scheme*, no similar criterion exists which does not impose strong restrictive hypotheses on the functor. And with sufficiently special functors, one can try to deduce *a posteriori* that the algebraic space is a scheme. This is possible if, for example, the functor has a group structure and the base is a field. We are thus able to recover the theorem of Murre [12] on representability of group functors, in an improved form (cf. § 4). One can also recover Grothendieck's theorem [13] on unramified functors, with the above restriction on the base scheme S.

The applications of the last two sections, to Hilbert and Picard functors, are perhaps the most important. We show that the Hilbert functor is representable for every map $f: X \to S$ of algebraic spaces which is locally of finite type, and that the Picard functor is representable for every map f which is flat, proper, and "cohomologically flat in dimension zero" (EGA III, 7.8.1). The proofs are very abstract, making minimal use of projective methods. Such methods occur only implicitly, in that the standard results on cohomology which we use rely on reduction to the projective case for their proofs. These results, approximately the contents of (EGA III_1), have now been extended to the context of algebraic spaces by Knutson [8].

Examples exist for both of the above functors which show that they need not be represented by schemes, even if f is a map of schemes (cf. [1, § 6]). Our results are thus in a natural form. However, they do not completely contain the previous ones, as for instance Grothendieck's theorem (FGA, 221) on existence and local projectivity of the Hilbert scheme when f is projective. To obtain it, one will still have to prove the projectivity, and we do not know how easily this can be done. It would be nice to have a short, direct proof.

Even when projective methods are available, the use of Theorem (1.6) may result in simplified proofs. Consider the problem of constructing a Picard scheme for a projective scheme X over a field k: The classical proofs (e.g. [11]) begin with a big family of Cartier divisors. One then proceeds by taking sections for the linear equivalence relation, to obtain a local modular family. Now using (1.6), we can deduce the existence of such a local modular family immediately from standard deformation theory and the omnipresent Grothendieck existence theorem. The use of Cartier divisors is thus eliminated from the proof, making it methodologically simpler. Moreover our proof extends without change to the case that X is any proper algebraic space over k. In contrast, the previous proofs for a proper scheme (of Grothendieck and Murre [12]) resort to Chow's lemma to reduce to the projective case, where Cartier divisors take over.

Our treatment of representability of functors is based on the work of Grothendieck and Murre. In addition, it is a pleasure to acknowledge the many fruitful discussions which we had with Grothendieck and with Stolzenberg while working on this paper.

1. Statement of the algebraization theorem

We begin by reviewing some terminology. Denote by S a scheme or an algebraic space which is locally of finite type over a field or over an excellent (EGA IV, 7.8.5) Dedekind domain. We consider (contravariant) functors

$$F\colon (\text{schemes}/S)^\circ \longrightarrow (\text{sets})\ . \tag{1.1}$$

If $X = \operatorname{Spec} A$ is an affine S-scheme, we will often write $F(A)$ instead of $F(X)$.

Let $s \in S$ be a point whose residue field $k(s)$ is of finite type over $\mathcal{O}_S$, let k' be a finite field extension of $k(s)$, and let

$$\xi_0 \in F(k') \tag{1.2}$$

be an element. By *infinitesimal deformation* of ξ_0 we mean a pair (A, η) where A is an artinian local $\mathcal{O}_S$-algebra with residue field k', and $\eta \in F(A)$ is an element which induces $\xi_0 \in F(k')$ by functorality. A *formal deformation* of ξ_0 is a pair $(\bar{A}, \{\xi_n\})$, where $\bar{A}$ is a complete noetherian local $\mathcal{O}_S$-algebra with residue field k',

$$\xi_n \in F(\bar{A}/\mathfrak{m}^{n+1}) \qquad n = 0, 1, \cdots$$

is a compatible system of elements (i.e., ξ_n induces ξ_{n-1} in $F(\bar{A}/\mathfrak{m}^n)$), and ξ_0 is the element (1.2). Here as always, $\mathfrak{m}$ denotes the maximal ideal.

A pair $(\bar{A}, \bar{\xi})$, with $\bar{A}$ as above and $\bar{\xi} \in F(\bar{A})$ inducing ξ_0, will be called an *effective formal deformation*. We will also refer to a given formal deformation $(\bar{A}, \{\xi_n\})$ as *effective* if there exists an element $\bar{\xi} \in F(\bar{A})$ inducing $\{\xi_n\}$.

A formal deformation $(\bar{A}, \{\xi_n\})$ is said to be *versal* (respectively, *universal*) if it has the following property: Let (B', η') be an infinitesimal deformation of ξ_0, and say that the $(n+1)^{\text{st}}$ power of the maximal ideal of B' is zero. Let B be a quotient of B', and denote by $\eta \in F(B)$ the element induced by η'. Then every map

$$\bar{A}/\mathfrak{m}^{n+1} \longrightarrow B \text{ sending } \xi_n \longmapsto \eta$$

can be embedded (resp., can be uniquely embedded) in a commutative diagram of $\mathcal{O}_S$-algebras

$$(1.3) \qquad \begin{array}{ccc} \bar{A}/\mathfrak{m}^{n+1} & \longrightarrow & B' \\ & \searrow & \downarrow \\ & & B \end{array} \quad \text{sending} \quad \begin{array}{ccc} \xi_n & \longmapsto & \eta' \\ & \searrow & \downarrow \\ & & \eta \end{array} .$$

Thus the existence of a formal universal deformation is a slightly weaker property than the pro-representability (FGA, 195) of F at the point in question. To get pro-representability, we would have to take into account also artinian algebras A whose residue fields are finite extensions of k'. Since the ring which pro-represents F is already determined by infinitesimal extensions, if it exists, we get a result in (1.6) which is nominally at least somewhat stronger by considering formal deformations.

The functor of smooth proper schemes, analogous to the one considered by Kuranishi [9] in the analytic case, admits a formal versal deformation, but it will not be effective in general. An elementary example of non-effectivity is given in [1, (5.9)].

Schlessinger's theorem [14] is the basic abstract criterion for existence of a formal versal deformation. For effectivity, the existence theorem for formal sheaves (EGA III, 5) applies to many relative functors, such as of the Hilbert and Picard types. For such functors, one can use Grothendieck's theorem to conclude that for any complete local $\mathcal{O}_S$-algebra $\bar{A}$ the map

$$(1.4) \qquad F(\bar{A}) \longrightarrow \lim_{\leftarrow} F(\bar{A}/\mathfrak{m}^{n+1})$$

is bijective (cf. §§ 6, 7). This is more than we need. To conclude effectivity, it would be enough to know that the map (1.4) has a *dense* image. For, then there is a $\xi' \in F(\bar{A})$ which induces $\xi_1 \in F(\bar{A}/\mathfrak{m}^2)$. Since $(\bar{A}, \{\xi_n\})$ is versal, we can apply (1.3) to lift the identity map on $\bar{A}/\mathfrak{m}^2$ successively to diagrams

$$
\begin{array}{ccc}
\bar{A}/\mathfrak{m}^{n+1} & \xrightarrow{\varphi_n} & \bar{A}/\mathfrak{m}^{n+1} \\
\downarrow c & & \downarrow c \\
A/\mathfrak{m}^n & \xrightarrow{\varphi_{n-1}} & A/\mathfrak{m}^n
\end{array}
\quad \text{sending} \quad
\begin{array}{ccc}
\xi_n & \longmapsto & \xi'_n \\
\downarrow & & \downarrow \\
\xi_{n-1} & \longmapsto & \xi_{n-1}
\end{array}
\tag{1.5}
$$

for each n, where ξ'_n is the element induced by ξ', and where c is the canonical map. Hence there is an $\mathcal{O}_S$-map $\varphi\colon \bar{A} \to \bar{A}$ with the property that it sends $\{\xi_n\} \mapsto \{\xi'_n\}$, and such that $\varphi \equiv$ identity (modulo $\mathfrak{m}^2$). Now any map of a complete local ring to itself which is congruent to the identity (modulo $\mathfrak{m}^2$) is an automorphism. (This is well known.) Thus φ is one, and hence the inverse image $F(\varphi^{-1})(\xi') \in F(\bar{A})$ of ξ' is the required element.

This suggests another possible approach to effectivity: If the functor F is a quotient of a representable functor X of finite type over S, so that there is a surjective map $\varphi\colon X \to F$, say as sheaves for the etale topology, then one will be able to conclude density of the image of (1.4) from an approximation theorem for formal solutions of equations similar to [2, (1.2)]. (We have the necessary theorem only in characteristic zero for the moment.) For the application, it would of course be enough that X cover F locally at ξ_0. The important point is that no hypothesis other than surjectivity is made on the map φ, so that the scheme X itself may be very bad looking. Our theorem might also extend to the analytic case for functors of such a type.

THEOREM 1.6. (Existence of algebraization). *Assume that the functor F (1.1) is locally of finite presentation over S, and suppose that an effective formal versal deformation $(\bar{A}, \bar{\xi})$ of ξ_0 (1.2) exists. Then there is an S-scheme X of finite type, a closed point $x \in X$ with residue field $k(x) = k'$, and an element $\xi \in F(X)$, such that the triple (X, x, ξ) is a versal deformation of ξ_0. More precisely, there is an isomorphism $\hat{\mathcal{O}}_{X,x} \approx \bar{A}$ such that ξ induces ξ_n in $F(\bar{A}/\mathfrak{m}^{n+1})$ for each n. The isomorphism in unique if $(\bar{A}, \bar{\xi})$ is universal.*

It does not follow that the element of $F(\bar{A})$ induced by ξ is $\bar{\xi}$.

This is because the formal deformation $\{\xi_n\}$ may not determine the element $\bar{\xi}$ uniquely. A simple example, of two lines with an infinite order contact, is given in [1, (5.3)]. The example also shows that the algebraization need not be unique. But if $\bar{\xi}$ is uniquely determined by $\{\xi_n\}$, for instance if the map (1.4) is *injective*, then we obtain

THEOREM 1.7. (Uniqueness). *With the notation of* (1.6), *suppose the element* $\bar{\xi} \in F(\bar{A})$ *uniquely determined by the set* $\{\xi_n\}$ *of its truncations. Then the triple* (X, x, ξ) *is unique up to local isomorphism for the etale topology, in the following sense: If* (X', x', ξ') *is another algebraization, then there is a third one* (X'', x'', ξ'') *and a diagram*

$$\begin{array}{ccc} & X'' & \\ {\scriptstyle f}\swarrow & & \searrow{\scriptstyle f'} \\ X & & X' \end{array} \quad \text{sending} \quad \begin{array}{ccc} & x'' & \\ \swarrow & & \searrow \\ x & & x' \end{array}, \quad \begin{array}{ccc} & \xi'' & \\ \swarrow & & \searrow \\ \xi & & \xi' \end{array},$$

and where f, f' *are etale morphisms.*

The existence of an algebraization follows very simply from the approximation theorem [3, (1.12)] for values of a functor in the "unobstructed case," i.e., if the ring $\bar{A}$ is a power series ring $\bar{A} = \hat{\mathcal{O}}[[x_1, \cdots, x_n]]$, where $\hat{\mathcal{O}}$ is the completion of a local ring of S. Assume more generally that $\bar{A}$ is already known to be the completion of a local ring of an S-scheme X of finite type, at a point of finite type x. Then in the notation of [3, § 2], there is an etale neighborhood (X', x') of x in X and an element $\xi' \in F(X')$, such that $\xi' \equiv \bar{\xi} (\text{mod } \mathfrak{m}^2)$, i.e., such that ξ' induces ξ_1 in $F(\bar{A}/\mathfrak{m}^2)$. The completion $\hat{\mathcal{O}}_{X', x'}$ is still isomorphic to the ring $\bar{A}$. As above (cf. (1.5)), it follows from versality that there is an automorphism $\varphi: \bar{A} \rightarrow \bar{A}$, congruent to the identity (mod $\mathfrak{m}^2$), and sending $\xi_n \mapsto \xi'_n$ for each n (ξ'_n is the element induced by ξ'). Thus (X', x', ξ') is the required algebraization.

To prove existence in the general case in the next section, we approximate the ring $\bar{A}$ at the same time as the value $\bar{\xi}$ of the functor. We obtain an algebraic ring A' and an element $\xi' \in F(A')$. By versality, there is a map $\varphi: \bar{A} \rightarrow \hat{A}'$ sending $\xi_n \mapsto \xi'_n$ for each n. What has to be done is to control the structure of A' enough so that such a map, if it exists, is necessarily an isomorphism. There are certainly several ways to do this. For instance if $\bar{A}$ were an integral domain, it would be enough to control the dimension of A', which can be done rather easily. We use a stratification which may conceivably be of some independent interest.

2. Proofs of Theorems (1.6) and (1.7)

We retain the notation of § 1. First of all, to prove (1.6), it is no loss of generality to assume that the ground space S is the spectrum of an excellent Dedekind domain R. For, by assumption, S is of finite type over Spec R, where R is an excellent Dedekind domain or a field. In case R is a field, we replace it by the polynomial ring $R[t]$, where t acts on S as zero. Consider the functor

$$\mathbf{F}\colon (\text{Schemes}/R)^0 \longrightarrow (\text{Sets}) ,$$

where $\mathbf{F}(X)$ is defined as the set of pairs (f, ξ), $f\colon X \to S$ an R-map and $\xi \in F(X^f)$. Here X^f denotes the scheme X with its S-structure induced by f. If X is an S-scheme, any element $\xi \in F(X)$ determines canonically the element (f_X, ξ) of $\mathbf{F}(X)$, where $f_X\colon X \to S$ is the structure map of X. It is immediately seen that $(\bar{A}, \bar{\xi})$ is an effective versal deformation of ξ_0 (1.2) if and only if $(\bar{A}, (f_{\bar{A}}, \bar{\xi}))$ is an effective versal deformation of $(f_{k'}, \xi_0)$. Thus we may as well suppose $S = \operatorname{Spec} R$. Moreover, it is easily seen that we may suppose that the image of the point Spec k' in S is a closed point.

Denote by $\mathfrak{p}$ the prime ideal of R corresponding to the point $s \in S$, and by $\hat{R}$ the $\mathfrak{p}$-completion of R, which is a complete discrete valuation ring. The ring $\bar{A}$ has residue field k' finite over $R/\mathfrak{p} = k$, and hence has a structure of finite module over a power series ring $\hat{R}[[x_1, \cdots, x_n]]$. We begin by putting $\bar{A}$ in a special form, using a suitable choice of the elements x_i.

Let B be an $R[d_1, \cdots, d_n]$-algebra and M a finite B-module. Denote

$$B_\nu = B/(d_1, \cdots, d_\nu)B$$
$$M_\nu = M/(d_1, \cdots, d_\nu)M .$$

Definition 2.1. A *preparation* for M consists of

(a) Finite R-modules $L_\nu = L_\nu(M)$, $v = 0, \cdots, n$ which are sums of modules of the type $R/\mathfrak{p}^e$ and of free modules,

(b) For each $\nu = 0, \cdots, n - 1$, B_ν-homomorphisms $u_\nu = u_\nu(M)$ and $v_\nu = v_\nu(M)$ such that the diagrams

$$\begin{array}{ccc} M_\nu & \xrightarrow{d_{\nu+1}} & M_\nu \\ {\scriptstyle u_\nu}\uparrow & \searrow{\scriptstyle v_\nu} & \uparrow{\scriptstyle u_\nu} \\ L_\nu \otimes_R B_\nu & \xrightarrow{d_{\nu+1}} & L_\nu \otimes_R B_\nu \end{array}$$

commute, and

(c) An isomorphism $u_n = v_n^{-1}\colon L_n \otimes_R B_n \xrightarrow{\sim} M_n$.

The diagam (b) asserts in particular that M_ν is isomorphic to the "standard" module $L_\nu \otimes_R B_\nu$ outside the locus $\{d_{\nu+1} = 0\}$ in Spec B_ν.

PROPOSITION 2.2. *Let $\{M^\alpha\}$ be a finite set of $\hat{R}[[x_1, \cdots, x_n]]$-modules of finite type. Then there exist elements*

$$d_1, \cdots, d_n \in \hat{R}[[x_1, \cdots, x_n]]$$

such that $(\mathfrak{p}, d_1, \cdots, d_n)$ generate an ideal primary to the maximal ideal, and such that each M^α admits a preparation as $\hat{R}[[d_1, \cdots, d_n]]$-module.

PROOF. Since $\hat{R}[[d_1, \cdots, d_n]]/(d_1, \cdots, d_n) = \hat{R}$, every finite module over this ring is isomorphic to a suitable module L_n. Therefore condition (c) of (2.1) can always be fulfilled. Let s be an integer between 0 and $n - 1$, and suppose that we have already found $d_1, \cdots, d_n$ so that the condition (b) can be fulfilled for $0 \leqq \nu < s$ and for each α. We are going to find a new set $d'_1, \cdots, d'_n$ for which the conditions can be fulfilled for each $\nu \leqq s$ and each α, and with $d_i = d'_i$ for $i \leqq s$. This will suffice by induction.

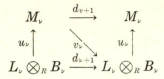

We write $B = \hat{R}[[d_1, \cdots, d_n]]$, so that $B_\nu = \hat{R}[[d_{\nu+1}, \cdots, d_n]]$. The problem is to choose an element d'_{s+1} appropriately. Now the localization of the ring B_S at the prime ideal generated by $\mathfrak{p}$ is a discrete valuation ring, and hence every finite module is isomorphic to a suitable standard module $L_s \otimes_R B_s$, locally at that point of Spec B_S. Therefore there is an element $z \in B_S$ not divisible by $\mathfrak{p}$, so that M_s^α becomes isomorphic to an $L_s^\alpha \otimes_R B_s$ when z is inverted. We may choose a single z which works for all α. Then it follows immediately (from the fact that Hom for finite modules commutes with localization) that a diagram (2.1) (b) exists for $\nu = s$ if we replace d_{s+1} by a suitable power of z, or more precisely, a representative d'_{s+1} in $\hat{R}[[d_1, \cdots, d_n]]$ of such a power. Since z is not divisible by $\mathfrak{p}$, the set $\{\mathfrak{p}, d'_1, \cdots, d'_{s+1}\}$ ($d'_i = d_i$ for $i \leqq s$) extends to a set of generators $(\mathfrak{p}, d'_1, \cdots, d'_n)$ for an ideal primary to the maximal ideal. Denote by B' the ring $\hat{R}[[d'_1, \cdots, d'_n]]$. By [15, IV-37, Prop. 22], B is a finite and *free* B'-module, say of rank r. Thus the diagrams of B_ν-modules for $\nu = 0, \cdots, s$

$$\begin{array}{ccc} M_\nu^\alpha & \xrightarrow{d'_{\nu+1}} & M_\nu^\alpha \\ \uparrow & \searrow & \uparrow \\ L_\nu^\alpha \otimes_R B_\nu & \xrightarrow[d'_{\nu+1}]{} & L_\nu^\alpha \otimes_R B_\nu \end{array}$$

yield diagrams of B'_ν-modules

$$\begin{array}{ccc} M_\nu^\alpha & \xrightarrow{d'_{\nu+1}} & M_\nu^\alpha \\ \uparrow & \searrow & \uparrow \\ L_\nu'^\alpha \otimes_R B'_\nu & \xrightarrow{d'_{\nu+1}} & L_\nu'^\alpha \otimes_R B'_\nu \end{array} ,$$

where $L_\nu'^\alpha = (L_\nu^\alpha)^r$. This completes the proof of the proposition.

For any $\hat{R}[[x_1, \cdots, x_n]]$ (resp. $\hat{R}[[d_1, \cdots, d_n]]$)-module M, let $I^{-1}(M) = 0$, and for $\nu = 0, \cdots, n$

$$I^\nu(M) = \{m \in M \mid \dim [\operatorname{supp}(m) \cap V(\mathfrak{p})] \leqq \nu\} ,$$

where $\operatorname{supp}(m)$ is the support in Spec $\hat{R}[[x]]$ (resp. Spec $\hat{R}[[d]]$) and where $V(\mathfrak{p})$ is the locus $\{\mathfrak{p} = 0\}$ in that space. Thus $I^n(M) = M$. Note that $I^\nu(M)$ will not be changed if we replace $\hat{R}[[x_1, \cdots, x_n]]$ by a subring $\hat{R}[[d_1, \cdots, d_n]]$ over which it is finite.

Returning to the ring $\bar{A}$ of (1.6), we apply the above proposition to prepare the $\hat{R}[[x]]$-modules $\bar{A}$, $I^\nu(\bar{A})$, $\nu = 0, \cdots, n-1$, with a suitable choice of $d_1, \cdots, d_n$.

LEMMA 2.3. *Consider the functor G on $R[d_1, \cdots, d_n]$-algebras which to an algebra B associates the set of isomorphism classes of the following data.*

(a) *A (commutative) B-algebra $\mathfrak{A}$ which is a finite prepared B-module with $L_\nu(\mathfrak{A}) = L_\nu(\bar{A})$, and an element $\eta \in F(\mathfrak{A})$.*

(b) *Finite $\mathfrak{A}$-modules $\mathfrak{I}^\mu$ which are prepared with $L_\nu(\mathfrak{I}^\mu) = L_\nu(I^\mu(\bar{A}))$ for $\mu = 0, \cdots, n$, and $\mathfrak{A}$-homomorphisms*

$$\mathfrak{I}^\mu \longrightarrow \mathfrak{A} .$$

This functor is locally of finite presentation.

PROOF. It is clear that to give a finite set of B-modules and maps satisfying a finite number of commutation relations is a problem locally of finite presentation (cf. (EGA IV, 8)). We apply the relative criterion of [3, (2.8)]. Thus we may assume that we have already given the prepared modules $\mathcal{A}$, $\mathcal{I}^{\mu}$. The additional structure of commutative algebra on $\mathcal{A}$ is given by homomorphisms $\mathcal{A} \otimes_B \mathcal{A} \to \mathcal{A}$ and $B \to \mathcal{A}$ satisfying certain identities, hence is described by a functor locally of finite presentation. The same is true of the $\mathcal{A}$-homomorphisms $\mathcal{I}^{\mu} \to \mathcal{A}$. Finally, if all the other data are already given, then the additional structure of the element $\eta \in F(\mathcal{A})$ is locally of finite presentation since F is, which proves the lemma.

Now to prove Theorem (1.6), we apply the approximation theorem [3, (1.12)] for values of a functor to the functor G of Lemma (2.3). Thus for any N we can approximate the value $\bar{\Delta}$ of G given by the chosen preparation of $\bar{A}$, $I^{\mu}(\bar{A})$, and the element $\bar{\xi} \in F(\bar{A})$ to order N on some etale neighborhood (Y, y), $Y = \operatorname{Spec} B$, of the "origin" $(\mathfrak{p}, d_1, \cdots, d_n)$ of $\operatorname{Spec} R[d_1, \cdots, d_n]$. Say the approximation is by the element Δ consisting of the $\mathcal{O}_Y$-algebra $\mathcal{A}$, the element $\eta \in F(\mathcal{A})$, and the $\mathcal{A}$-modules $\mathcal{I}^{\mu}$. The fact that it is an approximation to order N means that $\bar{\Delta}$ and Δ induce the same element of $G(R[d]/(\mathfrak{p}, d)^{N+1})$.

We write $X = \operatorname{Spec} \mathcal{A}$. The scheme X has a unique point x above y (for $N \geqq 0$), and $(\mathfrak{p}, d)$ generates an ideal primary to the maximal ideal there. We claim that if N is large enough then the triple (X, x, η) is the required algebraization.

Now the approximation gives an isomorphism of truncations

$$\mathcal{A}/(\mathfrak{p}, d)^{N+1} \approx \bar{A}/(\mathfrak{p}, \mathrm{d})^{N+1} \tag{2.4}$$

identifying the elements induced by η, ξ. Thus by versality of $(\bar{A}, \bar{\xi})$, we can lift the projection map $\bar{A} \to \mathcal{A}/(\mathfrak{p}, d)^{N+1}$ successively to maps

$$\bar{A} \longrightarrow \mathcal{A}/((\mathfrak{p}, d)^{C+1} \quad \text{sending} \quad \bar{\xi} \longmapsto \eta(C)$$

for $C = N, N+1, \cdots$, where $\eta(C)$ denotes the truncation of η (mod $(\mathfrak{p}, d)^{C+1}$). Hence we obtain a local R-homomorphism

$$\psi: \bar{A} \longrightarrow \hat{\mathcal{A}} \tag{2.5}$$

where $\hat{\mathcal{A}}$ is the completion of $\mathcal{A}$ at x. This homomorphism is not $\mathcal{O}_Y$-linear. However we claim that it is an *R-isomorphism* if $N \gg 0$, and this will complete the proof.

First of all, it is clear that ψ is *surjective*, provided at least that $N \geqq 1$. For since $\bar{A}$ is complete, it is enough to show that

ψ induces a surjection onto $\mathfrak{A}/\mathfrak{m}^2$, and $\mathfrak{m}^2 \subset (\mathfrak{p}, d)^2$, whence (2.4) applies.

Let $K = \ker \psi$, and suppose $K \neq (0)$. Say that $K \subset I^{n-\nu}(\bar{A})$, but $K \not\subset I^{n-\nu-1}(\bar{A})$, where $\nu \geqq 0$. Now $I^{\mu}(\bar{A})$ is just the set of elements $a \in \bar{A}$ such that the dimension of the locus $\operatorname{supp}(a) \cap V(\mathfrak{p})$ in Spec $\bar{A}$ is at most μ. Thus it is clear that $a \notin I^{n-\nu}(\bar{A})$ implies $\psi(a) \notin I^{n-\nu}(\hat{\mathfrak{A}})$. Therefore the map $I^{n-\nu}(\bar{A}) \to I^{n-\nu}(\hat{\mathfrak{A}})$ is surjective, i.e.,

$$0 \longrightarrow K \longrightarrow I^{n-\nu}(\bar{A}) \longrightarrow I^{n-\nu}(\hat{\mathfrak{A}}) \longrightarrow 0$$

is exact. Since by assumption $K \not\subset I^{n-\nu-1}(\bar{A})$, it follows that $I^{n-\nu}(\bar{A})$ and $I^{n-\nu}(\hat{\mathfrak{A}})$ are *not* isomorphic at the generic point P of the locus $\{\mathfrak{p} = 0\}$ of Spec $\hat{B}_\nu$ ($\hat{B}_\nu = \hat{R}[[d]]/(d_1, \cdots, d_\nu)$). If $\nu = 0$, this contradicts the fact that $L_0(\bar{A}) = L_0(\mathfrak{A})$. Suppose $\nu > 0$. Then since $L_\nu(I^{n-\nu}(\bar{A})) = L_\nu(\mathcal{I}^{n-\nu})$, it implies that the given map

$$\hat{\mathcal{I}}^{n-\nu} \longrightarrow \hat{\mathfrak{A}}\,, \tag{2.6}$$

which necessarily carries $\hat{\mathcal{I}}^{n-\nu}$ to $I^{n-\nu}(\hat{\mathfrak{A}})$, is not injective, locally at the point P. Thus to complete the proof, it suffices to prove injectivity of (2.6) locally at P for every ν, $1 \leqq \nu \leqq n$, and for $N \gg 0$.

Now since for $\hat{B} = \hat{R}[[d]]$ the module $L_\nu \otimes_R \hat{B}_\nu$ of (2.1) (b) is $d_{\nu+1}$-torsion free if $L_\nu \neq 0$, and since $v_\nu u_\nu = d_{\nu+1}$, it follows that u_ν is injective. Thus $d_1\bar{A}$ is contained in the standard module $L_0(\bar{A}) \otimes_R \hat{B}_0$ for which $I^{n-1} = 0$, whence $I^{n-1}(\bar{A}) \cap d_1\bar{A} = 0$, i.e., the canonical map

$$I^{n-1}(\bar{A}) \longrightarrow \bar{A}_1$$

is injective. Continuing this reasoning, with $\bar{A}_1$, etc. $\cdots$, one finds

$$I^{n-\nu}(\bar{A}) \cap (d_1, \cdots, d_\nu)\bar{A} = 0 \qquad \text{for } \nu = 1, \cdots, n\,,$$

hence that the canonical map

$$\varepsilon_\nu\colon I^{n-\nu}(\bar{A}) \longrightarrow \bar{A}_\nu$$

is injective. We claim that therefore the "nearby" map

$$\varepsilon'_\nu\colon \hat{\mathcal{I}}^{n-\nu} \longrightarrow \hat{\mathfrak{A}}_\nu = \hat{\mathfrak{A}}/(d_1, \cdots, d)\hat{\mathfrak{A}}$$

is injective outside the locus $\{d_{\nu+1} = 0\}$ of Spec $\hat{B}_\nu$, if $N \gg 0$. This will suffice.

To prove this property for ε'_ν, it suffices to show that, in the notation of (2.1) (b), the composition

$$\varphi'_\nu = v_\nu(\hat{\mathfrak{A}}) \circ \varepsilon'_\nu \circ u_\nu(\hat{\mathcal{I}}^{n-\nu}) ,$$

which is a map

$$\varphi'_\nu \colon L_\nu(I^{n-\nu}(\bar{A})) \otimes_R \hat{B}_\nu \longrightarrow L_\nu(\bar{A}) \otimes_R \hat{B}_\nu ,$$

is injective. But this is a map of standard modules which is congruent to the map $\varphi_\nu = v_\nu(\bar{A}) \circ \varepsilon_\nu \circ u_\nu(I^{n-\nu}(\bar{A}))$ (modulo $(\mathfrak{p}, d)^{N+1}$), and φ_ν is injective because ε_ν is. The injectivity of φ'_ν for large N is therefore clear, as follows: We interpret elements of L_ν as vectors,

$$\begin{aligned} L_\nu(I^{n-\nu}(\bar{A})) &= \{(a_1, \cdots, a_\alpha) \mid a_i \in R && \text{for } i = 1, \cdots, \alpha' , \\ &\qquad\qquad a_i \in R/\mathfrak{p}^{e_i} && \text{for } i = \alpha' + 1, \cdots, \alpha\} \\ L_\nu(\bar{A}) &= \{(b_1, \cdots, b_\beta) \mid b_i \in R && \text{for } i = 1, \cdots, \beta' , \\ &\qquad\qquad b_i \in R/\mathfrak{p}^{f_i} && \text{for } i = \beta' + 1, \cdots, \beta\} . \end{aligned}$$

Then since a generator p of $\mathfrak{p}$ is not a zero-divisor in $\hat{B}_\nu$, the map φ_ν is given as

$$(0, \cdots, 0, 1_i, 0, \cdots, 0) \longmapsto (c_{i1}, \cdots, c_{i\beta}) ,$$

where

$$\begin{aligned} c_{ij} &\in B_\nu && \text{if } i \leqq \alpha' \\ c_{ij} &= 0 && \text{if } i > \alpha' \text{ and } j \leqq \beta' \\ c_{ij} &= \mathfrak{p}^{r_{ij}}\gamma_{ij} && \text{if } i > \alpha' \text{ and } j > \beta' , \end{aligned}$$

where $r_{ij} = \max(0, f_j - e_i)$, and $\gamma_{ij} \in \hat{B}_\nu$ is determined modulo a suitable power of $\mathfrak{p}$. Now the injectivity of φ_ν is just equivalent with the assertions that the matrix (c_{ij}) $i = 1, \cdots, \alpha', j = 1, \cdots, \beta'$ have a non-zero α'-rowed determinant in B_ν, and that the matrix $(\bar{\gamma}_{ij})$ $i = \alpha' + 1, \cdots, \alpha, j = \beta' + 1, \cdots, \beta$ have an $(\alpha - \alpha')$-rowed determinant which is not congruent zero (modulo $\mathfrak{p}$), where $\bar{\gamma}_{ij} = \gamma_{ij}$ if $f_j \geqq e_i$, and is zero if $f_i < e_i$. The congruence class of this determinant is uniquely determined by φ_ν. We have thus expressed the injectivity by the non-vanishing of certain elements, which is preserved in φ'_ν if $N \gg 0$. This completes the proof.

PROOF OF THEOREM 1.7. Let (X, x, ξ), (X', x', ξ') be two algebraizations of $(\bar{A}, \bar{\xi})$. Then we have an isomorphism $\varphi \colon \bar{A} \longrightarrow \hat{\mathcal{O}}_{X,x}$ sending ξ_n to the truncation of ξ for each n. By assumption, the truncations $\{\xi_n\}$ determine $\bar{\xi}$ uniquely. Thus φ sends $\bar{\xi}$ to the element $\hat{\xi} \in F(\hat{\mathcal{O}}_{X,x})$ induced by ξ. We interpret $\xi \in F(X)$ as a map of functors $X \longrightarrow F$.Then, applying the same reasoning also to X', it follows that the diagram

$$\begin{array}{ccccc} & & \operatorname{Spec}\bar{A} & & \\ & \swarrow & \Big| & \searrow & \\ X & & \bar{\xi}\Big| & & X' \\ & {\scriptstyle\xi}\searrow & \downarrow & \swarrow{\scriptstyle\xi'} & \\ & & F & & \end{array}$$

commutes, and thus defines an element $\bar{\varepsilon} \in [X \times_F X'](\bar{A})$. Now since filtering direct limits commute with fibered products, it is clear from the definition [3, (1.5) and § 2], that a fibered product of functors locally of finite presentation is again locally of finite presentation. We identify $\bar{A}$ with the completion $\hat{\mathcal{O}}_{X,x}$ *via* φ and apply the approximation theorem [3, (1.12)] to the formal element $\bar{\varepsilon} \in [X \times_F X'](\bar{A})$. Thus there is an etale neighborhood (X'', x'') of x in X and an element $\varepsilon'' \in [X \times_F X'](X'')$ approximating $\bar{\varepsilon}$, and ε'' defines a commutative diagram

$$\begin{array}{ccccc} & & X'' & & \\ & {\scriptstyle f}\swarrow & & \searrow{\scriptstyle f'} & \\ X & & & & X' \\ & \searrow & & \swarrow & \\ & & F & & \end{array} \quad \text{sending} \quad \begin{array}{ccccc} & & x'' & & \\ & \swarrow & & \searrow & \\ x & & & & x \end{array}$$

Since $\hat{\mathcal{O}}_{X'',x} \approx \bar{A} \approx \hat{\mathcal{O}}_{X,x}$, one sees immediately that the map f is etale at x''. Similarly, f' is etale at x''. Thus we may replace X'' by an open neighborhood of x'' to make these maps etale. Then (X'', x'', ξ'') is as required, where ξ'' denotes the composed map $X'' \longrightarrow F$.

3. A criterion for representability

Let S be as in § 1, i.e., of finite type over a field or over an excellent Dedekind domain. We recall (cf. [1], [8]) that a locally separated (respectively, separated) *algebraic space* Y, locally or finite type over S, is a functor

$$Y\colon (\text{Schemes}/S)^\circ \longrightarrow (\text{Sets})$$

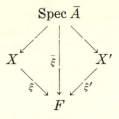

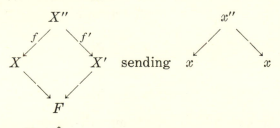

which is a sheaf for the etale topology [4], [5, exp. IV] and which is "locally representable" in the sense that there is an S-scheme locally of finite type U and a map $U \to Y$ which is represented by etale surjective maps. By this we mean more precisely that for every S-scheme V and map $V \to Y$, the product functor $U \times_Y V$ is represented by a subscheme (resp. closed subscheme) of $U \times_S V$, and the projection $U \times_Y V \to V$ is etale and surjective.

With this notation, the algebraic space Y is just the quotient U/R as sheaf for the etale topology, where $R = U \times_Y U \subset U \times_S U$. The scheme R is an equivalence relation on U whose projection maps are etale. Conversely, given an S-scheme U locally of finite type and any subscheme S (resp. closed subscheme) $R \subset U \times_S U$ which is a categorical equivalence relation and whose projections are etale, the quotient U/R as etale sheaf is an algebraic space as above [1, § 1].

We want to derive a criterion for a functor F to be an algebraic space, or by abuse of language, to be represented by an algebraic space, and we need a few preliminary remarks. Let X be an S-scheme and $\xi\colon X \to F$ a map (i.e., $\xi \in F(X)$). Let x be a point of X.

Definition 3.1. We say that ξ is *formally etale* at x if for every commutative diagram of solid arrows

$$\begin{array}{ccc} X & \xleftarrow{\ f_0\ } & Z_0 \\ {\scriptstyle \xi}\big\downarrow & \nwarrow^{f} & \big\downarrow \\ F & \longleftarrow & Z \end{array} \tag{3.2}$$

where Z is the spectrum of an artinian $\mathcal{O}_S$-algebra, Z_0 is a closed subscheme of Z defined by a nilpotent ideal, and f_0 is a map sending Z_0 to x set theoretically, there exists a unique dotted arrow f making the diagram commute.

LEMMA 3.3. *Let X be an S-scheme and let $\xi\colon X \to F$ be a map. Assume that ξ is relatively representable by morphisms locally of finite presentation*, i.e., *that for every scheme Z and map $Z \to F$ the fibered product $X \times_F Z$ is representable and is locally of finite presentation over Z. Let $x \in X$. Then ξ is formally etale at x if and only if the following condition holds. Let Y be an F-scheme, and let C be the inverse image of x in $X \times_F Y$. Then the projection $X \times_F Y \to Y$ is etale at every point of C.*

PROOF. The criterion (3.2) is evidently of the type which is preserved under base change, i.e., which holds for $X \times_F Y \to Y$ at any point of C. By (EGA IV, 17.14.2), formally etale for a map of schemes locally of finite presentation implies etale. Thus the condition of the proposition is verified. Conversely, suppose condition

(3.3) holds. Then with the notation of (3.2), the map $\mathrm{pr}_2\colon X \times_F Z \to Z$ is etale at the image of Z_0 under the graph of f_0, viewed as a map $\Gamma_{f_0}\colon Z_0 \to X \times_F Z$. Hence the section Γ_{f_0} extends uniquely to a section of pr_2, which is the graph of the required map f.

Let F be a functor (1.1). We follow Grothendieck's definition (FGA, 195) of pro-representability of F, in which the functor is restricted to the category of $\mathcal{O}_S$-algebras A which are *artinian* and of *finite type*. Thus the assertion that F is pro-representable means that there exists data as follows:

(a) an index set I,

(b) an $\mathcal{O}_S$-field of finite type k^p, and an element $\xi_0^p \in F(k^p)$ for each $p \in I$, and

(c) a formal deformation $(\bar{A}^p, \{\xi_n^p\})$ of ξ_0^p for each $p \in I$.

Here $\bar{A}^p$ is assumed to be a complete *noetherian* $\mathcal{O}_S$-algebra. The data has the property that for every artinian local $\mathcal{O}_S$-algebra B of finite type and every element $\eta \in F(B)$, there is a unique $p \in I$ and map $\bar{A}^p \to B$ sending $\{\xi_n^p\} \mapsto \eta$. Thus $(\bar{A}^p, \{\xi_n^p\})$ is in particular a universal formal deformation of ξ_0^p.

Effective pro-presentability is defined in the evident way as pro-representability together with the existence of an element $\bar{\xi}^p$ in $F(\bar{A}^p)$ inducing $\{\xi_n^p\}$ for each $p \in I$.

Suppose F pro-representable. Then we call the elements $p \in I$ the *points* of finite type of F. It is clear that if F is represented by an algebraic space X, then this notion is equivalent with the one given in [1, (2.3)] for a point x of X whose residue field $k(x)$ is of finite type over $\mathcal{O}_S$, and that the residue field identifies canonically with the field k^p *via* $\xi_0^p\colon \mathrm{Spec}\, k^p \to F$. Moreover, the ring $\bar{A}^p$ is canonically identified with the completion $\hat{\mathcal{O}}_{X,x}$ of the local ring at x [1, (2.5)].

We can now easily derive the following rather formal result.

THEOREM 3.4. *Let F be a contravariant functor from S-schemes to sets. Then F is a locally separated algebraic space (respectively, a separated algebraic space) locally of finite type over S if and only if the following conditions hold:*

[0] (sheaf axiom). *F is a sheaf for the etale topology.*

[1] (finiteness). *F is locally of finite presentation.*

[2] (pro-representability). *F is effectively pro-representable.*

[3] (relative representability). *Let X be an S-scheme of finite type, and let $\xi, \eta \in F(X)$. Then the condition $\xi = \eta$ is represented*

by a subscheme (*resp. a closed subscheme*) *of* X.

[4] *Let* X *be an* S*-scheme of finite type, and let* $\xi\colon X \to F$ *be a map. If* ξ *is formally etale at a point* $x \in X$, *then it is formally etale in a neighborhood of* x.

We number the sheaf condition [0] because it is just the natural condition which allows us to extend F to a functor on the category of algebraic spaces over S, so that the question of representability as algebraic space is reasonable. Conditions [1], [2] need no comment, and we will discuss [4] later.

The condition of relative representability [3] is well known to be equivalent, in the presence of [1], with the condition that for any pair of maps $\xi\colon X \to F$, $\eta\colon Y \to F$, the fibred product $X \times_F Y$ is represented by a subscheme (resp. closed subscheme) of $X \times_S Y$. For, a trivial limit argument using [1] shows that it suffices to consider schemes X, Y of finite type over S. Then $X \times_F Y$ can be described as the kernel of the pair of maps $\mathrm{pr}_1, \mathrm{pr}_2\colon X \times_S Y \to F$. Hence [3] implies the representability of the fibred product. Conversely, we can describe the kernel of a pair $\xi, \eta\colon X \to F$ as the object K making the diagram

$$\begin{array}{ccc} K & \longrightarrow & X \\ \downarrow & & \downarrow{\scriptstyle\Delta} \\ X \times_F X & \longrightarrow & X \times_S X \end{array}$$

cartesian.

PROOF OF THEOREM 3.4. *Necessity*. Suppose $F = Y$ is an algebraic space as in (3.4). Then condition [0] holds by definition. Condition [3] is given explicitly in [1, (1.5)]. To verify [1], we interpret Y as a quotient U/R of a scheme by an etale equivalence relation. Then a map f of an affine scheme Z to Y is represented by an etale covering Z' of Z of finite presentation, and a diagram

$$\begin{array}{ccccc} Z'' & \rightrightarrows & Z' & \longrightarrow & Z \\ \downarrow & & \downarrow{\scriptstyle f'} & & \\ R & \rightrightarrows & U & & \end{array}$$

where the upper and lower squares commute, and with $Z'' = Z' \times_Z Z'$. Two maps f_1, f_2 are equal if and only if they can be represented by maps $f_i'\colon Z' \to U$ for some Z' as above in such a way that the map $(f_1', f_2')\colon Z' \to U \times_S U$ factors through R. It is clear that this defines a functor locally of finite presentation.

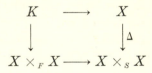

Consider condition [2]. Let $y \in Y$ be a point [1, (2.3)] of finite type. By [1, (2.4)] there is an etale neighborhood U of y in Y which is a scheme, and it is immediately seen that the completion of the local ring of y on U effectively pro-represents F at y, in the obvious sense. One could also avoid using [1, (2.4)] as follows: Choose an etale map $U \to Y$ which covers y, let $\{u_1, \cdots, u_n\}$ be the points of U lying over y, and let $\{v_1, \cdots, v_m\}$ be the points of $U \times_Y U$ lying over y. Let $\bar{A}'$ be the completion of the semi-local ring of U at $\{u_i\}$, and $\bar{A}''$ be the completion of the semi-local ring of $U \times_Y U$ at $\{v_j\}$. Then Spec $\bar{A}''$ defines an etale finite equivalence relation on Spec $\bar{A}'$, which has a quotient Spec $\bar{A}$, and $\bar{A}'$ is finite and etale over $\bar{A}$ [5, exp. V, 4.1]. By the sheaf condition [0], the map $\bar{\xi}'$: Spec $\bar{A}' \to Y$ factors through a map $\bar{\xi}$: Spec $\bar{A} \to Y$, and it is easily seen that $(\bar{A}, \bar{\xi})$ pro-represents F at y.

The property of being etale is local for the etale topology. Hence a map $X \to Y$ is represented by maps which are etale above $x \in X$ if and only if the map $X \times_Y U \to U$ is etale at some point lying over x. Since the projection $X \times_Y U \to X$ is etale, hence open, this is an open condition. If $X \to Y$ is formally etale at x, then $X \times_Y U \to U$ is etale at every point above x (3.3), hence $X \to Y$ is etale [1, § 2] in a neighborhood of x. This shows that [4] holds for Y.

Sufficiency. Suppose conditions [0]–[4] hold for F. We need to find a map $U \to F$ which is represented by etale surjective maps. Let p be a point of F of finite type, and let $(\bar{A}^p, \bar{\xi}^p)$ be the formal deformation pro-representing F at p. Since F is locally of finite presentation by condition [1], it follows from Theorem (1.6) that the pair $(\bar{A}^p, \bar{\xi}^p)$ is algebraizable, say by (X^p, x^p, ξ^p), where X^p is an S-scheme of finite type, etc. . .

LEMMA 3.5. *The map* $\xi^p : X^p \to F$ *is formally etale at* x^p.

PROOF. It is clear from the definition of pro-representability that (3.2) holds whenever Z is of finite type over S. We can now verify the condition of (3.3) as follows. Since F is locally of finite presentation, it suffices to treat the case that Y is of finite type over S. Then it is enough to prove that the projection $X \times_F Y \to Y$ is etale at *closed* points of C, which are of finite type over S. By (EGA IV. 17.14.2), etaleness is implied by the condition analogous to (3.2) for Z of finite type over S, which follows from (3.2) by base change.

Thus by condition [4], we may replace X^p by an open neighborhood of x^p so that $\xi^p : X^p \to F$ becomes formally etale at every point. Then since $X^p \to F$ is relatively representable by condition [3], it follows from Lemma (3.3) that it is represented by etale maps, i.e., that $X^p \times_F Z \to Z$ is etale for every F-scheme Z.

We claim that we can take for U the co-product of the X^p for all points p of F of finite type. In fact, let Z be any scheme, which we may assume of finite type. Then the projection $U \times_F Z \to Z$ is etale, and $U \times_F Z$ is a (closed) subscheme of $U \times_S Z$, by condition [3]. To show the projection surjective, it suffices to show that every point z of Z of finite type over S is in the image, since these points are dense. But by definition of pro-representability, the map $z \to F$ factors through Spec $\bar{A}^p$ for some point of finite type p, hence lifts to an X^p. Thus $X^p \times_F Z$ covers z, as required. This completes the proof.

The conditions [0]–[3] of the above theorem seem natural to us, but condition [4] is unpleasant since ways to verify it do not immediately come to mind. It can be expanded slightly, as follows.

Definition 3.6. Let F be a functor, X a scheme of finite type over S, and $\xi: X \to F$ a map which is represented by morphisms of finite presentation. We say that ξ is *etale* at a point $x \in X$ if for every S-scheme Y of finite type and map $Y \to F$, there is an open neighborhood U of x in X so that $U \times_F Y \to Y$ is etale.

By *geometric* discrete valuation ring, we mean one which is a localization of an $\mathcal{O}_S$-algebra of finite type, and with residue field of finite type over $\mathcal{O}_S$.

THEOREM 3.7. *Condition* [4] *of Theorem* (3.4) *may be replaced by the following two conditions*:

[4a] *Let A be a geometric discrete valuation ring, and let A'_K be an artinian local $\mathcal{O}_S$-algebra whose residue field is the field of fractions K of A. Let $\xi_1 \in F(A)$, $\xi_2 \in F(A'_K)$ be elements which induce the same element of $F(K)$. Then there is an $\mathcal{O}_S$-algebra A', an element $\xi \in F(A')$, and a diagram*

$$\begin{array}{ccc} & A' & \\ {\scriptstyle f}\swarrow & & \searrow{\scriptstyle g} \\ A & & A'_K \end{array} \quad \textit{sending} \quad \begin{array}{ccc} & \xi & \\ \swarrow & & \searrow \\ \xi_1 & & \xi_2 \end{array},$$

where f is surjective and with nilpotent kernel, and where g is the localization at the generic point of Spec $A' \approx$ Spec A.

[4b] *Let X be an S-scheme of finite type and $\xi: X \to F$ a map.*

If ξ is etale at a point $x \in X$, then it is etale in a neighborhood of x.

These two conditions are of completely different natures. Condition [4a] concerns a definite finite set of data and is usually easy to verify in practice. On the other hand, condition [4b] just asserts that, in the notation of definition (3.6), *the neighborhood U may be chosen independently of $Y \to F$*, and this cannot be tested on a finite set of data. A typical example (as in [1, (5.10)]) where [4b] does not hold is a union of the x-axis and of infinitely many vertical lines in the plane:

(3.8) 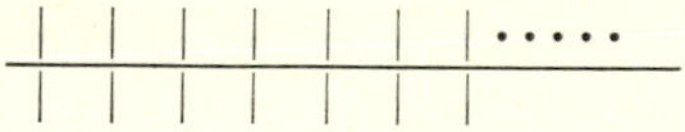

Examples illustrating the other conditions are also given if [1].

But [4b] will be automatic in many cases, for instance if the functor F is normal.

THEOREM 3.9. *With the notation of* (3.4) *and* (3.7), *suppose that condition* [2] *holds, and that the complete local rings $\bar{A}^p$ which pro-represent F all have geometrically unibranch spectra* (EGA IV, 6.15, *or* EGA0, 23.2.2) *and are free of embedded components. Then condition* [4b] *can be eliminated,* i.e., *F is represented by an algebraic space if and only if* [0]–[3] *and* [4a] *hold. If in addition the ground scheme S is of finite type over a field or over an excellent Dedekind domain having infinitely many points, and if the rings $\bar{A}^p$ are all normal and of a fixed Krull dimension d, then it is enough that* [0]–[3] *hold.*

PROOF OF THEOREM 3.7. *Necessity.* The necessity of [4b] is proved exactly as is the necessity of [4]. Suppose $F = Y$ is an algebraic space, and consider the situation of [4a]. It is clear that some ring A' exists admitting a diagram as in [4a], and which is a localization of an $\mathcal{O}_S$-algebra of finite type. This is a simple exercise which we omit. Let $Z' = \operatorname{Spec} A'$, and consider the scheme-theoretic closure W of the graph Γ_2 of ξ_2 in $Z' \times_S Y$. Since closure commutes with etale extensions, this notion extends to algebraic spaces. Clearly W is quasi-finite over Z', and has the graph Γ_1 of ξ_1 as subscheme. The graph Γ_1 is closed in an open set W' of W having the same underlying set as Γ_1. Thus $\Gamma_1 = W'_{\text{red}}$, and so W' is a scheme [1, (3.2)] and is finite over Z'. Hence the required map ξ exists if we replace A' by $\Gamma(W', \mathcal{O}_{W'})$.

Sufficiency. We proceed as in the proof of (3.4). Let (X^p, x^p, ξ^p) be an algebraization of the formal deformation $(\bar{A}^p, \bar{\xi}^p)$ which pro-represents F at the point p. Then using [4a], we will show

LEMMA. *The map* $\xi^p \colon X^p \to F$ *is etale at* x^p.

Since by [4b] etaleness is an open condition, the proof may then be completed as for (3.4).

Now we already know that ξ^p is formally etale at x^p, by (3.5), and so it suffices to prove the following lemma:

LEMMA 3.10. *Let F be a functor* (1.1) *satisfying conditions* (3.4) [1], [3] *and* (3.7) [4a], *let X be an S-scheme of finite type, and let $\xi \colon X \to F$ be a map which is formally etale at a point $x \in X$ of finite type. Then ξ is etale at x.*

PROOF. Let Z be an S-scheme of finite type, and $Z \to F$ a map. By (3.4) [3], the functor $X \times_F Z$ is represented by a scheme of finite type over S. We want to show that the map $X \times_F Z \to Z$ is etale above some neighborhood of x. Since ξ is formally etale at x, the map is etale at every point lying over x.

Setting $Z = X$, we find in particular that $X \times_F X$ is etale over X at the point (x, x) of the diagonal Δ. Hence there is a neighborhood of x in X above which Δ is open and closed in $X \times_F X$. We may replace X by this neighborhood, so that Δ is open and closed in $X \times_F X$. Since this property is preserved under pull-backs, it follows from [3] that $\xi \colon X \to F$ is representable by *unramified* maps (EGA IV, 17.4.2) of finite presentation. We now assume that this is so.

Returning to our map $Z \to F$, set $W = X \times_F Z$, and let

$$\begin{array}{ccc} W & \xrightarrow{g} & Z \\ {\scriptstyle f}\downarrow & & \downarrow \\ X & \xrightarrow{\xi} & F \end{array}$$

be the canonical cartesian diagram. We know that g is an unramified map. Let $C \subset W$ be the closed set of points at which g is not etale. What we have to prove is that x is not in the closure of $f(C)$. Suppose the contrary. Since f is of finite type, the image $f(C)$ is a constructible set. Therefore it contains a point $q \in X$ whose closure $\bar{q}$ in the local scheme $X_x = \operatorname{Spec} \mathcal{O}_{X,x}$ is of dimension one. Let $A_0 = \mathcal{O}_{X,x}/\mathcal{I}(\bar{q})$. This is a local integral domain of dimension one, localized from an $\mathcal{O}_S$-algebra of finite type.

Choose a point $w \in C$ lying over q, whose residue field K is finite over $k(q)$. Put $z = g(w)$. Then K is a finite extension of $k(z)$. To derive a contradiction, it is permissible to replace Z by a finite flat extension, locally at z, since non-etaleness is preserved under such an extension. Doing this suitably reduces us to the case that w and z have the same residue field K.

Let Z_n be the n-th order infinitesimal neighborhood of z in Z, and W_n the component of $W \times_Z Z_n$ containing w. Then $W_n \to Z_n$ is unramified and without residue field extension, hence is a closed immersion. Since g is assumed not to be etale at w, we have $W_n \neq Z_n$ for sufficiently large n. Write $Z_n = \operatorname{Spec} A'_K$, and let $\eta_2\colon Z_n \to F$ be induced by the inclusion of Z_n in the F-scheme Z.

Choose one of the finite set of discrete valuation rings A which are localizations of the integral closure of A_0 in K. Let $\eta_1\colon \operatorname{Spec} A \to F$ be the canonical map. Then η_1, η_2 agree on $\operatorname{Spec} K$ by construction. Hence we may apply condition [4a]. Let $\eta \in F(A')$ be the element whose existence is implied by that condition.

Consider the cartesian diagram obtained from η.

$$\begin{array}{ccc} X \times_F (\operatorname{Spec} A) & \longrightarrow & \operatorname{Spec} A \\ \downarrow & & \downarrow \\ X \times_F (\operatorname{Spec} A') & \longrightarrow & \operatorname{Spec} A' . \end{array}$$

Here the horizontal maps are of finite type and the vertical ones are obtained by dividing by the ideal of nilpotent elements of A'. The upper horizontal arrow has a section s given by the canonical lifting of η_1 to $\operatorname{Spec} A \to X$, and the image of the closed point *via* this section lies above our point x. Thus (3.3) both horizontal maps are etale at this image point. Therefore s extends to a section s' for the bottom arrow. This means that η lifts to a map $\operatorname{Spec} A' \to X$. Since it agrees with $\eta_2\colon \operatorname{Spec} A'_K = Z_n \to F$ in F, it gives a lifting of the inclusion of Z_n in Z to $Z_n \to W$, and by construction, this map sends $\operatorname{Spec} K$ to w, i.e., has as image the point w. Since $W_n \neq Z_n$ this is a contradiction, and completes the proof of the lemma.

PROOF OF THEOREM 3.9. Proceeding as in the proofs of (3.4), (3.7), let (X^p, x^p, ξ^p) be an algebraization of the formal deformation $(\bar{A}^p, \bar{\xi}^p)$ pro-representing F at p. It suffices to show that ξ^p is etale in a neighborhood of x^p, and as in the proof of (3.7), we may suppose $X^p \to F$ represented by unramified maps.

In the first case, we already know, dropping the superscript p, that ξ is etale (3.6) at x, by Lemma (3.10). Since the spectrum of the completion $\bar{A}^p$ of the local ring at x is geometrically unibranch, X is geometrically unibranch in a neighborhood of x (EGA IV, 9.7.10, and 7.8.3). Hence we may suppose X irreducible. Then we claim that ξ is etale, and it suffices to check this at closed points of X. Let x' be such a point, with image q in F, and let (X^q, x^q, ξ^q) be an algebraization of $(\bar{A}^q, \bar{\xi}^q)$. As for p, we may suppose ξ^q represented by unramified maps, and moreover we may suppose X^q geometrically unibranch and free of embedded components.

Consider the diagram

$$\begin{array}{ccc} X \times_F X^q = W & \xrightarrow{g} & X^q \\ {\scriptstyle f}\downarrow & & \downarrow{\scriptstyle \xi^q} \\ X & \xrightarrow{\xi} & F \end{array} \quad . \tag{3.11}$$

Since (3.10) ξ^q is etale at x^q, we may suppose f etale, again replacing X^q if necessary by a neighborhood of x^q. Moreover, g is unramified, and is etale above a neighborhood of x in X. Since f is etale and X irreducible, it follows that g is etale at all generic points of W. Then g must be etale. For, an unramified map is locally a closed immersion for the etale topology on the domain and range. Hence we are reduced to proving that a closed immersion which is etale at generic points of the domain, and whose range is geometrically unibranch and free of embedded components is, a local isomorphism. This is trivial.

Thus the maps g, f, ξ^q are etale at the points w, w, x^q respectively, where $w \in W$ lies over x' and x^q. This implies that ξ is etale at x', as is seen by pulling back to any F-scheme Z and applying (EGA IV, 17.7.1).

To prove the second part of (3.9) we remark to begin with that, under those hypotheses, the points of finite type of any scheme X of finite type over S are all closed points. Moreover, if A is an integral domain of finite type over S, then the Krull dimension of its local rings at closed points is constant. This is well known and easy to see.

Let $\xi: X \to F$ be as above. Then X is normal at x (EGA IV, 7.8.3), hence we may replace X by an open neighborhood of x which is entire, and we claim that then ξ is etale. It suffices to show that ξ is formally etale at every closed point $x' \in X$. Let x' be any closed point with image q in F, and let (X^q, x^q, ξ^q) be as above. We choose X^q so that it is entire and unramified over F, too. The Krull dimensions of X and X^q are equal to d, by assumption.

With the notation of (3.11), let $w \in W$ be the (unique) point lying over x' and x^q. Since (3.5) ξ^q is formally etale at x^q, f is etale at w (3.3). Hence $\dim_w(W)=d$. Since g unramified and X is entire, this implies that g is etale at W. It follows that ξ is formally etale at x'.

4. The theorem of Murre

Our version is the following.

THEOREM 4.1. *Let $S = \operatorname{Spec} k$, where k is a field, and let*

$$F\colon (\text{Schemes}/S)^0 \longrightarrow (\text{Groups})$$

be a functor. Then F is represented by a group scheme locally of finite type over S if and only if conditions [0]–[3] *of* (3.4) *hold.*

This result is an improvement over Murre's original version [12] in two respects. First of all we do not assume the groups abelian; secondly, we have eliminated the last of Murre's conditions, which concerns existence of a module for a map of a curve to F. The theorem of Matsumura and Oort [10] is also contained in (4.1).

The necessity of conditions [0]–[3] is contained in (3.4). The proof of sufficiency is in two parts: We will show that when F is a group functor, condition [4] of (3.4) can be eliminated, so that F is representable by an algebraic space. When this is done, the theorem follows from

LEMMA 4.2. *Let X be an algebraic space locally of finite type over a field k, which is a group object* (i.e., *is an "algebraic group"*). *Then X is a scheme.*

We remark that the ground field k in (4.2) can actually be replaced by any artinian ring. This follows immediately from (4.2) for the case k is a field and [1, (3.2)]. Applying this fact to (3.4), one obtains

COROLLARY 4.3. *Let $S=\operatorname{Spec} A$, where A is an artinian ring, and let F:* $(\text{Schemes}/S)^0 \to (\text{groups})$ *be a functor. Then F is represented by a group scheme locally of finite type over S if and only if conditions* [0]–[4] *of* (3.4) *hold.*

However, condition [4] is necessary when A is not a field, as is easily seen from examples similar to (3.8).

Suppose conditions [0]–[3] of (3.4) hold. To prove that F is an algebraic space, let $(\bar{A},\xi)$ be the effective formal deformation which pro-represents F at the identity element 1 of $F(k)$. Then the group law on F makes the formal scheme Spf $\bar{A} = \mathfrak{X}$ into a formal group. We claim that Spec $\bar{A}$ is geometrically unibranch and free of embedded components. This is certainly a general fact about formal groups, but in order to avoid applying some results about complete local rings which are not in the spirit of this discussion, we will deduce it from the algebraization as follows: Let (X, x, ξ) be an algebraization of the pair $(\bar{A}, \xi)$, which exists by (1.6). It is equivalent (EGA IV, 7.8.3) to show X geometrically unibranch and free of embedded components in a neighborhood of x, and to do this it is permissible to replace X by an etale neighborhood and to make a finite ground field extension $k \to k^*$; if we replace F by the induced functor F^*, etc., nothing essential will be changed.

Let $Y_1, \cdots, Y_r$ be the reduced irreducible subschemes of X corresponding to the associated primes of $\mathcal{O}_X$ passing through x. Replacing X by an etale neighborhood of x if necessary, we may suppose the Y_i are analytically irreducible at x. Then (after a ground field extension $k \to k^*$) the irreducible associated subschemes of $X \times X$ passing through (x, x) will be the $Y_i \times Y_j$. Denote by $\mathfrak{Y}_i$ the formal subscheme of $\mathfrak{X}$ corresponding to Y_i. It follows that the $\mathfrak{Y}_i \times \mathfrak{Y}_j$ are the associated irreducible subschemes of $\mathfrak{X} \times \mathfrak{X}$. Hence they must be permuted by the map

(4.4) $$\mathfrak{X} \times \mathfrak{X} \longrightarrow \mathfrak{X} \times \mathfrak{X}$$

defined functorially by

$$(\alpha, \beta) \longmapsto (\alpha, \alpha\beta) .$$

Using the canonical maps $\mathfrak{X} \to \mathfrak{X} \times \mathfrak{X}$, one sees immediately that $\mathfrak{Y}_i \times \mathfrak{Y}_j$ must be carried to itself by this map, and that $\mathfrak{Y}_i \subset \mathfrak{Y}_j$. Since this is true for each i, j, we have $r = 1$. Thus $\mathfrak{X}$ has no embedded components, and so X has none at x.

Again replacing k by a finite extension if necessary, we may suppose that the normalization of $(X \times X)_{\text{red}}$ is $\tilde{X} \times \tilde{X}$, where $\tilde{X}$ is the normalization of X_{red}. Let $Y \subset X$ be the closed subset of points where the map $\tilde{X} \to X_{\text{red}}$ is not an isomorphism. Then $\mathfrak{Y} \times \mathfrak{X} \cup \mathfrak{X} \times \mathfrak{Y}$ is an intrinsically defined closed subscheme of $\mathfrak{X} \times \mathfrak{X}$, hence must be carried into itself by the map (4.4). Such a $\mathfrak{Y}$ must be either empty or all of $\mathfrak{X}$. Since $\mathfrak{Y} = \mathfrak{X}$ is not possible, we have $\mathfrak{Y} = \varnothing$, whence X_{red} is normal, as required.

It follows that for every point p of F, Spec $\bar{A}^p$ (notation as in

§ 3) is geometrically unibranch and free of embedded components. For, to verify this it is permissible to make a finite field extension $k \longrightarrow k^*$, and one reduces oneself in this way to the case $k(p) = k$. Then the translation by p in F *via* the group law gives an isomorphism $\bar{A} \approx \bar{A}^p$.

We use a modification of the argument used in the proof of (3.9). Let (X, x, ξ) denote an algebraization of a formal deformation $(\bar{A}^p, \bar{\xi}^p)$ pro-representing F at p. Since Spec $\bar{A}^p$ is geometrically unibranch and free of embedded components, we may replace X by a neighborhood of x which is irreducible, geometrically unibranch, and free of embedded components. Moreover, we may suppose as in § 3 that ξ is represented by unramified maps. Let $x' \in X$ be any closed point with image q in F, and revert to the notation of (3.11). Then (3.5) ξ^q is formally etale at x^q, hence f is etale at the unique point $w \in W$ lying over x', x^q. We claim that g is etale at w. This will imply that ξ is formally etale at x', and hence that ξ is represented by etale maps, which will complete the proof.

It is enough to treat the case that the points x, x' are rational points. For, if we denote by a * a finite change of ground field $k \longrightarrow k^*$, then it suffices to show g^* is etale at any of the points of W^* lying over w. We can choose points $\tilde{x}$, $\tilde{x}'$ in X^* lying over x, x' respectively which are on the same irreducible component $\tilde{X}$ of X^*. Then let $\tilde{q}$ denote the image of $\tilde{x}'$ in F^*, etc... We are left with a situation analogous to the original one, and if k^* is chosen large enough, the points $\tilde{x}$, $\tilde{x}'$ will be rational points.

Now suppose x, x' are rational points. Since there is no residue field extension, and since g is unramified, the map $\varepsilon\colon \hat{\mathcal{O}}_{X^q,x^q} = \bar{A}^q \longrightarrow \hat{\mathcal{O}}_{W,w}$ is surjective, and we want to prove it bijective. Since X is irreducible, and f is etale, we have

$$\dim \hat{\mathcal{O}}_{W,w} = \dim_w W = \dim_x X = \dim \bar{A} = \dim \bar{A}^q .$$

Since Spec $\bar{A}^q$ is geometrically unibranch and free of embedded components, the map ε will be injective if it is so locally at the generic point of Spec $\bar{A}^q$. It therefore suffices to show that the lengths of $\bar{A}^q$ and $\hat{\mathcal{O}}_{W,w}$ at the generic points of their spectra are equal. Let us denote by $l(Z)$ the length of an irreducible scheme Z at its generic point. Then since f is etale, we have

$$l(X) = l(\operatorname{Spec} \mathcal{O}_{W,w}) = l(\operatorname{Spec} \hat{\mathcal{O}}_{W,w}) .$$

Hence

$$l(\operatorname{Spec} \hat{\mathcal{O}}_{W,w}) = l(X) = l(\operatorname{Spec} \bar{A}) = l(\operatorname{Spec} \bar{A}^q)\,,$$

which completes the proof.

It remains to prove Lemma (4.2). Now the connected component X^0 of X containing the identity is a subgroup of finite type over k. This is seen by the usual argument; if $U \subset X^0$ is a nonempty open subset of finite type, then $U \times U \to X^0$ is surjective. The cokernel X/X^0 is an etale group scheme [1, (3.3)]. Since algebraic groups are quasi-projective, there is no problem descending the ground field. Hence it suffices to prove X is a scheme in case X/X^0 is completely decomposed, and then X is a sum of copies of X^0. Thus we need only show X^0 is a scheme, which reduces us to the case that X is of finite type.

By [1, (3.1)] there is a dense open subset $U \subset X$ which is a scheme. This open subset has a birational law of composition on it, induced by the group law on X. Hence we may apply the theorem of Weil [16, p. 54, Th. 15] on construction of a group from birational data to complete the proof. The general case of this theorem, allowing nilpotent elements, is treated in [5, exp XVIII].

5. Further expansion of the conditions for representability

Let the ground scheme S be as in § 1. By *infinitesimal extension* of an $\mathcal{O}_S$-algebra A, we mean a surjective map of $\mathcal{O}_S$-algebras $A' \to A$ whose kernel is a finitely generated nilpotent ideal. The notion of infinitesimal extension $X \hookrightarrow X'$ of schemes is defined analogously.

Let F (1.1) be a functor. We will call *deformation situation* a collection of data

$$(A' \longrightarrow A \longrightarrow A_0,\ M,\ \xi_0) \tag{5.1}$$

where $A' \to A \to A_0$ is a diagram of infinitesimal extensions, A_0 is a noetherian $\mathcal{O}_S$-integral domain, $M = \ker (A' \to A)$ is a finite A_0-module, i.e., is annihilated by $\ker (A' \to A_0)$, and $\xi_0 \in F(A_0)$ is an element. The set of deformation situations is made into a category in the obvious way: A map of (5.1) to a deformation situation $(B' \to B \to B_0, N, \eta_0)$ consists of a commutative diagram

$$\begin{array}{ccccc} A' & \longrightarrow & A & \longrightarrow & A_0 \\ \big\downarrow{\scriptstyle f'} & & \big\downarrow{\scriptstyle f} & & \big\downarrow{\scriptstyle f_0} \\ B' & \longrightarrow & B & \longrightarrow & B_0 \end{array} \quad \text{sending} \quad \begin{array}{c} \xi_0 \\ \big\downarrow \\ \eta_0 \end{array}\,.$$

It induces an f_0-linear map $g: M \longrightarrow N$.

Definition (5.2). *A deformation theory* for F consists of

(a) A functor associating to every triple (A_0, M, ξ_0) (where A_0 is an $\mathcal{O}_S$-integral domain, M is an A_0-module of finite type, and $\xi_0 \in F(A_0)$) an A_0-module

$$D = D(A_0, M, \xi_0) ,$$

and to every map of triples $(A_0, M, \xi_0) \longrightarrow (B_0, N, \eta_0)$ (meaning a homomorphism $f_0: A_0 \longrightarrow B_0$ sending $\xi_0 \mapsto \eta_0$ and an f_0-linear map $g: M \longrightarrow N$) an f_0-linear map

$$\varphi: D(A_0, M, \xi_0) \longrightarrow D(B_0, N, \eta_0) .$$

If $f_0 =$ identity and $M = N$, so that the two triples are equal, then we require that φ be an A_0-linear function of g.

(b) For every deformation situation (5.1) an operation of the additive group of $D(A_0, M, \xi_0)$ on $F_{\xi_0}(A')$ such that two elements are in the same orbit under this operation if and only if they have the same image in $F_{\xi_0}(A)$. We require that this operation be compatible with the above map φ, when we are given a morphism of deformation situations.

Here $F_{\xi_0}(A')$ denotes the subset of $F(A')$ of elements whose image in $F(A_0)$ is ξ_0.

Note added in proof. Given an A-module M, let $A[M]$ denote the A-algebra $A \oplus M$, in which $MM = 0$. A deformation situation (5.1) yields obvious isomorphisms

$$A' \times_A A' \approx A'[M] \approx A_0[M] \times_{A_0} A' ,$$

hence induced maps

$$F(A') \times_{F(A)} F(A') \xleftarrow{\ u\ } F(A'[M]) \xrightarrow{\ v\ } F(A_0[M]) \times_{F(A_0)} F(A) .$$

Suppose u surjective and v bijective, for every deformation situation. Then one obtains a deformation theory with $D(A_0, M, \xi_0) = F_{\xi_0}(A_0[M]$, as in Schlessinger [14, 2.17]: The map $A'[M] \longrightarrow A'$ Sending $(a', m) \mapsto a' + m$ and the isomorphism v define an operation of D on $F_{\xi_0}(B')$, and the surjectivity of u implies that this operation is transitive on elements lying over a given $\xi \in F_{\xi_0}(A)$. I do not known a reasonable example of a deformatian theory other than the ones obtained in this way.

A natural condition to put on F which implies the above conditions on u, v is the following:

For every deformation situation (5.1) and every map $B \to A$, where B is an infinitesimal extension of A_0, the map

$$(*) \qquad F(A' \times_A B) \longrightarrow (F(A') \times_{F(A)} F(B)$$

is bijective.

This is a straight-forward extension of Schlessinger's conditions, and it includes [4′](b) and [5′](a) of theorem (5.3) as well. Although our hypotheses in (5.3) are weaker, it seems pedantic to me now to insist on such minor points. The remaining conditions of (5.3) and the bijectivity of (*) are also necessary for the representability of F.

THEOREM (5.3). *Let F be a functor* (1.1) *on S-schemes, and suppose a deformation theory* (5.2) *given for F. Then F is represented by a locally separated (resp. separated) algebraic space, locally of finite type over S, if the following conditions hold.*

[0′] (sheaf axiom). *F is a sheaf for the* fppf-*topology* (cf. [5, exp. 4]).

[1′] (finiteness). *F is locally of finite presentation.*

[2′] (commutation with inverse limits). *Let $\bar{A}$ be a complete noetherian local $\mathcal{O}_S$-algebra with residue field of finite type over S. Then the canonical map*

$$F(\bar{A}) \longrightarrow \lim_{\leftarrow} F(\bar{A}/\mathfrak{m}^n)$$

is injective, and has a dense image.

[3′] (separation condition). (a) *Let A_0 be a geometric* (3.6) *discrete valuation ring, and let K, k be the field of fractions and residue field of A_0 respectively. If $\xi, \eta \in F(A_0)$ are elements which induce the same element in $F(K)$ and in $F(k)$ (resp. the same element in $F(K)$), then $\xi = \eta$.*

(b) *Let A_0 be an $\mathcal{O}_S$-integral domain of finite type, and let $\xi, \eta \in F(A_0)$. Suppose that there is a dense set $\mathcal{S}$ of points of* Spec A_0 *of finite type, such that ξ and η are equal* (i.e., *induce the same element*) *in $F(k(s))$ for all $s \in \mathcal{S}$. Then $\xi = \eta$ on a non-empty open subset of* Spec A_0.

[4′] (conditions on the deformation theory). (a) *The module $D(A_0, M, \xi_0)$ commutes with localization in A_0, and is a finite module when M is free of rank one.*

(b) *D operates freely on $F_{\xi_0}(A')$ if M is of length one.*

(c) *Let A_0 be an $\mathcal{O}_S$-integral domain of finite type. There is a non-empty open set U of* Spec A_0, *such that for every closed* *point* $s \in U$, *we have*

$$D \otimes_{A_0} k(s) = D(k, M \otimes_{A_0} k(s), \xi_{0s})$$

where ξ_{0s} denotes the element of $F(k(s))$ induced by ξ.

[5′] (conditions on the obstructions). (a) *Consider a deformation situation* (5.2) *with* A_0 *of finite type and* M *of length one, and a diagram*

$$\begin{array}{ccccc} B' & \longrightarrow & B & & \\ \downarrow & & \downarrow & & \\ A' & \longrightarrow & A & \longrightarrow & A_0 \end{array}$$

of infinitesimal extensions of A_0, *with* $B' = A' \times_A B$. *If* $b \in F(B)$ *is an element lying over* ξ_0 *whose image* $a \in F(A)$ *lifts to* $F(A')$, *then* b *can be lifted to* $F(B')$.

(b) *Consider data of the form* (5.2), *with* A_0 *a geometric discrete valuation ring with field of fractions* K, *and* M *free of rank one. Denote by* A_K, A'_K *the localizations of* A, A' *respectively at the generic points of their spectra. If an element* $\xi \in F(A)$ *has the property that its image in* $F(A_K)$ *lifts to* $F(A'_K)$, *then its image in* $F(A_0 \times_K A_K)$ *lifts to* $F(A_0 \times_K A'_K)$.

(c) *With the notation of* (5.2), *suppose* A_0 *of finite type and* M *free of rank* n, *and let* $\xi \in F(A)$. *Let* K *be the field of fractions of* A_0 *and denote by a subscript* K *the localization at the generic point of* Spec A_0. *Suppose that for every one-dimensional quotient* M^*_K *of* M_K *the lifting of* ξ_K *to* $F(A^*_K)$ *is obstructed, where* $A'_K \to A^*_K \to A_K$ *is the extension determined by* M^*_K. *Then there is a non-empty open set* U *of* Spec A_0 *such that for every quotient* ε *of* M *of length one with support in* U, *the lifting of* ξ *to* $F(A_\varepsilon)$ *is obstructed. Here* $A' \to A_\varepsilon \to A$ *denotes the resulting extension.*

What we have done in this theorem is to expand the conditions [2]–[4] of (3.4). Note also that we have strengthened the sheaf axiom [0′]. Although the list of conditions is long, each condition can be verified relatively easily in the cases of Hilbert and Picard functors, using known results on cohomology.

We remark that since F is locally of finite presentation by [1′], it suffices to construct a deformation theory for situations (5.2) where A_0 is of finite type over $\mathcal{O}_S$. One obtains the modules and operations in the general case by passage to the limit.

The proof of (5.3) is less formal than that of (3.4). We first prove that [2] and [4] of (3.4) hold under the additional hypothesis that the relative representability condition (3.4) [3] is already known to hold. Since [0] and [1] are included in (5.3), we may then apply (3.4) to conclude representability. To complete the proof, we will apply the theorem in this special case to deduce the relative representability [3].

To prove (3.4) [2], it suffices to prove F pro-representable by a sum of complete noetherian local rings. As we have already remarked (1.4), the effectivity follows from condition [2′]. To do this, we want to apply the criterion of Schlessinger [14]. But Schlessinger's criterion does not take into account residue field extensions, and so we have to do a little extra work. The question of residue field extensions has also been considered by Levelt (unpublished).

Let k' be an $\mathcal{O}_S$-field of finite type, and $\xi'_0 \in F(k')$. Put $Z' =$ Spec k'. Then by (3.4) [3], the fibered product $Z' \times_F Z'$ of $\xi'_0 \colon Z' \to F$ with itself is represented by a subscheme R of $Z' \times_S Z'$. This subscheme is a flat, finite (by the *Nullstellensatz*) equivalence relation on $Z' =$ Spec k', and thus admits a quotient ([5], exp. V, 4.1) $Z = Z'/R$ which is necessarily the spectrum of a field k. Moreover, we have (*loc. cit.*) $Z' \times_F Z' =$ Spec $(k' \otimes_k k')$. By the sheaf condition [0′], the element $\xi'_0 \in F(Z)$ is induced by some $\xi_0 \in F(Z)$, and the map $\xi_0 \colon Z \to F$ is a monomorphism. Thus every point ξ'_0: Spec $k' \to F$ factors through one which is a monomorphism, and this monomorphism is determined, up to canonical isomorphism.

Let ξ_0: Spec $k \to F$ be such a monomorphism, with k an $\mathcal{O}_S$-field of finite type. It suffices to show that F is pro-representable at ξ_0. By this we mean the following: Consider the functor which, to an artinian local $\mathcal{O}_S$-algebra Λ of finite type, associates the set of elements $\eta \in F(\Lambda)$ for which there is a map (necessarily unique) $k \to k'$, sending $\xi_0 \to \eta_0$. Here k' denotes the residue field of Λ, and $\eta_0 \in F(k')$ the element induced by η. This functor is pro-representable. Now the ring $\bar{A}$ pro-representing F at ξ_0 will be uniquely determined, if it exists, by those Λ having residue field k. It therefore suffices to find an $\bar{A}$ having the universal property for every Λ whose residue field is contained in a given finite extension k^* of k: the uniqueness will show that any of the $\bar{A}$ has the universal property for every Λ.

Let $s \in S$ be the image of Spec k. We may suppose k^* a normal extension of $k(s)$ which is a tensor product of extensions given by a single element. Then there exists a flat extension S^* of S, finite in a neighborhood of s, such that there is a single point s^* of S^* lying over s, and that $k(s^*) = k^*$. Let $S^{**} = S^* \times_S S^*$, let F^* be the S^*-functor $F \times_S S^*$, and let F^{**} be the S^{**}-functor $F \times_S S^{**}$.

We obtain a finite set $\xi_0^{*(1)}, \cdots, \xi_0^{*(r)} \in F^*(k^*)$ of points determined by the residue fields of $k \otimes_{\mathcal{O}_S} \mathcal{O}_{S*}$.

LEMMA (5.4). *There is a universal formal deformation* (cf.§1) $(\bar{A}^{*(i)}, \{\xi_n^{*(i)}\})$ *of* $\xi_0^{*(i)}$ *for each* i.

PROOF. We apply the criterion of Schlessinger [14] to the functor F^*. Since the deformation theory is given by finite modules [4′] (a), it follows that

$$\mathbf{F}^*_{\xi_0^{*(i)}}(k^*[\varepsilon]) = F_{\xi_0^{(i)}}(k^*[\varepsilon]) \qquad (\varepsilon^2 = 0)$$

is a finite dimensional k^*-vector space [14], where $\xi_0^{(i)} \in F(k^*)$ is the image of $\xi_0^{*(i)}$. It remains to show that F^* commutes with fibered products as follows. Let $A' \to A$ be a surjective map of local artinian $\mathcal{O}_{S*}$-algebras, with residue field k^*, and kernel of length one, and let $B \to A$ be another local artinian $\mathcal{O}_{S*}$-algebra. We have to show that

$$F^*(A' \times_A B) \xrightarrow{\sim} F^*(A') \times_{F^*(A)} F^*(B) .$$

More precisely, we have to show this isomorphism when F^* is replaced by $F^*_{\xi_0^{*(i)}}$, but the above is stronger. In fact, it is just the assertion that

$$F(A' \times_A B) \xrightarrow{\sim} F(A') \times_{F(A)} F(B) , \tag{5.5}$$

if we denote by the same letter an $\mathcal{O}_{S*}$-algebra and its underlying $\mathcal{O}_s$-algebra structure.

Let $a' \in F(A')$, $b \in F(B)$ be elements with the same image $a \in F(A)$. Then b lifts to some $b'' \in F(B')$, by condition [5′](a). We use the notation of [5′](a), so that $A_0 = k^*$. It is immediately seen that $B' \to B$ is surjective with kernel of length one, isomorphic to M, since $A' \to A$ is. By [4′](b), the module $D = D(A_0, M, \xi_0)$ operates freely. Hence there is a unique element $d \in D$ with $d\bar{b}^* = a'$, where $\bar{b}^*$ is the image of b^* in $F(A')$. Thus $b' = db^*$ is the unique element of $F(B')$ lying over a' and b, which proves the lemma.

Now let $\bar{A}^*$ be the product of the rings $\bar{A}^{*(i)}$. There is also a finite set of points of $F^{**}(k^*)$ determined by the residue fields of

$k \otimes_{\mathcal{O}_S} \mathcal{O}_{S^{**}}$, and the reasoning of (5.4) shows that a formal universal deformation exists for each of these. Denote the product of their rings by $\bar{A}^{**}$. Since $F^{**} = F^* \times_{S^*} S^{**}$ in two ways, the universal property of $\bar{A}^*$ implies that in fact $\bar{A}^{**} \approx \bar{A}^* \otimes_{\mathcal{O}_{S^*}} \mathcal{O}_{S^{**}}$ in two ways. It follows easily that Spec $\bar{A}^{**}$ is a finite flat equivalence relation on Spec $\bar{A}^*$, and therefore [5, exp. V, 4.1] admits a quotient, say Spec $\bar{A}$, over which Spec $\bar{A}^*$ is finite and flat, and $\bar{A}$ is necessarily a complete local $\mathcal{O}_S$-algebra. We claim that it has the required universal property.

Let Λ be an artinian local $\mathcal{O}_S$-algebra with residue field $k' \subset k^*$, and let $Z = \operatorname{Spec} \Lambda$. Using the universal elements of Spec $\bar{A}^*$ and Spec $\bar{A}^{**}$, we obtain a diagram

$$\begin{array}{ccccc} \operatorname{Hom}_S(Z, \operatorname{Spec} \bar{A}) & \longrightarrow & \operatorname{Hom}_{S^*}(Z^*, \operatorname{Spec} \bar{A}^*) & \rightrightarrows & \operatorname{Hom}_{S^{**}}(Z^{**}, \operatorname{Spec} \bar{A}^{**}) \\ & & \downarrow \alpha^* & & \downarrow \alpha^{**} \\ F(Z) & \longrightarrow & F^*(Z^*) & \rightrightarrows & F^{**}(Z^{**}) \end{array} \tag{5.6}$$

of which the first row is exact since S^* is faithfully flat over S in a neighborhood of s (FGA, 190–08), and the second is exact by the sheaf axiom [0′].

Since k^* is a normal extension of $k(s)$, the residue fields of $Z^* = Z \times_S S^*$ and $Z^{**} = Z \times_S S^{**}$ are all canonically isomorphic to k^*. Thus the universal property of $\bar{A}^*$ shows that the image of α^* is the set of elements ξ^* lying over ξ_0, i.e., such that the map $k \to \Lambda^*_{\text{red}}$ sends ξ_0 to the image of ξ^* in $F(\Lambda^*_{\text{red}}) = F^*(\Delta^*_{\text{red}})$. The analogous assertion holds for $\bar{A}^{**}$. Thus the exactness of (5.6) gives the required universal property of $\bar{A}$. This shows that condition [2] of (3.4) holds.

We will verify conditions [4a] and [4b] of (3.7) separately. Let us denote the geometric discrete valuation ring of [4a] by A_0, retaining the rest of the notation of that condition: Choose some $\mathcal{O}_S$-algebra A' inducing A'_K, and which is an infinitesimal extension of A. We may assume A' torsion free. Performing the extension step by step, we are reduced to the following situation. Consider a diagram of extensions $A' \to A \to A_0$ with $M = \ker(A' \to A)$ free of rank 1 over A_0, an element $\xi \in F(A)$, and a lifting ξ^* of ξ_K to $F(A'_K)$. Here we denote by a subscript K the localization at the generic point of Spec $A_0 = \operatorname{Spec} A'$. We have to lift ξ to an element $\xi' \in F(A')$ such that $\xi^* = \xi'_K$, but we may replace A', A by other (torsion free) infinitesimal extensions of A_0 with the same rings of fractions.

First of all, we may suppose that ξ lifts in some way to $F(A')$, say to an η'. For, by condition [5′](b), the image of ξ in $F(A_0 \times_K A_K)$ lifts to $F(A_0 \times_K A'_K)$. Now the product $A_0 \times_K A'_K$ contains A' and is generated over that subring by pairs $(0, n)$, where n is in the nilradical of A'_K. Hence, viewing this product as a subring af A'_K, we have $A_0 \times_K A'_K = \lim_{\overrightarrow{\nu}} A'[t'^{-\nu}N']$, where N' is the nilradical of A', and $t' \in A'$ is some non-zero-divisor in the maximal ideal. Similarly, $A \times_K A_K = \lim_{\rightarrow} A[t'^{-\nu}N]$, where N is the nilradical of A. Since F is locally of finite presentation by [1′], the image of ξ lifts to $F(A'[t'^{-\nu}N'])$ for some ν, and so we may replace A, A' by $A[t'^{-\nu}N]$, $A'[t'^{-\nu}N']$ respectively.

Let $D = D(A_0, M, \xi_0)$. By [4′](a), we have

$$D \otimes_{A_0} K = D(K, M \otimes_{A_0} K, \xi_{0K}) .$$

Hence there is an element $d \in D$ and an integer n so that

$$t^{-n} d(\eta'_K) = \xi^* \qquad \text{in } F(A'_K),$$

where t is a local parameter of A_0. Let $m \in M$ be a generator of this module, and let $\tilde{A}' \supset A'$ be the ring obtained by adjoining $t^{-n}m$. Then we have a diagram

$$\begin{array}{ccccc} \tilde{A}' & \longrightarrow & A & \longrightarrow & A_0 \\ \cup & & \| & & \| \\ A' & \longrightarrow & A & \longrightarrow & A_0 , \end{array}$$

and we may identify $\ker(\tilde{A}' \to A)$ with M *via* multiplication by t^n in the natural inclusion. Thus D operates on $F_{\xi_0}(\tilde{A}')$ in such a way that the map $F_{\xi_0}(A') \to F_{\xi_0}(\tilde{A}')$ is compatible with multiplication by t^n in D. Denoting by $\sim$ the image in $F(\tilde{A}')$, we have

$$t^n \xi^* = (d\eta')_K = (\widetilde{d\eta'})_K = (t^n d\tilde{\eta}')_K ,$$

whence

$$d\tilde{\eta}'_K = \xi^* .$$

Thus we may replace A' by $\tilde{A}'$, and then $d\tilde{\eta}' = \xi'$ is the required element.

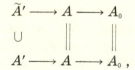

Now to verify [4b] of (3.7), let $\xi: X \to F$ be a morphism, with X affine and of finite type over S. We want to show that the condition of being etale at a point is an open condition. Suppose ξ etale at x. Then as in the proof of (3.7), we may, replacing X if necessary by an open neighborhood of x, assume ξ represented by unramified maps.

It is clear from the definition (3.6) that the condition of being etale is preserved under generalizations of points. Suppose it is not an open condition. Then there will be an irreducible closed subscheme Y of X and a dense set $\mathfrak{S}$ of points of Y, which we may assume of finite type, such that ξ is not etale at any point of $\mathfrak{S}$ but is etale at the generic point y of Y. Replacing X by an open neighborhood of y if necessary, we may suppose the points of $\mathfrak{S}$ closed. For, this is automatic if $\mathfrak{S}$ is of finite type over a field or over a Dedekind domain with infinitely many primes. If the Dedekind domain R has only finitely many primes, then either y lies over a closed point of Spec R, in which case $\mathfrak{S}$ consists of closed points, or it lies over the generic point, in which case we can localize X there.

Let $x \in X$ be any closed point, with image q in F, and let $X^q \to F$ be a map which is etale at a lifting $x^q \in X^q$ of q. Such a map exists as in (3.10), with $k(x^q) = k(q)$. With the notation of (3.11), g is unramified, and we may assume f etale. If g is etale at the unique point $w \in W$ lying over x and x^q, then it is immediately seen that $\xi\colon X \to F$ is etale at x.

LEMMA (5.7). *Let $\varphi\colon U \to V$ be an unramified map of finite type of noetherian schemes which is not etale at a closed point $u \in U$. Then there is an infinitesimal extension $U \hookrightarrow U_\varepsilon$ of U with $\mathcal{O}_U \approx \mathcal{O}_{U_\varepsilon}/(\varepsilon)$, ε an ideal of length one with support in u, and an unramified extension $\varphi_\varepsilon \to V$ of φ to U_ε.*

PROOF. Suppose first that V is the spectrum of a henselian ring. Then the connected component of U containing u is finite over V. We may replace U by that connected component. There is a unique connected finite etale extension $\tilde{V} \to V$ with residue field equal to $k(u)$, and φ factors by a closed immersion $i\colon U \hookrightarrow \tilde{V} \to V$. Since φ is not etale, i is not an isomorphism, and it is clear that an extension U_ε exists in this case, as a subscheme of $\tilde{V}$ containing U.

In general, there exists an etale neighborhood V' of $v = \varphi(u)$ and an infinitesimal extension $\varphi'_\varepsilon\colon U'_\varepsilon \to V'$ of $U' = U \times_V V'$. Now since the extension is of length one and $k(u) = k(u')$, the descent data for this extension φ'_ε relative to the etale base change $V' \to V$ (locally above v) is given automatically by the descent data for φ'. This follows immediately from the fact that φ is unramified:

one obtains an etale equivalence relation $(U' \times_U U')_\varepsilon = R_\varepsilon \rightrightarrows U'_\varepsilon$. Thus the required extension φ_ε with $U_\varepsilon = U'_\varepsilon/R_\varepsilon$ exists as an algebraic space. Since U is a scheme, so is U_ε [1, (3.2)].

Assume ξ not etale at x. Then applying the lemma to the map g, one obtains an unramified infinitesimal extension $g_\varepsilon: W_\varepsilon \to X^q$ of g of length one at the point w. Since $k(w) = k(x)$, this extension is induced from an infinitesimal extension X_ε of X of length one at the point x (same reasoning as above). By the sheaf axiom [0′], ξ extends to a map $\xi_\varepsilon: X_\varepsilon \to F$. This map is automatically unramified since g_ε is. We have thus proved

LEMMA (5.8). *If the unramified map $\xi: X \to F$ is not etale at the closed point x, then there is an unramified extension $\xi_\varepsilon: X_\varepsilon \to F$ of ξ of length one at x.*

Returning to our previous notation, let $Y \subset X$ be the irreducible closed subscheme and $\mathfrak{S}$ the dense set of closed points of Y at which ξ is not etale. We may assume $X = \operatorname{Spec} A$ lies over some affine etale over of S, say Spec $\mathcal{O}$. Now all the extensions X_ε of X of length one may be embedded in a suitable affine Space $\mathbf{E}_0^n$ compatibly with a given embedding of X. This is trivial. If say $A = \mathcal{O}[x_1, \cdots, x_n]/J$, and $X_\varepsilon = \operatorname{Spec} A_\varepsilon$ (X_ε is automatically affine), then the map $\mathcal{O}[x] \to A$ can be lifted to A_ε. Provided we add one extra x_ν with the relation $x_\nu = 0$ in A, we can make the lifting surjective, since the ideal ε is of length one.

We now consider all the X_ε as embedded in a given affine space. We will say that X_ε is induced by an infinitesimal extension Y_ε of Y of length one if $X_\varepsilon = X \cup Y_\varepsilon$ in the scheme-theoretic sense. This means that if J, J_ε, I, I_ε are the ideals of X, X_ε, Y, Y_ε respectively, in $\mathcal{O}[x]$, so that J/J_ε and I/I_ε are of length one, then we have

$$I_\varepsilon \cap J = J_\varepsilon .$$

When this is so, Y_ε is a closed subscheme of X_ε.

LEMMA (5.9). *There exists a closed subscheme $\tilde{Y}$ of X with underlying space that of Y, such that every extension X_ε of X of length one with support in Y is induced by an extension $\tilde{Y}_\varepsilon$ of $\tilde{Y}$.*

PROOF. An extension X_ε of length one is determined by a surjective homomorphism

$$J \longrightarrow k(x)$$

where $x \in X$ is the support of ε. To induce X_ε from a $\tilde{Y}_\varepsilon$, it suffices to extend this homomorphism to the ideal $\tilde{I}$ of $\tilde{Y}$. Let $\mathfrak{m}$ be the maximal ideal of X at x. Then the above map $J \longrightarrow k(x)$ factors through $J/\mathfrak{m}J$, and it is clear that the extension will always exist if the canonical map $J/\mathfrak{m}J \rightarrow \tilde{I}/\mathfrak{m}\tilde{I}$ is injective, i.e., if

$$\mathfrak{m}\tilde{I} \cap J = \mathfrak{m}J\,. \tag{5.10}$$

We try $\tilde{I} = I^n + J$. Then

$$\mathfrak{m}\tilde{I} \cap J = \mathfrak{m}(I^n + J) \cap J = (\mathfrak{m}I^n + \mathfrak{m}J) \cap J\,.$$

It is easily seen that

$$(\mathfrak{m}I^n + \mathfrak{m}J) \cap J = (\mathfrak{m}I^n \cap J) + \mathfrak{m}J\,.$$

Thus (5.10) holds if and only if

$$(\mathfrak{m}I^n \cap J) \subset \mathfrak{m}J\,.$$

Now by assumption, $x \in Y$, i.e., $\mathfrak{m} \supset I$. Since for large n we have $I^n \cap J \subset IJ$ by Artin-Rees, it follows that

$$(\mathfrak{m}I^n \cap J) \subset (I^n \cap J) \subset IJ \subset \mathfrak{m}J\,,$$

as required.

We replace Y by the $\tilde{Y}$ of the above lemma. Thus we may assume that every extension X_ε is induced by some extension Y_ε of Y, and we obtain unramified maps

$$\eta_\varepsilon\colon Y_\varepsilon \longrightarrow F \tag{5.11}$$

induced by the map ξ_ε of (5.8) at every point x of the dense set $\mathfrak{S} \subset Y$. We denote by $\eta\colon Y \rightarrow F$ the map induced by ξ.

LEMMA (5.12). *The map $\eta_\varepsilon\colon Y_\varepsilon \rightarrow F$ cannot be extended in any way to a map $\eta'\colon Y' \rightarrow F$, where $Y_\varepsilon \hookrightarrow Y'$ is an infinitesimal extension such that the kernel of $\mathcal{O}_{Y'} \rightarrow \mathcal{O}_Y$ is torsion free. Moreover, η_ε cannot be so extended in any neighborhood of the point $x =$* supp (ε) *on Y.*

PROOF. Let $\eta'\colon Y' \rightarrow F$ be such an extension, where we replace Y by an arbitrary neighborhood of x. Consider the diagram

$$\begin{array}{ccc}
X \times_F Y & \xrightarrow{g} & Y \\
\downarrow & & \downarrow \\
X \times_F Y_\varepsilon & \xrightarrow{g_\varepsilon} & Y_\varepsilon \\
\downarrow & & \downarrow \\
X \times_F Y' & \xrightarrow{g'} & Y' \\
\downarrow & & \downarrow \\
X & \xrightarrow{\quad\xi\quad} & F\,.
\end{array}$$

The maps g, g_ε, g' are all unramified. Since Y is a subscheme of X, it follows that $X \times_F Y$ is isomorphic to the diagonal of $Y \times_F Y$ locally at the point (x, x). Hence g is an isomorphism locally at that point.

We claim that moreover $X \times_F Y \to X \times_F Y_\varepsilon$ is an isomorphism at (x, x). In fact, this map is induced by the base change $Y_\varepsilon \to X_\varepsilon$ from $X \times_F X \to X \times_F X_\varepsilon$. Since $X \not\approx X_\varepsilon$, we know that

$$X \times_F X_\varepsilon \not\approx X_\varepsilon \times_F X_\varepsilon$$

at (x, x). But it is immediately seen that the infinitesimal extension $X \times_F X \hookrightarrow X_\varepsilon \times_F X_\varepsilon$ is of length one at (x, x). Hence necessarily $X \times_F X \approx X \times_F X_\varepsilon$, which proves the assertion. It follows that g_ε is not etale at (x, x).

Since $X \to F$ is etale at the general point y of Y, the map g' is etale above y on X. Thus this map is an immersion locally at (x, x) and is generically etale in a neighborhood of that point. Since the kernel of $\mathcal{O}_{Y'} \to \mathcal{O}_Y$ is torsion free and since g is an isomorphism locally at (x, x), it follows that g' is an isomorphism there. This contradicts the fact that g_ε is not etale at (x, x), and completes the proof of (5.12).

Now let Y' be any infinitesimal extension of Y such that each Y_ε can be realized as a subscheme of Y'. Write $Y = \operatorname{Spec} A$, $Y' = \operatorname{Spec} A'$, $Y_\varepsilon = \operatorname{Spec} A_\varepsilon$, and $Y_0 = \operatorname{Spec} A_0$, where $A_0 = A_{\text{red}}$. Since the extensions A_ε are of length one, we may suppose that ker $(A' \to A)$ is an A_0-module M, so that we have a deformation situation

$$(A' \longrightarrow A \longrightarrow A_0, M, \eta_0) .$$

Replacing A_0 etc$\cdots$ by a suitable localization, we may further assume M free, say of rank $n \geqq 1$.

Consider one dimensional quotients M_K^*, of the localization M_K of M at the generic point, as in condition [5′] (c). Since the set $\mathfrak{S}$ is dense, and we are given the extensions (5.11), this condition implies that η_K extends in some way to an $\eta_K^* \in F(A_K^*)$. Replacing Y by a non-empty affine open, we may suppose (by [1′]) that there is a one-dimensional quotient M^* of M such that η extends to an $\eta^* \in F(A^*)$. Let $Y^* = \operatorname{Spec} A^*$.

Case 1. There is a dense set of points $x \in \mathfrak{S}$ for which the extension Y_ε is not a subscheme of Y^*. In this case, $n > 1$, and we replace Y by Y^* and Y_ε by Y_ε^*, proceeding by induction on n. For we then have a diagram of type of [5′](a),

$$\begin{array}{ccccc} A_\varepsilon^* & \longrightarrow & A^* & & \\ \downarrow & & \downarrow & & \\ A_\varepsilon & \longrightarrow & A & \longrightarrow & A_0 \end{array}$$

where $A_\varepsilon^* = A_\varepsilon \times_A A^*$ is an infinitesimal extension of A^* of length one, and one sees as with (5.5) that $F(A_\varepsilon^*) = F(A_\varepsilon) \times_{F(A)} F(A')$. Thus the element η_ε extends to an element $\eta_\varepsilon^* \in F(A_\varepsilon^*)$.

Case 2. Y_ε is a subscheme of Y^* for a dense set of points $x \in S$. Then we may replace Y' by Y^*, which reduces us to the case $n = 1$, and that η lifts in some way to $\eta' \in F(A')$.

We apply condition [4′](c). Let $x \in \mathfrak{S}$ be a point, which we may assume to be in the open set U of [4′](c). If $\mathfrak{m}'$, $\mathfrak{m}_\varepsilon$, $\mathfrak{m}$ are the maximal ideals at x of A', A_ε, A respectively, then for large enough n we have a diagram

$$A_\varepsilon/\mathfrak{m}_\varepsilon^{n+1} \xrightarrow{f} A/\mathfrak{m}^{n+1} \longrightarrow k(x) ,$$

where the kernel of f is of length one. Denote by $(\eta_\varepsilon)_n$, $(\eta'_\varepsilon)_n$ the elements of $F(A_\varepsilon/\mathfrak{m}_\varepsilon^{n+1})$ induced by η_ε, η' respectively. By [4′] (c), there is an element $d^0 \in D \otimes k(x)$ which carries $(\eta'_\varepsilon)_n$ to $(\eta_\varepsilon)_n$. Let $d \in D$ represent d^0, and replace η' by $d\eta'$. Then $(\eta_\varepsilon)_n = (\eta'_\varepsilon)_n$, and we claim that if $\eta'_\varepsilon \in F(A_\varepsilon)$ is the element induced by η', then $\eta_\varepsilon = \eta'_\varepsilon$, which will contradict Lemma (5.12) and complete the proof of [4b]. But by (3.4)[3], the condition $\eta_\varepsilon = \eta'_\varepsilon$ is represented by a subscheme of Spec A_ε. This subscheme contains Spec $(A_\varepsilon/\mathfrak{m}_\varepsilon^{n+1})$ and Spec A, hence must be all of Spec A_ε. This proves [4b] of (3.7).

To complete the proof of Theorem (5.3), it remains to prove that the relative representability condition (3.4)[3] holds for F. Let $\xi, \eta: X \to F$ be as in that condition, and consider the functor

$$N\colon (\text{schemes}/X)^0 \longrightarrow (\text{sets})$$

which expresses the condition $\xi = \eta$, i.e., such that for an X-scheme Z,

$$N(Z) = \begin{cases} \varnothing & \text{if the elements of } F(Z) \text{ induced by } \xi, \eta \text{ are not equal} \\ \{\varnothing\} & \text{if they are equal.} \end{cases}$$

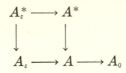

Now condition (3.4)[3] holds trivially for the functor N, since two elements of $N(Z)$ are always equal. Moreover, a deformation theory for N is given by the zero-modules. Thus we may apply what has already been proved to this functor. To prove it represented by an algebraic space Y locally of finite type over X, it suffices to verify conditions [0′]-[5′] of (5.3). Then $Y \to X$ is a monomorphism, whence locally quasi-finite, and so Y is a scheme [1, (3.3)].

Condition [3′] (b) will imply that Y is of finite type over X, and not only locally so. For, to prove this it suffices to write Y as union of locally closed sets which are noetherian topological spaces. Let $\bar{Y} \subset X$ be the closure of the image of Y. It follows from [3′] (b) that there is a non-empty open set U of X so that $Y \cap U$ covers $\bar{Y} \cap U$. Thus $Y \cap U$ is noetherian. By noetherian induction, we may assume $Y \cap (X - U)$ is also noetherian, which proves the assertion.

Finally, condition [3′](a) will imply that Y is a subscheme (resp. a closed subscheme) of X: It suffices to verify this at points $x_0 \in X$ of finite type which are in the image of Y. For, if $x_0 \notin \bar{Y}$, it is trivially true. If $\bar{Y} - Y$ were not closed, then there would be a point x_0 in the image of Y and in the closure of $\bar{Y} - Y$. Clearly Y cannot be an immersion there. Thus proving $Y \to X$ an immersion at the image points will also show $\bar{Y} - Y$ closed, hence that the map is everywhere an immersion.

Let x_0 be the image of $y_0 \in Y$. The property of being an immersion is local for the etale topology, and since $Y \to X$ is a monomorphism, hence unramified, it is a closed immersion locally for the etale topology on Y. Thus if we replace X by an etale neighborhood of x_0, we may suppose $Y \approx Y_1 \amalg Y_2$ where Y_1 is a closed subscheme of X containing x_0, and where Y_2 does not cover x_0. If the closure $\bar{Y}_2$ of Y_2 does not cover x_0, we are through. Suppose it does. Then since the image is constructible, it will contain a point which is locally of dimension one. Thus there will be a geometric discrete valuation ring A_0 and a map Spec $A_0 \to X$, such that its generic point maps to the image of Y_2 and its closed point maps to x_0. This contradicts [3′] (a).

It remains to verify conditions [0′]-[5′] of (5.3) for the functor N. Condition [3′] is contained in the above, and conditions [0′]-[2′] and [4′] are trivial. Consider [5′](a). In this condition, we have a diagram as in [5′](a) of $\mathcal{O}_X$-algebras. What has to be shown is that if the two elements $\xi, \eta \in F(X)$ are equal in $F(A')$ and in $F(B)$, then they are also equal in $F(B')$. Let $D = D(A_0, M, \xi_0)$ be the deformation module for F. Since $\xi_B = \eta_B$, there is an element $d \in D$ such that $d\xi_{B'} = \eta_{B'}$ in $F(B')$. But $\xi_{A'} = \eta_{A'}$ by assumption. Since by [4′](b) for F, D operates freely on $F_{\xi_0}(A')$, it follows that $d=0$, whence $\xi_{B'} = \eta_{B'}$.

Consider [5′](b). The assertion is this: Let a deformation situation (5.2) of $\mathcal{O}_X$-algebras be given, where A_0 is a geometric discrete valuation ring and M is free of rank one. Suppose that the elements $\xi, \eta \in F(X)$ are equal in $F(A)$ and in $F(A'_K)$. Then they are equal in $F(A_0 \times_K A'_K)$.

Denote by $\xi', \eta' \in F(A')$ the elements induced by ξ, η. Let $d \in D$ be such that $d(\xi') = \eta'$. Since $\xi'_K = \eta'_K$ and since D commutes with localization, the element induced by d in $D \otimes K$ leaves ξ'_K fixed, hence is zero by [4′] (b). Thus d is a torsion element of D; say $t^n d = 0$. Let $m \in M$ be a generator, and put $\tilde{A}' = A'[t^{-n}m]$. Then the kernel of $\tilde{A}' \to A$ is free of rank one, generated by $t^{-n}m$, and so D operates on $F_{\xi_0}(\tilde{A}')$ in such a way that the map $F_{\xi_0}(A') \to F_{\xi_0}(\tilde{A}')$ is compatible with multiplication by t^n in D. Hence we have

$$\tilde{\xi}' = (t^n d)\tilde{\xi}' = (\widetilde{d\xi'}) = \tilde{\eta}' \qquad \text{in } F(\tilde{A}') .$$

Since $\tilde{A}' \subset A_0 \times_K A'_K$, it follows that *a fortiori* $\xi = \eta$ in

$$F(A_0 \times_K A'_K) .$$

The assertion of [5′] (c) for N is this. Let a deformation situation (5.1) of $\mathcal{O}_X$-algebras of finite type be given, with M free of rank n, and suppose $\xi = \eta$ in $F(A)$. Suppose that $\xi \neq \eta$ in $F(A^*_K)$ for every one-dimensional quotient M^*_K of M_K. Then there is a non-empty open set U of Spec A such that $\xi \neq \eta$ in $F(A_\varepsilon)$ for every quotient ε of M of length one with support in U.

Now since D is a linear functor of M and M is free, it follows that

$$D = D(A_0, M, \xi_0) = \Delta^n \qquad \text{where } \Delta = D(A_0, A_0, \xi_0) .$$

Let $\xi', \eta' \in F(A')$ be the elements induced by ξ, η. Then there is an element $d = (\delta_1, \cdots, \delta_n) \in \Delta^n$ such that $d\eta' = \xi'$. Since the operation of $D(K, M^*_K, \xi_{0K}) \approx \Delta_K$ on $F_{\xi_0}(A^*_K)$ is free by [4′] (b), the assumption that $\xi \neq \eta$ in $F(A^*_K)$ for every M^*_K is equivalent with the linear independence of the images of $\delta_1, \cdots, \delta_n$ in Δ_K. Clearly there is an open set U so that the images of $\{\delta_i\}$ in $D \otimes k$ are linearly independent for every residue field k at a closed point u of U. We may suppose this U in the open set of condition [4′] (c). Then one sees easily that it is possible to find a map of deformation situations

$$\begin{array}{ccccccc} A' & \longrightarrow & A & \longrightarrow & A_0 & \quad M & \quad \xi_0 \\ \downarrow & & \downarrow & & \downarrow & \quad \downarrow & \quad \downarrow \\ \bar{A} & \longrightarrow & \bar{A} & \longrightarrow & k & \quad M \otimes k & \quad \xi_{0k} \end{array}$$

where $\bar{A} = A/\mathfrak{m}^r$ for sufficiently large r, $\mathfrak{m}$ being the maximal ideal at u. By [4′] (c), we have $D(k, M \otimes k, \xi_{0k}) = D \otimes k$. Hence the linear independence of the image of $\{\delta_i\}$ in $D \otimes k$ implies that $\xi \neq \eta$ in $F(\bar{A}_\varepsilon)$ for every quotient ε of $M \otimes k$ of length one. Thus certainly $\xi \neq \eta$ in $F(A_\varepsilon)$, if A_ε is the induced extension of A. This completes the proof of [5′] (c) for N, and proves Theorem (5.3).

6. Application to Hilbert functors

We adopt the notation of (FGA, 221). Our result is follows. Let $f\colon X \to S$ be a map locally of finite presentation of algebraic spaces, and let $\mathcal{F}$ be a coherent sheaf on X. Denote by Q the functor Quot $(\mathcal{F}/X/S)$ of Grothendieck, *viz.*, for every S-scheme Z, $Q(Z)$ is the set of isomorphism classes of quotients $\mathcal{G}$ of $\mathcal{F}_Z$ which are finitely presented, flat over Z, and with support proper over Z. Here $\mathcal{F}_Z$ denotes the coherent sheaf $\mathcal{O}_Z \otimes_{\mathcal{O}_S} F$ on $X_Z = Z \times_S X$.

THEOREM 6.1. *With the above notation, the functor Q is represented by an algebraic space locally of finite presentation over S. If f is separated, so is Q.*

COROLLARY 6.2. $\mathrm{Hilb}_{X/S}$ *is represented by an algebraic space locally of finite presentation over S.*

PROOF. If X' is an open subset of X and Q' is the functor Quot $(\mathcal{F}'/X'/S)$ (where $\mathcal{F}' = \mathcal{F} \mid X'$), then there is a canonical map $Q' \to Q$ which is represented by open immersions. For, Q' identifies with the subfunctor of Q of those quotient sheaves whose support lies in the inverse image of X'. Let $\mathcal{G} \in Q(Z)$ be a quotient of $\mathcal{F}_Z$. Then $C = \text{support } (G) \cap (X - X')_Z$ is closed and proper, and therefore $f(C)$ is closed in Z. Thus the condition of having support in X'_Z is an open condition, as required. Now it is immediately seen that Q is the union of the subfunctors Q', as X' runs over the (filtering) family of open subspaces of X of finite presentation. Thus we are reduced to proving representability for a finitely presented algebraic space. It also suffices to prove the separation assertion in that case.

It is clear by descent theory that Q is a sheaf for the fppf-topology. Thus the problem of representing Q is (by definition of algebraic space) local on S for the etale topology, and so we may assume S affine. Then the data $f: X \to S$, $\mathcal{F}$ are induced from some analogous data $f_0: X_0 \to S_0$, $\mathcal{F}_0$ *via* a map $S \to S_0$, where S_0 is an affine scheme of finite type over Spec **Z**. We are therefore reduced to the case $S = S_0$, i.e., that S is itself affine and of finite type over Spec **Z**. So we may apply Theorem (5.3).

The deformation theory for Q is described by Grothendieck in (FGA, 221–21). Since certain aspects are only tacitly given there, we will review the situation briefly for the convenience of the reader. Let

$$(A' \longrightarrow A \longrightarrow A_0, M, \mathcal{G}_0)$$

be a deformation situation (5.1). We use the notation $\mathcal{F}'$, $\mathcal{F}$, $\mathcal{F}_0$ for the sheaves induced on $X_{A'}$, X_A, X_{A_0} respectively, where $X_A = X \times_S \operatorname{Spec} A$, etc$\cdots$. Thus $\mathcal{G}_0$ is a quotient $\mathcal{F}_0/\mathcal{H}_0$ of $\mathcal{F}_0$ which is A_0-flat and with proper support. Consider a given quotient $\mathcal{G}$ of F inducing $\mathcal{G}_0$. Then $\mathcal{G}$ automatically has proper support, hence the requirement to be in $Q(A)$ is just that it be A-flat. We assume this so.

LEMMA 6.3. *Let $\mathcal{G}'$ be a quotient of $\mathcal{F}'$ inducing $\mathcal{G}$. Then $\mathcal{G}'$ is A'-flat,* i.e., $\mathcal{G}' \in Q_G(A)$, *if and only if the canonical sequence*

$$0 \longrightarrow \mathcal{G}_0 \otimes M \longrightarrow \mathcal{G}' \longrightarrow \mathcal{G} \longrightarrow 0\,,$$

which is automatically right exact, is exact. Here

$$\mathcal{G}_0 \otimes M = \mathcal{G}_0 \otimes_{A_0} M \approx \mathcal{G}' \otimes_{A'} M\,.$$

PROOF. If we tensor $\mathcal{G}'$ with the sequence $0 \to M \to A' \to A \to 0$, we, obtain

$$0 \longrightarrow \operatorname{Tor}_1^{A'}(\mathcal{G}', A) \longrightarrow \mathcal{G}_0 \otimes M \longrightarrow \mathcal{G}' \longrightarrow \mathcal{G} \longrightarrow 0\,.$$

Thus our sequence is exact if and only if $\operatorname{Tor}_1^{A'}(\mathcal{G}', A) = 0$. Now apply [6, exp. IV Th. 5.6].

It follows that for any $\mathcal{G}' \in Q_{\mathcal{G}}(A')$, we obtain an exact diagram

(6.4)
$$\begin{array}{ccccccccc}
 & & & & 0 & & 0 & & \\
 & & & & \downarrow & & \downarrow & & \\
 & & & & \mathcal{H}' & \dashrightarrow & \mathcal{H} & \dashrightarrow & 0 \\
 & & & & \downarrow & & \downarrow & & \\
0 & \longrightarrow & M\mathcal{F}' & \longrightarrow & \mathcal{F}' & \longrightarrow & \mathcal{F} & \longrightarrow & 0 \\
 & & \downarrow & & \downarrow & & \downarrow & & \\
0 & \dashrightarrow & \mathcal{G}_0 \otimes M & \dashrightarrow & \mathcal{G}' & \dashrightarrow & \mathcal{G} & \dashrightarrow & 0 \\
 & & \downarrow & & \downarrow & & \downarrow & & \\
 & & 0 & & 0 & & 0 & & ,
\end{array}$$

where $\mathcal{H}'$, $\mathcal{H}$ are the relevant kernels, and where the solid arrows are the given ones.

Now according to Grothendieck, the deformation theory for Q is given by the modules

(6.5) $$D(A_0, M, \mathcal{G}_0) = \mathrm{Hom}_{\mathcal{O}_{X_0}}(\mathcal{H}_0, \mathcal{G}_0 \otimes M) ,$$

and $D(A_0, M, \mathcal{G}_0)$ operates freely on $Q_{\mathcal{G}_0}(A')$. The operation can be described as follows: Denote by

$$\mathcal{E} = \mathcal{F}' \times_{\mathcal{F}} \mathcal{H}$$

the kernel of the canonical map $\mathcal{F}' \to \mathcal{G}$. Then given a diagram (6.4), the image of $\mathcal{E}$ in $\mathcal{G}'$ lies in $\mathcal{G}_0 \otimes M$. Thus $\mathcal{G}'$ determines an element $\varphi_{\mathcal{G}'}$ of

(6.6) $$\mathrm{Hom}_{\mathcal{O}_{X'}}(\mathcal{E}, \mathcal{G}_0 \otimes M) .$$

Since $\ker \varphi_{\mathcal{G}'} = \mathcal{H}'$ and since $\mathcal{H}'$ determines $\mathcal{G}'$, so does $\varphi_{\mathcal{G}'}$. Now from the map $\mathcal{E} \to \mathcal{H} \to 0$, we obtain an injection

$$\mathrm{Hom}(\mathcal{H}, \mathcal{G}_0 \otimes M) \hookrightarrow \mathrm{Hom}(\mathcal{E}, \mathcal{G}_0 \otimes M) ,$$

whence a free operation of

$$\mathrm{Hom}_{\mathcal{O}_{X'}}(\mathcal{H}, \mathcal{G}_0 \otimes M) = \mathrm{Hom}_{\mathcal{O}_{X_0}}(\mathcal{H}_0, \mathcal{G}_0 \otimes M)$$

on (6.6) by translation. An elementary verification shows that the subset $Q_{\mathcal{G}}(A')$ is a principal homogeneous space under this operation, as required.

LEMMA 6.7. *The obstruction to lifting $\mathcal{G}$ to an element $\mathcal{G}' \in Q(A')$ is the element of $\mathrm{Ext}^1_{X_{A'}}(\mathcal{H}, \mathcal{G}_0 \otimes M)$ determined from the extension*

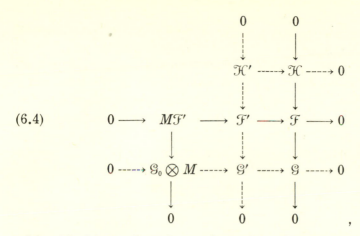

$$e\colon 0 \longrightarrow M\mathcal{F}' \longrightarrow \mathcal{F}' \longrightarrow \mathcal{F} \longrightarrow 0$$

via diagram (6.4).

PROOF. To find $\mathcal{G}'$ is equivalent with finding an element of $\mathrm{Ext}^1_{X_{A'}}(\mathcal{G}, \mathcal{G}_0 \otimes M)$ whose image in $\mathrm{Ext}^1_{X_{A'}}(\mathcal{F}, \mathcal{G}_0 \otimes M)$ is the same as that of e. Call this image $\bar{e}$. By the exact sequence of Ext for $0 \to \mathcal{H} \to \mathcal{E} \to \mathcal{G} \to 0$, such a $\mathcal{G}'$ exists if and only if $\bar{e}$ is mapped to zero in $\mathrm{Ext}^1_{X_{A'}}(\mathcal{H}, \mathcal{G}_0 \otimes M)$.

We now proceed to verify the conditions of Theorem (5.3) for Q. Conditions [0′] and [1′] are routine, and [2′] follows immediately from the existence theorem for formal sheaves (EGA III, 2; [8]).

[3′] (a) Since it is A_0-flat and hence torsion free, a quotient $\mathcal{G} \in Q(A_0)$ is determined uniquely by its generic fibre $\mathcal{G}_K = \mathcal{G} \otimes_{A_0} K$ and by its set-theoretic support, with the notation of [3′] (a). This is clear. Now if $f\colon X \to S$ is separated, then $\mathrm{supp}(\mathcal{G})$ is closed, and hence is just the closure of $\mathrm{supp}(\mathcal{G}_K)$. Thus $\mathcal{G}$ is determined by $\mathcal{G}_K$ in that case. In general, the support is determined by the two fibres $\mathcal{G}_K$, $\mathcal{G}_k$, hence $\mathcal{G}$ is determined by these two fibres, as was to be shown.

[3′] (b) Let $\mathcal{G}_1, \mathcal{G}_2 \in Q(A_0)$, and say $\mathcal{G}_i = \mathcal{F}/\mathcal{H}_i$ Put $\mathcal{E} = \mathcal{F}/(\mathcal{H}_1 + \mathcal{H}_2)$. Then $\mathcal{G}_1 = \mathcal{G}_2$ if and only if the map $\mathcal{G}_1 \to \mathcal{E}$ is an isomorphism. The formation of the sheaf $\mathcal{E}$ commutes with base change. Thus [3′] (b) reduces to the following: If a map $\varphi\colon \mathcal{G} \to \mathcal{E}$ of coherent sheaves has the property that $\varphi \otimes k(s)$ is an isomorphism for a dense set $\mathcal{S}$ of points s of Spec A_0, then φ is an isomorphism above an open set of Spec A_0. Or, if a coherent sheaf D on X_{A_0} has the property that $D \otimes k(s)$ is zero for all $s \in \mathcal{S}$, then D is zero above a non-empty open. This well known and trivial.

[4′] The finiteness theorem (EGA III, 3; [8]) for cohomology proves that $D(A_0, M, \mathcal{G}_0)$ is a finite module, since $\underline{\mathrm{Hom}}_{\mathcal{O}_{X_0}}(\mathcal{H}_0, \mathcal{G}_0 \otimes M)$ has proper support. It is clear that D commutes with localization. Thus [4′] (a), (b) hold. To verify [4′] (c), we need the following two lemmas.

LEMMA 6.8. *Let $\mathcal{F}, \mathcal{G}$ be coherent sheaves on an algebraic space X of finite type over $S =$ Spec A_0, where A_0 is a non-empty integral domain. Let q be an integer. There is a non-empty open set U of S such that for each $s \in U$ the canonical map*

$$\underline{\mathrm{Ext}}^q_X(\mathcal{F}, \mathcal{G})_s \longrightarrow \underline{\mathrm{Ext}}^q_{X_s}(\mathcal{F}_s, \mathcal{G}_s)$$

is an isomorphism, where the subscript s denotes restriction to *the fibre.*

PROOF. The problem is local on X for the etale topology. Hence we may suppose X affine. Now choose a free resolution of $\mathcal{F}$ and apply (EGA IV, 9.4.2, 3).

LEMMA 6.9. *Let* $X \xrightarrow{f} Y$ *be a map of S-algebraic spaces of finite type, where* $S = \operatorname{Spec} A_0$ *is the spectrum of a noetherian integral domain, and let* $\mathcal{F}$ *be a coherent sheaf on* X *with support proper over* Y. *Let* q *be an integer. Denote by a subscript* s *the restriction to the fibre at* $s \in S$. *There is a non-empty open* $U \subset S$ *such that for every* $s \in U$ *the canonical map*

$$(R^q f_* \mathcal{F})_s \longrightarrow R^q f_{s*} \mathcal{F}_s$$

is an isomorphism. In particular, setting $Y = S$, *there is a non-empty open* $U \subset S$ *so that*

$$H^q(X, \mathcal{F}) \otimes_{A_0} k(s) \xrightarrow{\sim} H^q(X_s, \mathcal{F}_s)$$

for all $s \in U$.

We know of no reference for this lemma. However, its proof involves standard techniques *via* reduction to the projective case, and we will leave the proof to the reader. Since in our situation the sheaf $\underline{\operatorname{Hom}}\,(\mathcal{H}_0, \mathcal{G}_0 \otimes M)$ has proper support, condition [4′] (c) follows immediately from the two lemmas.

[5′] Consider condition [5′] (a). With the notation of that condition, let $\mathcal{F}_A$ denote the sheaf induced by $\mathcal{F}$ on X_A, etc. If we are given elements $\mathcal{G}_{A'} \in Q(A')$ and $\mathcal{G}_B \in Q(B)$ inducing $\mathcal{G}_A$ in $Q(A)$, we obtain an exact diagram of $\mathcal{O}_{X_{A'}}$-modules

$$\begin{array}{ccccccccc} 0 & \longrightarrow & \mathcal{G}_0 \otimes M & \longrightarrow & \mathcal{G}_{A'} \times_{\mathcal{G}_A} \mathcal{G}_B & \longrightarrow & \mathcal{G}_B & \longrightarrow & 0 \\ & & \| & & \downarrow & & \downarrow & & \\ 0 & \longrightarrow & \mathcal{G}_0 \otimes M & \longrightarrow & \mathcal{G}_{A'} & \longrightarrow & \mathcal{G}_A & \longrightarrow & 0\,, \end{array}$$

and a canonical map $\mathcal{F}_{A'} \to \mathcal{G}_{A'} \times_{\mathcal{G}_A} \mathcal{G}_B$. It is easily seen that this map is surjective. Hence $\mathcal{G}_{A'} \times_{\mathcal{G}_A} \mathcal{G}_B$ is the required quotient of $\mathcal{F}_{A'}$.

In the situation of [5′] (b), denote by $i\colon X_{A'_K} \to X_+ = X_{A_0 \times_K A'_K}$ the inclusion. The functor i_* is exact on quasi-coherent sheaves, and since i^* is also exact, i_* carries injectives to injectives. Hence we have for any coherent $\mathcal{A}$, $\mathcal{B}$

$$\operatorname{Ext}^1_{X_{A'_K}}(i^*\mathcal{A}, \mathcal{B}) \approx \operatorname{Ext}^1_{X_+}(\mathcal{A}, i_*\mathcal{B})\,.$$

Now the obstruction to lifting an element $\mathcal{G} \in Q(A)$ to $Q(A_0 \times_K A'_A)$ lies (6.7) in $\operatorname{Ext}^1_{X_+}(\mathcal{H}_0, \mathcal{G}_0 \otimes \tilde{M})$, where $\tilde{M}$ is a free rank one K-

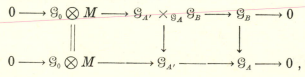

module, viewed as an A_0-module. Thus $\mathcal{G}_0 \otimes \tilde{M} \approx i_*(\mathcal{G}_{0K} \otimes \tilde{M})$, and it follows that the obstruction to lifting to $Q(A_0 \times_K A'_K)$ and to $Q(A'_K)$ are the same.

It remains, finally, to verify condition [5'] (c) of (5.3). In that condition, the module M is free of rank n. Hence the obstruction to lifting a $\mathcal{G} \in Q(A)$ to $Q(A')$ lies in

$$\mathrm{Ext}^1_{X_{A'}}(\mathcal{H}, \mathcal{G}_0 \otimes M) \approx \mathrm{Ext}^1_{X_{A'}}(\mathcal{H}, \mathcal{G}_0)^n = E^n .$$

Let the element be $e = (e_1, \cdots, e_n)$, and let the extension be

$$(6.10) \qquad 0 \longrightarrow \mathcal{G}_0 \otimes M \longrightarrow \mathcal{E} \longrightarrow \mathcal{H} \longrightarrow 0 .$$

The hypothesis that the lifting of $\mathcal{G}_K$ to $Q(A^*_K)$ be obstructed for every one-dimensional quotient M^*_K of M_K is clearly equivalent with the linear independence of the elements $\{e_i\}$ in E_K. It follows that there exists a non-empty open $U \subset \mathrm{Spec}\, A_0$ such that the images of the e_i in $E \otimes k(s)$ are linearly independent for all $s \in U$.

Consider a quotient $\varepsilon \approx k(s)$ of M of length one with support $s \in \mathcal{S}$, and the extension

$$(6.11) \qquad 0 \longrightarrow \mathcal{G}_0 \otimes \varepsilon \longrightarrow \mathcal{E}_\varepsilon \longrightarrow \mathcal{H} \longrightarrow 0$$

induced from (6.10). To split (6.11) as an X_{A_ε}-sequence is the same as to split it as an $X_{A'}$-sequence.

We claim that if we restrict U to a sufficiently small open set, then

$$(6.12) \qquad \underline{\mathrm{Ext}}^q(\mathcal{H}, \mathcal{G}_0 \otimes k(s)) \approx \underline{\mathrm{Ext}}^q(\mathcal{H}, \mathcal{G}_0) \otimes k(s)$$

for all $s \in U$, and all $q \leqq q_0$. It will then follow from the spectral sequence relating local and global Ext and Lemma (6.9) that, again restricting U if necessary,

$$\mathrm{Ext}^1(\mathcal{H}, \mathcal{G}_0 \otimes k(s)) \approx \mathrm{Ext}^1(\mathcal{H}, \mathcal{G}_0) \otimes k(s) = E \otimes k(s)$$

for all $s \in U$. Thus for s in a sufficiently small U, the sequence (6.11) does not split. This proves [5'] (c).

Now the assertion (6.12) is local on X_{A_0} for the etale topology, hence we may assume X_{A_0} affine. Let $\mathcal{L}\cdot$ be a free resolution of $\mathcal{H}$. Then $\underline{\mathrm{Hom}}(\mathcal{L}\cdot, \mathcal{G}_0 \otimes k(s)) = \underline{\mathrm{Hom}}(\mathcal{L}\cdot, \mathcal{G}_0) \otimes k(s)$. Since $\underline{\mathrm{Ext}}^q(\mathcal{H}, \mathcal{G}_0 \otimes k(s)) = H^q(\underline{\mathrm{Hom}}(\mathcal{L}\cdot, \mathcal{G}_0 \otimes k(s))$, we may apply (EGA IV, 9.4.3).

7. Application to the Picard functor

Throughout this section, $f: X \to S$ will denote a flat, proper map of algebraic spaces. The *relative Picard functor* $P = \mathrm{Pic}\, X/S$

is defined as in Grothendieck (FGA, 232): For every $g\colon S' \to S$,

$$P(S') = [\operatorname{Pic} X/S](S') = H^0(S', R^1f'_*\mathbf{G}_m) \,, \tag{7.1}$$

where f' is the map $f' = S' \times_S f\colon X' \to S'$, and $\mathbf{G}_m$ denotes the the sheaf of units (here on X'). The symbol $R^1f'_*$ must be interpreted as the derived functor of the direct image functor on sheaves for the fppf-topology. Thus $\operatorname{Pic} X/S$ is a functor from $(\text{schemes}/S)^0$ to (sets).

Recall (EGA III, 7.8.1) that f is called cohomologically flat in dimension zero if for every $S' \to S$ we have

$$\mathcal{O}_{S'} \otimes_{\mathcal{O}_S} (f_*\mathcal{O}_X) \approx f'_*\mathcal{O}_{X'} \,. \tag{7.2}$$

THEOREM 7.3. *Let $f\colon X \to S$ be a proper, flat map of algebraic spaces which is finitely presented and cohomologically flat in dimension zero. Then the relative Picard functor* $\operatorname{Pic} X/S = P$ *is represented by an algebraic space locally of finite presentation over S.*

PROOF. We need the results of (EGA III, 7). They are stated there only for noetherian schemes, but go over without change to the context of algebraic spaces, and the noetherian hypotheses are eliminated by standard limit considerations (EGA IV, 8).

The conditions that f be proper and flat are locally of finite presentation on S, by (EGA IV, 8). So is the cohomological flatness condition. In fact, it follows from (EGA III, 7.7.6) that there is a coherent sheaf $\mathcal{Q}$ on S and a functorial isomorphism

$$f'_*\mathcal{O}_{X'} \approx \underline{\operatorname{Hom}}_{\mathcal{O}_{S'}}(\mathcal{O}_{S'} \otimes_{\mathcal{O}_S} \mathcal{Q}, \mathcal{O}_{S'}) \tag{7.4}$$

for every $S' \to S$. Moreover (EGA III, 7.8.4), f is cohomologically flat in dimension zero if and only if $\mathcal{Q}$ is locally free. This is a condition locally of finite presentation. Thus the given data may, locally for the etale topology on S, be descended to an affine scheme S_0 of finite type over $\operatorname{Spec} \mathbf{Z}$, and so we may assume S is such a scheme. We may therefore apply Theorem (5.3).

It is perfectly possible to verify the conditions of (5.3) in the general case. However, we found that notationally inconvenient passages back and forth between various topologies were involved, and so we prefer to make two preliminary reductions at this stage. We leave the details of these reductions to the reader. First of all, let

$$\bar{S} = \underline{\operatorname{Spec}}_{\mathcal{O}_S} f_*\mathcal{O}_X \,.$$

Since f is cohomologically flat, it follows from (EGA III, 7.7.6) that $\bar{S}$ is finite and flat over S. Let the canonical maps be

$$\begin{array}{ccc} X & & \\ & \searrow^{\bar{f}} & \\ {\scriptstyle f}\big\downarrow & & \bar{S} \\ & \swarrow_{\pi} & \\ S & & \end{array} \quad .$$

Then it can be shown that $\bar{f}$ is flat, proper, and cohomologically flat in dimension zero. Moreover, we have

$$\pi_* \operatorname{Pic} X/\bar{S} = \operatorname{Pic} X/S \ ,$$

and

$$\bar{f}_* \mathcal{O}_X = \mathcal{O}_{\bar{S}} \ .$$

Now the direct image functor π_* carries algebraic spaces to algebraic spaces. This is the functor $\prod_{\bar{S}/S}$ in Grothendieck's notation (FGA, 195–12), and the fact is easily checked. Thus it suffices to to show $\operatorname{Pic} X/\bar{S}$ is representable, which reduces us to the case that

(7.5) $$f_* \mathcal{O}_X = \mathcal{O}_S \ .$$

Next, there is a finitely presented faithfully flat map $g: S' \to S$ such that X'/S' has a section. In fact, one can take $g = f$. It follows from [1, (7.2)] that the problem of representing $\operatorname{Pic} X/S$ is local on S for the fppf-topology. Thus we may replace S by S', which reduces us to the case that X/S has a section.

We now make these two assumptions. Then it follows from the Leray spectral sequence for the map $f': X' \to S'$ ($S' \to S$ arbitrary) that (FGA, 232, 2.4)

(7.6) $$P(S') = \operatorname{Pic} X'/\operatorname{Pic} S' \ .$$

Here $\operatorname{Pic} Z = H^1(Z, \mathbf{G}_m)$ is the group of invertible sheaves on Z, which is the same when calculated for the fppf or for the Zariski topology [4, exp. IX, (3.3)]. Thus we may now interpret the symbol $R^1 f'_*$ of (7.1) as the first derived functor of the direct image f'_* on the category of sheaves for the Zariski topology, and from now on all our calculations will be made in that topology. We need only remember that P is actually a sheaf for the fppf-topology, i.e., that condition [0′] of (5.3) holds for P.

The deformation theory for P is given as usual by the exponential map. For every deformation situation (5.1), we get an exact sequence of abelian sheaves on $X_{A'}$

$$(7.7) \qquad 0 \longrightarrow M_X \longrightarrow \mathcal{O}_{X_{A'}} \longrightarrow \mathcal{O}_{X_A} \longrightarrow 0 ,$$

where M_X denotes the coherent sheaf $M \otimes_{A_0} \mathcal{O}_{X_{A_0}}$ on X_{A_0}. The exponential map $x \mapsto 1 + x$ identifies this kernel with the kernel of the map ε,

$$(7.8) \qquad 0 \longrightarrow M_X \longrightarrow \mathcal{O}^*_{X_{A'}} \xrightarrow{\varepsilon} \mathcal{O}^*_{X_{A'}} \longrightarrow 0 .$$

Here $\mathcal{O}^*_Z = \mathbf{G}_{mZ}$.

Now since f is cohomologically flat in dimension zero, the map $f'_*\mathcal{O}_{X_{A'}} \to f'_*\mathcal{O}_{X_A}$ is surjective. Therefore $f'_*\mathcal{O}^*_{X_{A'}} \to f'_*\mathcal{O}^*_{X_A}$ is also surjective, and so (7.8) yields an exact sequence of sheaves on Spec A',

$$(7.9) \quad 0 \longrightarrow R^1f'_*M_X \longrightarrow R^1f'_*\mathcal{O}^*_{X_{A'}} \longrightarrow R^1f'_*\mathcal{O}^*_{X_A} \longrightarrow R^2f'_*M_X .$$

Here we have denoted by f' the map induced from f on Spec A'. Since Spec A' is affine, we have

$$H'(X_{A_0}, M_X) \approx H^0(X_{A'}, R^1f'_*M_X) ,$$

and so (7.9) yields

$$(7.10) \qquad 0 \longrightarrow H^1(X_{A_0}, M_X) \longrightarrow P(A') \longrightarrow P(A) .$$

Thus the A_0-module

$$(7.11) \qquad D(A_0, M, \xi_0) = H^1(X_{A_0}, M_X)$$

gives a deformation theory for P. Note that D does not depend on ξ_0.

We now proceed to verify the conditions of (5.3). Condition [0′] holds by definition, and [1′] is clear from (7.6).

[2′] Let A be a complete noetherian local $\mathcal{O}_S$-algebra. Since Pic (Spec A) = 0, it follows from (7.6) that $P(X_A) = \mathrm{Pic}\, X_A$. The interpretation of elements of Pic X_A as isomorphism classes of invertible sheaves and the existence theorem for formal sheaves [8, 4.3] imply easily that $P(A) \xrightarrow{\sim} \varprojlim P(A/\mathfrak{m}^n)$.

[3′] (a) We have Pic $(A_0) = 0$, whence by (7.6) the problem is the following. Let $\mathfrak{L}$ be an invertible sheaf on X_{A_0} having the property that $\mathfrak{L} \otimes K$ and $\mathfrak{L} \otimes k$ are trivial. Then $\mathfrak{L}$ is trivial, i.e., is isomorphic to $\mathcal{O}_{X_{A_0}}$. We are indebted to Raynaud for supplying the following proof, which is much simpler than our original argument. Let $\mathfrak{Q}_{\mathfrak{L}}$ be the A_0-module of (EGA III, 7.7.6) such that for all $g\colon S' \to \mathrm{Spec}\, A_0$,

$$f'_*(\mathcal{O}_{S'} \otimes_{A_0} \mathfrak{L}) \approx \underline{\mathrm{Hom}}_{\mathcal{O}_{S'}}(\mathcal{O}_{S'} \otimes_{A_0} \mathfrak{Q}_L, \mathcal{O}_{S'}) .$$

Then since the module $\mathfrak{Q}$ of (7.4) is free, and since $\mathfrak{L} \otimes K$ and $\mathfrak{L} \otimes k$ are trivial, it follows that $\mathfrak{Q}_{\mathfrak{L}} \otimes K$ and $\mathfrak{Q}_{\mathfrak{L}} \otimes k$ have the same rank, hence that $\mathfrak{Q}_{\mathfrak{L}}$ is free. Therefore by (EGA III, 7.8.4), $\mathfrak{L}$ is cohomologically flat in dimension zero. Now the fact that $\mathfrak{L} \otimes k$ is trivial means that this sheaf has a nowhere zero global section $\bar{\alpha}$ on X_k. By cohomological flatness, $H^0(X_k, \mathfrak{L} \otimes k) = H^0(X_{A_0}, \mathfrak{L}) \otimes k$. Hence we can lift $\bar{\alpha}$ to a section α of $\mathfrak{L}$. This section is then nowhere zero. Hence $\mathfrak{L}$ is trivial.

(b) Again, what has to be proved is that if an invertible sheaf $\mathfrak{L}$ on X_{A_0} has the property that $\mathfrak{L} \otimes k(s)$ is trivial for all $s \in \mathfrak{S}$, then $\mathfrak{L}$ is trivial above a non-empty open set U. By (6.9), we may suppose $H^0(X_{k(s)}, \mathfrak{L} \otimes k(s)) = H^0(X_{A_0}, \mathfrak{L}) \otimes k(s)$ for all $s \in \mathfrak{S} \cap U$. Choose a section α of $\mathfrak{L}$ above a sufficiently small neighborhood of one of the points s such that its residue $\bar{\alpha}$ is nowhere zero on $X_{k(s)}$. Its locus of zeros is a closed set in X_U, hence has a closed image C in U not containing s. The complement of C in U is as required.

[4′] The assertions of (a), (b) are trivial from the definition of D and (7.10), and (c) follows immediately from (6.9).

[5′] (a) With the notation of this condition, it is clear that the obstruction for lifting a is purely local at the support of M. Thus by (7.9) it is an element of $H^2(X_k, M_X)$, where k is the residue field of A_0 at the point supporting M. This is the same as the obstruction to lifting b.

(b) Here the obstruction to lifting ξ is by (7.9) an element $\alpha \in H^2(X_{A_0}, M_X) = H^0\,(\text{Spec}\, A', R^2 f'_* M_X)$. By assumption, ξ is zero in $H^2(X_K, M_X \otimes_{A_0} K)$. Now the kernel of the map $A_0 \times_K A'_K \to A_0 \times_K A_K$ is $M \otimes_{A_0} K$, viewed as an A_0-module. Thus the obstruction to lifting ξ to $P(A_0 \times_X A'_K)$ is the image of ξ in

$$H^2(X_{A_0}, (M \otimes_{A_0} K)_X) \approx H^2(X_K, M_X \otimes_{A_0} K)\;,$$

which vanishes by assumption.

(c) Let $M = (A_0)^n$. Then the obstruction to lifting locally is an element $\alpha = (\alpha_1, \cdots, \alpha_n)$ of $H^2(X_{A_0}, \mathcal{O})^n$, by (7.9). To say this does not vanish on any one-dimensional quotient of M_K at the generic point is just to say that the elements $\alpha_i \in H^2(X_{A_0}, \mathcal{O})$ are linearly independent there. Then it is clear that in a neighborhood of the generic point the elements $\alpha_i \otimes k \in H^2(X_{A_0}, \mathcal{O}) \otimes k \approx H^2(X_k, \mathcal{O})$ (cf. (6.9)) will also be linearly independent, for every residue field k of a point of that neighborhood.

But the obstruction to lifting to an A_ε lies in $H^2(X_{A_0}, \varepsilon \otimes \mathcal{O}) \approx$

$H^2(X_k, \mathcal{O})$, and to write $A' \to A_\varepsilon \to A$ is the same as to write ε as a one-dimensional quotient of $M \otimes k$. Hence the vanishing of such an obstruction corresponds to the linear dependence of the elements $\alpha_i \otimes k$. This proves [5′] (c), and completes the proof of (7.3).

MASSACHUSETTS INSTITUTE OF TECHNOLOGY

REFERENCES

[1] M. ARTIN, *The implicit function theorem in algebraic geometry*, (to appear).

[2] ———, *On the solutions of analytic equations*, Invent. Math. **5** (1968), 277–291.

[3] ———, *Algebraic approximation of structures over complete local rings*, (to appear).

[4] ———, A. GROTHENDIECK, and J-L. VERDIER, Séminaire de géométrie algébrique 1963–64; Cohomologie étale des schémas, I.H.É.S. (mimeographed notes).

[5] M. DEMAZURE and A. GROTHENDIECK, Séminaire de géométrie algébrique 1963–64; Schémas en groups, I.H.É.S. (mimeographed notes).

[6] A. GROTHENDIECK, Séminaire de géométrie algébrique 1960–61, I.H.É.S. (mimeographed notes).

[7] K. KODAIRA and D. SPENCER, *On deformations of complex analytic structures*, Ann. of Math. **67** (1958) 328–466.

[8] D. KNUTSON, Algebraic spaces, Ph. D. Thesis, M.I.T., September, 1968 (to appear).

[9] M. KURANISHI, *On the locally complete families of complex analytic structures*, Ann. of Math. **75** (1962), 536–577.

[10] H. MATSUMURA and F. OORT, *Representatibility of group functors, and automorphisms of algebraic schemes*, Invent. Math. **4** (1967), 1–25.

[11] T. MATSUSAKA, *On the algebraic construction of the Picard variety*, Jap. J. Math. **21** (1951), 217–235.

[12] J. P. MURRE, *On contravariant functors from the category of preschemes over a field into the category of abelian groups*, Pub. Math I.H.É.S. No. 23, 1964.

[13] ———, Representation of unramified functors, Applications, Séminaire Bourbaki, 17 (1964–65) No. 294 (mimeographed notes).

[14] M. SCHLESSINGER, *Functors of Artin rings*, Trans. Amer. Soc. **130** (1968), 205–222.

[15] J-P. SERRE, Algèbre Locale, multiplicités, Lecture Notes in Mathematics No. 11, Berlin, Springer, 1965.

[16] A. WEIL, Variétés abéliennes et courbes algébriques, Hermann, Paris, 1948.

EGA J. DIEUDONNÉ and A. GROTHENDIECK, *Éléments de géométrie algébrique*, Pub. Math. I.H.É.S. (1960–).

FGA A. GROTHENDIECK, Fondements de la géométrie algébrique, Extraits du Séminaire Bourbaki 1957–1962 (mimeographed notes).

(Received September 30, 1968)

The signature of fibre-bundles*

By M. F. ATIYAH

1. Introduction

For a compact oriented differentiable manifold X of dimension $4k$ the signature (or index) of X is defined as the signature of the quadratic form in $H^{2k}(X; \mathbf{R})$ given by the cup product. Thus

$$\text{Sign}(X) = p - q$$

where p is the number of $+$ signs in a diagonalization of the quadratic form and q is the number of $-$ signs. If dim X is not divisible by 4 one defines Sign (X) to be zero. Then one has the multiplicative formula

$$\text{Sign}(X \times Y) = \text{Sign}(X) \cdot \text{Sign}(Y) .$$

In [5] it was proved that this multiplicative formula continues to hold when $X \times Y$ is replaced by a fibre bundle with base X and fibre Y *provided that the fundamental group of X acts trivially on the cohomology of Y.*

In this paper we exhibit examples which show that this restriction on the action of $\pi_1(X)$ is necessary, and that the signature *is not multiplicative in general fibre-bundles*. Our examples are actually in the lowest possible dimension namely when dim $X =$ dim $Y = 2$. The total space Z of the bundle has dimension 4 and non-zero signature, whereas Sign $(X) =$ Sign $(Y) = 0$ (because their dimensions are not divisible by 4). Of course if one wants an example in which base and fibre have dimensions divisible by 4 it suffices to take Z^2, which is fibered over X^2 with fibre Y^2; we have

$$\text{Sign}(Z^2) = \text{Sign}(Z)^2 \neq 0$$
$$\text{Sign}(X^2) = \text{Sign}(X)^2 = 0$$
$$\text{Sign}(Y^2) = \text{Sign}(Y)^2 = 0 .$$

* Partially supported by AF-AFOSR-359-66.

Since this paper was submitted, my attention has been drawn to a very similar paper of Kodaira J. Anal. Math. **19** (1967), 207-215. Although this decreases the originality of the present paper, it enhances the appropriateness of the dedication.

Our 4-manifold Z will actually arise as a complex algebraic surface and the projection $\pi: Z \rightarrow X$ will be holomorphic (for some complex structure on X). The fibres $Y_x = \pi^{-1}(x)$ will therefore be algebraic curves but the complex structure will vary with x, so that Z is not a holomorphic fibre bundle. This is an essential feature of the example as we shall explain in § 3.

If T, E denote the Todd genus and Euler characteristic respectively, one has the following simple relations (for curves and surfaces)

$$2T(X) = E(X)\,, \qquad 2T(Y) = E(Y)$$
$$4T(Z) = \operatorname{Sign}(Z) + E(Z)\,.$$

Since E is always multiplicative for fibre bundles these relations imply

$$T(Z) = T(X)T(Y) + \frac{1}{4}\operatorname{Sign}(Z)\,.$$

Thus the non-vanishing of Sign (Z) is equivalent to the non-multiplicativity

$$T(Z) \neq T(X)T(Y) \tag{1.1}$$

of the Todd genus.

If $d \in H^2(Z)$ denotes the first Chern class of the tangent bundle T_π along the fibres of Z, the total Pontrjagin class $p(Z)$ is given by

$$\begin{aligned} p(Z) &= p(T_\pi)\cdot\pi^* p(X) \\ &= 1 + d^2\,. \end{aligned}$$

The Hirzebruch formula for the signature therefore gives

$$\operatorname{Sign}(Z) = \frac{d^2}{3}[Z]\,. \tag{1.2}$$

Thus the crucial property of our examples will be that

$$d^2 \neq 0\,. \tag{1.3}$$

In the next section we shall construct the surface Z and show that (1.3) holds. In § 3 we shall explain the connection with moduli of algebraic curves. Finally in § 4 we shall investigate in general the effect of the fundamental group on the signature of a fibre-bundle. We shall see that this is closely related to the homomorphism

$$H^*(B_G; \mathbf{Q}) \longrightarrow H^*(B_\Gamma; \mathbf{Q})$$

induced by a homomorphism of the discrete group Γ into the real Lie group G.

2. Construction of the surface Z

We first choose a curve C with a fixed-point-free involution τ. In other words C is a double covering of a curve C': these exist as soon as the genus $g' \geqq 1$, but we shall take $g' \geqq 2$. Note that the genus g of C is $2g' - 1$ and so $g \geqq 3$. We now take X to be the covering of C given by the homomorphism

$$\pi_1(C) \longrightarrow H_1(C; \mathbf{Z}) \longrightarrow H_1(C; Z_2) \cong Z_2^{2g} ,$$

where g is the genus of C. It has the property that any double covering of C becomes trivial when lifted to X. Thus, if $f: X \rightarrow C$ is the covering map, the induced homomorphism with mod 2 coefficients

$$f_1^*: H^1(C; Z_2) \longrightarrow H^1(X; Z_2)$$

is zero.

Consider now, in $X \times C$, the graphs Γ_f, $\Gamma_{\tau f}$

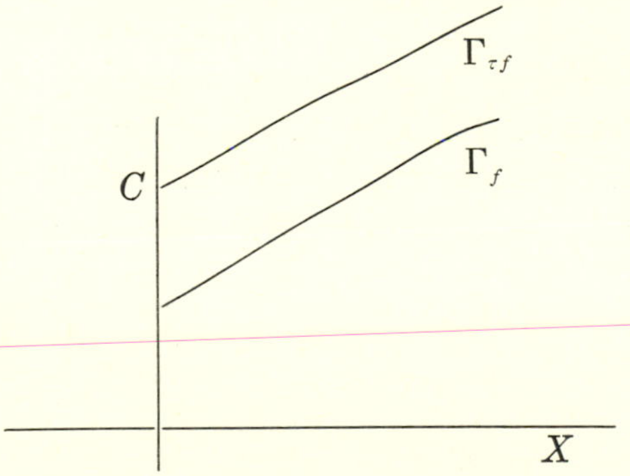

Our choice of f was to ensure the following property of these graphs:

LEMMA 2.1. *The homology class in $H^2(X \times C; \mathbf{Z})$ defined by the curve $\Gamma_f + \Gamma_{\tau f}$ is even.*

PROOF. Let $\gamma_f \in H^2(X \times C; Z_2)$ be the mod 2 reduction of the class of Γ_f. We have to show

$$\gamma_f + \gamma_{\tau f} = 0 .$$

Now, if we use the Künneth formula and Poincaré duality (for mod 2 coefficients) to identify

$$H^*(X \times C) \cong \mathrm{Hom}\big(H^*(C), H^*(X)\big)$$

it is well known that

$$\gamma_f = \textstyle\sum_p f_p^*$$

where f_p^* is the homomorphism induced by f in $H^p(\ ; Z_2)$. By our choice of f we have

$$\begin{aligned} f_1^* &= 0 \\ f_2^* &= \deg f (\mathrm{mod}\ 2) = 0 \\ f_0^* &= 1 \qquad\qquad \text{(identifying } H^0(C) \text{ and } H^0(X) \text{ with } Z_2) . \end{aligned}$$

Similarly

$$\begin{aligned} (\tau f)_1^* &= f_1^* \tau_1^* = 0 \\ (\tau f)_2^* &= 0 \\ (\tau f)_0^* &= 1 . \end{aligned}$$

Putting these together we get

$$\gamma_f + \gamma_{\tau f} = \textstyle\sum_p \big(f_p^* + (\tau f)_p^*\big) = 0 .$$

Let us next recall that, if A is a non-singular curve on an algebraic surface M, and if the homology class of A in $H^2(M; \mathbf{Z})$ is even, then we can construct a double covering $\widetilde{M}$ of M *ramified along* A. To see this[1] let L be the holomorphic line-bundle defined by A. We shall first show that we can find a holomorphic line-bundle $\widetilde{L}$ with $\widetilde{L}^2 \cong L$. Consider the exact cohomology sequence

$$\longrightarrow H^1(M; \mathcal{O}) \longrightarrow H^1(M; \mathcal{O}^*) \xrightarrow{\ \delta\ } H^2(M; \mathbf{Z}) \longrightarrow H^2(M; \mathcal{O}) \longrightarrow$$

arising from the exact sequence of sheaves

$$0 \longrightarrow \mathbf{Z} \longrightarrow \mathcal{O} \xrightarrow{\ \exp 2\pi i\ } \mathcal{O}^* \longrightarrow 0$$

($\mathcal{O}$ the sheaf of germs of holomorphic functions, $\mathcal{O}^*$ the multiplicative sheaf of germs of non-zero holomorphic functions). Since $H^1(M; \mathcal{O})$ and $H^2(M; \mathcal{O})$ are both complex vector spaces, it follows easily from this exact sequence that, for $k \in \mathbf{Z}$,

$$\delta(l) \text{ divisible by } k \text{ in } H^2(M; \mathbf{Z}) \Longrightarrow l \text{ divisible by } k \text{ in } H^1(M; \mathcal{O}^*) .$$

Taking l to be the class of L and $k = 2$ we see that A being even

[1] The argument which follows applies quite generally to any complex manifold M (not necessarily compact) with a non-singular divisor A defining an even class in $H^2(M; \mathbf{Z})$.

implies L is even, i.e. there exists a holomorphic line-bundle $\widetilde{L}$ with $\widetilde{L}^2 \cong L$. If s is now a section of L vanishing at A the ramified covering $\widetilde{M}$ is just the inverse image of $s(M) \subset L$ under the squaring map $\widetilde{L} \to L$. It is clear that $\widetilde{M}$ is algebraic.

If $\widetilde{A} \subset \widetilde{M}$ is the branch curve (mapped isomorphically onto A by $\widetilde{M} \to M$) we have

$$\widetilde{A}^2 = A^2/2 \tag{2.2}$$

for the self-intersection numbers. This follows from the fact that the normal bundles of $A, \widetilde{A}$ in $M, \widetilde{M}$ are just the restrictions $L \mid A$, $\widetilde{L} \mid \widetilde{A}$. If B is any curve on M having no component in common with A, and if $\widetilde{B}$ is its inverse image in $\widetilde{M}$, we have

$$A \cdot B = \widetilde{A} \cdot \widetilde{B} . \tag{2.3}$$

If a, b denote the corresponding cohomology classes (2.3) follows from the formulas

$$\{\pi^*(a)\pi^*(b)\}[\widetilde{M}] = 2(a \cdot b)[M]$$
$$\pi^*(b) = \widetilde{b}$$
$$\pi^*(a) = 2\widetilde{a} .$$

Returning now to the curve $A = \Gamma_f + \Gamma_{\tau f}$ on the surface $M = X \times C$ we see from Lemma (2.1) that a ramified double cover $\widetilde{M}$ exists. This is our surface Z referred to in § 1. It is clear that the composite projection

$$Z \longrightarrow X \times C \longrightarrow X$$

has as fibres the double coverings of C ramified at pairs of points of our involution τ. Thus the genus h of Y is given by

$$2 - 2h = 2(2 - 2g) - 2$$

or

$$h = 2g .$$

It remains now to calculate the first Chern class d of the tangent bundle T_π along the fibres of Z and to show that $d^2 \neq 0$.

Let ω be a holomorphic differential on C, and let Ω, $\widetilde{\Omega}$ be its lift to $X \times C$, Z respectively. Then $\widetilde{\Omega}$ is a holomorphic section of the dual T_π^* of T_π so that

$$d = -(\widetilde{\Omega})$$

where $(\widetilde{\Omega})$ denotes the class of the zeros of $\widetilde{\Omega}$. But clearly

$$(\widetilde{\Omega}) = \widetilde{p}^*(w) + (\widetilde{\Gamma}_f) + (\widetilde{\Gamma}_{\tau f})$$

where $\tilde{p}: Z \to C$ is the projection. Hence, computing self-intersection numbers, and using (2.2) and (2.3) we get

$$\begin{aligned} d^2[Z] &= 2\tilde{p}^*(\omega)\cdot(\tilde{\Gamma}_f) + 2\tilde{p}^*(\omega)\cdot(\tilde{\Gamma}_{\tau f}) + (\tilde{\Gamma}_f)^2 + (\tilde{\Gamma}_{\tau f})^2 \\ &= 2p^*(\omega)\cdot(\Gamma_f) + 2p^*(\omega)\cdot(\Gamma_{\tau f}) + \frac{1}{2}(\Gamma_f)^2 + \frac{1}{2}(\Gamma_{\tau f})^2 \\ &= 2(2g-2)\deg f + 2(2g-2)\deg \tau f + \frac{1}{2}(\Gamma_f)^2 + \frac{1}{2}(\Gamma_{\tau f})^2 \\ &= 8(g-1)\deg f + (\Gamma_f)^2 \,. \end{aligned}$$

Here p denotes the projection $X \times C \to C$ and we used the involution $1 \times \tau$ on $X \times C$ to take Γ_f into $\Gamma_{\tau f}$ and so derive the equality $(\Gamma_f)^2 = (\Gamma_{\tau f})^2$. Finally, if I denotes the identity map of C, we have

$$\begin{aligned} (\Gamma_f)^2 &= \deg f\cdot(\Gamma_I)^2 \\ &= \deg f\cdot(2-2g) \,. \end{aligned}$$

Putting this in the formula above for $d^2[Z]$ we therefore obtain

$$\begin{aligned} d^2[Z] &= 8(g-1)\deg f - 2(g-1)\deg f \\ &= 6(g-1)\deg f \\ &= 6(g-1)\cdot 2^{2g} \,. \end{aligned}$$

Thus the signature of our surface Z is given by

$$\text{(2.4)} \qquad \operatorname{Sign}(Z) = (g-1)2^{2g+1} = -E(X)$$

and (since $g \geqq 3$) this is non-zero as required.

Remarks. (1) An alternative method of calculating Sign (Z) has been pointed out to me by F. Hirzebruch. This is to use the following general formula for the signature of a ramified double covering $\tilde{M}$ of M:

$$\operatorname{Sign}(\tilde{M}) = 2\operatorname{Sign}(M) - \operatorname{Sign}(\tilde{S}\cdot\tilde{S}) \,.$$

Here $\tilde{S} \subset \tilde{M}$ is the ramification submanifold (of codimension 2) and $\tilde{S}\cdot\tilde{S}$ denotes a "self-intersection manifold" of $\tilde{S}$ in $\tilde{M}$. This formula is an easy consequence of the general G-signature theorem of [1] with G of order two (see [1; (6.15)]) and, applied in our case, it gives

$$\begin{aligned} \operatorname{Sign} Z &= 0 - (\tilde{\Gamma}_f^2 + \tilde{\Gamma}_{\tau f}^2) \\ &= -\Gamma_f^2 = -E(X) \,. \end{aligned}$$

(2) The formula (2.4) shows that there exist algebraic

surfaces with arbitrarily large signature, contrary to a conjecture of Zappa. In fact Borel in [3] produced counter-examples to this conjecture which are somewhat similar to our examples. In both cases the surfaces are classifying spaces for a discrete group Γ, that is the universal covering surface is contractible. In Borel's example the universal covering is the unit ball B^2 in $\mathbf{C}^2$, but in our examples the universal covering is definitely not B^2. One way to distinguish the two cases is to observe that the Borel surfaces are rigid [4] whereas our surface Z has moduli. In fact it is easy to see (using the footnote in § 2) that, as we vary the original curve C' in a local family C'_t, we obtain a family Z_t and that the moduli of C'_t (i.e., the holomorphic periods) give rise to non-trivial moduli for Z_t. We may also note that the universal covering of Z cannot be the product $B^1 \times B^1$ of two unit discs because, if it were, the argument of Borel [3] would imply Sign $(Z) = 0$. Since B^2 and $B^1 \times B^1$ are the only homogeneous bounded domains in $\mathbf{C}^2$ it follows that the universal covering of Z is not a homogeneous bounded complex domain. Note finally that the fundamental group $\pi_1(Z)$ is a split extension with $\pi_1(Y)$ as subgroup and $\pi_1(X)$ as quotient.

(3) The smallest non-zero value of Sign (Z) in (2.4) occurs when $g = 3$. This gives Sign $(Z) = 2^8$.

(4) Since the fibre Y_x of Z is, by construction, a double covering of C ramified at x and $\tau(x)$ it is fairly clear that the "moduli" of Y_x vary non-trivially with $x \in X$. In the next section we shall in fact see that this is always the case for any family of curves $Z \to X$ for which the total space Z has non-vanishing signature. Conversely, appealing to facts about the space of moduli of curves, one could deduce the existence of families $Z \to X$ with Sign $(Z) \neq 0$. This was my original approach to the problem. The specific construction by ramified double coverings was suggested to me by I. R. Šafarevic.

(5) Despite the algebro-geometric aspect of the preceding remark it should be emphasized that the complex structure of the fibering $Z \to X$ is purely auxiliary. We could have constructed Z as a ramified covering of $X \times C$ without using the complex structure.

3. Relation with moduli

If we apply the Grothendieck Riemann-Roch theorem to the map $Z \xrightarrow{\pi} X$ we get

$$\text{ch}(1 - \mathcal{H}) = \pi_*\left(1 + \frac{d}{2} + \frac{d^2}{12}\right) \tag{3.1}$$

where $\mathcal{H}$ is the vector bundle over X whose fibre $\mathcal{H}_x$ is $H^1(Y_x, \mathcal{O}_{Y_x})$: its dual $\mathcal{H}^*$ is the bundle whose fibre $\mathcal{H}_x^*$ is the space of holomorphic differentials on Y_x. Equating the two-dimensional terms in (3.1) we get

$$c_1(\mathcal{H}^*) = -c_1(\mathcal{H}) = \pi_*\left(\frac{d^2}{12}\right). \tag{3.2}$$

Thus the essential feature of our example is that $c_1(\mathcal{H}^*) \neq 0$.

The cohomology class $c_1(\mathcal{H}^*)$ can also be interpreted as follows. In the differentiable fibre bundle $Z \to X$ the fundamental group $\pi_1(X)$ acts on the cohomology $H^1(Y; \mathbf{Z})$ preserving the symplectic form given by the cup product. Thus we get a homomorphism

$$\alpha: \pi_1(X) \longrightarrow \text{Sp}(2h; \mathbf{Z})$$

and hence a homomorphism

$$\beta: \pi_1(X) \longrightarrow \text{Sp}(2h; \mathbf{R}).$$

This induces a homomorphism in the cohomology of the classifying spaces

$$\beta^*: H^*(B_G; \mathbf{Q}) \longrightarrow H^*(X; \mathbf{Q})$$

where $G = \text{Sp}(2h; \mathbf{R})$. Now the maximal compact subgroup of G is isomorphic to the unitary group $U(h)$ and so there is a universal complex vector bundle V of dimension h over B_G. One can verify that the bundle $\beta^*(V)$ over X induced by β is isomorphic to $\mathcal{H}^*$ (or $\mathcal{H}$, depending on how we choose V) and so

$$c_1(\mathcal{H}^*) = \beta^*(c_1(V)).$$

Since $c_1(\mathcal{H}^*) \neq 0$ and since β factors through α it follows that *the inclusion*

$$\text{Sp}(2h; \mathbf{Z}) \longrightarrow \text{Sp}(2h; \mathbf{R})$$

induces a non-zero homomorphism in the rational cohomology of the classifying spaces (in dimension 2). Note that in our example $h = 2(2g' - 1)$ where $g' \geqq 2$, so that the smallest value is 6.

Remark. For a discrete subgroup Γ of a real Lie group G, one might ask under what circumstances the homomorphism

$$H^*(B_G; \mathbf{Q}) \longrightarrow H^*(B_\Gamma; \mathbf{Q})$$

is non-trivial. When G/Γ is compact one can use the theory of harmonic forms to investigate this question. Our example here, with $G = \mathrm{Sp}(2h; \mathbf{R})$, $\Gamma = \mathrm{Sp}(2h; \mathbf{Z})$ is of a different type since G/Γ has finite volume but is not compact.

Let D denote the Siegel upper half-plane $\mathrm{Sp}(2h; \mathbf{R})/U(h)$ and let $\Gamma\backslash D$ denote the space obtained by dividing by the action of the discrete group $\Gamma = \mathrm{Sp}(2h; \mathbf{Z})$. Then the holomorphic family of curves $Z \to X$ defines a holomorphic map

$$\varphi: X \longrightarrow \Gamma\backslash D .$$

Except for the presence of fixed points of Γ on D the space $\Gamma\backslash D$ would be a classifying space for Γ. In any case, for real cohomology, the fixed points cause no trouble[2] and we can identify

$$\varphi^*: H^*(\Gamma\backslash D; \mathbf{Q}) \longrightarrow H^*(X; \mathbf{Q})$$

with the homomorphism

$$\alpha^*: H^*(B_\Gamma; \mathbf{Q}) \longrightarrow H^*(X; \mathbf{Q})$$

induced by $\alpha: \pi_1(X) \to \Gamma$. This shows that we cannot get examples of families Z with $c_1(\mathcal{H}^*) \neq 0$ unless φ is non-trivial, that is unless the holomorphic structure of the fibres varies.

4. General remarks

We shall now discuss in general the effect of the fundamental group of the base on the signature of a fibre bundle. When our manifolds are not complex we cannot use holomorphic differentials as in § 3 but, choosing a riemannian metric, we can use harmonic forms. With this modification we shall see that the discussion of § 3 can be paralleled in the general case.

Let $\pi: Z \to X$ be a differentiable fibre bundle with fibre Y. We assume X, Y, Z are compact oriented and that X, Y have even[3] dimension. We fix riemannian metrics on the fibres, for example by taking a riemannian metric on Z and giving Y_x the induced metric.

Let $\dim Y = 2k$ and consider the bundle H^k over X given by the real cohomology of the fibres in dimension k. This is the bundle associated to the action of $\Gamma = \pi_1(X)$ on $H^k(Y; \mathbf{R})$. This action

[2] We can replace Γ by a subgroup Γ_0 of finite index acting freely on D.

[3] When the dimensions are odd one always has

$$\mathrm{Sign}(X) = \mathrm{Sign}(Y) = \mathrm{Sign}(Z) = 0 .$$

preserves the bilinear form given by the cup product. Thus for k odd we have a homomorphism

$$\beta\colon \Gamma \longrightarrow \mathrm{Sp}\,(2n; \mathbf{R})$$

and for k even we have

$$\beta\colon \Gamma \longrightarrow O(p, q; \mathbf{R}) .$$

Here $2n = \dim H^k(Y; \mathbf{R})$ for k odd, and $\sum_1^p x_i^2 - \sum_1^q x_j^2$ is the diagonalization of the quadratic form on $H^k(Y; R)$ for k even.

If we identify $H^k(Y_x; \mathbf{R})$ with the space of harmonic k-forms on Y_x we get an inner product in the bundle H^k and so a reduction of its structure group to a maximal compact subgroup. For k odd this is $U(n)$ and for k even it is $O(p) \times O(q)$. For k odd we thus have an associated n-dimensional complex vector bundle V over X, and for k even we have two real vector bundles W^+, W^- of dimensions p, q respectively. These can be defined directly in terms of the inner product on $H^k(Y_x; \mathbf{R})$ as follows. The $*$-operator on the harmonic k-forms satisfies $*^2 = (-1)^k$. Thus for k odd it defines a complex structure on the bundle H^k — this is V — and for k even it defines a decomposition $H^k = H^k_+ \oplus H^k_-$ into the ± 1-eigenspaces — these are W^+, W^-. We now define the signature of the map π to be

$$\begin{aligned} \mathrm{Sign}\,(\pi) &= V^* - V \in K(X) && (k \text{ odd}) \\ &= W^+ - W^- \in KO(X) && (k \text{ even}) . \end{aligned}$$

From the classifying space description of the bundles V, W^+, W^- it is clear that[4] $\mathrm{ch}\,(\mathrm{Sign}\,(\pi))$ is induced from a universal characteristic class

$$\mathrm{ch}\,(\mathrm{Sign}) \in H^*(B_G; \mathbf{Q}) = H^*(B_K; \mathbf{Q})$$

by the composite map

$$X \longrightarrow B_\Gamma \longrightarrow B_G$$

where $G = \mathrm{Sp}\,(2n; \mathbf{R})$ or $O(p, q; \mathbf{R})$ according as k is odd or even and K is a maximal compact subgroup. The map $X \to B_\Gamma$ is the classifying map of the universal covering of X and $B_\Gamma \to B_G$ is induced by the homomorphism $\beta\colon \Gamma \to G$.

On the other hand from the harmonic form description of V, W^+, W^- one can show that

[4] For a real bundle ch is defined as the Chern character of the complexification.

$$(4.1)\qquad \begin{aligned} V^* - V &= \operatorname{index} D^+ \in K(X) && k \text{ odd} \\ W^+ - W^- &= \operatorname{index} D^+ \in KO(X) && k \text{ even} \end{aligned}$$

where D^+ is a family of elliptic operators along the fibres of π. For $x \in X$ the operator D_x^+ is the signature operator defined in [1; § 6]. The vector space $\operatorname{Ker} D_x^+$ has a constant dimension and is the fibre of a bundle $\operatorname{Ker} D^+$ over X (complex for k odd, real for k even). Similarly for $\operatorname{Coker} D^+$ and the index of the family D^+ is defined to be the element of $K(X)$ or $KO(X)$ (as k is odd or even) given by

$$\operatorname{index} D^+ = \operatorname{Ker} D^+ - \operatorname{Coker} D^+ .$$

Note that (for k even say) $\operatorname{Ker} D^+$ contains W^+ as a sub-bundle and similarly $\operatorname{Coker} D^+$ contains W^-. The equality (4.1) holds because the remaining parts cancel, that is

$$\operatorname{Ker} D^+/W^+ \cong \operatorname{Coker} D^+/W^- .$$

The index theorem for families of elliptic operators [2] applied to D^+ gives

$$(4.2)\qquad \operatorname{ch}(\operatorname{index} D^+) = \pi_*(\widetilde{\mathfrak{L}}T_\pi) \in H^*(X; \mathbf{Q})$$

where, for any real vector bundle E of dimension $2k$, $\widetilde{\mathfrak{L}}(E)$ is that function of the Pontrjagin classes of E defined by

$$\widetilde{\mathfrak{L}}(E) = \prod_{i=1}^{k} \frac{x_i}{\tanh x_i/2} .$$

(As usual the Pontrjagin classes of E are interpreted as the elementary symmetric functions in $x_1^2, \cdots, x_k^2$.) Combined with (4.1) we therefore obtain

$$(4.3)\qquad \operatorname{ch}\big(\operatorname{Sign}(\pi)\big) = \pi_*\big(\widetilde{\mathfrak{L}}(E)\big) .$$

For the special case considered in § 3 formula (4.3) coincides essentially with (3.2) which was derived from the Grothendieck Riemann-Roch theorem.

The Hirzebruch index theorem gives the signature of Z in terms of $\widetilde{\mathfrak{L}}(Z)$; i.e., $\widetilde{\mathfrak{L}}$ of the tangent bundle of Z,

$$\begin{aligned} \operatorname{Sign}(Z) &= \widetilde{\mathfrak{L}}(Z)[Z] \\ &= \{\widetilde{\mathfrak{L}}(T_\pi)\cdot\pi^*\widetilde{\mathfrak{L}}(X)\}[Z] \\ &= \{\pi_*\widetilde{\mathfrak{L}}(T_\pi)\cdot\widetilde{\mathfrak{L}}(X)\}[X] \\ &= \{\operatorname{ch}\big(\operatorname{Sign}(\pi)\big)\cdot\widetilde{\mathfrak{L}}(X)\}[X] && \text{by (4.3)} . \end{aligned}$$

This last expression for Sign (Z) shows clearly why the signature is not multiplicative in general. The deviation from multiplicativity is in a sense measured by the (positive-dimensional components of) the cohomology class ch $(\text{Sign}(\pi))$. As we have seen this is induced from a universal characteristic class in B_G *via* the fundamental group of X.

OXFORD UNIVERSITY and INSTITUTE FOR ADVANCED STUDY

REFERENCES

[1] M. F. ATIYAH and I. M. SINGER, *The index of elliptic operators*: III, Ann. of Math. **87** (1968),

[2] ———, *The index of elliptic operators*: IV, (to appear).

[3] A. BOREL, *Compact Clifford-Klein forms of symmetric spaces*, Topology **2** (1963), 111-122.

[4] E. CALABI and E. VESENTINI, *On compact locally symmetric Kähler manifolds*, Ann. of Math. **71** (1960), 472-507.

[5] S. S. CHERN, F. HIRZEBRUCH, and J-P. SERRE, *The index of a fibered manifold*, Proc. Amer. Math. Soc. **8** (1957), 587-596.

(Received May 28, 1968)

On Hensel's lemma and exponential sums*

By WALTER L. BAILY, JR.

Introduction

It is our purpose to present here a result generalizing Hensel's lemma [1, 2, 5], together with some related ideas which may be applied to the evaluation of certain exponential sums, connected with the Fourier coefficients of Eisenstein series. It is the latter application which motivated our original investigations. However, we think that the generalization of Hensel's lemma may be considered of some interest in itself.

Part of what we do here is based on ideas suggested by P. Cartier and we wish to acknowledge that fact here. When we indicated to him an interest in certain ideas of Siegel [6, § 7], he told us of some results he had been able to prove regarding numbers of points on varieties modulo powers of a prime ideal. Presumably some of these ideas are not unfamiliar to others; in particular, it would seem that they should be implicit in the language of schemes. However, our methods are quite elementary, and having been unable to find the explicit formulation of the results in this form in the literature, we beg to impose on the reader's indulgence to present them here in detail.

In closing our prefatory remarks, we wish most gratefully to dedicate this note to my former teacher, Professor K. Kodaira, in appreciation of his past years of patient instruction and kind guidance.

Part I. A generalization of Hensel's lemma and related matters

Let k be an algebraic number field and let V be a non-singular, irreducible, affine variety of dimension d and codimension d', imbedded in the affine space of dimension n, and defined over k. Then

* The author wishes to acknowledge research support from NSF Grant GP-6654 during the preparation of this article.

$d + d' = n$. If v is any place of k, we may take as universal domain the algebraic closure $\Omega = \Omega_v$ of the completion k_v of k at v. If V had any singularities, then it would have singularities which are algebraic over k. Therefore, the statement that V is non-singular has nothing to do with the choice of universal domain containing k. It is convenient to take advantage of this fact in order not to be specific about our choice of universal domain at certain places of the discussion.

Let $\mathfrak{o}$ be the maximal order (of integers) of k and let the ideal of V have generators $f_1, \cdots, f_N \in \mathfrak{o}[X_1, \cdots, X_n]$.

Let $\mathfrak{p}$ be any prime ideal of k, $k_\mathfrak{p}$, the completion of k at the place of $\mathfrak{p}$, $\mathfrak{o}_\mathfrak{p}$, the maximal compact subring of $k_\mathfrak{p}$, and denote by $\mathfrak{p}$ again the maximal ideal of $\mathfrak{o}_\mathfrak{p}$. Let π be a prime of element of $\mathfrak{p}$ in k, so that $\mathfrak{p} = \pi \cdot \mathfrak{o}_\mathfrak{p}$. By our assumption, all coefficients of f_i are integral at $\mathfrak{p}$, $i = 1, \cdots, N$.

If I is a subset of $\{1, \cdots, N\}$, and J, a subset of $\{1, \cdots, n\}$ such that $|I| = |J|$, where $|A|$ denotes the cardinality of A, we let Δ_{IJ} denote the determinant of the square matrix

$$\left\{\frac{\partial f_i}{\partial X_j}\right\}_{i \in I, j \in J},$$

with the rows and columns in natural order. The condition that V be non-singular is then equivalent to the requirement that for every $a \in V$, there exist sets I and J as above with $|I| = |J| = d'$ such that $\Delta_{IJ}(a) \neq 0$. This property of V implies [8, Thm. 2]: If $a \in V$, there exists a subset I of $\{1, \cdots, N\}$ with $|I| = d'$ such that for every $l = 1, \cdots, N$ we have

$$f_l = \sum_{i \in I} r_{li} f_i , \tag{1}$$

where r_{li} are rational functions defined at a.

Moreover, for the given prime $\mathfrak{p}$, $V_{\mathfrak{o}_\mathfrak{p}} = V \cap \mathfrak{o}_\mathfrak{p}^n$ is compact, because V is closed and $\mathfrak{o}_\mathfrak{p}^n$ is compact in the metric topology on $k_\mathfrak{p}^n$. We let q be the number of elements in $\mathfrak{o}_\mathfrak{p}/\mathfrak{p}$ and take the usual norm $|\ |_\mathfrak{p}$ on $k_\mathfrak{p}$ defined by $|a|_\mathfrak{p} = q^{-\mathrm{ord}_\mathfrak{p} a}$ for $a \in k_\mathfrak{p}^* = k_\mathfrak{p} - \{0\}$, $|0|_\mathfrak{p} = 0$. Then we have

LEMMA 1. *For each $\mathfrak{p}$, there exists a positive integer $\nu = \nu_\mathfrak{p}$ such that if $a \in \mathfrak{o}_\mathfrak{p}^n$ and $f_i(a) \equiv 0(\mathfrak{p}^\nu)$, $i = 1, \cdots, N$, then there exist I and J with $|I| = |J| = d'$ such that $\Delta_{IJ}(a) \not\equiv 0(\mathfrak{p}^\nu)$. For all but a finite number of $\mathfrak{p}$, we may take $\nu_\mathfrak{p} = 1$.*

PROOF. Suppose that for each $\nu \in \mathbf{Z}$, $\nu > 0$, we could find $a_\nu \in \mathfrak{o}_\mathfrak{p}^n$ satisfying $f_i(a_\nu) \equiv \Delta_{IJ}(a_\nu) \equiv 0(\mathfrak{p}^\nu)$ for all i, I, J. Then from the compactness of $\mathfrak{o}_\mathfrak{p}^n$, it follows that the sequence $\{a_\nu\}$ has a limit point a in $\mathfrak{o}_\mathfrak{p}^n$. It is evident from continuity that $f_i(a) = \Delta_{IJ}(a) = 0$ for all i, I, J, thus a is a singular point of V, which is a contradiction. This proves the existence of $\nu_\mathfrak{p}$ for all prime ideals $\mathfrak{p}$.

We know that the coefficients of the polynomials f_i and Δ_{IJ} are in $\mathfrak{o}_\mathfrak{p}$. The last part follows then from the proof of [4, Prop. 22, § 12] in taking the F of that proposition to be the (empty) set of singularities of V, in letting the polynomials G_i of that proposition be the polynomials f_i, and in making other obvious substitutions of notation.

Henceforth, if $\mathfrak{p}$ is any fixed prime ideal, we denote by a bar over any object, for which it makes sense, the reduction of that object modulo $\mathfrak{p}$.

Let (S_*, N_*) be a pair consisting of a set S_* of elements $f_1, \cdots, f_N$ of $k_\mathfrak{p}[X_1, \cdots, X_n]$ which generate the ideal of V in $k_\mathfrak{p}[X_1, \cdots, X_n]$ and of a non-empty, open, compact subset N_* of $k_\mathfrak{p}^n$. Such a pair is called a Hensel pair for V if there exist a positive integer M_* and a non-negative integer μ_* with the following properties: If $a \in N_*$ satisfies $f_i(a) \equiv 0(\mathfrak{p}^m)$ for a positive integer $m \geqq \max(M_*, 2\mu_* + 1)$, then there exists $b \in N_*$ such that $b \equiv a(\mathfrak{p}^{m-\mu_*})$ and such that $f_i(b) \equiv 0(\mathfrak{p}^{m+1})$, $i = 1, \cdots, N$. The Hensel pair is said to be of order (M_*, μ_*).

THEOREM 1. *Let S_0 be a fixed finite set of ideal generators $\{f_i^{(0)}\}_{i \in I_0}$ of the ideal $\mathfrak{J}$ of V having coefficients in k. Let $\mathfrak{p}$ be a prime ideal of k such that the polynomial coefficients of all $f_i^{(0)}$ are integral at $\mathfrak{p}$. Then there exist a positive integer $M = M_\mathfrak{p}$, a non-negative integer $\mu = \mu_\mathfrak{p}$, and a finite collection $\{S_\lambda, N_\lambda\}_{\lambda \in \Lambda}$ of Hensel pairs of order (M, μ) for V with the following properties*: 1) *$N_\lambda \subset \mathfrak{o}_\mathfrak{p}^n$ for $\lambda \in \Lambda$*; 2) *if m is a positive integer $\geqq$* $\max(M, 2\mu + 1)$, *and if $f_i(a) \equiv 0(\mathfrak{p}^m)$, $i \in I_0$, then $a \in N_\lambda$ and $f^{(\lambda)}(a) \equiv 0(\mathfrak{p}^m)$ when $f^{(\lambda)} \in S_\lambda$, for some $\lambda \in \Lambda$—in particular, $V_{\mathfrak{o}_\mathfrak{p}} \subset \bigcup_\lambda N_\lambda$. For all but a finite number of prime ideals $\mathfrak{p}$, we can choose $\Lambda = \{0\}$, $N_0 = \mathfrak{o}_\mathfrak{p}^n$, $\mu_\mathfrak{p} = 0$, and $M_\mathfrak{p} = 1$.*

PROOF. First suppose $V_{\mathfrak{o}_\mathfrak{p}}$ is non-empty. Take $I_0 = \{1, \cdots, N\}$ and let $f_i^{(0)} = f_i$, $i \in I_0$. If $a_0 \in V_{\mathfrak{o}_\mathfrak{p}}$, there exists a subset I of I_0 with $|I| = d'$, which for simplicity we may assume to be $\{1, \cdots, d'\}$, such that for $l > d'$, we have [8, Th. 2]

$$f_l = \sum_{i=1}^{d'} r_{li} f_i \ ,$$

where r_{li} are rational functions, defined over k and holomorphic at a_0. Define

$$f'_l = f_l - \sum_{i=1}^{d'} r_{li}(a_0) f_i \ , \qquad l > d' \ .$$

Then

$$f'_l = \sum_{i=1}^{d'} r'_{li} f_i \ , \tag{4}$$

and $r'_{li}(a_0) = 0$. Since r'_{li} is continuous at a_0 in the $\mathfrak{p}$-adic topology, there exists a neighborhood N_{a_0} of a_0 in affine space such that $| r'_{li} |_{\mathfrak{p}} < 1$ on N_{a_0}. Take $S_{a_0} = \{f_1, \cdots, f_{d'}, f'_{d'+1}, \cdots, f'_N\}$. Define $f'_i = f_i$ for $i \leqq d'$. In order to have $f(b) \equiv 0(\mathfrak{p}^{m+1})$, for $b \in N_{a_0}$ and for all $f \in S_{a_0}$, it will be sufficient to have this hold for $f = f_1, \cdots, f_{d'}$. Since $f_i(a_0) = 0$, $i = 1, \cdots, d'$, we have

$$\frac{\partial f'_l}{\partial X_j}(a_0) = \sum_{i=1}^{d'} r'_{li}(a_0) \frac{\partial f_i}{\partial X_j}(a_0) = 0 \ , \qquad l > d' \ .$$

Since the rank of the matrix $(\partial f'_i / \partial X_j(a_0))$ is d', the same will be true if we restrict i to be $\leqq d'$. Then by an $\mathfrak{o}_{\mathfrak{p}}$-unimodular change of coordinates and an $\mathfrak{o}_{\mathfrak{p}}$-unimodular linear transformation of the system of functions $f_1, \cdots, f_{d'}$, we may assume that the matrix $(\partial f_i / \partial X_j)$ will satisfy

$$\frac{\partial f_i}{\partial X_j}(a_0) = \begin{cases} \pi^{\nu_i} & \text{if } i = j = 1, \cdots, d', \text{ with } \nu_1 \leqq \cdots \leqq \nu_{d'}; \\ 0 & \text{otherwise}; \end{cases} \tag{5}$$

and will have the same form modulo $\mathfrak{p}^{\mu+1}$ in N_{a_0} if N_{a_0} is chosen small enough for any fixed μ. Using Taylor's expansion, we have for $y \in \mathfrak{o}_{\mathfrak{p}}^n$, $a \in N_{a_0}$, and $i \leqq d'$:

$$f_i(a + \pi^{m-\mu} y) = f_i(a) + \pi^{m-\mu} \sum_{j=1}^{n} \frac{\partial f_i}{\partial X_j}(a) y_j + T \ , \tag{6}$$

where T consists of terms divisible by $\pi^{2(m-\mu)}$ because each f_i has integral coefficients. If $m \geqq 2\mu + 1$, we have $2(m - \mu) \geqq m + 1$. Hence, to solve for $b = a + \pi^{m-\mu} y$ such that $f_i(b) \equiv 0(\mathfrak{p}^{m+1})$, $i = 1, \cdots, d'$, it is sufficient to solve

$$\pi^{m-\mu} \sum_{j=1}^{n} \frac{\partial f_i}{\partial X_j}(a) y_j \equiv -f_i(a)(\mathfrak{p}^{m+1}) \ , \qquad i = 1, \cdots, d' \ , \tag{7}$$

for $y = (y_1, \cdots, y_n)$. If we take $\mu \geqq \max_i \nu_i$, and use the fact that $f_i(a) \equiv 0(\text{mod } \mathfrak{p}^m)$, then these congruences are consistent modulo the given powers of $\mathfrak{p}$, and we can solve for $y \in \mathfrak{o}_{\mathfrak{p}}^n$. It is then elementary to prove that the set of solutions y modulo $\mathfrak{p}$ is a d-dimensional vector space over the finite field F_q of q elements.

Let $\mathcal{P}$ be a finite set of prime ideals of k such that all f_i have integral coefficients if $\mathfrak{p} \notin \mathcal{P}$, and such that if $\nu_{\mathfrak{p}}$ is the least positive integer satisfying the conditions of Lemma 1, and if $\mathfrak{p} \notin \mathcal{P}$, then $\nu_{\mathfrak{p}} = 1$. Let $\mathfrak{p} \notin \mathcal{P}$ and let $a \in V_{\mathfrak{o}_{\mathfrak{p}}}$. If $a' \in N_1(a)$, and if $f_1, \cdots, f_{d'}$ are such that the matrix $(\partial f_i/\partial X_j(a))$, $1 \leqq i \leqq d'$, $1 \leqq j \leqq n$, has rank d' modulo $\mathfrak{p}$, then from [2] it follows that every solution a'' of $f_1(a'') = \cdots = f_{d'}(a'') = 0$, obtained from a' as a limit of successive approximations by solving the congruences (7), is actually a point of $V_{\mathfrak{o}_{\mathfrak{p}}}$. Moreover, in this case, we have $\nu_1 = \cdots = \nu_{d'} = 0$. Therefore, if $\mathfrak{p} \notin \mathcal{P}$, we may choose $\mu_{\mathfrak{p}} = 0$, $M_{\mathfrak{p}} = 1$, $N_0 = \mathfrak{o}_{\mathfrak{p}}^n$, and the collection $\{S_\lambda\}_{\lambda \in \Lambda}$ to consist of S_0 alone. In addition, if m is any positive integer, and if $a' \in \mathfrak{o}_{\mathfrak{p}}^n$ satisfies $f_i(a') \equiv 0(\mathfrak{p}^m)$, $i = 1, \cdots, N$, then by [2], $a' \in N_1(a)$ for some $a \in V_{\mathfrak{o}_{\mathfrak{p}}}$; in that case, the set of congruence classes modulo $\mathfrak{p}$ of solutions y of the system of congruences $f_i(a' + \pi^m y) \equiv 0(\mathfrak{p}^{m+1})$ (which of course become linear when we ignore terms of higher order in y, which are $\equiv 0(\mathfrak{p}^{2m})$) forms a vector space of dimension d over F_q. This completes the proof of the theorem for $\mathfrak{p} \notin \mathcal{P}$.

What we have proved to this point is the following:

(A) For each $a_0 \in V_{\mathfrak{o}_{\mathfrak{p}}}$, there exists a neighborhood

$$N_{M(a_0)}(a_0) = \{a \in \mathfrak{o}_{\mathfrak{p}}^n \mid a \equiv a_0(\mathfrak{p}^{M(a_0)})\}$$

of a_0 in $k_{\mathfrak{p}}^n$ and a non-negative integer $\mu = \mu(a_0)$ such that if $a \in N_{M(a_0)}(a_0)$ satisfies $f_i(a) \equiv 0(\mathfrak{p}^m)$, $i = 1, \cdots, N$, for $m \geqq 2\mu + 1$, then there exists $b \equiv a(\mathfrak{p}^{m-\mu})$ satisfying $f_i(b) \equiv 0(\mathfrak{p}^{m+1})$. The set of all b satisfying this, modulo $\mathfrak{p}^{m-\mu+1}$, forms a d-dimensional vector space over F_q. If, moreover, $m \geqq \mu + M(a_0)$, then also $b \in N_{M(a_0)}(a_0)$.

(B) If $\mathfrak{p} \notin \mathcal{P}$, all the assertions in (A) are satisfied if we take $M(a_0) = 1$ and $\mu = 0$.

To complete the proof for $\mathfrak{p} \in \mathcal{P}$, we proceed as follows. Let a run through all points of $V_{\mathfrak{o}_{\mathfrak{p}}}$, and for each a choose $M(a)$ and $\mu(a) \geqq M(a)$ as in (A); let $N_{M(a)}(a) = N(a)$. By the compactness of $V_{\mathfrak{o}_{\mathfrak{p}}}$, there exist a finite number of points $\{a_\lambda\}$ of $V_{\mathfrak{o}_{\mathfrak{p}}}$ such that $\{N(a_\lambda)\}$ is a covering of $V_{\mathfrak{o}_{\mathfrak{p}}}$, and then the union of all $N(a_\lambda)$ is a neighborhood of $V_{\mathfrak{o}_{\mathfrak{p}}}$. Since $\mathfrak{o}_{\mathfrak{p}}^n$ is compact, it is clear that there exists a positive integer $M = M_{\mathfrak{p}}$ such that if $a \in \mathfrak{o}_{\mathfrak{p}}^n$, and if $f_i(a) \equiv 0(\mathfrak{p}^M)$, $i \in I_0$, then $a \in \bigcup_\lambda N(a_\lambda)$. Let $S = S_a$ be defined as earlier in the proof. It is then clear that the conclusions of our theorem are satisfied if we choose this $M \geqq \max_\lambda M(a_\lambda)$, and put $\mu = \max \mu(a_\lambda)$.

Now, if $V_{\mathfrak{o}_\mathfrak{p}}$ is empty, it follows from the compactness of $\mathfrak{o}_\mathfrak{p}^n$ that there exists $M > 0$ such that the system

$$f_i^{(0)}(a) \equiv 0(\mathfrak{p}^M) , \qquad i \in I_0 ,$$

has no solutions $a \in \mathfrak{o}_\mathfrak{p}^n$. Hence, the claims of the theorem are vacuously fulfilled.

COROLLARY 1. (*of the proof*). *If* $a \in V_{\mathfrak{o}_\mathfrak{p}}$, $m \in \mathbf{Z}$, $m > 0$, *define*

$$V_m(a) = \{x \in V_{\mathfrak{o}_\mathfrak{p}} \mid x \equiv a(\mathfrak{p}^m)\} .$$

Then for each $\mathfrak{p}$, *there exists a positive integer* $m_\mathfrak{p}$ *such that for* $m \geqq m_\mathfrak{p}$, *the quotient of* $V_m(a)$ *by the equivalence relation of congruence modulo* $\mathfrak{p}^{m+1}$ *is isomorphic to* F_q^d *for all* $a \in V_{\mathfrak{o}_\mathfrak{p}}$. *For all but a finite number of* $\mathfrak{p}$, *we may take* $m_\mathfrak{p} = 1$.

The proof of this is contained in the proof of the theorem.

Henceforth, for simplicity, we agree to take $m_\mathfrak{p} \geqq M_\mathfrak{p}$. This is legitimate, since by the theorem we may take $M_\mathfrak{p} = 1$ for all but a finite number of $\mathfrak{p}$.

From now on, if $a \in V_{k_\mathfrak{p}}$, we denote by V_a the tangent space to V at a, viewed as a linear subspace of the ambient affine space of V.

COROLLARY 2. *Let* M *and* μ *be as in the theorem and suppose* $V_{\mathfrak{o}_\mathfrak{p}}$ *is not empty. Let* $a \in V_{\mathfrak{o}_\mathfrak{p}}$ *and let* m *be a positive integer such that* $m \geqq \max(M/2, 2\mu + 1)$. *Let* N_m *be the set of points* x *in the affine space* (*over* $k_\mathfrak{p}$) *such that* $x \equiv a(\mathfrak{p}^m)$. *Then there is a one-to-one correspondence between the congruence classes of points of* $N_m \cap V$ *and of points of* $N_m \cap V_a$ *modulo* $\mathfrak{p}^{2m-\mu}$.

PROOF. This follows at once by applying Taylor's expansion to $f(a + \pi^m y)$ for f in the ideal of V with coefficients in $\mathfrak{o}_\mathfrak{p}$ in two cases: once, when $a + \pi^m y \in V$, and once when $a + \pi^m y \in V_a$ and using the results of the theorem. (We note, for convenience in the proof, that we may always assume the jacobian matrix $(\partial f_i/\partial X_j(a))$ to be in elementary divisor form (5).)

We denote the ambient affine space of V by E and let $E_\mathfrak{p}$ denote the set of points of E which are rational over $k_\mathfrak{p}$. Let Λ be a $k_\mathfrak{p}$-lattice [7, p. 28] in $E_\mathfrak{p}$ and let $h\colon E_\mathfrak{p} \to k_\mathfrak{p}$ be a $k_\mathfrak{p}$-linear mapping such that $h(\Lambda) \subset \mathfrak{o}_\mathfrak{p}$; and for each $a \in V_{\mathfrak{o}_\mathfrak{p}}$ let h_a be the restriction of h to V_a and define

$$|h_a| = \sup_{\lambda \in \Lambda \cap V_a} |h_a(\lambda)|_{\mathfrak{p}} .$$

Then we have

PROPOSITION 1. *The real-valued function Ψ defined by $\Psi(a) = |h_a|$ for $a \in V_{\mathfrak{o}_{\mathfrak{p}}}$ is a continuous function of a. If $h \not\equiv 0$, if h^0 is the hyperplane annihilated by h, and if h^0 is not tangent to V at any point of $V_{\mathfrak{o}_{\mathfrak{p}}}$, then Ψ is strictly positive and is bounded away from zero on $V_{\mathfrak{o}_{\mathfrak{p}}}$.*

PROOF. Since $V_{\mathfrak{o}_{\mathfrak{p}}}$ is compact, the second part of the proposition will obviously follow, once we have proved the continuity of Ψ. Of course, we may assume from the outset that $h \not\equiv 0$. We can [7, Th. 1, p. 29] always find a basis $\beta_1^a, \cdots, \beta_n^a$ of Λ such that $\Lambda = \sum_{i=1}^n \mathfrak{o}_{\mathfrak{p}} \beta_i^a$ and such that $\beta_1^a, \cdots, \beta_d^a$ is a basis of V_a. If we write $h_a(\sum_{i=1}^d \omega_i \beta_i^a) = \sum_{i=1}^d c_i^a \omega_i$, then clearly

$$|h_a| = \sup_{1 \leq i \leq d} |c_i^a|_{\mathfrak{p}} .$$

Therefore it will be enough to show we may choose the basis $\{\beta_i^a\}$ for a in a sufficiently small neighborhood of any pre-assigned point such that the functions c_i^a are continuous there.

We take a fixed basis $\gamma_1, \cdots, \gamma_n$ of Λ in $E_{\mathfrak{p}}$ and can then write

$$\beta_i^a = \sum_j d_{ij}(a)\gamma_j .$$

It is a trivial calculation to show that to obtain c_i^a as continuous functions of a, it will be sufficient to make $d_{ij}(a)$ continuous. The linear space V_a is defined by the system of equations

$$\sum_{j=1}^n \frac{\partial f_i}{\partial X_j}(a)(X_j - a_j) = 0 , \qquad i = 1, \cdots, N .$$

Each partial derivative $\partial f_i / \partial X_j$ is a continuous function on V. We fix a_0 and subject $\{f_1, \cdots, f_N\}$ and $\{X_1, \cdots, X_n\}$ to $\mathfrak{p}$-adic, integral, linear, unimodular transformations to bring the matrix $(\partial f_i / \partial X_j(a_0))$ to elementary divisor form (5); as such, it must have rank r equal to $d' = n - d$. Let $\pi^{\nu_1}, \cdots, \pi^{\nu_r}$ be the elementary divisors with $\nu_1 \leq \cdots \leq \nu_r$. If $m > \nu_r$ and $a \equiv a_0$ $(\mathfrak{p}^m)$, then

$$\frac{\partial f_i}{\partial X_j}(a) \equiv \frac{\partial f_i}{\partial X_j}(a_0)(\mathfrak{p}^m) .$$

It is then easy to see that elementary row operations on the left and column operations on the right can be chosen to depend continuously on a such that these transform the matrix $(\partial f_i / \partial X_j(a))$

into the matrix $(\partial f_i/\partial X_j(a_0))$. In this way, a basis $\{\beta_i^a\}$ can be transformed into a basis $\{\beta_i^{a_0}\}$ in a continuous manner for a close enough to a_0. From this, a locally continuous choice of the functions $d_{ij}(a)$ is easily seen to exist, as asserted. This completes the proof.

The hypotheses being as above, assume that $h \not\equiv 0$ and that h^0 is nowhere tangent to V. Then define $|h|_V = \inf_{a \in V_{\mathfrak{r}\mathfrak{p}}} |h_a|$. Since $|h|_V > 0$, we have $|h|_V = q^{-m}$ for some $m \geqq 0$. If $m = 0$, we say that h is primitive relative to V at $\mathfrak{p}$. If $V = E$, of course this coincides with the usual notion of primitive linear polynomial. We call π^m the content of h relative to V at $\mathfrak{p}$ and write $m = m(h, V)$. Since Ψ takes its values in the finite set $\{q^{-m} \mid 0 \leqq m \leqq m(h, V)\}$, it is locally constant on $V_{\mathfrak{o}_\mathfrak{p}}$. Therefore there exists a non-negative integer κ and a finite number of points a_i of $V_{\mathfrak{o}_\mathfrak{p}}$ such that the neighborhoods $\{N_\kappa(a_i)\}$ form a disjoint covering of $V_{\mathfrak{o}_\mathfrak{p}}$ and such that in each neighborhood $N_\kappa(a_i)$, Ψ is constant and equal, say, to q^{-m_i}. We agree henceforth to choose κ as small as possible having this property and write $\kappa = \kappa(h, V)$. In particular, if h is primitive relative to V, we have $\kappa = 0$.

We now let ε be a character on $k_\mathfrak{p}/\mathfrak{o}_\mathfrak{p}$ which is not trivial on $\mathfrak{p}^{-1}$. The values of ε on $\mathfrak{p}^{-1}$ will be q^{th} roots of unity.

THEOREM 2. *Let h be a non-zero linear form on $E_\mathfrak{p}$ such that $h(\Lambda) \subset \mathfrak{o}_\mathfrak{p}$ and such that h^0 is nowhere tangent to V. Let $m_0 = m(h, V)$, $\kappa = \kappa(h, V)$, and let $m_\mathfrak{p}$ be as in Corollary 1 of Theorem 1. Then if $\nu \in \mathbf{Z}$ satisfies $\nu \geqq \kappa + m_\mathfrak{p} + 2m_0 + 1 + \mu$, we have*

$$\sum_{a \in V_{\mathfrak{o}_\mathfrak{p}} \bmod \mathfrak{p}^\nu} \varepsilon(\pi^{-\nu} h(a)) = 0 . \tag{8}$$

In particular, for all but a finite number of $\mathfrak{p}$, if h is primitive relative to V at $\mathfrak{p}$, it is sufficient to take $\nu \geqq 2$.

PROOF. If $V_{\mathfrak{o}_\mathfrak{p}}$ is empty, the sum has no terms and so is formally zero. Suppose, on the other hand, that $V_{\mathfrak{o}_\mathfrak{p}}$ is not empty. We cover $V_{\mathfrak{o}_\mathfrak{p}}$ with disjoint neighborhoods $\{N_\kappa(a_i)\}$ such that in $N_\kappa(a_i)$, Ψ is constant and equal to q^{-m_i}. It is then sufficient to show that the sum in (8) extended over the a modulo $\mathfrak{p}^\nu$ in each neighborhood is zero. Fix i. We have $m_0 \geqq m_i$. We let a run over a complete set $\{a_\iota\}$ of representatives of residue classes modulo $\mathfrak{p}^{\nu - m_i - 1}$ in $N_\kappa(a_i)$. Then $\nu - m_i - 1 \geqq \nu - m_0 - 1 > \kappa$, so that $\{N_{\nu - m_i - 1}(a_\iota)\}$ is a disjoint covering of $N_\kappa(a_i)$. Also, $\nu - m_i - 1 \geqq \nu - m_0 - 1 \geqq m_\mathfrak{p}$, so that if we define $\Lambda_a = V_a \cap \Lambda$, the set of residue classes

$$N_{\nu - m_i - 1}(a_\iota)/\mathfrak{p}^{\nu - m_i}$$

is isomorphic to $\Lambda_{a_\iota}/\mathfrak{p} \cong F_q^d$ by the inverse of the mapping which assigns to each residue class y of Λ_{a_ι} modulo $\mathfrak{p}$ the residue class of $\pi^{\nu-m_i-1}y$ modulo $\mathfrak{p}^{\nu-m_i}$. If $a \in N_{\nu-m_i-1}(a_\iota)$, and if $b \equiv a(\mathfrak{p}^m)$, $\nu - m_i \leqq m \leqq \nu$, then $b = a + \pi^m z$ and, since h is linear, $h(b) = h(a) + h(\pi^m z)$. We can easily prove from Taylor's formula that there exists $z' \in V_a$ such that $z \equiv z'(\mathfrak{p}^{m-\mu})$, so that

$$h(b) \equiv h(a) + h_a(\pi^m z')\,(\mathfrak{p}^{2m-\mu}) .$$

Then

$$2m - \mu - \nu \geqq 2(\nu - m_i) - \mu - \nu = \nu - 2m_i - \mu > 0 ,$$

and

$$\pi^{-\nu}h_a(\pi^m z') = \pi^{m-\nu}\textstyle\sum_j c_j^a z_j .$$

Since $m - \nu \geqq -m_i$, and since, by definition of m_i, we have $c_j^a \in \mathfrak{p}^{m_i}$ for all j, it follows that $\pi^{m-\nu}\sum_j c_j^a z_j \in \mathfrak{o}_\mathfrak{p}$, so that

$$\varepsilon\big(\pi^{-\nu}h(b)\big) = \varepsilon\big(\pi^{-\nu}h(a)\big) .$$

Using Corollary 1 of Theorem 1, one can prove by induction that if $a \in N_{\nu-m_i-1}(a_\iota)$, then the number of residue classes modulo $\mathfrak{p}^\nu$ in $N_{\nu-m_i}(a)$ is q^{dm_i}. Now let $a_j \in N_{\nu-m_i-1}(a_\iota)$ run over a complete set of residue classes modulo $\mathfrak{p}^{\nu-m_i}$ in $N_{\nu-m_i-1}(a_\iota)$. The calculations just carried out give us

$$(9) \quad \sum_{\substack{a \in N_{\nu-m_i-1}(a_\iota)\\ \mathrm{mod}\,\mathfrak{p}^\nu}} \varepsilon\big(\pi^{-\nu}h(a)\big) = q^{dm_i}\varepsilon\big(\pi^{-\nu}h(a_\iota)\big)\cdot\textstyle\sum_y \varepsilon\big(\pi^{-\nu}h(y)\big) ,$$

where the sum on the right is over y with $y + a_\iota \in N_{\nu-m_i-1}(a_\iota)$ modulo $\mathfrak{p}^{\nu-m_i}$; and by the same calculations, the sum on the right hand side becomes

$$(10) \qquad \sum_{y'=(y'_1,\cdots,y'_d)\in F_q^d} \varepsilon\big(\pi^{-1}\textstyle\sum_{j=1}^d \pi^{-m_i}c_j y'_j\big) ,$$

where the coordinates are taken with reference to a lattice basis $\beta_1^{a_\iota}, \cdots, \beta_d^{a_\iota}$ as before, so that

$$h_{a_\iota}\big(\textstyle\sum_{j=1}^d y_j\beta_j^{a_\iota}\big) = \sum_{j=1}^d c_j y_j .$$

By the definition of m_i and κ, we see that each c_j is divisible by π^{m_i}, and there exists a j such that $\pi^{-m_i}c_j$ is a unit of $\mathfrak{o}_\mathfrak{p}$. Using this fact together with the definition of ε, we see finally that the sum in (10) is zero. This completes the proof of the first assertion of the theorem. As for the second, if $\mathfrak{p}$ does not belong to the finite set of prime ideals such that $m_\mathfrak{p} > 1$, and if h is primitive relative to V at $\mathfrak{p}$, then $\mu = \kappa = m_0 = 0$ and

$$\kappa + m_{\mathfrak{p}} + 2m_0 + 1 + \mu = 2 .$$

LEMMA 2. *With notation as in Theorem 2, assume that the multiplicative group* $k_{\mathfrak{p}}^* = k_{\mathfrak{p}} - \{0\}$ *operates on the variety* V *in such a way that*

(1) $h(x.a) = xh(a)$ *for* $x \in k_{\mathfrak{p}}^*$ *and* $a \in V_{k_{\mathfrak{p}}}$,

(2) $x.V_{\mathfrak{o}_{\mathfrak{p}}} = V_{\mathfrak{o}_{\mathfrak{p}}}$ *for* x *belonging to the group* $U_{\mathfrak{p}}$ *of units of* $\mathfrak{o}_{\mathfrak{p}}$, *and*

(3) *if* $y, z \in V_{\mathfrak{o}_{\mathfrak{p}}}$, $y \equiv z(\mathfrak{p}^\nu)$, *then* $x.y \equiv x.z(\mathfrak{p}^\nu)$ *for all positive integers* ν *and* $x \in U_{\mathfrak{p}}$.

Then the sum appearing in (8) *is a rational number for each non-negative rational integer* ν.

PROOF. Since the group $U_{\mathfrak{p}}$ operates on $V_{\mathfrak{o}_{\mathfrak{p}}}$ and, for any positive rational integer ν, on the set $V_{\mathfrak{o}_{\mathfrak{p}}}/\mathfrak{p}^\nu$ of residue classes of points of $V_{\mathfrak{o}_{\mathfrak{p}}}$ modulo $\mathfrak{p}^\nu$, it follows that $V_{\mathfrak{o}_{\mathfrak{p}}}/\mathfrak{p}^\nu$ consists of finitely many orbits of $U_{\mathfrak{p}}$. It is then obviously sufficient to prove that for any $b \in \mathfrak{o}_{\mathfrak{p}}$, the sum

$$\sum_{u \in U_{\mathfrak{p}} \bmod \mathfrak{p}^\nu} \varepsilon(\pi^{-\nu}ub)$$

is rational. This sum lies in the field F of $q^{\nu\text{th}}$ roots of unity, and the effect of any element of the Galois group of F on this sum is to raise each term to the mth power, where m is a residue class prime to $\mathfrak{p}$ modulo q^ν; since multiplication by m induces a translation of $U_{\mathfrak{p}}$ onto itself, the terms of the sum are permuted among themselves by that element of the Galois group, and so the sum is a rational number.

LEMMA 3. *Let* P *be a polynomial in* n *variables* $X_1, \cdots, X_n$ *with coefficients from* $\mathfrak{o}_{\mathfrak{p}}$ *and let*

$$a = (a_1, \cdots, a_n) \in \mathfrak{o}_{\mathfrak{p}}^n .$$

Let m *and* ν *be positive integers with* $\nu \leqq m$. *If there exists* $y \in \mathfrak{o}_{\mathfrak{p}}^n$ *such that* $P(a + \pi^m y) \not\equiv P(a)(\mathfrak{p}^{m+\nu})$, *then at least one of the partial derivatives* $\partial P/\partial X_i$ *must satisfy*

$$\frac{\partial P}{\partial X_i}(a) \not\equiv 0(\mathfrak{p}^\nu) .$$

If P *is linear, the restriction* $\nu \leqq m$ *is unnecessary.*

PROOF. This is a trivial consequence of the Taylor expansion

$$P(a + \pi^m y) = P(a) + \pi^m \sum_{i=1}^{n} \frac{\partial P}{\partial X_i}(a) y_i + \pi^{2m}(\cdots) \,,$$

where the terms $(\cdots)$ are in $\mathfrak{o}_{\mathfrak{p}}$.

Part II. Applications

Let k be an algebraic number field, and let W be a finite-dimensional vector space with a fixed k-structure. Denote by $\mathfrak{o}$ the ring of integers in k. Let W_1 be a one-dimensional vector space with a k-structure and put $W' = W \oplus W_1$. By a k-lattice in W we mean [7, p. 83] an $\mathfrak{o}$-module of finite type in W_k containing a basis of W. If Λ is a k-lattice in W, then $\Lambda' = \Lambda \oplus \mathfrak{o}$ is one in W'. Let E and E_1 denote the algebras of endomorphisms of W and W_1 respectively and define $E' = E \oplus E_1$. The given k-structures and k-lattices in W and W' define k-structures and k-lattices in E and E' in a natural way, and we denote the lattices by Λ^E and $\Lambda^{E'}$. If $\mathfrak{p}$ is a prime ideal in k, we denote the $k_{\mathfrak{p}}$-lattice [7, p. 28] obtained from any of the above, in taking the tensor product with $\mathfrak{o}_{\mathfrak{p}}$, by affixing the subscript $\mathfrak{p}$.

Let G be a linear algebraic subgroup of $GL(W)$ defined over k. In order to facilitate application of our preceding results, we may replace G by a linear algebraic group G' which, as an affine variety, contains all finite specializations of all its points as follows. The group G' is contained in E' and is the image of G in $GL(W \oplus W_1)$ under the mapping φ defined by

$$\varphi(g)(w, w_1) = \left(g.\, w, (\det g)^{-1} \cdot w_1\right) .$$

Define $G_{\mathfrak{o}}$, $G'_{\mathfrak{o}}$, $G_{\mathfrak{o}_{\mathfrak{p}}}$, and $G'_{\mathfrak{o}_{\mathfrak{p}}}$ as the stabilizers in the respective groups of the lattices Λ, Λ', $\Lambda_{\mathfrak{p}}$, and Λ' in the respective vector spaces. It is obvious that we have $\varphi(G_{\mathfrak{o}}) = G'_{\mathfrak{o}}$ and $\varphi(G_{\mathfrak{o}_{\mathfrak{p}}}) = G'_{\mathfrak{o}_{\mathfrak{p}}}$. Imbed $k_{\mathfrak{p}}$ in $E_{k_{\mathfrak{p}}*}$ as the set of multiplications by scalars and assume that $k_{\mathfrak{p}}^* \subset G$. Allow $x \in k_{\mathfrak{p}}^*$ to act on G by left translations, and on G' by $x.\varphi(g) = \varphi(xg)$. Clearly the group $U_{\mathfrak{p}}$ acts on $G_{\mathfrak{o}_{\mathfrak{p}}}$ and on $G'_{\mathfrak{o}_{\mathfrak{p}}}$.

Let τ be a k-linear functional on W such that $\tau(\Lambda) \subset \mathfrak{o}$. We may extend τ to a $k_{\mathfrak{p}}$-linear functional on $W_{k_{\mathfrak{p}}}$ for all $\mathfrak{p}$, such that $\tau(\Lambda_{\mathfrak{p}}) \subset \mathfrak{o}_{\mathfrak{p}}$. Fix τ, and if $w \in \Lambda_{\mathfrak{p}}$, define $h_w \colon E_{k_{\mathfrak{p}}} \to k_{\mathfrak{p}}$ by $h_w(e) = \tau(e.w)$. Extend h_w to $E'_{k_{\mathfrak{p}}}$ by defining it to be zero on $E_{1k_{\mathfrak{p}}}$.

PROPOSITION 2. *With notation as above, assume for all $\mathfrak{p}$ and all non-zero $w \in \Lambda_{\mathfrak{p}}$ that $h_w \not\equiv 0$ and that h_w^0 is nowhere tangent to G'. Assume, furthermore, that there exists a finite set $\mathcal{P}$ of* *prime ideals such that if* $\mathfrak{p} \notin \mathcal{P}$ *and if* w *is a primitive vector of* $\Lambda_{\mathfrak{p}}$, *then* h_w *is primitive relative to* G'. *Let* ε *be a character of* $k_{\mathfrak{p}}/\mathfrak{o}_{\mathfrak{p}}$ *which is non-trivial on* $\mathfrak{p}^{-1}$. *Under these hypotheses, there exists a positive integer* $\nu_{\mathfrak{p}}$ *for each* $\mathfrak{p}$ *with the property that for* $\nu > \nu_{\mathfrak{p}}$ *and all primitive* $w \in \Lambda_{\mathfrak{p}}$ *we have*

$$\sum_{g \in G_{\mathfrak{O}_{\mathfrak{p}}} \bmod \mathfrak{p}^{\nu}} \varepsilon(\pi^{-\nu} h_w(g)) = 0 .$$

For $\mathfrak{p} \notin \mathcal{P}$ *we may take* $\nu_{\mathfrak{p}} = 1$.

PROOF. This is an immediate consequence of the results of Part I, applied by taking $V = G'$, $h = h_w$, and using the fact that h_w vanishes on $E_{1\mathfrak{p}}$.

In the computation of the Fourier coefficients of Eisenstein series for the Siegel modular group $\mathrm{Sp}(n, \mathbf{Z})$, one must study [6] series of the following type:

$$S_p^g(T) = \sum_{u \in \mathfrak{S}_{Q_p} \bmod \mathfrak{S}_{Z_p}} \varepsilon((T, u))\kappa(u)^{-g} ,$$

where the symbols have the following meanings: $\varepsilon(a) = e^{2\pi i a}$, $\mathfrak{S}$ is the space of n by n symmetric matrices, g is an even natural integer greater than $n + 1$, T is a $\frac{1}{2}$-integral[1], positive definite, symmetric matrix, $(\ ,\ )$ is the inner product on $\mathfrak{S}$ defined by $(x, y) = \mathrm{tr}\,(xy)$ (which makes the $\mathbf{Z}$-module of $\frac{1}{2}$-integral symmetric matrices into the dual lattice of the lattice of integral symmetric matrices), and $\kappa(u)$ is the absolute value of the product of the reduced denominators at the rational prime p of the elementary divisors of u. Using gaussian sums, one may prove that each $S_p^g(T)$ is a rational number, and that for all but the finite number of p dividing $2 \det T$, $S_p^g(T)$ has a simple form, given in [6, § 7].

One may now prove the same thing another way, without recourse (directly) to gaussian sums, by using the results we have just proved. Namely, for each prime p, $GL(n, \mathbf{Z}_p)$ operates on $\mathfrak{S}_{Q_p}$; if $g \in GL(n, \mathbf{Z}_p)$, $\sigma \in \mathfrak{S}$, then $g(\sigma) = {}^t g \sigma g$. Let G be the image of $GL(n)$ in $GL(\mathfrak{S})$ under this representation. To apply the preceding proposition, we let $W = \mathfrak{S}$, $\tau(\sigma) = \mathrm{tr}\,(T . \sigma)$, and note that $\kappa(gu) = \kappa(u)$ for $g \in G_{Z_p}$. (The last statement is not quite trivial. To prove it, we may view u as the matrix, with respect to a certain lattice basis, of a linear transformation of an n-dimensional vector space X into itself. Then one has the following facts. (1) The product of the first k elementary divisors of u is the greatest common divisor of the coefficients of the matrix of the kth skew-

---

[1] $T = (t_{ij})$, where t_{ii} and $2t_{ij}$ are integers.

symmetric tensor representation of u; (2) the coefficients in the kth skew-symmetric tensor representation of gu are linear combinations of those of u; and (3) if $g \in G_{Z_p}$, then the coefficients of those linear combinations, referred to in (2) are in $\mathbf{Z}_p$.

From (1), (2), and (3), it follows that g acts as a unimodular transformation of the lattice of linear, integral, transformations of the kth exterior product of X with itself, so that the elementary divisors of u and of gu are the same, hence that $\kappa(u) = \kappa(gu)$.)

It is clear that for each positive integer m, $p^{-m}\mathfrak{S}_{Z_p}/\mathfrak{S}_{Z_p}$ is finite, and that if $\kappa(u) = p^m$, then

$$u \in p^{-m}\mathfrak{S}_{Z_p}/\mathfrak{S}_{Z_p} - p^{-[\frac{m}{n}]+1}\mathfrak{S}_{Z_p}/\mathfrak{S}_{Z_p} ,$$

where $[x]$ denotes the biggest integer in x; moreover, the set $p^{-m}\mathfrak{S}_{Z_p}/\mathfrak{S}_{Z_p} - p^{-(m-1)}\mathfrak{S}_{Z_p}/\mathfrak{S}_{Z_p}$, which we denote by $\mathfrak{S}_{p,m}$, is a union of orbits of G_{Z_p}. Finally, it is clear that G contains the group of scalar multiplications by elements of Q_p^*; if τ and h_w are defined as in the discussion preceding Proposition 2, then it is evident that the hypotheses of Lemma 2 are satisfied for h_w.

PROPOSITION 3. *For every positive integer* ν, *we define*

$$S^g_{p,\nu}(T) = \sum_{u \in \mathfrak{S}_{p,\nu} \bmod \mathfrak{S}_{Z_p}} \varepsilon\big((T, u)\big)\kappa(u)^{-g} .$$

Then for every prime number p there exists a least positive integer ν_p *such that*

$$S^g_{p,\nu}(T) = 0 \qquad \textit{for } \nu > \nu_p .$$

Each of the quantities $S^g_{p,\nu}(T)$, $S^g_p(T)$ *is a rational number and* $S^g_p(T)$ *is different from zero for sufficiently large g. For all but a finite number of p, we have* $\nu_p = 1$.

PROOF. If we take into account the remarks preceding the statement of this proposition, Proposition 2 and the discussion preceding it, and Lemma 2, we see that in order to prove all of the statements of the proposition, except the fact that $S^g_p(T)$ is different from zero for large g, it will be sufficient to prove that the linear functional τ, defined by $\tau(\sigma) = \operatorname{tr}(T.\sigma)$, satisfies the hypotheses of Proposition 2. We shall show how to verify these hypotheses if we take $\mathcal{P}$ to be the set of primes which divide 2 det T. If $g \in G_{Z_p}$, then $(T, gu) = (g^*T, u)$, where g^* is the adjoint of g with respect to the inner product (,) and must preserve the lattice of $\frac{1}{2}$-integral, symmetric matrices. If $gu = {}^t huh$, then $g^*T = hT\,{}^th$, where ${}^tA$ is the transpose of A. Since one considers sums over orbits of G_{Z_p} for each p, one may choose (to simplify calculations) $g \in G_{Z_p}$ such that

$$g^*T = D(\varepsilon_1 p^{\alpha_1}, \cdots, \varepsilon_n p^{\alpha_n}) , \qquad \text{if } p \neq 2 , \tag{11}$$

where each ε_i is a unit of $\mathbf{Z}_p$, all $\alpha_i \geqq 0$, and $D(\cdots)$ is the n by n diagonal matrix with $(\cdots)$ as diagonal entries; and if $p = 2$, one may obtain, for suitable $g \in G_{Z_2}$,

$$g^*T = D'(\varepsilon_1 2^{\alpha_1}, \cdots, \varepsilon_q 2^{\alpha_q}; 2^{\beta_1}\Phi_1, \cdots, 2^{\beta_r}\Phi_r) , \tag{12}$$

where $q + 2r = n$ and D' is the n by n matrix made up of blocks along the main diagonal, of which the first q are one by one, with units ε_i of $\mathbf{Z}_2$ and $\alpha_i \geqq 0$, and of which the last r are 2 by 2, with $\beta_j \geqq -1$ and

$$\Phi_j = \begin{pmatrix} \gamma_j & \delta_j \\ \delta_j & \eta_j \end{pmatrix} ,$$

where δ_j is a unit and $\gamma_j \equiv \eta_j \equiv 0$ (2)—in this case, the otherwise unspecified entries of D' are zero. If $p \neq 2$, $p \nmid \det T$, then all α_i in (11) are zero. Let T_p be the normalized form of g^*T given for each p by (11) or (12). Let w be a primitive vector in $\mathfrak{S}_{Z_p}$, and let m be a suitably large positive integer. We consider the following two types of elementary transformations:

(1) If a is a unit of $\mathbf{Z}_p$ and $\xi = 1 + ap^m$, let $E_{\xi, j}$ be the matrix obtained from the identity matrix by replacing the jth diagonal entry by ξ, and define $g_{\xi, j} \in G_{Z_p}$ by $g_{\xi, j}(X) = E_{\xi, j} X E_{\xi, j}$;

(2) with a again a unit of $\mathbf{Z}_p$, i, j two different elements of the set $\{1, \cdots, n\}$, let $E_{a, ij}$ be the matrix obtained from the identity matrix by replacing the entry at the intersection of the ith row and j^{th} column by ap^m, and define $g_{a, ij}$ by

$$g_{a, ij}(X) = {}^tE_{a, ij} X E_{a, ij} .$$

Then by elementary calculations, which we omit here, one may show that for any prime p there exists a positive integer ν_p, which is equal to one if $p \nmid 2 \det T$, with the property that, if w is any primitive vector in $\mathfrak{S}_{Z_p}$, then there exists an elementary transformation g of one of the two above types such that

$$\operatorname{tr}\big(T_p(gw - w)\big) \not\equiv 0(p^{m+\nu_p})$$

if m is large enough. By Lemma 3 and Corollary 2 of Theorem 1, this implies that h_w is $\not\equiv 0$, that h_w^0 is not tangent to G (or to G') at any integral point, and that h_w is primitive relative to G (or to G') if $p \nmid 2 \det T$. That $S_p^g(T) \neq 0$ for large g is obvious, since we have now proved $S_p^g(T)$ to be equal to an expression of the form

$$1 + \sum_{j=1}^{n\nu_v} a_j p^{-jg} ,$$

where a_j are rational numbers independent of g.

The explicit calculation of $S_p^g(T)$ for $p \nmid 2 \det T$ requires more calculation than space allows here. However, in principal it involves a calculation for a finite field that is analogous to computation of the usual Γ-integrals over the reals by means of the Babylonian reduction process. Of course, the end result is that of [6, § 7].

Our purpose in introducing this method of evaluating the series S_p^g has been to avoid the gaussian sums. There might be little point to this except that it makes it possible to calculate the Fourier series coefficients for Eisenstein series attached to a certain arithmetic group acting on an exceptional tube domain (of 27 dimensions). These coefficients, as will be proved elsewhere, are rational numbers, and have an Euler product expansion. A typical Euler factor for "good" p is

$$(1 - p^{-18g})(1 - p^{4-18g})(1 - p^{8-18g}) .$$

We think it might be interesting to find an algebraico-geometrical interpretation of these factors, similar to that in terms of representation densities of quadratic forms in the classical cases. It would appear that such an interpretation, if it exists, might be closely related to finding something analogous to gaussian sums for an exceptional Jordan algebra, the absence of which at present forced us to find another way of calculating the properties of the Fourier coefficients.

UNIVERSITY OF CHICAGO.

REFERENCES

[1] W. L. CHOW, *The criterion for unit multiplicity and a generalization of Hensel's lemma*, Am. J. Math. **80** (1958), 539-552.

[2] P. SAMUEL, "Remarques sur le Lemme de Hensel", in Proc. Internat. Cong. Math., v. **2**, pp. 63-64, North-Holland Pub. Co., Amsterdam, 1954.

[3] G. SHIMURA, *Reduction of algebraic varieties with respect to a discrete valuation of the basic field*, Am. J. Math. **77** (1955), 134-176.

[4] G. SHIMURA and Y. TANIYAMA, Complex Multiplication of Abelian Varieties and its Applications to Number Theory, Publications of

the Mathematical Society of Japan **6**, Kenkyusha Printing Co., Tokyo, 1961.

[5] C. L. Siegel, *Über die analytische Theorie der quadratischen Formen*, Annals of Math. **36** (1935), 527-606.

[6] ———, *Einführung in die Theorie der Modulfunktionen n-ten Grades*, Math. Ann. **116** (1939), 617-657.

[7] A. Weil, Basic Number Theory, Springer-Verlag, New York, 1967.

[8] O. Zariski, *The concept of a simple point of an abstract algebraic variety*, Trans. Am. Math. Soc. **62** (1947), 1-52.

(Received September 23, 1968)

An intrinsic characterization of harmonic one-forms

By EUGENIO CALABI*

I. Introduction

The author recalls hearing the question since about 1955, do the zeros of a harmonic 1-form in a compact riemannian manifold possess certain combinatorial properties that depend only on the differential topological structure of the manifold, and not on the particular riemannian metric. In the case of surfaces the Riemann-Roch theorem (or alternately the Hopf-Stiefel theorem) on the indices of the zeros of any harmonic 1-form on a closed surface provided a positive answer to the question, rendering an analogous conjecture at least plausible for perhaps some higher dimensional manifolds. The fact that such a conjecture fails completely is the main consequence of the results of this paper. The technique is to consider, in a given, compact, n-dimensional manifold X, a given 1-form α, necessarily closed, and satisfying some reasonable technical condition on the points where it vanishes (only non-degenerate zeros admitted). Then the condition for the existence of a riemannian metric **g** with respect to which α is also co-closed are considered and settled completely, in terms of local properties (in § 2) and a new semi-global one (transitivity, §§ 4 and 5). Then in the sixth and concluding section it is shown how a harmonic 1-form can be perturbed by a closed 1-form in a small neighborhood, yielding a new closed 1-form with two additional zeros, such that under a different choice of metric, the resulting 1-form is again co-closed.

The proofs have undergone several modifications and corrections; the author is indebted to P. Deligne for suggesting a substantial simplification in the proof of Theorem 1 (precisely, Lemma 2, § 5).

Throughout this paper the term "differential manifold" denotes a connected manifold without boundary (unless otherwise

* Research partly supported by NSF under Grant GP 6895 and by the Institut des Hautes Etudes Scientifiques, 91 Bures-sur-Yvette, France

specified) of class C^r with $r \geqq 2$ (tensor fields are then of class C^{r-1}). The usual difficulty in dealing with exterior calculus in non-orientable manifolds is bypassed by using de Rham's idea of "forms of odd kind". A linear differential form (1-form, or pfaffian form) is called integrable, if it is locally a scalar multiple of a closed form by a factor which is nowhere zero. The adjoint of an exterior j-form φ in a riemannian n-manifold with metric $\mathbf{g}$, is an $(n-j)$-form of odd kind, which is denoted by $*_{\mathbf{g}}\varphi$, since the operation of adjunction depends on the particular riemannian metric, and the latter is subject to modification in the present context.

II. Local theory

We shall derive first a set of local conditions on a closed 1-form α in a differential manifold $\mathbf{X}$ for the existence of a local riemannian metric $\mathbf{g}$ with respect to which α is co-closed. Given a closed 1-form α on $\mathbf{X}$, for any point $\mathbf{p} \in \mathbf{X}$ and any simply connected neighborhood U of $\mathbf{p}$ we shall denote by $\int_{\mathbf{p}} \alpha\,|_U$ the uniquely defined scalar function f on U satisfying the equations

$$df = \alpha \qquad \text{in } U\,,$$
$$f(\mathbf{p}) = 0\,.$$

An equivalent formulation of the problem is that of characterizing locally the differentiable functions f that admit a local riemannian metric $\mathbf{g}$, with respect to which they satisfy the Laplace-Beltrami equation; in terms of local coordinates $\mathbf{x} = (x^1, \cdots, x^n)$, set $\mathbf{g} = g_{ij}(\mathbf{x})dx^i dx^j$, $(g_{ij}) = \mathfrak{g}$,

$$(1) \qquad \Delta_{\mathbf{g}} f = -\sum_{i,j=1}^{n} (\det \mathfrak{g})^{-\frac{1}{2}} \frac{\partial}{\partial x^i}\Big((\det \mathfrak{g})^{\frac{1}{2}} g^{ij} \frac{\partial f}{\partial x^j}\Big) = 0\,;$$
$$(g^{ij}) = (g_{ij})^{-1} = \mathfrak{g}^{-1}\,.$$

In terms of the matrix valued function $G^{ij} = (\det \mathfrak{g})^{\frac{1}{2}}(g_{ij})^{-1}$, equation (1) is a first order, under-determined, linear homogeneous differential equation, whose coefficients are determined by the differential $df = \alpha$ and its first order jet (i.e., the second order jet of f, disregarding additive constants).

PROPOSITION 1. *Let α be a closed 1-form, not identically zero, in a differential manifold $\mathbf{X}$. In order for $\mathbf{X}$ to admit (locally or globally) a differentiable riemannian metric $\mathfrak{g}$ with respect to* *which α is co-closed, then α must satisfy at least the following conditions*

(i) *The zeros of α are all of finite order;*

(ii) *For any simply connected neighborhood U of any point $\mathbf{p} \in \mathbf{X}$ the integral $\int_{\mathbf{p}} \alpha|_U$ does not achieve a relative maximum or minimum value at any interior point of U.*

PROOF. Under the assumptions stated, the conditions on α are well-known properties of solutions of the Laplace-Beltami equation. Property (i) follows from the Carleman-Aronszajn estimates [1] leading to the unique continuation theorem for solutions of a strongly elliptic differential equation, while property (ii) restates the strong maximum principle of E. Hopf [2].

The conditions of Proposition 1 are not sufficient for the existence of a riemmannian metric $\mathbf{g}$ (example: on the manifold $\mathbf{R}$, the real number line, the differential x^2dx satisfies the two conditions everywhere, but cannot be co-closed with respect to any riemannian metric in any neighborhood of the point $x = 0$). Consider a local integral $f = \int_{\mathbf{p}} \alpha|_U$ of a closed 1-form: the zeros of α are synonymous with the critical points of f in the sense of Morse's theory; with this in mind we transfer to α the concepts attached in that theory to f.

Definition 1. Let α be a 1-form in an n-dimensional differential manifold $\mathbf{X}$. A zero point $\mathbf{p}$ of α in $\mathbf{X}$ is called *non-degenerate*, if the matrix of first order partial derivatives of α with respect to a local coordinate system $(x^1, \cdots, x^n)$ is non-singular at $\mathbf{p}$. If in addition α is closed, this matrix of partial derivatives is symmetric, and the non-degenerate quadratic form it defines at $\mathbf{p}$ (called the hessian form) on the tangent space is assigned an *index* $j = j_\alpha(\mathbf{p})$, defined to be the maximal dimension of a linear subspace of the tangent space at $\mathbf{p}$, in which the hessian form is negative definite.

We recall some of the classical, elementary properties of non-degenerate critical points of functions, as applied to α. The set of 1-forms (respectively, closed 1-forms or exact 1-forms) α all of whose zeros are non-degenerate is open and dense in Whitney's C^1-topology; non-degenerate zeros of 1-forms are always isolated; the index j of a non-degenerate zero a of closed 1-form $(0 \leqq j \leqq n)$ is invariant under transformation of the local coordinates in terms of which it is defined.

We shall specialize our attention now to 1-forms α on $\mathbf{X}$ all of whose zeros are non-degenerate: condition (i) of Proposition 1 is now trivially satisfied, since in that case each zero of α is of order exactly 1; condition (ii) means that the index j of each zero of α is restricted to the range $1 \leqq j \leqq n-1$, i.e., excluding the extreme values 0 and n. It it worth nothing here that a closed 1-form α on a *compact* manifold $\mathbf{X}$ satisfying condition (ii) can never be exact. The next proposition states that the two conditions of Proposition 1, applied to forms all of whose zeros are non-degenerate, are also sufficient for the existence of *local* metrics $\mathbf{g}$ with respect to which α is co-closed, i.e., for the existence of local solutions $\mathbf{g}$ of equation (1).

PROPOSITION 2. *Let* $\mathbf{X}$ *be an* n*-dimensional differential manifold and* α *a closed* 1*-form on* $\mathbf{X}$, *all of whose zeros are non-degenerate. Then there exists in some neighborhood* U *of any given point* $\mathbf{p} \in \mathbf{X}$ *a riemannian metric* $\mathbf{g}_{\mathbf{p}}$ *with respect to which* α *is co-closed, if and only if the index* j *of each zero of* α *is in the range* $1 \leqq j \leqq n-1$.

PROOF. The necessity of the condition on the index is obvious after Proposition 1; so we shall just prove the sufficiency.

Let $\mathbf{p} \in \mathbf{X}$. Suppose first that $\alpha(\mathbf{p}) \neq 0$; then there exists a simply connected neighborhood U_0 of $\mathbf{p}$ where $\alpha \neq 0$ everywhere and hence $f = \int_{\mathbf{p}} \alpha|_{U_0}$ has no critical points. By the implicit function theorem there exists a smaller neighborhood U of $\mathbf{p}$ with a local coordinate system $(x^1, \cdots, x^n)$ on U such that $f|_U = x^1$.

Put on U the local euclidean metric form

$$ds^2 = \sum_{i=1}^{n} (dx^i)^2 \; ; \tag{2}$$

then x^1 satisfies the Laplace equation, so that $\alpha|_U = dx^1$ is co-closed with respect to the metric (2).

Now suppose that $\alpha(\mathbf{p}) = 0$, and let j be the index of $\mathbf{p}$ as a zero point of α. Then from a lemma of Morse there exists a simply connected neighborhood U of $\mathbf{p}$ with a local coordinate system $(x^1, \cdots, x^n)$ with $x^i(\mathbf{p}) = 0$ such that

$$f = \int_{\mathbf{p}} \alpha|_U = -\frac{n-j}{2} \sum_{i=1}^{j} (x^i)^2 + \frac{j}{2} \sum_{i=j+1}^{n} (x^i)^2 \, . \tag{3}$$

Again we introduce the euclidean metric (2) with respect to this coordinate system and observe that the function f of the coordinates satisfies Laplace's equation, so that $\alpha|_U = df$ is co-closed with respect to (2). We write $\alpha|_U$ in terms of these coordinates for future use as follows:

$$
\begin{aligned}
\alpha = df &= -(n-j)\sum_{i=1}^{j} x^i dx_i + j\sum_{i=j+1}^{n} x^i dx^i \\
&= (j\sum_{i=1}^{n} - n\sum_{i=1}^{j})x^i dx^i \,.
\end{aligned} \tag{4}
$$

This completes the proof of Proposition 2.

It is inevitable to guess at this point whether or not the local conditions of Proposition 2 are sufficient for the existence of a global riemannian structure on **X** with respect to which α is co-closed, particularly since the relevant equation can be linearized locally by using the matrix $G^{ij} = (\det(g_{ij}))^{\frac{1}{2}}(g^{ij})$, $(g^{ij}) = (g_{ij})^{-1}$, as the "unknown" in equation (1). It is clear that when the dimensionality n of **X** is 1, the problem has a trivial answer, since the conditions imply that α can have no zeros anywhere. In the higher dimensional cases the analogous conjecture becomes harder to prove, mainly because it is, in general, false, as one sees from the following example, in which **X** is a closed surface of genus 2.

III. A counter-example

Consider in euclidean 3-space $\mathbf{R}^3$ the two unit circles defined respectively by $\{x^2 + y^2 = 1;\ z = 1\}$, $\{x^2 + y^2 = 1;\ z = -1\}$ in cartesian coordinates (x, y, z); join them by the arc of a circular helix, defined in terms of a parameter t by $\{x = \cos t,\ y = \sin t,\ z = t;\ -1 \leqq t \leqq 1\}$. Visualize a tubular neighborhood of the resulting graph, with the boundary smoothed out at the two "Y-joints". The surface **X** obtained from the boundary is closed, of genus 2, and does not meet the z-axis $\{x = y = 0\}$. We map the surface **X** onto the circle $\mathbf{R}/2\pi\mathbf{Z}$ by the map f, where f is defined in $\mathbf{R}^3$ outside the z-axis by $f(x, y, z) = \arg(x + y\sqrt{-1}) \pmod{2\pi}$ and let $\alpha = df$ induce on **X** the closed 1-form to be considered. Barring all unnecessary bumps on the surface, α has precisely two zeros, both non-degenerate, of index 1, one in each of the crotches of the Y-joints, and so satisfies the conditions of Proposition 2. On the other hand there can be no riemannian metric **g** on **X** with respect to which α is co-closed, for the reason given below. Suppose that **X** had a riemannian structure with respect to which α were co-closed, i.e., that the 1-form of odd kind $*\alpha$ were closed. Consider any integral curve of the exact differential equation

$\alpha = 0$, i.e., any component of $f^{-1}(\theta)$, $\theta \in \mathbf{R}/2\pi\mathbf{Z}$ these curves have a natural orientation on their normal bundle (positive side = transversal directions along which $\alpha > 0$), so that each component is a 1-cycle of odd kind; the restriction of $*\alpha$ to each component of such a 1-cycle of odd kind is strictly positive at each point, except at the zero points of α (where the curves have a singularity).

Let γ be the "middle" one of the three components of $f^{-1}(0)$, i.e., the 1-cycle of odd kind linked about the helix near the point (1, 0, 0); by the argument just given, $\int_\gamma *\alpha > 0$. On the other hand γ is homologous to 0, since it cuts the surface $\mathbf{X}$ into two components, so that by Stoke's theorem, if $*\alpha$ is to be closed $\int_\gamma *\alpha = 0$. This contradiction precludes the possibility that $*\alpha$ may be closed, no matter what metric is used in defining it. It is easy to produce similar examples also with surfaces of higher genus or manifolds of more dimensions.

The above example motivates the need for some global invariants attached to closed 1-forms α on a manifold, other than their de Rham cohomology class (periods of integrals).

IV. Transitive 1-forms

Definition 2. Let $\mathbf{X}$ be a differential manifold and α an integrable 1-form on $\mathbf{X}$. For any point $\mathbf{p} \in \mathbf{X}$, we define the *upland* (respectively, the *lowland*) in $\mathbf{X}$ relative to $\mathbf{p}$ with respect to α, denoted by $C_\alpha^+(\mathbf{p})$ (respectively, $C_\alpha^-(\mathbf{p})$) to be the set of points $\mathbf{q}$ for which there exists a non-trivial, differentiable, oriented path from $\mathbf{p}$ to $\mathbf{q}$, along which the form α is everywhere strictly positive (respectively, negative), except possibly at $\mathbf{p}$ itself. If, for every point $\mathbf{p} \in \mathbf{X}$ the set $C_\alpha^+(\mathbf{p})$ (respectively, $C_\alpha^-(\mathbf{p})$) is everywhere dense in $\mathbf{X}$, then α is called *transitive* on $\mathbf{X}$.

An easy result on the concepts just defined is the following.

PROPOSITION 3. *Given any differential manifold* $\mathbf{X}$ *with an integrable* 1-*form* α, *for each point* $\mathbf{p} \in \mathbf{X}$ *the upland and the lowland relative to* $\mathbf{p}$ *with respect to* α *are open subdomains of* $\mathbf{X}$; *their boundary sets consist of points where* α *vanishes and, otherwise, of non-singular, integral hypersurfaces of the total differential equation* $\alpha = 0$. *Furthermore, if all the zeros of* α *are non-degenerate, then* α *is transitive, if and only if there exists one point* $\mathbf{p} \in \mathbf{X}$ *such that both* $C_\alpha^+(\mathbf{p})$ *and* $C_\alpha^-(\mathbf{p})$ *are every-*

where dense in $\mathbf{X}$, *in which case they coincide with the set of points where* $\alpha \neq 0$.

PROOF. It is clear that $C_n^+(\mathbf{p})$ and $C_\alpha^-(\mathbf{p})$ can contain only points $\mathbf{q}$ such that $\alpha(\mathbf{q}) \neq 0$. In view of the proof of Proposition 2, for any such point $\mathbf{q}$ there exists a local coordinate system $\{U; x^1, \cdots, x^n\}$ in a neighborhood U of $\mathbf{q}$ such that $x^i(\mathbf{q}) = 0$ $(1 \leqq i \leqq n)$ and $\alpha|_U = e^f dx^1$, i.e., $x^1 = \int_{\mathbf{p}} e^{-f}\alpha|_U$. Without loss of generality we may assume that the range in $\mathbf{R}^n$ of such a coordinate system is an open, convex domain. It is then obvious that $C_\alpha^+(\mathbf{p}) \cap U$ (respectively, $C_\alpha^-(\mathbf{p}) \cap U$) is either empty or all of U or there exists a real number c such that the intersection is defined by the inequation $x^1 > c$ (respectively, $x^1 < c$). This shows that the uplands and lowlands are open domains and that all boundary points $\mathbf{q}$ such that $\alpha(\mathbf{q}) \neq 0$ are, in terms of the same local coordinate descriptions, defined locally by the equation $x^1 = c$, which represents locally the integral hypersurface of the differential equation $\alpha = 0$ near $\mathbf{q}$. There remains to show that, if for some $\mathbf{p} \in \mathbf{X}$ the sets $C_\alpha^+(\mathbf{p})$ and $C_\alpha^-(\mathbf{p})$ are both everywhere dense, and if all the zeros of α are non-degenerate, then α is transitive and both the upland and lowland relative to $\mathbf{p}$ coincide with the set $\{\mathbf{q} \mid \alpha(\mathbf{q}) \neq 0\}$.

Suppose that both $C_\alpha^+(\mathbf{p})$ and $C_\alpha^-(\mathbf{p})$ are everywhere dense; then for every $\mathbf{q}$ such that $\alpha(\mathbf{q}) \neq 0$, the whole coordinate neighborhood $\{U; x^1, \cdots, x^n\}$ of $\mathbf{q}$ with convex range and such that $\alpha|_U = e^f dx^1$ is contained in both $C_\alpha^+(\mathbf{p})$ and $C_\alpha^-(\mathbf{p})$, so that both domains coincide with the set where $\alpha \neq 0$. To show that α is transitive we have to consider an arbitrary point $\mathbf{q}$, and prove that $C_\alpha^+(\mathbf{q})$ and $C_\alpha^-(\mathbf{q})$ are again everywhere dense. First we see that, using the property that all zeros of α are non-degenerate, there exist oriented paths both from $\mathbf{p}$ to $\mathbf{q}$ and from $\mathbf{q}$ to $\mathbf{p}$, such that α along the path is everywhere strictly positive, except possibly at the end points, and that

(a) if α is non-zero at either $\mathbf{p}$ or $\mathbf{q}$, then the positivity condition on α can be retained at that end point as well;

(b) any path from $\mathbf{q}$ to $\mathbf{p}$ along which α is everywhere positive (respectively, negative) followed by another similar path from $\mathbf{p}$ to some given point $\mathbf{q}'$ can be smoothed out near $\mathbf{p}$ to give a positive (respectively, negative) path from $\mathbf{q}$ to $\mathbf{q}'$;

(c) if $\alpha(\mathbf{p}) \neq 0$, the smoothing process can be achieved without pulling the path away from $\mathbf{p}$ (a fact that is relevant to the corollary that follows).

These facts show that $C_\alpha^+(\mathbf{q}) \supset C_\alpha^+(\mathbf{p})$ and $C_\alpha^-(\mathbf{q}) \supset C_\alpha^-(\mathbf{p})$, completing the proof of the proposition.

COROLLARY. *Let* $\mathbf{X}$ *be a differential manifold and* α *transitive* 1-*form on* $\mathbf{X}$. *Then for each point* $\mathbf{p} \in \mathbf{X}$ *such that* $\alpha(\mathbf{p}) \neq 0$, *there passes a simple, oriented, closed, differentiable path* γ *along which* α *is strictly positive everywhere.*

PROOF. Since $C_\alpha^+(\mathbf{p})$ and $C_\alpha^-(\mathbf{p})$ are both open and everywhere dense, so is their intersection; let $\mathbf{q}$ be a point in the intersection; then, since α vanishes neither at $\mathbf{p}$ nor at $\mathbf{q}$, there exists oriented paths γ_1 from $\mathbf{p}$ to $\mathbf{q}$ and γ_2 from $\mathbf{q}$ to $\mathbf{p}$, along which α is strictly positive everywhere, including both end points. By the smoothing process mentioned at the end of the proof of Proposition 3, the combined closed path can be smoothed at both $\mathbf{p}$ and $\mathbf{q}$ to a differentiable path passing, in particular, still through $\mathbf{p}$. Finally, to obtain a simple path from this, in case it is not, let $\mathbf{q}'$ be the first self-intersection of the oriented, closed path after starting at $\mathbf{p}$, corresponding to the value t_1 of a global parameter, and let $t_2 > t_1$ be the last value of the parameter before the path becomes closed that corresponds again to $\mathbf{p}$. Skipping the segment from t_1 to t_2 we obtain again a closed path with a possible discontinuity of the tangent at $\mathbf{q}'$; this path can be once more smoothed out at $\mathbf{q}'$ locally without introducing new double points. The elimination of all double points requires at most finitely many repetitions of this process, yielding at the end a closed, oriented, simple path through $\mathbf{p}$ along which $\alpha > 0$. This completes the proof of the Corollary.

V. The main theorem

The results of the preceding sections lead up to the formulation of the main theorem.

THEOREM 1. *Let* $\mathbf{X}$ *be a compact, connected,* n*-dimensional, differential manifold and let* α *be a closed* 1-*form on* $\mathbf{X}$ *all of whose zeros are non-degenerate. Then a necessary and sufficient condition in order that* $\mathbf{X}$ *admit a riemannian metric* $\mathbf{g}$ *with respect to which* α *is co-closed is that* α *be transitive.*

Remark. The condition of transitivity on α implies immediately the local conditions of Proposition 2; in fact if $\mathbf{p}$ is a zero of α of index 0 (respectively n) then the set $C_\alpha^-(\mathbf{p})$ (respectively $C_\alpha^+(\mathbf{p})$) is clearly empty.

PROOF. We prove first that, if α is a harmonic 1-form, then it is transitive. Suppose, otherwise, that for a certain point $\mathbf{p} \in \mathbf{X}$ the set $C_\alpha^+(\mathbf{p})$ is not everywhere dense. Since the zeros of α are isolated, the boundary of $C_\alpha^+(\mathbf{p})$ consists, except for an at most finite set of points, of integral manifolds of the differential equation $\alpha = 0$. This set, with the transversal orientation assigning the positive "side" to the interior $C_\alpha^+(\mathbf{p})$, defines an $(n-1)$-cycle of odd kind γ, that is homologically trivial. Since we suppose that $*\alpha$ is closed, $\int_\gamma *\alpha = 0$; on the other hand the restriction of $*\alpha$ to the part of γ where $\alpha \neq 0$ is everywhere strictly positive and otherwise zero. Hence the boundary of $C_\alpha^+(\mathbf{p})$ is contained in the zero-set of α, which is at most a finite set. This means that $C_\alpha^+(\mathbf{p})$ is either everywhere dense (contrary to what we supposed to begin with) or empty; this last possibility is ruled out, since this can not be for any $\mathbf{p}$, since α has only non-degenerate zeros and under this condition it would mean that $\mathbf{p}$ is a zero of index n, contradicting E.Hopf's maximum principle.

We now assume that α is transitive and try to construct a riemannian metric $\mathbf{g}$ with respect to which α is co-closed. This can be accomplished after a few lemmas that we shall now formulate.

LEMMA 1. *Let* $\mathbf{X}$ *be a paracompact, n-dimensional manifold; let* α *be a* 1-*form on* $\mathbf{X}$, *all of whose zeros are non-degenerate, and let* ψ *be an* $(n-1)$-*form on X of odd kind satisfying the following conditions.*

(a) *At each point* $\mathbf{p} \in \mathbf{X}$ *where* $\alpha(\mathbf{p}) \neq 0$ *the n-form of odd kind* $\alpha \wedge \psi$ *is strictly positive;*

(b) *For each point* $\mathbf{p} \in \mathbf{X}$ *where* α *vanishes there exists in a neighborhood* $U_\mathbf{p}$ *of* $\mathbf{p}$ *a riemannian metric* $\mathbf{g}_\mathbf{p}$, *with respect to which* $\psi|_{U_\mathbf{p}} = *_{(\mathbf{g}_\mathbf{p})}\alpha|_{U_\mathbf{p}}$, *the neighborhoods* $U_\mathbf{p}$ *being pairwise disjunct.*

Then there exists on $\mathbf{X}$ *a riemannian metric* $\mathbf{g}$, *with respect to which* $*\alpha = \psi$. *Furthermore one can construct* $\mathbf{g}$ *so as to coincide with* $\mathbf{g}_\mathbf{p}$ *in any neighborhood* $\mathrm{V}_\mathbf{p}$ *of a point* $\mathbf{p}$ *where* α *vanishes, where* $\bar{V}_\mathbf{p} \subset U_\mathbf{p}$.

Remark. One could replace condition (b) by a weaker one

involving the first order jets of α and ψ respectively at each point where α vanishes (ψ being also there, of necessity, zero); the metric $\mathbf{g}$ constructed in this case has a prescribed zero order jet. The details of this, however, are extraneous to the main result of this paper, and are omitted for the sake of brevity.

PROOF. Let A^j and A^j_- denote the vector bundles over $\mathbf{X}$ of j-forms and of j-forms of odd kind respectively, and consider the differentiable vector bundle $\mathrm{Hom}\,(A^1, A^{n-1}_-)$. A vector sub-bundle of this bundle, denoted by $\mathrm{Hom}_S\,(A^1, A^{n-1}_-)$ will be called that of symmetric homomorphisms, defined as follows. An element $\tau_{\mathbf{p}} \in \mathrm{Hom}\,(A^1, A^{n-1}_-)$ over $\mathbf{p} \in \mathbf{X}$ is called symmetric, if and only if for any two 1-forms $\beta_{\mathbf{p}}$, $\gamma_{\mathbf{p}}$ at $\mathbf{p}$ we have

$$\beta_{\mathbf{p}} \wedge \tau_{\mathbf{p}}(\gamma_{\mathbf{p}}) = \gamma_{\mathbf{p}} \wedge \tau_{\mathbf{p}}(\beta_{\mathbf{p}}) \, .$$

Furthermore a symmetric homomorphism $\tau_{\mathbf{p}}$ at $\mathbf{p} \in \mathbf{X}$ is called positive definite (respectivery semi-definite), if for every $\beta_{\mathbf{p}} \neq 0$

$$\beta_{\mathbf{p}} \wedge \tau_{\mathbf{p}}(\beta_{\mathbf{p}}) > 0 \qquad\qquad (\text{resp. } \geqq 0) \, ,$$

where the inequality in the bundle of n-forms of odd kind has the obvious meaning. The bundle of positive definite, symmetric homomorphisms of A^1 into A^{n-1}_- is a differentiable, open sub-bundle of $\mathrm{Hom}_S\,(A^1, A^{n-1})$ whose fibres are convex: we denote this bundle by T. Since the fibres are contractible, given any differentiable cross-section τ of T over an open set $U \subset \mathbf{X}$, for any closed set $K \subset U$ there exists a global cross section of T over $\mathbf{X}$ that coincides with τ in K. A typical cross section of T is defined in terms of any riemannian structure $\mathbf{g}$ in $\mathbf{X}$ by the homomorphism $*_{(\mathbf{g})}$; indeed the map $\{\mathbf{g} \longrightarrow *_{\mathbf{g}}\}$ is a bundle homomorphism of the bundle of riemannian metrics into the bundle T. The interesting thing is that, for $n \neq 2$ this homomorphism is bijective. In fact, let $\omega^1_{\mathbf{p}}, \cdots, \omega^n_{\mathbf{p}}$ be a basis for A^1 at any point $\mathbf{p} \in \mathbf{X}$. In terms of this basis the components of a riemannian form at $\mathbf{p}_n$ are a symmetric, positive definite matrix $(g_{ij}) = \mathrm{g}$, where if $\beta = \sum_{i=1}^n b_i \omega^i$ is a 1-form, we have

$$|\beta|^2 = \textstyle\sum_{ij} g^{ij} b_i b_j \qquad \left((g^{ij}) = (g_{ij})^{-1}\right) ,$$

Consequently in terms of the riemannian form $\mathbf{g}$, we have

$$(5) \quad *\beta = \left(\det(\mathrm{g})\right)^{1/2} \textstyle\sum_{i,k=1}^n (-1)^{k-1} g^{ik} b_i \,|\omega^1 \wedge \cdots \vee^{(k)} \cdots \wedge \omega^n|\, \mathrm{sgn}\, \omega^k .$$

On the other hand, the most general homomorphism τ of A^1 into A^{n-1}_- is represented by a matrix (G^{ij}) in the formula

$$(6) \quad \tau(\beta) = \textstyle\sum_{i,k=1}^n (-1)^{k-1} G^{ik} b_i \,|\, \omega^1 \wedge \cdots \vee^{(k)} \cdots \wedge \omega^n \,|\, \mathrm{sgn}\, \omega^k$$

where τ is symmetric (respectively, positive definite or positive semi-definite) according to the applicability of the same properties to the matrix $(G^{ik}) = \mathfrak{G}$. From formulas (5) and (6) we see that, given a riemannian metric $\mathbf{g}$ represented by a matrix $\mathfrak{g}$, the matrix $\mathfrak{G}$ representing $*_{\mathbf{g}}$ is given by

$$\mathfrak{G} = \big(\det(\mathfrak{g})\big)^{1/2}\,\mathfrak{g}^{-1}\,,$$

whence

$$\det(\mathfrak{G}) = \big(\det(\mathfrak{g})\big)^{n-2/2}\,.$$

Thus one sees that, for $n = 2$, $\det(\mathfrak{G}) = 1$, showing that the homomorphism is not surjective on T, while for $n \neq 2$ the map $\{\mathbf{g} \rightarrow \tau = *_{\mathbf{g}}\}$ is bijective, its inverse being given by

$$\mathfrak{g} = \big(\det(\mathfrak{G})\big)^{1/n-2}\,\mathfrak{G}^{-1}\,.$$

Let (W_ν) be a locally finite, open cover of $\mathbf{X}$ with the following properties: for each ν either α vanishes in precisely one point $p \in W_\nu$ and $\bar{V}_{\mathbf{p}} \subset W_\nu \subset U_{\mathbf{p}}$, or else α vanishes nowhere in W_ν; W_ν is contractible and $\bar{W}_\nu$ is disjunct from the closed neighborhoods $\bar{V}_{\mathbf{p}}$ of every zero point $\mathbf{p}$ of α prescribed in the statement; let (φ_ν) be a differentiable partition of unity subordinate to the locally finite cover (W_ν). If $W_\nu \supset V_{\mathbf{p}}$ for some zero point of α, we restrict the given riemannian metric structure $\mathbf{g}_{\mathbf{p}}$ from U_ν and define a positive semi-definite, differentiable cross section τ_ν over $\mathbf{X}$ in $\mathrm{Hom}_S(A^1, A^{n-1}_-)$ by

$$\tau_\nu = \begin{cases} = \varphi_\nu *_{(\mathbf{g}_{\mathbf{p}})} & \text{in } W_\nu \\ = 0 & \text{in } X - W_\nu\,. \end{cases}$$

If W_ν is disjunct from all the neighborhoods $V_{\mathbf{p}}$, since W_ν is contractible, there exists a differentiable frame $(\omega_\nu^1, \cdots, \omega_\nu^n)$ for A^1 over W_ν satisfying the following.

$$\omega_\nu^1 = \alpha\,,$$
$$\omega_\nu^i \wedge \psi = 0 \qquad (i = 2, \cdots, n)\,,$$

and a "dual" basis $(\eta_\nu^1, \cdots, \eta_\nu^n)$, uniquely defined by

$$\eta_\nu^1 = \psi\,,$$
$$\omega_\nu^i \wedge \eta_\nu^k = \begin{cases} \alpha \wedge \psi > 0 & (i = k)\,, \\ 0 & (i \neq k)\,. \end{cases}$$

We then have again a positive semi-definite cross section τ_ν in

$\mathrm{Hom}_S(A^1, A_-^{n-1})$ defined by

$$\tau_\nu(\omega_\nu^i) = \begin{cases} = \varphi_\nu \eta_\nu^i & \text{in } W_\nu \\ = 0 & \text{outside .} \end{cases}$$

Finally the sum

$$\tau = \sum_\nu \tau_\nu$$

is a differentiable, differentiable cross section over **X** in the bundle T of positive definite homomorphisms in $H_S(A^1, A_-^{n-1})$. By the previous argument, if $n \neq 2$, τ determines a riemannian structure $\mathbf{g}$ on **X** such that $*_{\mathbf{p}}\alpha = \psi$ and $\mathbf{g} = \mathbf{g}_{\mathbf{p}}$ in each neighborhood $V_{\mathbf{p}}$ of any zero point $\mathbf{p}$ of α.

In the remaining case $n = 2$ the proof is somewhat simpler. We cover **X** by the two open sets U_0, U_1, where

$$U_0 = \bigcup_{\alpha(\mathbf{p})=0} U_p \ , \qquad U_1 = X - \bigcup_{\alpha(\mathbf{p})=0} \bar{V}_p$$

and let $\{\varphi_0, \varphi_1\}$ be a differentiable partition of unity subordinate to $\{U_0, U_1\}$. In U_0 we have the riemannian structure $\mathbf{g}_0$ obtained from the given metrics $\mathbf{g}_{\mathbf{p}}$ each defined in the pairwise disjunct $U_{\mathbf{p}}$. In U_1 we define the riemannian metric g_1 defined by

$$\mathbf{g}_1 = \alpha \otimes \alpha + \psi \otimes \psi \ ,$$

where $\mathbf{g}_1$ is interpreted as a cross section in the tensor product bundle $A^1 \otimes A^1 = A_-^1 \otimes A_-^1$. The riemannian metric

$$\mathbf{g} = \varphi_0 \mathbf{g}_0 = \varphi_1 g_1$$

can be readily verified to satisfy all the required properties stated by the lemma.

We return now to the proof of the main theorem. The preceding lemma reduces the problem, given α with the stated assumptions, to find a closed $(n-1)$-form of odd kind ψ in **X** such that, wherever $\alpha \neq 0$, $\alpha \wedge \psi > 0$ and at each zero point $\mathbf{p}$ of α there exists a local riemannian metric $\mathbf{g}_{\mathbf{p}}$ with respect to which $\psi = *_{(\mathbf{g}_{\mathbf{p}})}\alpha$. The latter property is satisfied as a consequence of Proposition 2, since we have already noted that there can be no zeros of α of index 0 or n. For each zero point $\mathbf{p}$ of α we choose a fixed coordinate neighborhood $\{U_{\mathbf{p}}; x^1, \cdots, x^n\}$ about $\mathbf{p}$ with $x^i(\mathbf{p}) = 0$, with respect to which α is locally represented by (4) (cf. Proposition 2); we assume furthermore that the neighborhoods $U_{\mathbf{p}}$ of the different zero points are pairwise disjunct. Let $\mathbf{p}_\nu (\nu = 1, 2, \cdots, N)$ be an enumeration of the zero points of α and let δ_ν be positive numbers such that the range of the coordinate domain $U_{\mathbf{p}_\nu}$ includes the ball defined by

$$\sum_{i=1}^{n} (x^i)^2 \leqq \delta_\nu^2$$

and let V_t, for any real t in the interval $0 < t \leqq 1$, denote the union of the disjunct neighborhoods of the different $\mathbf{p}_\nu$ defined in the respective coordinate systems by

$$\sum_{i=1}^{n} (x^i)^2 < t^2\delta_\nu^2 . \tag{7}$$

We next prove a lemma that is a sharper version of the corollary of Proposition 3, § 4.

LEMMA 2 (Deligne). *Let* $\mathbf{X}$ *be an* n*-dimensional differential manifold,* α *a closed* 1*-form that is transitive. For any point* $\mathbf{p} \in \mathbf{X}$ *such that* $\alpha(\mathbf{p}) \neq 0$ *there exists a closed* $(n-1)$*-form of odd kind* ψ *with compact support, such that* $\alpha \wedge \psi \geqq 0$ *everywhere in* $\mathbf{X}$ *and* $(\alpha \wedge \psi)(\mathbf{p}) > 0$.

PROOF. Let $\mathbf{p} \in \mathbf{X}$ be a point where α does not vanish. According to the corollary of Proposition 3 there exists a simple closed, differentiable, oriented curve in $\mathbf{X}$ passing through $\mathbf{p}$, along which α is strictly positive everywhere. Let γ be such a curve: we construct a tubular neighborhood W of γ, such that the local integral manifolds defined by the equation $\alpha = 0$ (that are transversal to γ at each point) are everywhere transversal to the boundary of W and intersect W in an open $(n-1)$-ball meeting γ in exactly one point. Then W is a differentiable fibre bundle over γ, the fibres being the integral manifolds of α. By virtue of a theorem of Mazur the structure group of this bundle is reducible to the identity or the cyclic group of order 2 generated by a reflection along a hypersurface, according as to whether or not W is orientable. In other words the open submanifold W of $\mathbf{X}$ with the 1-form α in it is either isomorphic to the product $S^1 \times E^{n-1}$, or to the n-dimensional Möbius band, with fibre E^{n-1} and base S^1, where $S^1 = \mathbf{R}/c\mathbf{Z}$ for some positive constant $c = \int_\gamma \alpha$ and E^{n-1} is the interior of the unit euclidean ball in $\mathbf{R}^{n-1}$. Let π denote the projection of W onto S^1, with fibres along the integral manifolds of the equation $\alpha = 0$, let S^1 be parametrized by $x^1 \bmod c\mathbf{Z}$ so that $\pi^* dx^1 = \alpha$ and E^{n-1} be locally parametrized by $(x^2, \cdots, x^n) \in \mathbf{R}^{n-1}$ satisfying $\sum_{i=2}^{n} (x^i)^2 < 1$ and the group action, if required, represented by the involution $(x^2, x^3, \cdots, x^n) \to (-x^2, x^3, \cdots, x^n)$, and suppose that γ corresponds to $\{x^2 = x^3 = \cdots = x^n = 0\}$ and in γ the point $\mathbf{p}$ corresponds to $x^1 = 0 \bmod c\mathbf{Z}$.

Let $h(t)$ be an even, non-negative valued, differentiable function on $\mathbf{R}$ which is identically equal to 1 for $|t| \leqq \frac{1}{3}$ and identically zero for $|t| \geqq \frac{2}{3}$. Then the $(n-1)$-form $\psi_{\mathbf{p}}$ of odd kind can be represented by the equations

$$\begin{cases} \psi_p = h\left(\sum_{i=2}^n (x^i)^2\right) |dx^2 \wedge \cdots \wedge dx^n| (\operatorname{sgn} dx^1) & \text{inside } W; \\ \psi_{\mathbf{p}} = 0 & \text{in } X - W . \end{cases}$$

Clearly $\psi_{\mathbf{p}}$ is closed and, since

$$\alpha \wedge \psi_{\mathbf{p}} = h\left(\sum_{i=2}^n (x^i)^2\right) |dx^1 \wedge \cdots \wedge dx^n| \geqq 0$$

everywhere in W and vanishes outside, it satisfies the properties required and completes thereby the proof of the lemma.

It is to be noted that the existence of such a $\psi_{\mathbf{p}}$ for every $\mathbf{p}$ with $\alpha(\mathbf{p}) \neq 0$ is equivalent to α being transitive, once one knows that the zeros of α are reasonable, i.e., for instance all non-degenerate. It would make matters easier in the sequel if we could produce an analogous $\psi_{\mathbf{p}}$, *mutatis mutandis* when $\alpha(\mathbf{p}) = 0$, if $\mathbf{p}$ is a non-degenerate zero of a transitive α; however this is not the case up to now.

We complete the proof of the theorem as follows. Consider the open neighborhood V_1 defined by (7), for $t = 1$, of the set of all zeros of α. To each $\mathbf{p} \in \mathbf{X} - V_1$ we assign a closed $(n-1)$-form of odd kind $\psi_{\mathbf{p}}$ with the properties asserted in Lemma 2; such a form satisfies $\alpha \wedge \psi_{\mathbf{p}} > 0$ in a neighborhood of $\mathbf{p}$ but $\psi_{\mathbf{p}} = 0$ identically in a neighborhood V_t of all the zeros of α for some positive $t \leqq 1$, depending on $\mathbf{p}$. Let $U_{\mathbf{p}}$ denote the open set of points where $\psi_{\mathbf{p}} \neq 0$, whence also by the construction $\alpha \wedge \psi_{\mathbf{p}} > 0$. Since $\mathbf{X} - V_1$ is compact, there exists a finite set of points $\mathbf{q}_1, \cdots, \mathbf{q}_M \in \mathbf{X} - V_1$ such that $\bigcup_{\mu=1}^M U_{\mathbf{p}_\mu} \supset \mathbf{X} - V_1$; in this case there is a positive $\delta_0 < 1$ such that $\bigcup_{\mu=1}^M U_{q_\mu} \subset \mathbf{X} - V_{\delta_0}$. Thus the form $\psi' = \sum_{\mu=1}^M \psi_{\mathbf{q}_\mu}$ has the following properties:

(i) ψ' is closed and vanishes identically in V_{δ_0},

(ii) $\alpha \wedge \psi' > 0$ at all points where $\psi' \neq 0$,

and is bounded away from zero in $\mathbf{X} - V_{\delta'}$ for some δ' in the interval $\delta_0 < \delta' < 1$.

Now in the neighborhood of each zero point $\mathbf{p}_\nu$ of α $(\nu = 1, \cdots, N)$ we carry out the following construction. Let $V_{\delta,\nu}$ be the component of V_δ containing $\mathbf{p}_\nu$ and refer to the same local coordinate $x^1, \cdots, x^n$ in $V_{1,\nu}$ used in (7), satisfying the relations (3) and (4) with respect to α, where j is the index of $\mathbf{p}_\nu$. Let η_ν be a non-negative, differentiable function on $\mathbf{X}$ which is identically 1 in $V_{\delta',\nu}$ and vanishing identically outside $V_{1,\nu}$; consider the local euclidean metric $\mathbf{g}_\nu$ defined by (2) in $V_{1,\nu}$ and with respect to it we form the $(n-2)$-form of odd kind

$$\begin{cases} \chi_\nu = \eta_\nu \sum_{j+1 \leqq i \leqq n}^{1 \leqq h \leqq j,} (-1)^{h+i} x^h x^i \\ \qquad \times |dx^1 \wedge \cdots \vee^h \cdots \wedge dx^j \wedge dx^{j+1} \wedge \cdots \vee^i \cdots \wedge dx^n| \\ \qquad\qquad \times \operatorname{sgn}(dx^h \wedge dx^i) \qquad \text{in } V_{1,\nu}, \\ \chi_\nu = 0 \qquad \text{outside } V_{1,\nu}. \end{cases}$$

A direct computation shows that in $V_{\delta',\nu}$ (i.e., in a set where $\eta = 1$ identically, $d\chi_\nu = *_{(\mathbf{g}_\nu)}\alpha$. Hence we can let $\psi_\nu = d\chi_\nu$ globally, since $\psi)$ vanishes identically outside $V_{1,\nu}$; let

$$\psi'' = \sum_{\nu=1}^{N} \psi_\nu .$$

Then the $(n-1)$-form ψ'' of odd kind has the following properties:

(i) ψ'' is exact;

(ii) $\alpha \wedge \psi'' \geqq 0$ in $V_{\delta'} = \bigcup_{\nu=1}^{N} V_{\delta',\nu}$;

(iii) ψ'' vanishes outside V_1;

(iv) there exists a local riemannian structure $\mathbf{g}''$ (the union of the local euclidean metrics $\mathbf{g}_\nu$) in $V_{\delta'}$ with respect to which $\psi'' = *_{\mathbf{g}''}\alpha$.

Combining the above properties of ψ'' with those previously established for ψ', we see that for a sufficiently large, positive constant K, the form

$$\psi = K\psi' + \psi'' ,$$

which coincides with ψ'' in V_{δ_0}, satisfies $\alpha \wedge \psi > 0$ everywhere else. In other words ψ satisfies the conditions stated in Lemma 1 for the existence of a metric $\mathbf{g}$ on $\mathbf{X}$ for which $\psi = *_{\mathbf{g}}\alpha$. Since ψ is closed, α is co-closed with respect to this metric. Theorem 1 is thereby proved.

It was mentioned earlier that, given a closed 1-form α all of whose zeros are non-degenerate, in order that α be transitive it is necessary (but not sufficient) that each zero of α have index different from 0 and n. One can show by a method of continuity that transitivity, in a compact manifold $\mathbf{X}$, can be deduced from the additional condition that each zero of α have index different from 1 as well (or alternately different from $n - 1$). In particular, every closed 1-form with no zeros whatever is transitive.

VI. An example

We show that one can construct some rather pathological-looking, closed 1-forms that are transitive. In particular, there exist in a 3-dimensional (or higher) torus a riemannian metric with respect to which some harmonic 1-forms can have isolated zeros. We shall construct in euclidean 3-space a smooth function $f(x, y, z)$ with the following properties:

(i) for $x^2 + y^2 + z^2 > M^2$, M sufficiently large, $f(x, y, z) = x$ identically;

(ii) $f(x, y, z)$ is an even function of both y and z;

(iii), $f(x, y, z)$ has exactly two critical points, $(x, y, z) = (\pm c, 0, 0)$, both non-degenerate and of index respectively 1 and 2.

An example of such a function is the following: let $h(t)$ be an even function of one real variable, of class C^∞, identically equal to 1 for $|t| \leqq 1$, identically zero, for $|t| \geqq M$, non-negative and with derivative less than $\dfrac{2}{M}$ in absolute value throughout. Then, for M sufficiently large, letting $r = (x^2 + y^2 + z^2)^{1/2}$,

$$f(x, y, z) = \frac{x(x^2 + y^2 + z^2 - 1) + (y^2 - z^2)h(r)}{h(r) + (x^2 + y^2 + z^2 - 1)\big(1 - h(r)\big)}$$

the resulting function has the desired properties, with the constant $c = 3^{-1/2}$. Let $\mathbf{X}$ be an n-dimensional compact riemannian manifold ($n \geqq 3$) and α a non-trivial harmonic 1-form on $\mathbf{X}$. Choose a point $\mathbf{p} \in \mathbf{X}$ such that $\alpha(\mathbf{p}) \neq 0$ and let $\{U, x^1, \cdots, x^n\}$ be a local coordinate system in a neighborhood U of $\mathbf{p}$ with $x^i(\mathbf{p}) = 0$, such that $\alpha|_U = dx^1$, and choose two positive integers j, k such that $j + k = n - 1$; we denote by x' the subset of coordinates $(x^2, \cdots, x^{j+1})$ and by x'' the coordinates $(x^{j+2}, \cdots, x^n)$ and by y, z respectively the sums $\big(\sum_{i=2}^{j+1} (x^i)^2\big)^{\frac{1}{2}}$ and $\big(\sum_{i=j+2}^{n} (x^i)^2\big)^{\frac{1}{2}}$. If the local coordinate system includes in its range the ball

$$\textstyle\sum_{i=1}^{n} (x^i)^2 = (x^1)^2 + y^2 + z^2 < A \,,$$

we can perturb α by replacing it inside U by

$$\frac{A}{M} df\Big(\frac{Mx^1}{A}, \frac{My}{A}, \frac{Mz}{A}\Big) \,,$$

which coincides with α near the boundary of U, and hence can be continued outside by the given α. The new differential has the same zeros, of course, as α outside U, while inside U, where α had no zeros, the new closed 1-form has precisely two new zeros, both non-degenerate, corresponding to the two critical points of $f(x, y, z)$; their indices are respectively k and $k+1$. A careful check of the transversal trajectories of the new form inside U with respect to the original metric as compared with those of α preserves the transitivity property. Therefore it is possible to increase the number of zeros of a closed form by two at a time, and still preserve the existence of a riemannian metric with respect to which the form is harmonic. In particular any n-torus ($n \geqq 3$) admits riemannian metric with respect to which harmonic 1-forms may have any given, even number of isolated zeros.

UNIVERSITY OF PENNSYLVANIA AND I.H.E.S., Bures-sur-Yvette

REFERENCES

[1] N. ARONSZAJN, *A unique continuation theorem for solutions of elliptic partial differential equations or any qualities of second order*, J. Math. Pures Appl. (9) **36** (1957), 235-249.

[2] E. HOPF, *Elementare Bemerkungen über die Lösungen partieller Differentialgleichungen zweiter Ordnung vom elliptischen Typus*, Sitzungsber. der Preuss. Akad. der Wissenschaften, **19**, (1927), 147-152.

(Received October 15, 1968)

Intrinsic norms on a complex manifold

S. S. CHERN, HAROLD I. LEVINE AND LOUIS NIRENBERG*

1. Introduction

We propose to define in this paper certain norms (or more precisely, semi-norms) on the homology groups of a complex manifold. They will be invariants of the complex structure and do not increase under holomorphic mappings. Their definitions depend on the bounded plurisubharmonic functions on the manifold and are modelled after the notion of harmonic length introduced by H. Landau and R. Osserman [8] for Riemann surfaces (= one-dimensional complex manifolds). It is possible to extend the definition to certain families of chains. In particular we get in this way an intrinsic pseudo-metric on the manifold which is closely related to that of Caratheodory [3] and to one recently introduced by S. Kobayashi [7].

To define one of these semi-norms (possibly the most significant one among those to be introduced below), let M be a complex manifold of complex dimension n. Let

$$d^c = i(\bar{\partial} - \partial)\,, \tag{1}$$

so that d^c is a differential operator of degree one on smooth complex-valued exterior differential forms and maps a real form into a real form. Let $\mathcal{F}$ be the family of plurisubharmonic functions u of class C^2 on M satisfying the condition $0 < u < 1$. To a homology class γ of M with real coefficients we set

$$\begin{aligned} N\{\gamma\} &= \sup_{u\in\mathcal{F}} \inf_{T\in\gamma} |\, T[d^c u \wedge (dd^c u)^{k-1}]\,| \\ &\qquad\qquad \text{if } \dim\gamma = 2k-1\,, \\ N\{\gamma\} &= \sup_{u\in\mathcal{F}} \inf_{T\in\gamma} |\, T[du \wedge d^c u \wedge (dd^c u)^{k-1}]\,| \\ &\qquad\qquad \text{if } \dim\gamma = 2k\,, \end{aligned} \tag{2}$$

* The first author was partially supported by NSF grant GP-8623. The second author was partially supported by NSF grant GP-6761. The third author was partially supported by Air Force Office of Scientific Research, Grant AF-49(638)-1719.

where T runs over all currents (in the sense of de Rham [10]) of γ. Our main theorem (cf. § 3) asserts that $N\{\gamma\}$ is always finite. The following properties are easily verified.

$$N\{a\gamma\} = |a| N\{\gamma\}, \qquad a \in R; \tag{3}$$

$$N\{\gamma_1 + \gamma_2\} \leqq N\{\gamma_1\} + N\{\gamma_2\}, \qquad \dim \gamma_1 = \dim \gamma_2. \tag{4}$$

Furthermore, under a holomorphic mapping, $N\{\gamma\}$ is non-increasing. These properties make it a useful tool in the study of holomorphic mappings.

Unfortunately in the case when M is compact, the family $\mathcal{F}$ will consist only of constants and the seminorm will be identically zero. We will show, however, that our definition can be refined to give a meaningful invariant which, in the case of compact Riemann surfaces, is "equivalent" to the extremal length of Ahlfors-Beurling [2]. As is well known, the latter, together with the classical topological invariants, gives a complete system of conformal invariants to compact Riemann surfaces, in the sense of the following theorem of Accola [1]. Let $f: M \to M'$ be a diffeomorphism between two compact Riemann surfaces under which corresponding homology classes of curves have the same extremal length. Then f is a conformal equivalence.

2. A lemma on bounded plurisubharmonic functions in C_n

Note: Just the corollary of this paragraph is used in the proof of our Theorem 1. The lemma itself is used only in Remark 3 following Theorem 2.

Let C_n be the complex number space of dimension n with the coordinates z^k, $1 \leqq k \leqq n$. We denote its volume element by

$$dV = \left(\frac{i}{2}\right)^n \bigwedge_k dz^k \wedge d\bar{z}^k. \tag{5}$$

For a real-valued smooth function v defined in an open subset of C_n its partial derivatives will be denoted by

$$v_j = \frac{\partial v}{\partial z^j}, \quad v_{\bar{k}} = \frac{\partial v}{\partial \bar{z}^k}, \quad v_{j\bar{k}} = \frac{\partial^2 v}{\partial z^j \partial \bar{z}^k}, \text{ etc.} \qquad 1 \leqq j, k \leqq n. \tag{6}$$

We recall that such a function is called *plurisubharmonic* if the hermitian matrix $(v_{j\bar{k}})$ is positive semi-definite; v is called *pluriharmonic* if $(v_{j\bar{k}}) = 0$ or $dd^c v = 0$.

LEMMA. *Let v be a plurisubharmonic negative-valued function of class C^2 in a polydisc*

$$\Delta: |z^i| < r_i , \qquad 1 \leqq i \leqq n ,$$

in C_n. Let Δ_1 be a compact subpolydisc

$$|z^i| \leqq \rho_i < r_i , \qquad 1 \leqq i \leqq n .$$

Then there is a constant A independent of v, depending only on the numbers ρ_i, r_i, $1 \leqq i \leqq n$, such that for the integral of any $r \times r$ minor of $(-v_{j\bar{k}}/v)$ over Δ_1 we have the estimate

$$\int_{\Delta_1} \text{abs.} \begin{vmatrix} v_{i_1\bar{k}_1} & \cdots & v_{i_1\bar{k}_r} \\ & \cdots & \\ v_{i_r\bar{k}_1} & \cdots & v_{i_r\bar{k}_r} \end{vmatrix} \frac{1}{|v|^r} dV \leqq A . \tag{7}$$

We prove the lemma by induction on r. For $r = 1$ we prove a stronger form of (7), namely,

$$\int_{\Delta_1} \left(\left| \frac{v_{i\bar{k}}}{v} \right| + \frac{1}{2} \left| \frac{v_k}{v} \right|^2 \right) dV \leqq A , \qquad 1 \leqq i, k \leqq n . \tag{7'}$$

Since $(v_{i\bar{k}})$ is a positive semi-definite matrix, we have $2|v_{i\bar{k}}| \leqq v_{i\bar{i}} + v_{k\bar{k}}$. Hence it suffices to prove (7′) for $i = k$. Let $\zeta \geqq 0$ be a C^∞ function with support in Δ and equal to one in Δ_1. Then, by Green's theorem we have

$$\int \left(\frac{v_{k\bar{k}}}{-v} + \left| \frac{v_k}{v} \right|^2 \right) \zeta^2 dV = \int \left(\frac{v_k}{-v} \right)_{\bar{k}} \zeta^2 dV = 2 \int \frac{v_k}{v} \zeta \zeta_{\bar{k}} dV$$

$$\leqq \int \left(\frac{1}{2} \left| \frac{v_k}{v} \right|^2 \zeta^2 + 2 |\zeta_{\bar{k}}|^2 \right) dV ,$$

the integrations being over Δ. It follows that

$$\int \left(\frac{v_{k\bar{k}}}{-v} + \frac{1}{2} \left| \frac{v_k}{v} \right|^2 \right) \zeta^2 dV \leqq 2 \int |\zeta_{\bar{k}}|^2 dV = A,$$

which yields (7′).

To proceed by induction we suppose the truth of (7) for $r-1$. Since v is plurisubharmonic, the absolute value of a general $r \times r$ minor of $(v_{i\bar{k}})$ is less than or equal to the maximum of the principal minors of order r. It therefore suffices to prove (7) for a principal minor, and we can restrict ourselves to the case $i_1 = k_1 = 1, \cdots, i_r = k_r = r$. We set

$$\Psi = 2^{-2r} \left(\frac{i}{2} \right)^{n-r} \wedge_{k>r} dz^k \wedge d\bar{z}^k$$

and choose ζ as before. Then we have

$$I_r = \int_{\Delta_1} \begin{vmatrix} v_{1\bar 1} & \cdots & v_{1\bar r} \\ & \cdots & \\ v_{r\bar 1} & \cdots & v_{r\bar r} \end{vmatrix} \frac{dV}{(-v)^r} = \int_{\Delta_1} \left(\frac{dd^c v}{-v}\right)^r \wedge \psi \leqq \int_{\Delta} \zeta \left(\frac{dd^c v}{-v}\right)^r \wedge \psi$$

$$= -r \int_{\Delta} \frac{dv \wedge d^c v \wedge (dd^c v)^{r-1}}{(-v)^{r+1}} \zeta\psi - \int_{\Delta} \frac{d\zeta \wedge d^c v \wedge (dd^c v)^{r-1}}{(-v)^r} \wedge \psi ,$$

where the last equality follows from Green's theorem. Since

$$d\zeta \wedge d^c v - dv \wedge d^c \zeta = i(\partial + \bar\partial)\zeta \wedge (\bar\partial - \partial)v - i(\partial + \bar\partial)v \wedge (\bar\partial - \partial)\zeta$$
$$= 2i(\bar\partial\zeta \wedge \bar\partial v - \partial\zeta \wedge \partial v) ,$$

it contains no term of type (1,1). It follows that

$$\int_{\Delta} \frac{d\zeta \wedge d^c v \wedge (dd^c v)^{r-1}}{(-v)^r} \wedge \psi = \int_{\Delta} \frac{dv \wedge d^c \zeta \wedge (dd^c v)^{r-1}}{(-v)^r} \wedge \psi ,$$

and by Green's theorem this is equal to

$$-\frac{1}{r-1} \int_{\Delta} \frac{dd^c \zeta \wedge (dd^c v)^{r-1}}{(-v)^{r-1}} \wedge \psi .$$

On the other hand, one sees easily that $dv \wedge d^c v \wedge (dd^c v)^{r-1} \wedge \psi$ is a non-negative multiple of dV. Hence we get

$$I_r \leqq \frac{1}{r-1} \int_{\Delta} dd^c \zeta \wedge \left(\frac{dd^c v}{-v}\right)^{r-1} \wedge \psi ,$$

and the desired inequality (7) follows from induction hypothesis.

COROLLARY. *Let the polydiscs Δ, Δ_1 be defined as in the lemma. Let u be a C^2-plurisubharmonic function in Δ with $0 < u < 1$. Then there is a constant B independent of u such that for any $r \times r$ minor of $(u_{i\bar k})$ we have the estimate*

$$(8) \qquad \int_{\Delta_1} \left[\text{abs.} \begin{vmatrix} u_{i_1\bar k_1} & \cdots & u_{i_1\bar k_r} \\ & \cdots & \\ u_{i_r\bar k_1} & \cdots & u_{i_r\bar k_r} \end{vmatrix} + \sum_k |u_k|^2 \right] dV \leqq B .$$

To deduce this corollary from the lemma we set $v = u - 1$. Then $0 < -v < 1$ and

$$\text{abs.} \begin{vmatrix} u_{i_1\bar k_1} & \cdots & u_{i_1\bar k_r} \\ & \cdots & \\ u_{i_r\bar k_1} & \cdots & u_{i_r\bar k_r} \end{vmatrix} \leqq \frac{1}{(-v)^r} \text{abs.} \begin{vmatrix} v_{i_1\bar k_1} & \cdots & v_{i_1\bar k_r} \\ & \cdots & \\ v_{i_r\bar k_1} & \cdots & v_{i_r\bar k_r} \end{vmatrix} .$$

Thus the corollary follows from the Lemma and (7′).

For $r = 1$ the corollary was proved by P. Lelong [9].

The lemma proved above has a real analogue whose proof is *similar. Let v be a negative convex function of class C^2 in a domain $\mathfrak{D}$ in* $\mathbf{R}^n$, i.e., *a function whose hession matrix is positive semidefinite,*

$$\sum_{i,j} v_{x^i x^j} \xi_i \xi_j \geqq 0 .$$

For any subdomain K with compact closure in $\mathfrak{D}$, there is a constant A, independent of v, such that

$$\int_K \left\{ \left| \frac{v_{x^k}}{v} \right|^2 + \frac{1}{|v|^r} \left| \text{any } r \times r \text{ minor of } (v_{x^i x^j}) \right| \right\} dV \leqq A .$$

3. Semi-norms and their properties

Using the definition (2) we shall prove the theorem.

THEOREM 1. *Let M be a complex manifold and γ a homology class with real coefficients. Then $N\{\gamma\}$ is finite.*

To prove the theorem, let T be a closed current belonging to γ. By a theorem of de Rham [10, §. 15], there exist operators RT, AT whose supports belong to an arbitrarily small neighborhood of the support of T such that

$$RT = T + bAT + AbT , \tag{9}$$

where b is the boundary operator of currents. The operator R is a regularizing operator, constructed by convolution with a smooth kernel, which is given by

$$RT[\varphi] = \int_M \varphi \wedge \psi , \tag{10}$$

where ψ is a closed C^∞-form with support in a neighborhood of the support of T. Since $bT = 0$, there exists in every homology class a regular current and it suffices to show that $\int_M \varphi \wedge \psi$ has a finite upper bound independent of u, where

$$\varphi = d^c u \wedge (dd^c u)^{k-1} , \qquad \dim \gamma = 2k - 1 ,$$
$$\varphi = du \wedge d^c u \wedge (dd^c u)^{k-1} , \qquad \dim \gamma = 2k .$$

Consider first the case $\dim \gamma = 2k - 1$. By Green's theorem we have

$$\int_M \varphi \wedge \psi = - \int_M u (dd^c u)^{k-1} \wedge d^c \psi .$$

Since ψ is a fixed C^∞-form, it follows from our Corollary in § 2 that this integral is bounded in absolute value by a constant independent of $u \in \mathcal{F}$.

If $\dim \gamma = 2k$, we have, since $d\psi = 0$,

$$\int_M \varphi \wedge \psi = -\int_M u(dd^c u)^k \wedge \psi .$$

The existence of an upper bound for the absolute value of this integral again follows from our Corollary.

The following theorem is an immediate consequence of our definition.

THEOREM 2. *Let $H_l(M, R)$ be the l-dimensional homology group of M with real coefficients. Then $N\{\gamma\}$, $\gamma \in H_l(M, R)$, defines a semi-norm on the real vector space $H_l(M, R)$*, i.e.,

$$\begin{aligned} &N\{a\gamma\} = |a| N\{\gamma\} , && a \in R ; \\ &N\{\gamma_1 + \gamma_2\} \leqq N\{\gamma_1\} + N\{\gamma_2\} , && \gamma_1, \gamma_2 \in H_l(M, R) . \end{aligned} \tag{11}$$

Moreover, under a holomorphic mapping $f: M \to P$ we have

$$N_M\{\gamma\} \geqq N_P\{f_* \gamma\} , \tag{12}$$

where f_* is the induced homomorphism on the homology classes, and the semi-norms are taken in M and P respectively.

Remark 1. If the closed currents T_1, T_2 of γ are such that $T_1[\varphi] \neq T_2[\varphi]$, there exists a real number t with

$$\{tT_1 + (1-t)T_2\}[\varphi] = 0 ,$$

while the current $tT_1 + (1-t)T_2$ still belongs to γ. Therefore in the definition (2) we need only consider functions u for which φ is closed. These functions u satisfy respectively the differential equations

$$(dd^c u)^k = 0 , \qquad \dim \gamma = 2k - 1 , \tag{13}$$

$$du \wedge (dd^c u)^k = 0 , \qquad \dim \gamma = 2k . \tag{13a}$$

In § 5 we will give examples for which $N\{\gamma\}$ is a norm i.e., $N\{\gamma\} > 0$ when $\gamma \neq 0$.

Remark 2. The family, $\mathcal{F}$, of plurisubharmonic functions on M with values between 0 and 1 can be used to define another semi-norm which assigns to a homology class γ the number

$$N'\{\gamma\} = \sup_{u_i \in \mathcal{F}} \inf_{T \in \gamma} | T[d^c u_0 \wedge dd^c u_1 \wedge \cdots \wedge dd^c u_k] | ,$$

$$\text{if } \dim \gamma = 2k + 1$$

$$N'\{\gamma\} = \sup_{u_i \in \mathcal{F}} \inf_{T \in \gamma} | T[du_0 \wedge d^c u_1 \wedge dd^c u_2 \wedge \cdots \wedge dd^c u_k] | , \quad \text{if } \dim \gamma = 2k .$$

The proof of finiteness of $N'(\gamma)$ requires a slight modification of the preceding arguments. The analogue of the corollary needed here is

LEMMA. *Let the polydiscs Δ, Δ_1 be defined as in lemma of § 2. Let $u_1, u_2, \cdots, u_r$ be C^2 plurisubharmonic functions in Δ with $0 < u_i < 1$. Then there is constant C independent of the u_i, such that if $J = (j_1, \cdots, j_r)$ and $K = (k_1, \cdots, k_r)$ $1 \leqq j_1 < \cdots < j_r \leqq n$ and $1 \leqq k_1 < \cdots < k_r \leqq n$,*

$$\int_{\Delta_1} | U_{J\bar{K}} | \, dV \leqq C ,$$

where $U_{J\bar{K}}$ is the coefficient of $dz_{j_1} \wedge d\bar{z}_{k_1} \wedge \cdots \wedge dz_{j_r} \wedge d\bar{z}_{k_r}$ in $dd^c u_1 \wedge \cdots \wedge dd^c u_r$, and dV is the element of volume in C_n. To prove this we note first that the matrix with $(J, K)^{\text{th}}$ entry, $U_{J\bar{K}}$ is positive semi-definite (by induction on r), and so we need only consider the case $J = K = (1, 2, \cdots, r)$. Then using the notation and technique of the lemma of § 2, we have

$$\begin{aligned}
\int_{\Delta_1} | U_{J\bar{J}} | \, dV &= \int_{\Delta_1} dd^c u_1 \wedge \cdots \wedge dd^c u_r \wedge \psi \\
&\leqq \int_{\Delta} \zeta \cdot dd^c u_1 \wedge \cdots \wedge dd^c u_r \wedge \psi \\
&= - \int_{\Delta} d\zeta \wedge d^c u_1 \wedge dd^c u_2 \wedge \cdots dd^c u_r \wedge \psi \\
&= - \int_{\Delta} du_1 \wedge dd^c u_2 \wedge \cdots \wedge dd^c u_r \wedge d^c \zeta \wedge \psi \\
&= \int_{\Delta} u_1 \cdot dd^c u_2 \wedge \cdots \wedge dd^c u_r \wedge dd^c \zeta \wedge \psi .
\end{aligned}$$

Theorem 2 and Remark 1 following it are true for N', and $N' \geqq N$. The equations replacing (13) and (13a) are their multilinear versions.

(13′) $\quad dd^c u_0 \wedge \cdots \wedge dd^c u_k = 0 , \quad$ for N' on $H_{2k+1}(M, R)$.

(13′a) $\quad du_0 \wedge dd^c u_1 \wedge \cdots \wedge dd^c u_k = 0 , \quad$ for N' on $H_{2k}(M, R)$.

Another, possibly larger, semi-norm results if we change the definition of N' by allowing u_0 to be any C^2 function with $0 < u_0 < 1$, but still requiring $u_1, \cdots, u_k$ to be in $\mathcal{F}$. The other norms which we introduce may also be modified in a similar manner with the aid of $k + 1$ functions in place of one.

Remark 3. Another seminorm can be defined by the consideration of a different family of functions. Let $\mathcal{F}_1$ be the family of negative C^2-functions, defined locally up to a multiplicative positive constant, which are plurisubharmonic. For such a function v the forms

$$\frac{dv}{v}, \qquad \frac{d^c v}{v}, \qquad \frac{dd^c v}{v} \tag{14}$$

are well defined on M. With the aid of the functions of $\mathcal{F}_1$ we define, to a homology class γ,

$$\begin{aligned} N_1\{\gamma\} &= \sup_{v \in \mathcal{F}_1} \inf_{T \in \gamma} \left| T\left[\frac{d^c v \wedge (dd^c v)^{k-1}}{(-v)^k}\right] \right| , \\ &\qquad\qquad \text{if } \dim \gamma = 2k - 1 , \\ N_1\{\gamma\} &= \sup_{v \in \mathcal{F}_1} \inf_{T \in \gamma} \left| T\left[\frac{dv \wedge d^c v \wedge (dd^c v)^{k-1}}{(-v)^{k+1}}\right] \right| , \\ &\qquad\qquad \text{if } \dim \gamma = 2k . \end{aligned} \tag{15}$$

By applying the Lemma in § 2 it can be proved that $N_1\{\gamma\}$ is always finite, and is hence a semi-norm in the homology vector space $H_l(M, R)$ ($l = 2k - 1$ or $2k$). Unfortunately we know no example for which $N_1\{\gamma\}$ is not zero. In particular, if $\dim \gamma = 1$, then necessarily $N_1\{\gamma\} = 0$.

It may be observed that in the proof of Theorem 1, only the property of local boundedness of the functions u is utilized. We will therefore introduce wider families of functions and thereby generalize the semi-norms introduced above. Let $\mathfrak{U} = \{U_i\}$ be a locally finite open covering of M. We denote by $\mathcal{F}(\mathfrak{U})$ the family of plurisubharmonic C^2-functions $u_i: U_i \to R$ defined in each member of the covering which satisfy the following conditions:

(1) the oscillation of u_i in U_i is less than one;

(2) $du_i = du_j$ in $U_i \cap U_j \neq \emptyset$.

The latter defines a closed real one-form in M. Similarly, $d^c u_i$ and $dd^c u_i$ are also well defined in M. Without ambiguity, we can denote them without the indices. Analogous to (2) we define

$$\begin{aligned} N\{\gamma, \mathfrak{U}\} &= \sup_{u \in \mathcal{F}(\mathfrak{U})} \inf_{T \in \gamma} | T[d^c u \wedge (dd^c u)^{k-1}] | \\ &\qquad\qquad \text{if } \dim \gamma = 2k - 1 , \\ N\{\gamma, \mathfrak{U}\} &= \sup_{u \in \mathcal{F}(\mathfrak{U})} \inf_{T \in \gamma} | T[du \wedge d^c u \wedge (dd^c u)^{k-1}] | \\ &\qquad\qquad \text{if } \dim \gamma = 2k . \end{aligned} \tag{16}$$

Let π: $\tilde{M} \to M$ be the universal covering manifold of M and let $\tilde{U}$ be a fundamental domain on $\tilde{M}$. We denote by $\mathcal{F}(\tilde{U})$ the family of plurisubharmonic C^2-functions on $\tilde{M}$ such that their oscillation in $\tilde{U}$ is less than one and their differentials are well defined on M. We define

$$
(17)\qquad
\begin{aligned}
N\{\gamma, \tilde{U}\} &= \sup_{u \in \mathcal{F}(\tilde{U})} \inf_{T \in \gamma} | T[d^c u \wedge (dd^c u)^{k-1}] | , \quad \text{if } \dim \gamma = 2k-1 , \\
N\{\gamma, \tilde{U}\} &= \sup_{u \in \mathcal{F}(\tilde{U})} \inf_{T \in \gamma} | T[du \wedge d^c u \wedge (dd^c u)^{k-1}] | , \quad \text{if } \dim \gamma = 2k .
\end{aligned}
$$

We will suppose of the fundamental domain $\tilde{U}$ that each of its points is in the interior of the union of $\tilde{U}$ and a finite number of its translates by deck transformations.

By a partition of unity the proof of Theorem 1 also gives the following theorem.

THEOREM 3. *The $N\{\gamma, \mathfrak{U}\}$ and $N\{\gamma, \tilde{U}\}$ defined in* (16) *and* (17) *are finite and define seminorms in the homology vector spaces $H_l(M, R)$. Between them and $N\{\gamma\}$ there are the inequalities*

$$
(18)\qquad N\{\gamma\} \leqq N\{\gamma, \mathfrak{U}\} , \qquad N\{\gamma\} \leqq N\{\gamma, \tilde{U}\} .
$$

Addendum. For non-compact complex manifolds M and P and f a holomorphic map from M to P, we do not in general have an inequality analogous to (12) for the semi-norms, $N\{\cdot, \mathfrak{U}\}$ and $N\{\cdot, \tilde{U}\}$. However, let $\tilde{U}$ and $\tilde{V}$ be fundamental domains for the universal covering spaces $\tilde{M}$ and $\tilde{P}$ and let $\tilde{f}$ cover f. Then if $\tilde{f}(\tilde{U})$ is covered by a finite number T of deck-transforms of $\tilde{V}$, we have

$$
N_P\{f_*(\gamma), \tilde{V}\} \leqq T^k N_M\{\gamma, \tilde{U}\} , \qquad \text{for } \dim \gamma = 2k-1 \text{ or } 2k-2 .
$$

In particular, if M is compact we always have such inequalities. Similarly, let $\mathfrak{U}$ be a *simple* open covering of M, that is a locally finite open covering of M each member of which is simply connected, and let $\mathfrak{V}$ be an arbitrary, locally finite open covering of P. If for any element $U_i \in \mathfrak{U}$, $f(U_i)$ is covered by S or fewer elements of $\mathfrak{V}$, we again have

$$
N_P\{f_*(\gamma), \mathfrak{V}\} \leqq S^k N_M\{\gamma, \mathfrak{U}\} , \qquad \text{for } \dim \gamma = 2k-1 \text{ or } 2k-2 .
$$

Of course if M is compact and $\mathfrak{U}$ is simple, we have such inequalities.

4. Comparision of semi-norms.

Very little is known about the relations between the different semi-norms. If we apply the above Addendum to the case that $M = P$ is compact and f is the identity map, we find that the equivalence class of $N\{\cdot, \tilde{U}\}$ and the equivalence class of $N\{\cdot, \mathfrak{U}\}$ are independent of the choice of fundamental domain $\tilde{U}$ and simple covering U respectively. Both of these facts are implied by

THEOREM 4. *Let M be a compact complex manifold without boundary. Let $\mathfrak{U} = \{U_i\}$ be a finite open simple covering of M, and let $\tilde{U}$ be a fundamental domain in the universal covering manifold of M. Then the seminorms $N\{\gamma, \mathfrak{U}\}$ and $N\{\gamma, \tilde{U}\}$ are equivalent*, i.e., *there is a constant $C > 0$, independent of γ, such that*

$$C^{-1}N\{\gamma, \mathfrak{U}\} \leqq N\{\gamma, \tilde{U}\} \leqq CN\{\gamma, \mathfrak{U}\} . \tag{19}$$

It follows that for any two finite simple open coverings $\mathfrak{U}$ and $\mathfrak{V}$ the semi-norms $N\{\gamma, \mathfrak{U}\}$ and $N\{\gamma, \mathfrak{V}\}$ are equivalent.

In fact, if a point of U_i is lifted to a point of $\tilde{U}$, the lifting of U_i to $\tilde{M}$ is uniquely determined. It follows that if $u \in \mathcal{F}(\tilde{U})$, then the oscillation of u on each of the U_i is bounded by some constant c. Hence $c^{-1}u \in \mathcal{F}(\mathfrak{U})$, from which the second inequality of (19) follows, with $C = c^k$ or c^{k+1} according as $\dim \gamma = 2k - 1$ or $2k$. In a similar way it is easily seen that if $u \in \mathcal{F}(\mathfrak{U})$ then $c^{-1} u \in \mathcal{F}(\tilde{U})$ for some constant c and the first of (19) follows.

If M is a compact Riemann surface and γ is a one-dimensional homology class, the *extremal length* $\lambda(\gamma)$ of Ahlfors-Beurling is defined by

$$\lambda^{\frac{1}{2}}(\gamma) = \sup_\rho \frac{\inf_{C \in \gamma} \int_C \rho\,|dz|}{\left(\iint_M \rho^2 dxdy\right)^{\frac{1}{2}}} \tag{20}$$

where $\rho \geqq 0$ ranges over all lower semicontinuous densities which are not identically zero [2].

THEOREM 5. *Let M be a compact Riemann surface without boundary and $\mathfrak{U} = \{U_i\}$ a finite open simple covering of M. Then the semi-norms $N\{\gamma, \mathfrak{U}\}$ and $\lambda^{\frac{1}{2}}(\gamma)$ defined over the one-dimensional homology group $H_1(M, R)$ are equivalent*, i.e., *there is a constant $C > 0$, independent of $\gamma \in H_1(M, R)$, such that*

(21) $$C^{-1}\lambda^{\frac{1}{2}}(\gamma) \leqq N(\gamma, \mathfrak{U}) \leqq C\lambda^{\frac{1}{2}}(\gamma)\,.$$

We proceed to prove this theorem. For $u \in \mathcal{F}(\mathfrak{U})$ we set $\rho = |u_z|$. Let C be any closed curve belonging to the homology class γ. Since $d^c u = 2\,\mathrm{Im}\,(u_z d_z)$, we have

$$\int_C |d^c u| \leqq 2\int_C |u_z||d_z|\,,$$

and hence

$$\inf_{C \in \gamma} \int_C |d^c u| \leqq 2\lambda^{\frac{1}{2}}(\gamma)\left\{\iint_M |u_z|^2\,dxdy\right\}^{\frac{1}{2}}.$$

On U_i the oscillation of u is less than 1. It follows that on any compact subset K of U_i we can find a uniform bound for

$$\iint_K |u_z|^2\,dxdy$$

for all harmonic functions u. Hence there is a constant C_1 depending only on $\mathfrak{U}$ such that

$$\iint_M |u_z|^2\,dxdy \leqq C_1^2\,.$$

Thus

$$\inf_{C \in \gamma} \int_C |d^c u| \leqq 2C_1\lambda^{\frac{1}{2}}(\gamma)\,.$$

Since this holds for all $u \in \mathcal{F}(\mathfrak{U})$, the last inequality of (21) follows.

To prove the first inequality of (21) we make use of a theorem of Accola [1] which says that there is a harmonic one-form σ on M representing the homology class γ such that

(22) $$||\sigma||^2 = \lambda(\gamma)\,,$$

where $||\sigma||$ is the L_2-norm of σ. In U_i we write $\sigma = du$, where u is a harmonic function defined up to an additive constant. Then we have

$$\lambda(\gamma) = ||\sigma||^2 = 2\iint_M |u_z|^2\,dxdy\,.$$

On the other hand, by standard results on harmonic functions we have

$$|\operatorname{osc} u \text{ in } U_i| \leqq \text{const}\left(\iint_M |u_z|^2\,dxdy\right)^{\frac{1}{2}}.$$

Thus we may suppose u be so chosen that on each U_i,

$$0 < u < C_2\lambda^{\frac{1}{2}}(\gamma) = a ,$$

say, where C_2 is a constant. It follows that $u/a \in \mathcal{F}(\mathfrak{U})$. Since $d^c u$ is closed, we have, for a current $T \in \gamma$,

$$T\left(\frac{d^c u}{a}\right) = \frac{1}{a}\int_M d^c u \wedge \sigma = \frac{1}{a}\int_M \sigma^* \wedge \sigma$$

$$= \frac{1}{a}\,||\,\sigma\,||^2 = \frac{1}{C_2}\,\lambda^{\frac{1}{2}}(\gamma) .$$

Hence $\lambda^{\frac{1}{2}}(\gamma) \leqq C_2 N\{\gamma, \mathfrak{U}\}$, and the first inequality of (21) is proved.

Remark. From the comparison with the extremal length it seems natural to extend our definition to a family of chains of a fixed dimension. We could also take the integrals of the absolute values of the corresponding differential forms. For instance, let G be a family of chains of dimension $2k - 1$. We define

$$\hat{N}\{G\} = \sup_{u \in \mathcal{F}} \inf_{g \in G} \int_g |\, d^c u \wedge (dd^c u)^{k-1}\,| . \tag{23}$$

Unfortunately we are unable to prove that $\hat{N}\{G\}$ is finite, except the following case. *If $k = 1$ and G contains all curves homotopically equivalent to a closed curve, then $\hat{N}\{G\}$ is finite.*

5. Some examples

In C_n with the coordinates z^k, $1 \leqq k \leqq n$, we set

$$r = |\,z\,| = (\textstyle\sum_k |\,z^k\,|^2)^{\frac{1}{2}} . \tag{24}$$

We consider the annulus A_n defined by $1 < r < a$. The homology group $H_{2n-1}(A_n, Z)$ is free cyclic and we denote by γ its generator defined by the natural orientation of C_n.

By definition we find

$$\begin{aligned} d^c \log r &= \frac{-i}{2r^2} \textstyle\sum_k (\bar{z}^k dz^k - z^k d\bar{z}^k) , \\ dd^c \log r &= \frac{i}{r^4} \{r^2 \textstyle\sum_k dz^k \wedge d\bar{z}^k - (\sum_k \bar{z}^k dz^k) \wedge (\sum_k z^k d\bar{z}^k)\} . \end{aligned} \tag{25}$$

The differential form in the last expression is a real-valued two-form of type (1,1). It remains unchanged when z^k are multiplied by the same factor. This means that if we denote by

$$\psi\colon C_n - \{0\} \to P_{n-1}$$

the identification of the space of lines through 0 in C_n with the complex projective space P_{n-1} of dimension $n-1$, $dd^c \log r$ can be regarded as a form in P_{n-1}. The function $u = \log r/\log a$ satisfies in A_n the condition $0 < u < 1$. Since P_{n-1} is of real dimension $2n-2$, we have

$$(dd^c u)^n = 0 . \tag{26}$$

It follows that the integral

$$\int d^c u \wedge (dd^c u)^{n-1} = \frac{1}{(\log a)^n} \int d^c \log r \wedge (dd^c \log r)^{n-1} \tag{27}$$

over a cycle of the homology class γ depends only on γ. It is an easy computation that over the unit sphere in C_n the form

$$d^c \log r (dd^c \log r)^{n-1}$$

is equal to $(n-1)!\, 2^{n-1}$ times its volume element. Using the value of the volume of the unit sphere in C_n, we find that the integral (27) is equal to $(2\pi/\log a)^n$. By definition we have

$$N\{\gamma\} \geqq \left(\frac{2\pi}{\log a}\right)^n > 0 . \tag{28}$$

Thus $N\{\gamma\}$ is a norm on $H_{2n-1}(A_n, R)$.

Since the norm is non-increasing under a holomorphic mapping, we derive from this the theorem: *Let $f\colon A_n \to A_n$ be a holomorphic mapping. Then* $f_* \gamma = \pm \gamma$ or 0, *f_* being the induced homomorphism on homology*. This generalizes a theorem of M. Schiffer [12] and H. Huber [6] for $n = 1$.

We do not know the exact value of $N\{\gamma\}$. In the case $n = 1$ Landau and Osserman [8] showed that the equality sign holds in (28). Let $\tilde{U}$ be the domain: $0 \leqq \arg z < 2\pi$. We wish to show that for $n = 1$ we have

$$N\{\gamma\} = N\{\gamma, \tilde{U}\} = \frac{2\pi}{\log a} . \tag{29}$$

In fact, let u be a harmonic function in $\mathbf{F}(\tilde{U})$. Imagining the ring slit on the positive x-axis between 1 and a, the function u is well defined in $\tilde{U}$. It suffices to prove the inequality

$$\left| \int_{|z|=r} d^c u \right| = \left| \int_0^{2\pi} r u_r d\theta \right| \leqq \frac{2\pi}{\log a} ,$$

where $z = re^{i\theta}$. Let S denote the operator of averaging with respect to angle. Since u_θ is periodic, it is easily seen that $v(r) = Su$

is harmonic. Thus $v = c \log r$, with $c \leqq 1/\log a$, since $0 < v < 1$. It follows that

$$|rv_r| \leqq \frac{1}{\log a},$$

which is the inequality to be proved.

Our next example is concerned with the torus $M = S^1 \times S^1$ and with γ the homology class of the torus itself. For any current $T \in \gamma$ we have then

$$T[\varphi] = \int_M \varphi,$$

where φ is a C^∞ two-form. We consider M to be covered by the (x, y)-plane. Let $\tilde{U}$ be the fundamental domain consisting of the open square

$$0 < x < 1,\ 0 < y < 1$$

and the segments $y = 0$, $0 \leqq x < 1$ and $x = 0$, $0 \leqq y < 1$. We shall prove that $N\{\gamma, \tilde{U}\} = 1$.

For this purpose let $u \in \mathcal{F}(\tilde{U})$. Since du is well defined on M and M is compact, we have

$$\int_M dd^c u = 0.$$

Since u is plurisubharmonic, this implies $dd^c u = 0$, i.e., u is harmonic. Its derivatives with respect to x and y are single-valued harmonic functions on M and are therefore constants. Thus u is a linear function and we may take

$$u = az + \bar{a}\bar{z}. \tag{30}$$

For $T \in \gamma$ we have

$$T[du \wedge d^c u] = \int_{\tilde{U}} du \wedge d^c u = 4 \iint_{\tilde{U}} |u_z|^2\, dxdy = 4|a|^2.$$

Now the values of u at the corners of $\tilde{U}$ are 0, $a + \bar{a}$, $i(a - \bar{a})$, $a(1 + i) + \bar{a}(1 - i)$. The fact that the oscillation of u in $\tilde{U}$ is at most one means that

$$\begin{aligned} &2|\operatorname{Re} a| < 1, && 2|\operatorname{Im} a| < 1, \\ &2|\operatorname{Re} a(1+i)| < 1, && 2|\operatorname{Re} a(1-i)| < 1. \end{aligned} \tag{30a}$$

These imply $|a| < (1/2)$. Consequently we have

$$N\{\gamma, \tilde{U}\} = \sup_u 4|a|^2 = 1.$$

6. Intrinsic pseudo-metrics

In § 4 we remarked about the possibility of defining the semi-norm of a family of chains of a fixed dimension. The simplest case is the family γ of *curves* (rather than chains) having two given points $z, \zeta \in M$ as boundary and containing all curves homotopic to a given one in the family. To indicate that the notions so introduced will be pseudo-distances on M we will change our notation and repeat the definitions as follows:

$$\rho_\gamma(z, \zeta) = \sup_{u \in \mathcal{F}} \inf_{T \in \gamma} |T[d^c u]| , \tag{31}$$

$$\rho_\gamma(z, \zeta; \mathfrak{U}) = \sup_{u \in \mathcal{F}(\mathfrak{U})} \inf_{T \in \gamma} |T[d^c u]| , \tag{32}$$

$$\rho_\gamma(z, \zeta; \tilde{U}) = \sup_{u \in \mathcal{F}(\tilde{U})} \inf_{T \in \gamma} |T[d^c u]| . \tag{33}$$

We shall omit the subscript γ if the family consists of *all* (smooth) curves joining z and ζ and we shall denote $\rho_\gamma(z, \zeta)$ by $\rho_c(z, \zeta)$ if the family consists of all *chains* joining z and ζ. Clearly $\rho \leqq \rho_\gamma$ for the three definitions, with equality if M is simply connected. The definitions (32) and (33) refer respectively to a locally finite open covering $\mathfrak{U}$ of M and a fundamental domain $\tilde{U}$ in the universal covering manifold $\tilde{M}$ of M. In all three formulas T denotes a curve with $z - \zeta$ as boundary. As in the general case it suffices to restrict ourselves to functions u which satisfy the additional condition $dd^c u = 0$, i.e., which are pluriharmonic. From the definitions we have

$$\rho(z, \zeta) \leqq \rho(z, \zeta; \mathfrak{U}) ; \qquad \rho(z, \zeta) \leqq \rho(z, \zeta; \tilde{U}) . \tag{34}$$

Remark. If M is compact, then $\rho(z, \zeta) = 0$. However, the quantities in (32), (33) need not be zero. Consider for example the torus discussed in the end of the last section with $\tilde{U}$ defined as before. The harmonic functions in $\mathcal{F}(\tilde{U})$ are given by (30), where a satisfies the inequalities (30a). For z, ζ on the torus we have

$$T[d^c u] = \int_z^\zeta d^c u = 2 \operatorname{Im}[a(\tilde{\zeta} - \tilde{z})] , \tag{35}$$

where $\tilde{z}$ and $\tilde{\zeta}$ are points in the plane, the covering surface, lying over z and ζ respectively. N. Kerzman has found that in this case $\rho(z, \zeta; \tilde{U})$ is equal to the maximum of the horizontal and vertical distances from $\tilde{\zeta} - \tilde{z}$ to the sides of a period square containing $\tilde{\zeta} - \tilde{z}$.

THEOREM 6. *The quantities* $\rho(z, \zeta)$, $\rho(z, \zeta; \mathfrak{U})$, $\rho(z, \zeta; \tilde{U})$ *are pseudo-distances*, i.e., *they are finite and satisfy the triangle inequalities.*

The finiteness follows from the fact that the only functions u which enter into consideration are bounded pluriharmonic functions, so that their first partial derivatives are uniformly bounded on compact sets. The last statement follows immediately from definition.

We wish to compare our pseudo-metric with those of Caratheodory [3] and S. Kobayashi [7]. We recall their definitions as follows. Let D be the unit disk $|\tau| < 1$, whose hyperbolic distance we denote by $h(\tau_1, \tau_2)$, $\tau_1, \tau_2 \in D$. Then the Caratheodory pseudo-distance is defined by

$$c(z, \zeta) = \sup h\big(f(z), f(\zeta)\big) , \tag{36}$$

as f runs over the family of all holomorphic mappings $f: M \to D$.

To define the Kobayashi pseudo-distance let

$$f_i: D \to M , \qquad 1 \leqq i \leqq k ,$$

be holomorphic mappings which satisfy the conditions

$$z \in f_1(D) , \qquad \zeta \in f_k(D) ,$$
$$f_i(D) \cap f_{i+1}(D) \neq \varnothing , \qquad 1 \leqq i \leqq k-1 .$$

We choose $z_0 = z, z_1, \cdots, z_{k-1}, z_k = \zeta$, such that

$$z_i \in f_i(D) \cap f_{i+1}(D) , \qquad 1 \leqq i \leqq k-1 .$$

Let $a_i, b_i \in D$ be points satisfying

$$z_{i-1} = f_i(a_i), z_i = f_i(b_i) , \qquad 1 \leqq i \leqq k .$$

Then the Kobayashi pseudo-distance is defined by

$$d(z, \zeta) = \inf \textstyle\sum_{1 \leqq i \leqq k} h(a_i, b_i) , \tag{37}$$

where the infimum is taken with respect to all the choices made.

Kobayashi proved that $c(z, \zeta) \leqq d(z, \zeta)$. We will establish the theorem.

THEOREM 7[1]. *Between the pseudo-distances the following inequalities are valid*

$$c(z, \zeta) \leqq \frac{\pi}{2} \rho_c(z, \zeta) \leqq \frac{\pi}{2} \rho(z, \zeta) \leqq d(z, \zeta) . \tag{38}$$

[1] In our original proof we showed $c \leqq (2/\pi)\rho \leqq d$. Kerzman observed that our argument could be used to prove the more general result (38).

The first inequality becomes an equality if M is simply connected.

By the conformal mapping

$$\tau = \frac{i - \exp(\pi i w)}{i + \exp(\pi i w)}, \tag{39}$$

we map the unit disk D onto the infinite strip $S: 0 < u < 1$, $w = u + iv$. Under (39) the real axis of D corresponds to the line $\frac{1}{2} + iv$ and the origin of D to the point $w = \frac{1}{2}$. S has a hyperbolic metric induced from that of D by the mapping (39), which is given by

$$ds^2 = \frac{d\tau d\bar{\tau}}{(1 - \tau\bar{\tau})^2} = \frac{\pi^2 dw d\bar{w}}{2(1 - \cos 2\pi u)}. \tag{40}$$

Thus the hyperbolic distance on S between the points $\frac{1}{2}$ and $\frac{1}{2} + vi$ is

$$h_S\left(\frac{1}{2}, \frac{1}{2} + vi\right) = \frac{\pi}{2} |v|. \tag{41}$$

Since S admits the group of hyperbolic motions, we can normalize the holomorphic mappings $M \to S$ such that the image of z is the point $\frac{1}{2}$ and the image of ζ lies on the line $\operatorname{Re} w = \frac{1}{2}$. Hence the Caratheodory pseudo-distance can be redefined as follows.

$$c(z, \zeta) = \frac{\pi}{2} \sup_g |\operatorname{Im} g(\zeta)|, \tag{42}$$

where g runs over all holomorphic mappings $g: M \to S$ such that $g(z) = \frac{1}{2}$ and $\operatorname{Re} g(\zeta) = \frac{1}{2}$. If σ denotes the segment joining $g(z)$ to $g(\zeta)$, we have

$$\operatorname{Im} g(\zeta) = \int_\sigma dv = \int_\sigma d^c u = \int_z^\zeta d^c u.$$

where $u = \operatorname{Re} g$ is a pluriharmonic function which belongs to the family $\mathcal{F}$. Since $T[d^c u] = \operatorname{Im} g(\zeta)$ for any chain T this proves that

$$c(z, \zeta) \leqq \frac{\pi}{2} \rho_c(z, \zeta).$$

To prove equality consider a pluriharmonic function u in $\mathcal{F}$. If among all *chains* T, $\inf_T |T[d^c u]| \neq 0$, then the integral of $d^c u$ around any closed path vanishes. But then the function

$$v(z) = \int_{z_0}^z d^c u$$

is well defined independent of the path of integration and is therefore a conjugate pluriharmonic function of u. Hence $w = u + iv$ defines a holomorphic mapping $w: M \to S$. By (40) we have

$$\frac{\pi}{2} |dv| \leqq ds .$$

Since $dv = d^c u$, it follows that

$$\frac{\pi}{2} \rho_c(z, \zeta) \leqq c(z, \zeta) ,$$

and hence

$$c(z, \zeta) = \frac{\pi}{2} \rho_c(z, \zeta) .$$

If M is simply connected then to every pluriharmonic function u in $\mathcal{F}$ there exists a conjugate pluriharmonic function v defined up to an additive constant. It follows that for any chain T bounded by ζ and z, $T[d^c u]$ equals $v(\zeta) - v(z)$ and is therefore independent of the chain T. Consequently

$$\rho_c(z, \zeta) = \rho(z, \zeta) .$$

To prove the last inequality in (38) we use the fact that ρ is non-increasing under a holomorphic mapping. We will also follow the above notation in the definition of the Kobayashi pseudo-distance. Let l_i be the straight segment in D joining a_i to b_i and let $f_i(l_i) = L_i$, $1 \leqq i \leqq k$. Let u be a pluriharmonic function on M, with $0 < u < 1$. Then we have

$$\frac{\pi}{2} \left| \int_{L_i} d^c u \right| \leqq \frac{\pi}{2} \rho_0(a_i, b_i) = h(a_i, b_i) , \qquad 1 \leqq i \leqq k ,$$

where ρ_0 is our metric in D. The last equality follows from what we just proved, as D is simply connected. It follows that

$$\frac{\pi}{2} \left| \sum_i \int_{L_i} d^c u \right| \leqq \sum_i h(a_i, b_i) , \qquad 1 \leqq i \leqq k .$$

Now the right-hand side may be chosen as close to $d(z, \zeta)$ as we like, while the left-hand side is not smaller than $(\pi/2)(\rho(z, \zeta))$. This implies the desired inequality.

Remark. We do not know when $(\pi/2)\rho(z, \zeta) = d(z, \zeta)$, nor how $(\pi/2)\rho(z, \zeta; \tilde{U})$ compares with $d(z, \zeta)$. Using chains one may also introduce pseudo-metrics $\rho_c(z, \zeta; \mathfrak{U})$ and $\rho_c(z, \zeta; \tilde{U})$; but on a compact manifold these are zezo.

7. Remarks on the differential equations

The differential equations (13) and (13a) are, in general, over-determined systems of non-linear differential equations. For $k=1$, equation (13),

$$dd^c u = 0 ,$$

asserts that u is pluriharmonic, while in general, equation (13) means that the rank of the hessian $u_{i\bar{j}}$ is less than k. Almost nothing is known about the solvability of these equations. The case $k = n$ reduces to a single equation which is the complex analogue of the Monge-Ampère equation

$$\det \{u_{i\bar{j}}\} = 0 ;$$

it is non-linear degenerate elliptic in view of our requirement that the matrix $u_{i\bar{j}}$ be positive semi-definite.

It would be interesting to formulate boundary value problems for these equations. We remark that the equation $(dd^c u)^n = 0$ also arises as the Euler equation for a stationary point of the functional

$$I(u) = \int_M du \wedge d^c u \wedge (dd^c u)^{n-1} \tag{43}$$

under, perhaps, some boundary conditions. Consider for example the class B of C^2 plurisubharmonic functions which are required to equal one on some components of the (smooth) boundary of a compact manifold M, and zero on the others. If v is a member of B, let γ denote the $(2n-1)$-dimensional homology class of the level hypersurfaces $v =$ constant. Then we observe that for $T \in \gamma$, if v satisfies $(dd^c v)^n = 0$,

$$\int_T dv \wedge (dd^c v)^{n-1} = \int_M dv \wedge d^c v \wedge (dd^c v)^{n-1} = I(v) . \tag{44}$$

It is not difficult to verify that the functional I is convex and one is therefore tempted to conjecture that

$$N\{\gamma\} = \inf_{v \in B} I(v) . \tag{45}$$

If this is the case then $N\{\gamma\}$, which is defined as the supremum of a functional would also be characterized as the infimum of another, a situation that often arises in, so called, dual variational problems in the calculus of variations. The problem of minimizing $I(v)$ seems an interesting one. In the case of the annulus $1 < |z| < a$, the function $v_0 = \log|z|/\log a$ is indeed the minimizing function, since the convex functional I is stationary at v_0.

The differential equation (13) has a real analogue, which is

$$\operatorname{rank}\left(\frac{\partial^2 u}{\partial x^i \partial x^j}\right) \leqq k\,, \qquad 1 \leqq i, j \leqq n\,, \tag{46}$$

where $u(x^1, \cdots, x^n)$ is a real-valued C^2-function in the real variables $x^1, \cdots, x^n$. Equation (46) and its generalizations have been studied in connection with some geometrical problems (cf. [4], [5], [11]). In fact, if $u = u(x^1, \cdots, x^n)$ is considered as the equation of a non-parametric hypersurface in the euclidean $(n+1)$-space E^{n+1}, the left-hand side of (46) is called the index of relative nullity, being the rank of its second fundamental form. Hartman and Nirenberg [5, p. 912] proved that for $n = 2$, $k = 1$ (in which case condition (46) means that the surface has zero gaussian curvature) the surface is a cylinder if $u(x^1, x^2)$ is defined for all $(x^1, x^2) \in R^2$. For higher dimensions a similar result is not true, as shown by an example of Sacksteder [11]. In this respect we wish to refer to a general theorem of Hartman [4] concerned with sufficient conditions for an isometrically immersed submanifold in an euclidean space to be cylindrical.

UNIVERSITY OF CALIFORNIA, BERKELEY
BRANDEIS UNIVERSITY
COURANT INSTITUTE, NEW YORK UNIVERSITY.

REFERENCES

[1] R. D. M. ACCOLA, *Differentials and extremal length on Riemann surfaces.* Proc. Nat. Acad. Sci. USA **46** (1960), 540–543.

[2] L. V. AHLFORS and L. SARIO, Riemann Surfaces, Princeton University Press, Princeton, 1960.

[3] C. CARATHEODORY, *Über eine spezielle Metrik, die in der Theorie der analytischen Funktionen auftritt.* Atti. Pont. Acad. Sci. Nuovo Lincei **80** (1927), 135–141.

[4] P. HARTMAN, *On isometric immersions in euclidean space of manifolds with non-negative sectional curvatures.* Trans. Amer. Math. Soc. **115** (1965), 94–109.

[5] ——— and L. NIRENBERG, *On spherical image maps whose jacobians do not change sign.* Amer. J. Math. **81** (1959), 901–920.

[6] H. HUBER, *Über analytische Abbildungen von Ringgebieten in Ringgebiete.* Compos. Math. **9** (1951), 161–168.

[7] S. KOBAYASHI, *Invariant distances on complex manifolds and holomorphic mappings.* J. Math. Soc. Japan **19** (1967), 460–480.

[8] H. J. LANDAU and R. OSSERMAN, *On analytic mappings of Riemann surfaces*. J. Anal. Math. **7** (1959-60), 249-279.

[9] P. LELONG, *Sur les dérivées d'une fonction plurisousharmonique*. C. R. Acad. Sci. Paris **238** (1954), 2276-2278.

[10] G. DE RHAM, Variétés Différentiables, Actualités Sci. et Ind. No. 1222, Hermann, Paris, 1955.

[11] R. SACKSTEDER, *On hypersurfaces with no negative sectional curvatures*. Amer. J. Math. **82** (1960), 609-630.

[12] M. SCHIFFER, *On the modulus of doubly connected domains*. Quart. J. Math. **17** (1946), 197-213.

(Received September 9, 1968)

On the area of complex manifolds

BY G. DE RHAM

This is part of a Seminar given in 1957 at the Institute for Advanced Study, Princeton, where I had the privilege of many fruitful discussions with K. Kodaira. It contains a proof of a theorem of Lelong, according to which an anlytic set in a complex manifold defines a closed current. Another proof has been given by Federer [8].

1. Area of complex analytic submanifolds of the hermitian parabolic space

Let $z_\alpha = x_\alpha + iy_\alpha (\alpha = 1, 2, \cdots, n)$ be coordinates in C^n. The hermitian parabolic metric defined by

$$ds^2 = \sum_{\alpha=1}^n dz_\alpha d\bar{z}_\alpha = \sum_1^n dx_\alpha^2 + dy_\alpha^2$$

is nothing but the euclidean metric. Taken with this metric, C^n is the hermitian parabolic space. The parabolic motions of C^n are the linear (in general non-homogeneous) transformations on $z_1, \cdots, z_n$ which leave ds^2 invariant. They also leave invariant the associated 2-form

$$\omega = \frac{i}{2} \sum_1^n dz_\alpha \wedge d\bar{z}_\alpha = \sum_1^n dx_\alpha \wedge dy_\alpha$$

and its powers

$$\frac{\omega^p}{p!} = \sum_{\alpha_1 < \cdots < \alpha_p} dx_{\alpha_1} \wedge dy_{\alpha_1} \wedge \cdots \wedge dx_{\alpha_p} \wedge dy_{\alpha_p} .$$

The general term of this last sum is the element of the 2p-dimensional area in the complex coordinate plane $Oz_{\alpha_1} \cdots z_{\alpha_p}$. Therefore

If S is an oriented piece of a real $2p$-dimensional submanifold of C^n, the integral

$$J(S) = \int_S \frac{\omega^p}{p!}$$

is equal to the sum of the areas, taken with the signs corre-

sponding to the orientations of S, of the projections of S into the complex p-dimensional coordinate planes.

Now, *if S is complex analytic and suitably oriented, $J(S)$ is equal to the area of S.* It is sufficient to verify it in the case where S is a piece of a complex analytic plane; then, by a parabolic motion which leaves the area as well as the form ω^p invariant, this plane can be transformed into the coordinate plane $Oz_1z_2 \cdots z_p$ and the assertion becomes trivial.

If S is any piece of a real $2p$-dimensional submanifold of C^n, we have the Wirtinger inequality $J(S) \leqq$ area of S, for which a simple proof has been given by Federer [8]. But, as we will concern ourselves only with analytic complex submanifold, we do not need it. The result can be stated as follows.

THEOREM A. *The $2p$-dimensional area of a piece of a complex p-dimensional analytic submanifold of the parabolic hermitian space C^n is equal to the sum of the areas of its projections into the complex p-dimensional coordinate planes.*

Now, let us consider a domain $D \subset C^n$ and a complex analytic set V in D, i.e., a set, closed in D, which can be locally defined by holomorphic equations (see [2]). V is said to be of (complex) *dimension p at a point z_0*, if $n - p$ is the greatest dimension of the complex planes L passing through z_0, such that z_0 is an isolated point of $L \cap V$. If there exists a neighborhood U of z_0, such that $U \cap V$ can be defined by a system of $n - p$ equations

$$f_j(z) = 0 \qquad (j = 1, \cdots, n - p)\ ,$$

where the f_j are holomorphic functions in U whose differentials df_j are linearly independent, we say that V is regular p-dimensional at z_0. If V is of dimension p at each of its points, we say that V is of *pure dimension p*; in that case, V is the topological closure in D of the set of points W at which V is regular p dimensional. If V is of dimension $\leqq p$ at each of its points and of dimension p at some of its points, we say that V is of *dimension p*. The topological closure $V^{(k)}$ in D of the set of points at which V is regular k-dimensional, if not empty, is an analytic set in D of pure dimension k, and if V is of dimension p, we have

$$V = V^{(0)} \cup V^{(1)} \cup \cdots \cup V^{(p)}$$

(see [2]).

If V is of dimension p at z_0, there exists an algebraic cone of dimension p, with summit at z_0, which contains all the complex lines tangent to V at z_0, i.e., all the limits of a complex line passing through z_0 and a point z of V tending to z_0. This can be proved in the following way.

Let $f(z) = 0$ be one of the equations defining V in the neighborhood of z_0. Take a coordinate system $z = (z_1, \cdots, z_n)$ with origin at z_0 and let

$$f(z) = \sum_{k=m}^{\infty} P_k(z)$$

be the expansion of $f(z)$ in a power series in $z_1, \cdots, z_n$, where $P_k(z)$ is a homogeneous polynomial of degree k in $z_1, \cdots, z_n$ and $P_m(z)$ does not vanish identically. Then the equation $P_m(z) = 0$ defines an algebraic cone of dimension $n - 1$ which contains all the complex lines tangent to V at z_0. Let $q = n - p$. For $q = 1$, the assertion is proved; for $q > 1$, we proceed by induction. We can choose the coordinate system in such a way that the complex line $z_1 = \cdots = z_{n-1} = 0$ is not contained in the cone $P_m(z) = 0$. Then there exists a neighborhood U of z_0, such that the projection of $U \cap V$ into the $(n - 1)$-dimensional plane C^{n-1} defined by $z_n = 0$ is contained in a set $V' \subset C^{n-1}$ which is analytic and of dimension p at z_0. By induction hypothesis, there exists an algebraic cone C' of dimension p, in C^{n-1}, which contains all the complex lines tangent to V' at z_0. Then the set of all points of the cone $P_m(z) = 0$ whose projection in C^{r-1} lie in C' is an algebraic cone of dimension p which contains all the complex lines tangent to V at z_0.

The smallest algebraic cone C which contains all the complex lines tangent to V at z_0 will be called *the tangent cone of V at z_0*. It follows from the above that its dimension is $\leqq p$ and this is all we neeed. But, by a similar argument, it can be shown that its dimension is equal to p. Of course, if z_0 is a regular point of V, the tangent cone at z_0 is nothing but the tangent plane.

Let us consider the complex manifold M consisting of all $(n - p)$-dimensional complex planes L^{n-p} passing through z_0. In M, the L^{n-p} which contain at least one complex line of the cone C (of summit z_0 and of dimension p) form a closed subset of complex codimension 1. Therefore, the set E of all $L^{n-p} \in M$ such that $L^{n-p} \cap C = \{z_0\}$ is open and dense in M. It follows that *there exists unitary coordinate systems with origin at z_0* (unitary or orthonomal with respect to the hermitian parabolic metric), *all of whose* $(n - p)$-

dimensional coordinate planes belong to E. In the manifold of all unitary coordinate systems with origin at z_0 they form a set which is open and dense.

LEMMA 1. *Let V be an analytic set of pure dimension p in a domain $D \subset C^n$ and let W be the set of the regular points of V. Let $B(z, r)$ denote the open ball of center $z \in C^n$ and of radius r. Then, for $r \to 0$,*

$$\textit{area of } W \cap B(z, r) = O(r^{2p}) .$$

More precisely, given a compact set $K \subset D$, there are two positive numbers M and R such that the area of $W \cap B(z, r)$ is $\leqq Mr^{2p}$ for every $z \in K$ and for every $r \leqq R$.

This lemma is in Lelong [1, p. 254, Th. 5]. The case where $p = n - 1$ is already contained in Stoll [5, p. 151, Satz 10]. It can be deduced from Theorem A in the following way.

Suppose $z_0 \in V$. Take a unitary coordinate system with origin at z_0 such that the intersection of the tangent cone of V at z_0 with every $(n - p)$-dimensional coordinate plane consists of the single point z_0. Let L^{n-p} be any $(n - p)$-dimensional coordinate plane and L^p the complementary p-dimensional coordinate plane. Clearly, z_0 is an isolated point of $V \cap R^{n-p}$. We may suppose L^{n-p} is $\{z_1 = \cdots = z_p = 0\}$ and L^p is $\{z_{p+1} = \cdots = z_n = 0\}$. According to a well known theorem (see [2, p. 267]), there exist a ball $B(z_0, R_0)$ of center z_0 and $n - p$ polynomials $P_s(t)$ $(s = p + 1, \cdots, n)$, in which the coefficient of the highest power of t is equal to 1, and the other coefficients are holomorphic functions of $z_1, \cdots, z_p$ in $L^p \cap B(z_0, R_0)$, such that the coordinates of any point of $V \cap B(z_0, R_0)$ satisfy the equations $P_s(z_s) = 0$ $(s = p + 1, \cdots, n)$. Therefore, if $N_0 \geqq$ the product of the degrees of the polynomial $P_s(t)$, *any point of $L^p \cap B(z_0, R_0)$ is the projection of at most N_0 points of $V \cap B(z_0, R_0)$.*

We can take N_0 and R_0 in such a way that this statement will be true for any p-dimensional coordinate plane L^p of the chosen coordinate system. If $B(z, r) \subset B(z_0, R_0)$, the projection of $W \cap B(z, r)$ into L^p has at most N_0 sheets; therefore, its area is $\leqq N_0$ times the $2p$-dimensional area of a ball of radius r in L^p, i.e., $N_0(\pi^p/p!)\, r^{2p}$. On account of Theorem A, it follows that the area of $W \cap B(z, r)$ is $\leqq \binom{n}{p} N_0(\pi^p/p!)\, r^{2p}$.

Now, let K be a compact subset of D. To every point $z_0 \in V \cap K$ are associated a ball $B(z_0, R_0)$ and a positive integer N_0 with the above property (of course, R_0 and N_0 depends on z_0). According to the Borel-Lebesgue theorem, $V \cap K$ is covered by a finite number of these balls; let us denote them by $B_j (j = 1, \cdots, h)$ and let N be the greatest of the corresponding integers N_0. Since the sets $B_j (j = 1, \cdots, h)$ and $D - V$ form an open covering of K, there exists a positive number R such that every ball $B(z, r)$, with center $z \in K$ and radius $r \leqq R$, is contained in one of the B_j or in $D - V$. Setting $M = \binom{n}{p} N(\pi^p/p!)$, it follows that the area of $W \cap B(z, r)$ is $\leqq Mr^{2p}$, for every $z \in K$ and for every $r \leqq R$. q.e.d.

LEMMA 2. *Let D, V, and W be as in Lemma* 1, *and let R^q be a real q-dimensional linear subspace of C^n, K a compact subset of $D \cap R^q$, and $T(K, r)$ the set of all points whose distance to K is less than r. Then, for $r \rightarrow 0$,*

$$\textit{area of } W \cap T(K, r) = O(r^{2p-q}) .$$

Let us consider in R^q a net of (real q-dimensional) cubes of side equal to r. Let $N(r)$ be the number of the cubes of the net which meet K. Since the total (real q-dimensional) area (or volume) $N(r) r^q$ of these cubes is bounded with r, we have, for $r \rightarrow 0$, $N(r) = O(r^{-q})$.

For each of these cubes C, take a point $z \in C \cap K$. The ball $B = B(z, r + r\sqrt{q})$ of center z and of radius $(1 + \sqrt{q})r$ contains all the points whose distance to C is $< r$. Therefore, the union of the $N(r)$ balls B contains $T(K, r)$ and, because of Lemma 1, it follows that, for $r \rightarrow 0$, the area of $W \cap T(K, r)$ is

$$N(r)O(r^{2p}) = O(r^{2p-q}) . \qquad \text{q.e.d.}$$

2. Integration of a differential form on an analytic set

We consider again a domain $D \subset C^n$, an analytic set V in D of pure dimension p and the set W of the regular points of V. It follows from Lemma 1 that, for any compact subset K of D, the (real 2p-dimensional) area of $W \cap K$ is finite. Therefore, if φ is any differential form in C^n of degree $2p$, of class C^0 and with a compact support $K \subset D$, the integral $\int_W \varphi$ is convergent (as noticed by W. Stoll [6, p. 299, Satz 7]). If M is an upper bound of the sum of the moduli we have

$$(1) \qquad \left| \int_W \varphi \right| \leqq M \text{ area of } W \cap K .$$

Considered as a linear functional of φ, this integral is a *current* (see [3, Ch. III]), defined in D and associated to the analytic set V. Now, this current is closed; in other words

THEOREM. *For every differential form ψ of degree $2p-1$, of class C^1 and with a compact support $K \subset D$, we have*

$$\int_W d\psi = 0 \,.$$

This theorem is proved by P. Lelong [1]. The following proof (sketched at the Conference on The Theory of Analytic Functions, IAS, September 1957) differs slightly from that of Lelong.

The set $V_1 = V - W$ is an analytic set in D of dimension $\leqq p-1$ (as proved in [2]). Let W_1 be the set of all points at which V_1 is regular $(p-1)$-dimensional; then $V_2 = V_1 - W_1$ is an analytic set in D of dimension $\leqq p-2$. Let W_2 be the set of all points at which V_2 is regular $(p-2)$-dimensional; then $V_3 = V_2 - W_2$ is an analytic set in D of dimension $\leqq p-3$. In this way, we define the sets $V_k (k = 1, 2, \cdots, p+1)$ and $W_k (k = 1, 2, \cdots, p)$, such that V_k is analytic in D of dimension $\leqq p-k$, W_k is the set of all points at which V_k is regular $(p-k)$-dimensional and $V_{k+1} = V_k - W_k$. V_{p+1} is the empty set and W_k is a complex submanifold of dimension $p-k$ or the empty set.

Now, consider the following statement.

I_k. *If the support of ψ does not meet V_k, then*

$$\int_W d\psi = 0 \,.$$

If the support of ψ does not meet V_1, its intersection with W is compact and the restriction of ψ in the submanifold W has a compact support in W; therefore, because of a well known theorem (see [3, p. 26]), I_1 is true. On the other hand, since V_{p+1} is the empty set, I_{p+1} is what we want to prove. Hence, we only have to show that, *for $1 \leqq k \leqq p$, I_k implies I_{k+1}.*

Every point $z \in W_k$ has a neighborhood $U_z \subset D - V_{k+1}$ in which there are complex analytic local coordinates with the help of which $W_k \cap U_z$ is defined by linear equations. Since the sets $U_z (z \in K)$ and $D - V_k$ form an open covering of $D - V_{k+1}$, there exists, according to a well known theorem (see e.g. [3, p. 4]), a partition of unity in $D - V_{k+1}$, $1 = \sum \varphi_j$, such that each φ_j is a function of class C^∞ with a compact support contained in one of the sets U_z or in $D - V_k$, and which is locally finite (this means that any compact subsetof $D - V_{k+1}$ meet the supports of only a finite number of the φ_j). Hence, if we suppose (as in I_{k+1}) that the support of ψ is a compact subset of $D - V_{k+1}$, only a finite number of the forms $\psi_j = \varphi_j \psi$ are not identically zero, and since $\psi = \sum \psi_j$, we have only to prove our assertion for the forms ψ_j, whose supports are contained in one of the U_z or in $D - V_k$. In the last case, the assertion is true because I_k holds by hypothesis. Hence, replacing ψ_j by ψ, we may assume that the support of ψ is contained in an open set $U = U_z$ in which, by using suitable local coordinates, W_k is defined by linear equations. In other words, *we may assume that $W_k \cap U$ is contained in a complex $(p - k)$-dimensional plane L^{p-k}.*

Let $\rho(z)$ be the distance of $z \in U$ to L^{p-k}, let $f(t)$ be a function of class C^1 of a real variable t, such that $f(t) = 0$ for $t > 1$ and $f(t) = 1$ for $t < 1/2$, and let r be an arbitrary positive constant. We can write

$$\psi = f\left(\frac{\rho(z)}{r}\right)\psi + \left[1 - f\left(\frac{\rho(z)}{r}\right)\right]\psi .$$

The form $[1 - f(\rho(z)/r)]\psi$ is of class C^1 and since it vanishes for $\rho(z) < r/2$, its support does not meet V_k; hence, according to I_k, the integral of its differential over W vanishes and consequently

$$\begin{aligned}\int_W d\psi &= \int_W d\left[f\left(\frac{\rho(z)}{r}\right)\psi\right] \\ &= \frac{1}{r}\int_W f'\left(\frac{\rho(z)}{r}\right)d\rho \wedge \psi + \int_W f\left(\frac{\rho(z)}{r}\right)d\psi .\end{aligned}$$

The forms under the sign $\int$ in the right hand side vanish for $\rho(z) \geqq r$; therefore, they vanish outside of $T(K, r)$, where K is a suitably chosen compact subset of L^{p-k}. Since their coefficients are bounded (with a bound independent of r), by using inequality (1), we get

$$\left|\int_W d\psi\right| \leqq \left(\frac{A}{r} + B\right) \text{ area of } W \cap T(K, r) , \tag{2}$$

with two constants A and B. Now, the real dimension of $L^{p-k} \supset K$ being $2p - 2k$, Lemma 2 shows that

$$\text{area of } W \cap T(K, r) = O(r^{2k}) .$$

Since r is arbitrarily small and $k \geqq 1$, it follows that the left hand side of (2) is equal to zero. q.e.d.

UNIVERSITY OF LAUSANNE AND UNIVERSITY OF GENEVA

REFERENCES

[1] PIERRE LELONG, *Integration sur un ensemble analytique complexe*, Bull. Soc. Math. France, **85**, 239-262 (1957).

[2] R. REMMERT and K. STEIN, *Über die wesentlichen Singularitäten analytischer Mengen*, Math. Ann. **126**, 263-306 (1953).

[3] G. DE RHAM. Variétés differentiables. Paris, Hermann et Cie, 1955.

[4] ———, *Über mehrfache Integrale*, Abh. Math. Sem. Univ. Hamburg, **12** (1938), 313-339.

[5] W. STOLL, *Mehrfache Integrale auf complexen Mannigfaltigkeiten*, Math. Z. **57**, 116-154 (1952).

[6] ———, *Einige Bemerkungen zur Fortsetzbarkeit analytischer Mengen*, Math. Z. **60** (1954), 287-304.

[7] W. WIRTINGER, *Eine Determinantenidentität und ihre Anwendung auf analytische Gebilde in Euklidischer und Hermitescher Messbestimmung*, Monatsh. f. Math. u. Physik **44** (1936) 343-365.

[8] HERBERT FEDERER, *Some theorems on integral currents*, Trans. Amer. Math. Soc. **117** (1965), 43-67.

(Received Ausust 29, 1968)

On the deformation of sheaves of rings

BY MURRAY GERSTENHABER

Among the current "deformation theories", two exhibit remarkable formal similarities. The first, initiated in its modern form by Froelicher-Nijenhuis [1], and brilliantly elaborated by Kodaira-Spencer [8, a, b], (cf. also [10]), concerns itself with the variation of the complex analytic structure of a manifold X. Denoting by Θ the sheaf of germs of holomorphic tangent vectors, the group of "infinitesimal deformations" is shown to be $H^1(X, \Theta)$. There is a natural quadratic map $H^1(X, \Theta) \rightarrow H^2(X, \Theta)$ which assigns to every infinitesimal deformation the obstruction to prolonging it one step. If such prolongation is possible, then one meets another obstructions, which like all subsequent obstructions lies in $H^2(X, \Theta)$. If one can pass all obstructions then a significant analytic question remains. The formal deformation so constructed *a priori* need not converge and therefore need not define a genuine variation of the complex analytic structure of X. In the special case where $H^2(X, \Theta) = 0$ convergence is assured by the work of Kodaira-Nirenberg-Spencer [7] bnt the general conditions necessary to pass from a formal to a genuine deformation are still not completely understood.

The second deformation theory concerns itself, in particular, with varying the structure of an algebra A which, in the simplest case, is considered at all times to be a fixed vector space over a field [2]. There is once again a group of infinitesimal deformations, which is here $H^2(A, A)$, the second Hochschild cohomology group of A with coefficients in itself, considered as an A-bimodule. It is only an accident of notation that the present group of infinitesimal deformations appears to have a different dimension from that in the analytic case; the cocycles in both cases are functions of two variables. There is a quadratic map $H^2(A, A) \rightarrow H^3(A, A)$ which gives the first obstruction to any infinitesimal deformation, and all subsequent obstructions lie also in $H^3(A, A)$. If all obstructions can be passed then one does, in this case have a "deformation" of

A, at least when A is finite dimensional. Analytic questions analogous to those in the deformation of a manifold do not appear unless one puts more stringent conditions on the nature of the deformation.

The similarity between the analytic and algebraic deformation theories suggests that there is something more than a formal analogy between them. In the present note we shall outline how the deformation theory of rings can be extended to sheaves of rings, and we shall show how this provides a link between the algebraic and analytic theories. Our concern, for the moment, is only with formal questions, mainly the definition of infinitesimal deformations and their obstructions. The basic technical material is contained in [3].

1. Let $\mathcal{O}$ be a sheaf of rings over a space X. By a *deformation of* $\mathcal{O}$ we shall, for the present, mean a sheaf of rings $\tilde{\mathcal{O}}$ over X, filtered by sheaves of ideals,

$$\tilde{\mathcal{O}} = F^0\tilde{\mathcal{O}} \supset F^1\tilde{\mathcal{O}} \supset F^2\tilde{\mathcal{O}} \supset \cdots$$

with

$$F^i\tilde{\mathcal{O}} \cdot F^j\tilde{\mathcal{O}} \subset F^{i+j}\tilde{\mathcal{O}} , \qquad \bigcap F^i\tilde{\mathcal{O}} = 0 ,$$

and

$$\text{gr}\,\tilde{\mathcal{O}} \cong \mathcal{O}[t] . \tag{1}$$

Here $\text{gr}\,\tilde{\mathcal{O}} = \tilde{\mathcal{O}}/F^1\tilde{\mathcal{O}} \oplus F^1\tilde{\mathcal{O}}/F^2\tilde{\mathcal{O}} \oplus \cdots$, and $\mathcal{O}[t]$ is the sheaf whose stalk, $\mathcal{O}[t]_x$ at $x \in X$ is the polynomial ring $\mathcal{O}_x[t]$ in a variable t over $\mathcal{O}_x$. Both $\text{gr}\,\tilde{\mathcal{O}}$ and $\mathcal{O}[t]$ are in a natural way graded sheaves of rings, and in (1) it is understood that the isomorphism is one of graded sheaves.

It has not been assumed that the rings in question are necessarily commutative and it will prove important to investigate the case where X is an algebraic variety, $\mathcal{O}$ its sheaf of local rings and where $\tilde{\mathcal{O}}$ is *not* commutative. However, the basic example of a deformation in the present sense, arises in the following way. Suppose that we have a morphism $f\colon \mathcal{V} \to C$ from a variety $\mathcal{V}$ to a curve C, and a simple point $P \in C$ such that $X = f^{-1}P$ is a subvariety non-singular on $\mathcal{V}$ but which may itself be singular. Let $\mathcal{O}, \tilde{\mathcal{O}}$ be the sheaves of local rings on X and $\mathcal{V}$, respectively, and τ be a function on C vanishing to first order at P. Then $\tilde{\mathcal{O}}$ is filtered by the powers of $t = \tau \circ f$, and the associated graded ring is $\mathcal{O}[t]$. In this example, the rings are, of course, all commutative.

If $\tilde{\mathcal{O}}$ is a deformation of $\mathcal{O}$, then $\mathcal{O}_n = \tilde{\mathcal{O}}/F^{n+1}\tilde{\mathcal{O}}$ is again a filtered sheaf of rings, but now the $(n+1)^{\text{st}}$ filtering part is 0 and $\operatorname{gr}\mathcal{O}_n \cong \mathcal{O}[t]/(t^{n+1})$. There is further a sequence of epimorphisms

$$(2) \qquad \mathcal{O} = \mathcal{O}_0 \xleftarrow{\pi_1} \mathcal{O}_1 \xleftarrow{\pi_2} \mathcal{O}_2 \longleftarrow \cdots$$

in which $\ker \pi_n$ may be identified in a natural way with the sheaf of $\mathcal{O}$-modules $t^n\mathcal{O}$. By a *formal deformation* of a sheaf $\mathcal{O}$ we shall mean a sequence of epimorphisms of filtered sheaves, as in (2), with $\operatorname{gr}\mathcal{O}_n \cong \mathcal{O}[t]/(t^{n+1})$ and $\ker \pi_n \cong t^n\mathcal{O}$. To every actual deformation there is associated a formal deformation, but the converse may fail, in analogy with the analytic case where a formal deformation may conceivably not be associated with a genuine one. We shall, for the present, leave this question entirely aside. *A deformation-jet of order n* of $\mathcal{O}$ is a sequence

$$(2_n) \qquad \mathcal{O} = \mathcal{O}_0 \xleftarrow{\pi_1} \mathcal{O}_1 \xleftarrow{\pi_2} \mathcal{O}_2 \longleftarrow \cdots \xleftarrow{\pi_n} \mathcal{O}_n$$

with all the properties of a formal deformation except that it terminates at $\mathcal{O}_n$. An *infinitesimal deformation* of $\mathcal{O}$ is just a a deformation-jet of order 1.

We shall always tacitly assume that all $\mathcal{O}_i$ are in fact sheaves of unital algebras over some constant coefficient ring k (which may be a field) and that all π_i are unital. Note that if we have a morphism $k' \to k$ then $\mathcal{O}$ may also be considered as a sheaf of k' algebras but the concept of a "deformation of $\mathcal{O}$" will generally change. For example, if $\mathcal{O}$ has characteristic $p \neq 0$ and is considered as a sheaf of $\mathbf{Z}/p$-modules then all deformations still have characteristic p, but considered as a sheaf of $\mathbf{Z}$-modules, some deformations may have characteristic zero.

2. Since $t\mathcal{O}$ is isomorphic, as an $\mathcal{O}$-bimodule, to $\mathcal{O}$ itself, it follows from the definition that an infinitesimal deformation of the sheaf of rings $\mathcal{O}$ is a "singular extension" of sheaves of k-algebras,

$$(3) \qquad \mathbf{E}\colon 0 \longrightarrow \mathcal{O} \xrightarrow{i} \mathcal{O}_1 \xrightarrow{\pi_1} \mathcal{O} \longrightarrow 0 .$$

Under the usual Baer addition, the equivalence classes of such extensions form a group which will be denoted by $\mathcal{E}^2(\mathcal{O}, \mathcal{O})$. When $\mathcal{O}$ is commutative then the subgroup consisting of those (classes of) extensions in which $\mathcal{O}_1$ is commutative will be denoted $\mathcal{E}^2_c(\mathcal{O}, \mathcal{O})$.

An extension $\mathbf{E}$ is *locally trivial* or *locally split* if X has a covering by open "trivializing neighborhoods" $\mathcal{U}$ such that

$$\mathbf{E} \mid \mathcal{U}\colon 0 \longrightarrow \mathcal{O} \mid \mathcal{U} \xrightarrow{i \mid \mathcal{U}} \mathcal{O}_1 \mid \mathcal{U} \xrightarrow{\pi_1 \mid \mathcal{U}} \mathcal{O} \mid \mathcal{U} \longrightarrow 0$$

has a "splitting morphism" $s_{\mathcal{U}}\colon \mathcal{O} \mid \mathcal{U} \to \mathcal{O}_1 \mid \mathcal{U}$ (i.e., $(\pi_1 \mid \mathcal{U}) \circ s_{\mathcal{U}}$ is the identity map of $\mathcal{O} \mid \mathcal{U}$). The locally split extensions form a subgroup, $\mathcal{E}_l^2$ $(= \mathcal{E}_l^2(\mathcal{O}, \mathcal{O}))$ of $\mathcal{E}^2$. It is evident that $\mathcal{E}_l^2 \subset \mathcal{E}^2$. In the following "classical cases" it is known that $\mathcal{E}_c^2 = \mathcal{E}_l^2$, i.e., that every element of $\mathcal{E}_c^2$ is locally split.

Case 1. X a non-singular algebraic variety over a perfect field k, $\mathcal{O}$ the algebraic sheaf of local rings of X, considered as a sheaf of k-algebras. Here $\mathcal{O}_x$ is the k-algebra of germs of rational functions on X which are holomorphic at $x \in X$. The fact that here $\mathcal{E}_c^2 = \mathcal{E}_l^2$ is essentially Theorem 22 of Harrison's classical paper [6].

Case 2. X an analytic manifold, $\mathcal{O}$ the sheaf of local rings, considered as a sheaf of $\mathbf{C}$-algebras. Here the essential theory is due to Gray [5].

Recall now that if

$$0 \longrightarrow M \longrightarrow B \longrightarrow A \longrightarrow 0$$

is a split singular extension of a ring A by an A-module M, and if $s_1, s_2\colon A \to B$ are splitting morphisms, then $s_1 - s_2$ is a derivation of A into M. This observation can be extended to singular extensions of $\mathcal{O}$. In particular, if (3) is locally split, if $\mathcal{U}, \mathcal{V}$ are trivializing neighborhoods, and if $s_{\mathcal{U},\mathcal{V}}$ denotes the ristriction of $s_{\mathcal{U}}$ to $\mathcal{O} \mid \mathcal{U} \cap \mathcal{V}$, then $s_{\mathcal{U},\mathcal{V}} - s_{\mathcal{V},\mathcal{U}}$ is a derivation of $\mathcal{O} \mid \mathcal{U} \cap \mathcal{V}$ into itself. Denoting by Θ the sheaf of germs of derivations of $\mathcal{O}$, it follows that there is a k-module morphism

$$\mathcal{E}_l^2(\mathcal{O}, \mathcal{O}) \longrightarrow H^1(X, \Theta) .$$

By an argument paralleling one of Gray [5], one can show that this is an isomorphism. In those cases where every element of $\mathcal{E}_c^2$ is locally split, it follows that there is a natural isomorphism

$$\mathcal{E}_c^2(\mathcal{O}, \mathcal{O}) \xrightarrow{\sim} H^1(X, \Theta) .$$

In particular, in the classical cases, where X is a non-singular variety or a manifold, one obtains groups of infinitesimal deformations of the sheaf of local rings $\mathcal{O}$ (considered as algebras over the ground field) isomorphic to the classical group of infinitesimal deformations, $H^1(X, \Theta)$. But if X is singular, or if we allow that the characteristic may change, then the present concept of "infinitesimal deformation" is more general.

Let $\mathfrak{S}$ denote the sheaf of germs of singular extensions of $\mathcal{O}$ by itself. This is a sheaf of sets whose global sections, $\Gamma(\mathfrak{S})$, are the singular extensions of $\mathcal{O}$ by itself. The classes of the latter constitute $\mathcal{E}^2(\mathcal{O}, \mathcal{O})$. Now let Θ_1 denote the sheaf of *classes* of germs of singular extensions of $\mathcal{O}$ by itself. This is obtained from the pre-sheaf of "local Baer groups", the group associated with an open $\mathfrak{U}$ being the group of classes of singular extensions of $\mathcal{O}_{\mathfrak{U}}$ by itself. There is a natural onto map $\mathfrak{S} \rightarrow \Theta_1$, and hence a natural map $\Gamma(\mathfrak{S}) \rightarrow \Gamma(\Theta_1)$. Since equivalent elements of $\Gamma(\mathfrak{S})$ clearly have the same image there is a morphism

$$\mathcal{E}^2(\mathcal{O}, \mathcal{O}) \longrightarrow \Gamma(\Theta_1)$$

whose kernel is $\mathcal{E}^2_l(\mathcal{O}, \mathcal{O})$. In the commutative case, letting $\Theta_{1,c}$ denote the sheaf of germs of classes of commutative singular extensions of $\mathcal{O}$, this morphism carries $\mathcal{E}^1_c(\mathcal{O}, \mathcal{O})$ into $\Gamma(\Theta_{1,c})$. In the classical cases, where X is non-singular, we have $\Theta_{1,c} = 0$. But if X is singular and if $\mathbf{e} \in \mathcal{E}^2_c(\mathcal{O}, \mathcal{O})$, then its image in $\Gamma(\Theta_{1,c})$ measures how much of the infinitesimal deformation is due to infinitesimal change of the singularities.

3. If A is a ring and M an A-module then it is possible in a way paralleling that of Baer, to put an additive group structure on those equivalence classes of exact squences

$$\text{(4)} \qquad \mathbf{D}\colon 0 \longrightarrow M \longrightarrow N \xrightarrow{\rho} B \longrightarrow A \longrightarrow 0$$

with the following properties.

(i) N, B are rings, and all mappings are, in particular, ring morphisms

(ii) A operates on N, both left and right, in such a way that $A \oplus N$ is a ring with multiplication given by

$$(a, n)(a', n') = (aa', an' + na' + nn') ,$$

$a, a' \in A$; $n, n' \in N$. We say that N is an "A-structure".

(iii) $\rho(nn') = (\rho n)n' = n(\rho n')$.

(iv) $M \rightarrow N$ is a morphism of A-structures.

The construction of this group, denoted $\mathcal{E}^3(A, M)$, is given in [3], where it is observed that generalization is possible, in particular, to sheaves of rings. The necessary and sufficient condition that the sequence (4) represent the zero element of $\mathcal{E}^3(A, M)$ is that there exist a commutative diagram

$$\begin{array}{lcccccccccc}
\mathbf{S}: & & & 0 & \longrightarrow & N & \longrightarrow & \bar{B} & \longrightarrow & A & \longrightarrow & 0 \\
& & & & & \| & & \downarrow & & \| & & \\
\mathbf{D}: 0 & \longrightarrow & M & \longrightarrow & & N & \longrightarrow & B & \longrightarrow & A & \longrightarrow & 0
\end{array}$$

This has been called a "solution" for (4), [3]. The equivalence classes of solutions form a principal homogeneous space over $\mathcal{E}^2(A, M)$.

Suppose now that we have a singular extension

$$\mathbf{E}: 0 \longrightarrow \mathcal{O} \longrightarrow \mathcal{O}_1 \longrightarrow \mathcal{O} \longrightarrow 0$$

of $\mathcal{O}$ by itself. Then the composite sequence

$$\begin{array}{ccccccccccc}
\mathbf{E} \circ \mathbf{E}: 0 & \longrightarrow & \mathcal{O} & \longrightarrow & \mathcal{O}_1 & \xrightarrow{\rho} & \mathcal{O}_1 & \longrightarrow & \mathcal{O} & \longrightarrow & 0 \\
& & & & \searrow & & \nearrow & & & & \\
& & & & & \mathcal{O} & & & & & \\
& & & & \nearrow & & \searrow & & & & \\
& & & & 0 & & 0 & & & &
\end{array}$$

represents an element of $\mathcal{E}^3(\mathcal{O}, \mathcal{O})$, and represents the zero element if and only if there is a solution

$$\begin{array}{ccccccccccc}
& & & 0 & \longrightarrow & \mathcal{O}_1 & \longrightarrow & \mathcal{O}_2 & \longrightarrow & \mathcal{O} & \longrightarrow 0 \\
& & & & & \| & & \downarrow{\scriptstyle \pi_2} & & \| & \\
\mathbf{E} \circ \mathbf{E}: 0 & \longrightarrow & \mathcal{O} & \longrightarrow & & \mathcal{O}_1 & \longrightarrow & \mathcal{O}_1 & \xrightarrow{\pi_1} & \mathcal{O} & \longrightarrow 0
\end{array}$$

This is equivalent to the existence of a 2nd order deformation-jet

$$\mathcal{O} = \mathcal{O}_0 \xleftarrow{\pi_1} \mathcal{O}_1 \xleftarrow{\pi_2} \mathcal{O}_2$$

of $\mathcal{O}$. The class of $\mathbf{E} \circ \mathbf{E}$ is thus the obstruction to extending the infinitesimal deformation $\mathbf{E}$ to a second order deformation-jet.

The *locally trivial* elements of $\mathcal{E}^3$ are the classes of those $\mathbf{D}$ for which there exists a covering of X by trivializing neighborhoods $\mathfrak{U}$ on each of which $\mathbf{D} \mid \mathfrak{U}$ has a solution. Let Θ_1 denote, as before, the sheaf of germs of classes of singular extensions of $\mathcal{O}$. If $\mathbf{D}$ represents a locally trivial element of $\mathcal{E}^3(\mathcal{O}, \mathcal{O})$, and if $S_{\mathfrak{U}}$, $S_{\mathfrak{V}}$ are local solutions in trivializing neighborhoods $\mathfrak{U}, \mathfrak{V}$ respectively, then $(\mathbf{S}_{\mathfrak{U}} \mid \mathfrak{U} \cap \mathfrak{V}) - (\mathbf{S}_{\mathfrak{V}} \mid \mathfrak{U} \cap \mathfrak{V})$ is a singular extension of $\mathcal{O} \mid \mathfrak{U} \cap \mathfrak{V}$ by itself. Taking the class of this singular extension, we see that there is morphism

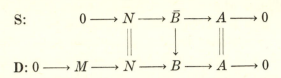

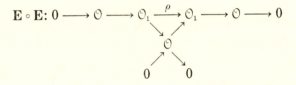

$$\theta\colon \mathcal{E}_l^3(\mathcal{O}, \mathcal{O}) \longrightarrow H^1(X, \Theta_1) .$$

Let $\mathbf{d}$ be the class of $\mathbf{D}$ in $\mathcal{E}_l^3(\mathcal{O}, \mathcal{O})$. If $\mathbf{D}$ has a global solution then the image of $\mathbf{d}$ in $H^1(X, \Theta_1)$ must vanish, but not conversely. For if the image is zero then we can conclude only that the *classes* of the local solutions, $\mathbf{S}_{\mathfrak{U}}$, of $\mathbf{D}$ agree on the intersection of two trivializing neighborhoods. Equivalently, for every pair of trivializing neighborhoods $\mathfrak{U}$, $\mathfrak{V}$ for $\mathbf{D}$ there is a morphism

$$\varphi_{\mathfrak{U}\mathfrak{V}}\colon \mathbf{S}_{\mathfrak{U}} \mid \mathfrak{U} \cap \mathfrak{V} \longrightarrow \mathbf{S}_{\mathfrak{V}} \mid \mathfrak{U} \cap \mathfrak{V} .$$

Suppose now that indeed $\mathbf{d} \in \ker \theta$. Then $\mathbf{D}$ has a global solution if and only if the local solutions $\mathbf{S}_{\mathfrak{U}}$ can be so chosen that on $\mathfrak{U} \cap \mathfrak{V} \cap \mathfrak{W}$ ($\mathfrak{W}$ a third trivializing neighborhood) we have

$$\varphi_{\mathfrak{V}\mathfrak{W}} \circ \varphi_{\mathfrak{U}\mathfrak{V}} = \varphi_{\mathfrak{U}\mathfrak{W}} .$$

Let us write $\mathbf{S}_{\mathfrak{U}}$ in the form

$$\mathbf{S}_{\mathfrak{U}}\colon 0 \longrightarrow \mathcal{O}_1 \mid \mathfrak{U} \longrightarrow \mathcal{O}_{2,\mathfrak{U}} \longrightarrow \mathcal{O} \mid \mathfrak{U} \longrightarrow 0 .$$

Recall now (cf. [3] and [4]) that the morphism $\varphi_{\mathfrak{U}\mathfrak{V}}$ is really nothing other than a morphism

$$\varphi_{\mathfrak{U}\mathfrak{V}}\colon \mathcal{O}_{2,\mathfrak{U}} \mid \mathfrak{U} \cap \mathfrak{V} \longrightarrow \mathcal{O}_{2,\mathfrak{V}} \mid \mathfrak{U} \cap \mathfrak{V}$$

which when adjoined to the following commutative diagram produces a diagram which still commutes.

$$\begin{array}{lcccccccccc}
\mathbf{S}_{\mathfrak{U}} \mid \mathfrak{U}\cap\mathfrak{V}\colon & & & 0 & \longrightarrow & \mathcal{O}_1 \mid \mathfrak{U}\cap\mathfrak{V} & \longrightarrow & \mathcal{O}_{2,\mathfrak{U}} \mid \mathfrak{U}\cap\mathfrak{V} & \longrightarrow & \mathcal{O} \mid \mathfrak{U}\cap\mathfrak{V} & \longrightarrow 0 \\
\downarrow & & & & & \| & & \downarrow & & \|{\scriptstyle \pi_{2,\mathfrak{U}} \mid \mathfrak{U}\cap\mathfrak{V}} & \\
\mathbf{D} \mid \mathfrak{U}\cap\mathfrak{V}\colon 0 & \longrightarrow & \mathcal{O} \mid \mathfrak{U}\cap\mathfrak{V} & & \longrightarrow & \mathcal{O}_1 \mid \mathfrak{U}\cap\mathfrak{V} & \longrightarrow & \mathcal{O}_1 \mid \mathfrak{U}\cap\mathfrak{V} & \longrightarrow & \mathcal{O} \mid \mathfrak{U}\cap\mathfrak{V} & \longrightarrow 0 \\
\downarrow & & & & & \| & & \uparrow & & \|{\scriptstyle \pi_{2,\mathfrak{V}} \mid \mathfrak{U}\cap\mathfrak{V}} & \\
\mathbf{S}_{\mathfrak{U}} \mid \mathfrak{U}\cap\mathfrak{V}\colon & & & 0 & \longrightarrow & \mathcal{O}_1 \mid \mathfrak{U}\cap\mathfrak{V} & \longrightarrow & \mathcal{O}_{2,\mathfrak{V}} \mid \mathfrak{U}\cap\mathfrak{V} & \longrightarrow & \mathcal{O} \mid \mathfrak{U}\cap\mathfrak{V} & \longrightarrow 0
\end{array}$$

The difference $\varphi_{\mathfrak{U}\mathfrak{V}} - \varphi'_{\mathfrak{U}\mathfrak{V}}$ between two such morphisms must vanish on $\mathcal{O}_1 \mid \mathfrak{U} \cap \mathfrak{V}$, hence may be viewed as a map $\mathcal{O} \mid \mathfrak{U} \cap \mathfrak{V} \to \mathcal{O}_{2,\mathfrak{V}} \mid \mathfrak{U} \cap \mathfrak{V}$. The composite of this with the morphism

$$\pi_{2,\mathfrak{V}} \mid \mathfrak{U} \cap \mathfrak{V}\colon \mathcal{O}_{2,\mathfrak{V}} \mid \mathfrak{U} \cap \mathfrak{V} \longrightarrow \mathcal{O}_1 \mid \mathfrak{U} \cap \mathfrak{V}$$

vanishes, so the image of $\varphi_{\mathfrak{U}\mathfrak{V}} - \varphi'_{\mathfrak{U}\mathfrak{V}}$ lies in the kernel of

$$\mathcal{O}_{2,\mathfrak{V}} \mid \mathfrak{U} \cap \mathfrak{V} \longrightarrow \mathcal{O}_1 \mid \mathfrak{U} \cap \mathfrak{V} ,$$

which is just $\mathcal{O} \mid \mathfrak{U} \cap \mathfrak{V}$. Therefore $\varphi_{\mathfrak{U}\mathfrak{V}} - \varphi'_{\mathfrak{U}\mathfrak{V}}$ may be viewed as a map $\mathcal{O} \mid \mathfrak{U} \cap \mathfrak{V} \to \mathcal{O} \mid \mathfrak{U} \cap \mathfrak{V}$, and it is easy to verify that it is in fact a derivation. Restricting attention to $\mathfrak{U} \cap \mathfrak{V} \cap \mathfrak{W}$, we see that on this set $\varphi_{\mathfrak{U}\mathfrak{W}} \circ \varphi_{\mathfrak{U}\mathfrak{V}} - \varphi_{\mathfrak{U}\mathfrak{W}}$ is a derivation of $\mathcal{O} \mid \mathfrak{U} \cap \mathfrak{V} \cap \mathfrak{W}$ into itself. These derivations define an element of $H^2(X, \Theta)$, so we have a morphism

$$\ker \theta \longrightarrow H^2(X, \Theta) .$$

It is easy to verify that this is an isomorphism, yielding an exact sequence

$$0 \longrightarrow H^2(X, \Theta) \longrightarrow \mathcal{E}^3_l(\mathcal{O}, \mathcal{O}) \longrightarrow H^1(X, \Theta_1) .$$

If we choose to consider everywhere only sheaves of commutative rings, then the group analogous to $\mathcal{E}^3$ will be denoted $\mathcal{E}^3_c$, the locally trivial elements will be denoted $\mathcal{E}^3_{c,l}$, and we have an exact sequence

$$0 \longrightarrow H^2(X, \Theta) \longrightarrow \mathcal{E}^3_{c,l}(\mathcal{O}, \mathcal{O}) \longrightarrow H^1(X, \Theta_{1,c}) .$$

In the classical cases, $\Theta_{1,c} = 0$, yielding a canonical isomorphism

$$\mathcal{E}^3_{c,l}(\mathcal{O}, \mathcal{O}) \cong H^2(X, \Theta) .$$

If a singular extension

$$\mathbf{E}\colon 0 \longrightarrow \mathcal{O} \longrightarrow \mathcal{O}_1 \longrightarrow \mathcal{O} \longrightarrow 0$$

represents an element $\mathbf{e}$ of $\mathcal{O}^2_c(\mathcal{O}, \mathcal{O})$ then the class $\mathbf{d}$ of $\mathbf{E} \circ \mathbf{E}$ is an element of $\mathcal{O}^3_{c,l}(\mathcal{O}, \mathcal{O})$ depending only on $\mathbf{e}$ and will be denoted $\mathbf{e} \circ \mathbf{e}$. For note, in any case, that if $\mathbf{E}$ is locally trivial, then so is $\mathbf{E} \circ \mathbf{E}$. One can verify that the quadratic map

$$H^1(X, \Theta) = \mathcal{E}^2_c(\mathcal{O}, \mathcal{O}) \longrightarrow \mathcal{E}^3_c(\mathcal{O}, \mathcal{O}) = H^2(X, \Theta)$$

defined by $\mathbf{e} \mapsto \mathbf{e} \circ \mathbf{e}$ assigns to every element of $H^1(H, \Theta)$ its usual obstruction in $H^2(X, \Theta)$. It will be evident now, by comparison with [3], that in the classical cases the formal theory of the higher obstructions proceeds for deformations of $\mathcal{O}$ just as it does classically for deformations of X.

4. The theory presented here has been purely formal and leaves open all of the really significant questions. Among these are

(a) In the classical cases, when does a deformation of $\mathcal{O}$, in the present sense, actually correspond to a deformation of X in the usual sense?

(b) When does a formal deformation correspond to an actual one?

(c) When does there exist a Kuranishi-type theory [9] yielding a space of moduli for deformations?

Observe that the present theory includes the possibility, even in the classical cases, of deforming the sheaf $\mathcal{O}$ of local rings into a filtered non-commutative sheaf; explicit examples exist and are easy to produce. This suggests that, in certain cases at least, it is meaningful to consider objects analogous in various respects to varieties or manifolds but whose local rings are non-commutative. In the case of a "variety", in place of the function field one will have a division ring. It is already clear from other phenomena that algebraic geometry must be expanded to include non-commutative cases. The present theory suggests that some of these cases come from deformation of more usual objects; this may be useful in providing examples.

UNIVERSITY OF PENNSYLVANIA

REFERENCES

[1] A. FROEHLICHER and A. NIJENHUIS, *A theorem on stability of complex structures*, Proc. Nat. Acad. Sci., USA **43** (1957), 239-241.

[2] M. GERSTENHABER, *On the deformation of rings and algebras*, Ann. of Math. **79** (1964), 59-103.

[3] ———, *On the deformation of rings and algebras*; II, ibid, **84** (1966), 1-19.

[4] ———, *The obstruction to an automorphism of a filtered ring*, Bull. A. M.S. **74** (1968), 659-697.

[5] J. W. GRAY, *Extensions of sheaves of algebras*, Illinois J. Math. **5** (1961), 159-174.

[6] D. K. HARRISON, *Commutative algebras and cohomology*, Trans. A. M. S. **104** (1962), 191-204.

[7] K. KODAIRA, L. NIRENBERG, and D. C. SPENCER, *On the existence of deformations of complex analytic structures*: I-II, Ann. of Math. **68** (1958), 450-459.

[8] K. KODAIRA and D. C. SPENCER, (a) *On deformations of complex analytic strvctures*; I-II, Ann. of Math. **67** (1958), 328-466; (b) *On deformations of complex analytic structures*: III. *Stability theorems for complex structures*, ibid, **71** (1960), 43-76 (for further references, see [10]).

[9] M. KURANISHI, *On the locally complete families of complex analytic structures*, Ann. of Math. **75** (1962), 536-577.

[10] D. C. SPENCER, *Deformations of structures on manifolds defined by transitive continuous pseudogroups*, Ann. of Math. **76** (1962), 306-445.

(Received December 16, 1968)

Die Azyklizität der affinoiden Überdeckungen

Von Lothar Gerritzen und Hans Grauert

Einleitung

In der komplexen Analysis hat die Methode der Steinschen Überdeckungen zu wichtigen Resultaten geführt. Man denke etwa an das 1953 von H. Cartan und J. P. Serre bewiesene "théorème de finitude", das aussagt, daß auf geschlossenen komplexen Mannigfaltigkeiten die Kohomologiegruppen mit Koeffizienten in einer kohärenten analytischen Garbe endlichdimensionale Vektorräume bilden. Beim Beweis dieses Satzes ist es wichtig zu wissen, daß auf einem Steinschen Raum alle diese Kohomologiegruppen für Dimensionen grösser als 0 verschwinden.

In der nichtarchimedischen Funktionentheorie treten an Stelle der Überdeckungen mit Steinschen Räumen gewisse Überdeckungen mit *affinoiden Räumen*. Auch in diesem Falle muß gezeigt werden, daß entsprechend gebildete Kohomologiegruppen verschwinden. Darüber hinaus braucht man, daß jede nulldimensionale Kohomologieklasse eine *affinoide Funktion* ist. In der komplexen Analysis ist dieser zweite Sachverhalt trivial richtig.

Wir werden in dieser Arbeit die beiden eben genannten Aussagen beweisen. Ihr Inhalt stellt gerade die Azyklizität der affinoiden Überdeckungen von affinoiden Räumen dar. Beim Beweis ist eine Untersuchung der *affinoiden Teilbereiche* der affinoiden Räume von Bedeutung. Ein wichtiges Resultat ergibt sich dabei: Die *affinoide Algebra* eines Teilbereiches ist eindeutig bestimmt!

Wegen der Hauptergebnisse kann die Tate'sche Definition des *rigiden analytischen* Raumes vereinfacht werden. Wir nennen solche Räume fortan *holomorphe Räume* und zeigen, daß jeder algebraische Raum (im Sinne von A. Weil) einen solchen holomorphen Raum bestimmt, daß es andererseits aber auch holomorphe Räume gibt, die nicht von einem algebraischen Raum herrühren (Hopfsche Mannigfaltigkeiten).

Das Hauptergebnis der Arbeit wurde von dem zweiten der

Autoren in [6] angekündigt und im Falle von einer Dimension in [4] bewiesen. In diesem Falle konnte ein schärferes Ergebnis gewonnen werden. Der erste Autor zeigte dann, daß man den Beweis wesentlich einfacher durchführen kann, wenn man auf die Verschärfung verzichtet (man vgl. [3]). Die Beweisidee aus [3] ist nun in diese Arbeit übernommen worden. Es ist dadurch gelungen, den Beweis wesentlich kürzer darzustellen, als ursprünglich für nötig gehalten wurde.

1. Affinoide Räume

1. Es sei k ein Körper, der mit einer nichtarchimedischen Bewertung versehen ist. Wir schreiben $|a|$ für den Wert von $a \in k$. Die Bewertung sei in multiplikativer Form, es gelte also das Gesetz $|a_1 \cdot a_2| = |a_1| \cdot |a_2|$, und k sei vollständig in Bezug auf die induzierte natürliche uniforme Struktur. Ferner sei die Bewertung nicht die triviale.—Wir setzen k nicht als algebraisch abgeschlossen voraus.

Es sei T_n die Algebra der strikt konvergenten Potenzreihen über k in n Unbestimmten. T_n ist bekanntlich ein noetherscher Ring und mit einer natürlichen Topologie versehen. Ist $I \subset T_n$ ein Ideal, so heißt jede Algebra A, die zu T_n/I isomorph ist, eine *affinoide Algebra*. Natürlich ist A wieder noethersch.

Wir bezeichnen fortan mit $\bar{k}$ den algebraischen Abschluß von k. Es gibt eine eindeutig bestimmte Fortsetzung der Bewertung von k nach $\bar{k}$. Diese Fortsetzung sei fortan durchgeführt. Wir schreiben wieder $|a|$ für den Wert von $a \in \bar{k}$. Es sei

$$E^n = \{(x_1, \cdots, x_n); x_\gamma \in \bar{k}, |x_\gamma| \leqq 1\} .$$

Die Bewertungstopologie in $\bar{k}$ macht E^n zu einem topologischen Raum.

T_n besteht gerade aus den Potenzreihen über k, die auf ganz E^n konvergieren. Die Grenzfunktionen heißen affinoide Funktionen. Wir dürfen T_n mit der k-Algebra dieser Funktionen identifizieren. Es sei $\mathcal{O}$ die Garbe der konvergenten Potenzreihen mit Koeffizienten aus $\bar{k}$ über E^n.

Ist $I \subset T_n$ ein Ideal, so erzeugen die Funktionen $f \in I$ eine Idealgarbe $J \subset \mathcal{O}$. Wir setzen $H = \mathcal{O}/J$. Der Träger von H besteht gerade aus den Punkten

$$X = \{x \in E^n : J_x \neq \mathcal{O}_x\} .$$

Wir denken uns fortan H auf X beschränkt. H ist eine Garbe von lokalen $\bar{k}$-Algebren über (dem topologischen Raum) X.

Die natürliche Abbildung $T_n \to \Gamma(X, H)$ führt, wie in [5] gezeigt wurde, gerade das Ideal I in Null über. Man hat deshalb eine Injektion $A = T_n/I \to \Gamma(X, H)$. Wir können also die affinoide Algebra A als k-Unteralgebra von $\Gamma(X, H)$ auffassen. Das Tripel (X, H, A) nennen wir einen in E^n *eingebetteten affinoiden Raum.*

Definition 1. *Ein affinoider Raum ist ein Tripel* (X, H, A), *bei dem folgendes gilt:*

(1) X *ist ein topologischer Raum,*

(2) H *ist eine Garbe von lokalen* $\bar{k}$*-Algebren auf* X,

(3) $A \subset \Gamma(X, H)$ *ist eine* k*-Unteralgebra,*

(4) (X, H, A) *ist isomorph zu einem in* E^n *eingebetteten affinoiden Raum.*

Wir nennen H die *Strukturgarbe,* A die *Strukturalgebra* unseres affinoiden Raumes. Die Elemente $f \in A \subset \Gamma(X, H)$ heißen wieder *affinoide Funktionen* auf X. Ist $x \in X$, so sei $f_x \in H_x$ das Bild von f in dem Halm H_x. Es sei $m_x \subset H_x$ das maximale Ideal, und $f(x)$ bezeichne das Quotientenbild von f_x in $H_x/m_x = \bar{k}$. Jeder affinoiden Funktion ist also eine (stetige) $\bar{k}$-wertige Funktion $[f]$ so zugeordnet, daß gilt: $f(x) = [f](x)$. Die Zuordnung $f \to [f]$ ist ein Homomorphismus von k-Algebren, der im allgemeinen nicht injektiv ist.

Sind X_1, X_2 affinoide Räume, so trägt das topologische Produkt $X_1 \times X_2$ eine natürliche Strukturgarbe H und eine natürliche Strukturalgebra A, die $X_1 \times X_2$ zu einem affinoiden Raum machen. Man erhält H und A auf folgende Weise. Zunächst wird X_γ in E^{n_γ} eingebettet. Die Ideale I_γ von X_γ lassen sich dann auch als Ideale in $E^{n_1+n_2}$ auffassen. Sie erzeugen ein Ideal I, das dann $X_1 \times X_2$ definiert. Die Konstruktion ist unabhängig von der Auswahl der Einbettungen.

Von besonderer Bedeutung ist der Fall $X \times E^n$. Die Strukturalgebra $A(X \times E^n)$ ist in diesem Falle isomorph zu $A\langle \xi_1, \cdots, \xi_n \rangle$, wobei $A\langle \xi_1, \cdots, \xi_n \rangle$ die k-Algebra der strikt konvergenten Potenzreihen in den Unbestimmten $\xi_1, \cdots, \xi_n$ über A bezeichnet. Die affinoide Algebra $A = A(X)$ trägt dabei eine Topologie, die in einer Darstellung $A = T_m/I$ von der Topologie in T_m erzeugt wird. Bekanntlich ist die Topologie in A unabhängig von dieser Darstellung.

Ist (X, H, A) ein affinoider Raum, $I \subset A$ ein Ideal, so erzeugt I eine Idealgarbe $J \subset H$. Es sei dann $X_1 = \{x \in X: J_x \neq H_x\}$; $H_1 =$

$H/J \mid X_1$; $A_1 = A/I$. Man kann A_1 wieder als Unteralgebra von $\Gamma(X_1, H_1)$ auffassen und (X_1, H_1, A_1) ist ein affinoider Raum (ein in X eingebetteter). Bezeichnet (X^*, H^*, A^*) einen weiteren affinoiden Raum und ist ein Isomorphismus von X_1 und X^* gegeben, so sagt man, daß X^* in X eingebettet ist.

2. Es sei fortan (X, H, A) ein affinoider Raum. Wir bezeichnen ihn einfach mit X, sofern keine Mißverständnisse entstehen.

Es sei S eine analytische Garbe über X, das ist eine Garbe von H-Moduln. Die Menge der Schnittflächen $\Gamma(X, S)$ ist dann ein A-Modul. Im allgemeinen ist $\Gamma(X, S)$ viel zu groß. Deshalb nennt man S erst dann eine *affinoide Garbe*, wenn ein A-Untermodul $M \subset \Gamma(X, S)$ fest vorgegeben ist. Insbesondere ist also das Paar (H, A) eine affinoide Garbe. Ebenso sind (qH, qA) affinoide Garben, wenn qH, qA für $q = 1, 2, 3, \cdots$ die q-fache direkte Summe bezeichnet. Unter einem *affinoiden Garbenmorphismus* einer affinoiden Garbe (S_1, M_1) in eine affinoide Garbe (S_2, M_2) versteht man einen analytischen Garbenmorphismus φ: $S_1 \rightarrow S_2$, der M_1 in M_2 abbildet. Eine (affinoide) Garbensequenz $(S_1, M_1) \rightarrow \cdots \rightarrow (S_l, M_l)$ heißt *exakt*, wenn die Sequenzen $S_1 \rightarrow \cdots \rightarrow S_l$; $M_1 \rightarrow \cdots \rightarrow M_l$ exakt sind. Schließlich nennt man eine affinoide Garbe (S, M) *kohärent*, wenn es auf X eine exakte Sequenz $(pH, pA) \rightarrow (qH, qA) \rightarrow (S, M) \rightarrow 0$ gibt. Dabei bezeichnet $0 = (0, 0)$ die Nullgarbe.—Eine affinoide Garbe auf X werden wir fortan wieder meistens einfach mit S bezeichnen.

Es sei $X' \subset X$ eine offene Teilmenge. Wir definieren:

Definition 2. *X' ist ein affinoider Teilbereich von X, wenn eine k-Unteralgebra $A' \subset \Gamma(X', H')$ mit $H' = H \mid X'$ gegeben ist, so daß*

(1) *(X', H', A') ein affinoider Raum ist,*

(2) *für jedes $f \in A$ die Beschränkung $f \mid X'$ in A' enthalten ist.*

Wir werden später zeigen:

SATZ 1. *A' ist durch die Teilmenge X' eindeutig festgelegt.*

Es macht also Sinn zu fragen, wann eine offene Teilmenge von X ein affinoider Teilbereich ist.

Der Durchschnitt endlich vieler affinoider Teilbereiche X_ν, $\nu = 1, \cdots, l$ von X ist stets wieder ein affinoider Teilbereich von X. Zunächst ist $X_* = X_1 \cap \cdots \cap X_l$ eine offene Teilmenge von X.

Die Strukturalgebra A_* von X_* erhält man sodann auf folgende Weise: Es sei X_ν in $X \times E^{n_\nu}$ so eingebettet, dass die Produktprojektion $X \times E^{n_\nu} \to X$ auf X_ν die Identität $X_\nu \to X$ ist (Das ist stets möglich!). Es seien $f_{\nu_1}, \cdots, f_{\nu_m}$ affinoide Funktionen auf $X \times E^{n_\nu}$, die das Ideal von X_ν erzeugen. Wir können die f_{ν_μ} auch als affinoide Funktionen auf $X \times E^{n_1+\cdots+n_l}$ auffassen. Sie erzeugen zusammen ein Ideal I. Dieses Ideal I definiert dann den affinoiden Raum X_*. Man zeigt leicht, daß die Konstruktion unabhängig von der Auswahl der Einbettung der X_ν und der affinoiden Funktionen f_{ν_μ} ist. X_* ist affinoider Teilbeich von X, X_ν und aller Durchschnitte $X_{\nu_1} \cap \cdots \cap X_{\nu_r}$ mit $\nu_\lambda \in \{1, \cdots, l\}$.

Es sei $S = (S, M)$ eine affinoide Garbe auf X und $X' \subset X$ ein affinoider Teilbereich. Unter der Beschränkung $S \mid X'$ wollen wir dann die affinoide Garbe (S', M') mit

$$S' = S \mid X' \quad \text{und} \quad M' = (M \mid X') \cdot A'$$

verstehen. Die *Cechschen Kohomologiegruppen* lassen sich nun auf üblichem Wege definieren:

Es seien $U_1, \cdots, U_l$ endlich viele affinoide Teilbereiche von X. Gilt $\bigcup U_\nu = X$, so heißt $\mathfrak{U} = \{U_\nu : \nu = 1, \cdots, l\}$ eine affinoide Überdeckung von X. Wir setzen $U_{\nu_0 \cdots \nu_l} = U_{\nu_0} \cap \cdots \cap U_{\nu_l}$. Dann ist $U_{\nu_0 \cdots \nu_l}$ ein affinoider Raum. Es sei nun S eine affinoide Garbe über X. Wir setzen $S \mid U_{\nu_0 \cdots \nu_l} = (S_{\nu_0 \cdots \nu_l}, M_{\nu_0 \cdots \nu_l})$. Unter einer Kokette ξ der Dimension l versteht man sodann eine Kollektion von Schnittflächen $\xi_{\nu_0 \cdots \nu_l} \in M_{\nu_0 \cdots \nu_l}$ mit $\nu_\lambda \in \{1, \cdots, l\}$. Da man durch die Beschränkung einen Homomorphismus $M_{\nu_0 \cdots \hat{\nu}_\lambda \cdots \nu_{l+1}} \to M_{\nu_0 \cdots \nu_{l+1}}$ hat[1], ist in dem k-Modul $C^l(\mathfrak{U}, S)$ aller l-dimensionalen Koketten die Korandoperation definiert. Man erhält auf bekannte Weise die Čechschen Kohomologiegruppen $H^l(\mathfrak{U}, S)$. Das Hauptergebnis dieser Arbeit ist der folgende:

SATZ 2. *Es sei S eine kohärente affinoide Garbe über X, es sei $\mathfrak{U}$ eine (endliche) affinoide Überdeckung von X. Dann gilt $H^l(\mathfrak{U}, S) = 0$ für $l = 1, 2, 3 \cdots$ und $H^0(\mathfrak{U}, S) = M$.*

Der im Satz 2 ausgedrückte Sachverhalt wird üblicherweise die *Azyklizität* der affinoiden Überdeckungen genannt.

3. Es seien (X, H, A) und (Y, G, B) zwei affinoide Räume. Unter einer *analytischen Abbildung* $\varphi: (X, H) \to (Y, G)$ versteht

[1] $\hat{\nu}_\lambda$ bedeutet, daß ν_λ ausgelassen werden soll.

man ein Paar von Abbildungen (φ_0, φ_1), bei dem

(1) $\varphi_0: X \to Y$ eine stetige Abbildung, und

(2) $\varphi_1: G \circ \varphi_0 \to H$ ein Homomorphismus von Garben von lokalen $\bar{k}$-Algebren ist.

Dabei bezeichnet $G \circ \varphi_0$ die durch φ_0 nach X geliftete Garbe. Natürlich ergibt φ_1 einen Homomorphismus $\Gamma(Y, G) \to \Gamma(X, H)$.

Definition 3. *Eine analytische Abbildung* $\varphi: (X, H) \to (Y, G)$ *heißt affinoid, wenn durch* φ *die Strukturalgebra* $B \subset \Gamma(Y, G)$ *in* $A \subset \Gamma(X, H)$ *abgebildet wird.*

Wir nennen eine analytische Abbildung $\varphi: (X, H) \to (Y, G)$ *bianalytisch*, wenn φ_0 und φ_1 topologische Abbildungen sind. In diesem Falle existiert als "Umkehrung" ein natürlicher Garbenhomomorphismus $\check{\varphi}_1: H \circ \varphi_0^{-1} \to G$. Wir nennen $(\varphi_0^{-1}, \check{\varphi}_1)$ die Umkehrung φ^{-1} von φ. Die Abbildung φ^{-1} ist ebenfalls analytisch. Sind φ, φ^{-1} affinoid, so heißt φ eine *biaffinoide* Abbildung. In diesem Falle ergibt φ einen Algebraisomorphismus $B \xrightarrow{\sim} A$. Eine biaffinoide Abbildung ist also nichts anderes als ein Isomorphismus des affinoiden Raumes (X, H, A) und des affinoiden Raumes (Y, G, B).

Definition 4. *Eine affinoide Abbildung* $\varphi: X \to Y$ *heißt eine Immersion, wenn in jedem Punkt* $x \in X$ *der Homomorphismus* $\varphi_1: (G \circ \varphi_0)_x \to H_x$ *surjektiv ist.*

Unter einer *eineindeutigen* (*injektiven*) *Immersion* verstehen wir eine Immersion $\varphi: X \to Y$, bei der φ_0 injektiv ist. Eine Immersion heißt *offen*, wenn φ_0 eine offene Abbildung und φ_1 in jedem Halm injektiv ist.

Ist X ein affinoider Teilbereich von Y und bezeichnet φ die identische Abbildung, so ist φ wegen der Eigenschaft $B \mid X \subset A$ eine affinoide Abbildung. Da auch φ_1 die Identität und mithin bijektiv ist, folgt, daß φ eine (offene, eineindeutige) Immersion darstellt.

Es seien nun X, Y affinoide Räume und $\varphi: X \to Y$ sei eine offene, eineindeutige Immersion. Dann ist $\varphi_0(X) = Y' \subset Y$ eine offene Teilmenge. Ist $G' = G \mid Y'$, so bildet φ den Raum (X, H) bianalytisch auf (Y', G') ab. Bezeichnet man noch mit B' das Bild von A in $\Gamma(Y', G')$, so ist (Y', G', B') ein affinoider Teilbereich von (Y, G, B) und φ ist eine biaffinoide Abbildung von (X, H, A) auf (Y', G', B'). In anderen Worten heißt has: Man kann X als affinoiden Teilbereich von Y auffassen.

Eine affinoide Abbildung $\varphi: X \to Y$ heißt *endlich*, wenn die

Strukturalgebra A bezüglich des Homomorphismus $B \to A$ ein endlicher B-Modul ist. Es sei nun φ eine endliche und eineindeutige Immersion von X in Y. Nach [5] ist das 0-te direkte Bild

$$\varphi_*(H, A) = (\varphi_*(H), \varphi_*(A))$$

eine kohärente affinoide Garbe über Y. Der Träger Y' von $\varphi_*(H)$ ist gerade die Menge $\varphi_0(X)$. Es sei (J, I) die Annullatorgarbe von $\varphi_*(H, A)$. Diese ist nach [5] kohärent, und es gilt

$$Y' = \{y \in Y: J_y \neq G_y\} .$$

Die Menge Y' wird nun durch I zu einem in (Y, G) eingebetteten affinoiden Raum. Es sei $G' = G/J \mid Y'$ die Strukturgarbe von Y' und $B' = B/I \subset \Gamma(Y', G')$ die Strukturalgebra. Da φ_1 surjektiv ist, erhalten wir einen Isomorphismus φ_1': $G' \circ \varphi_0 \to H$. Die Abbildung $\varphi' = (\varphi_0, \varphi_1')$ ist also eine bianalytische Abbildung von (X, H) auf (Y', G').

Man hat den natürlichen Homomorphismus

$$(G', B') \to (\varphi_*(H), \varphi_*(A)) .$$

Dieser ist lokal surjektiv, d.h. der Homomorphismus $G' \to \varphi_*(H)$ ist surjektiv. Dieses folgt, weil φ_1 die Garbe $G \circ \varphi_0$ auf H abbildet. Nach einem Resultat aus [5] folgt, daß auch die Abbildung $B' \to \varphi_*(A) \approx A$ surjektiv ist. Mithin ergibt φ' einen Homomorphismus von B' auf A. Dieser ist dann auch injektiv, weil φ' eine bianalytische Abbildung ist. Damit ist gezeigt, daß φ' eine biaffinoide Abbildung darstellt. Wir haben also den folgenden Satz hergeleitet:

SATZ 3. *Es sei φ: $X \to Y$ eine endliche und eineindeutige Immersion. Dann ist das Bild $\varphi(X) \subset Y$ ein in Y eingebetteter affinoider Raum (mit einer natürlich gegebenen Strukturgarbe und natürlichen Strukturalgebra). Man kann φ als eine biaffinoide Einbettungsabbildung $X \to \varphi(X)$ auffassen.*

2. Rationale Bereiche

1. Im folgenden sei X ein fester affinoider Raum und A die Algebra der affinoiden Funktionen auf X. Wir untersuchen spezielle affinoide Teilbereiche, die sog. rationalen Bereiche.

Es seien f_i, $0 \leqq i \leqq n$, affinoide Funktionen auf X ohne gemeinsame Nullstelle und $X' = \{x \in X: |f_i(x)| \leqq |f_n(x)|\}$; X' ist offen in X. Da die Funktion $f_n' = f_n \mid X'$ keine Nullstelle in X' hat, ist $f_i/f_n \mid X'$ eine analytische Funktion. Die Algebra A' der affinoiden Funktionen auf X' sei

$$A\left\langle\frac{f_i}{f_n}\right\rangle = A\left\langle\frac{f_0}{f_n}, \cdots, \frac{f_{n-1}}{f_n}\right\rangle$$
$$= A\langle\xi_i\rangle/(f_i - f_n\xi_i)$$
$$= A\langle\xi_0, \cdots, \xi_{n-1}\rangle/(f_0 - f_n\xi_0, \cdots, f_{n-1} - f_n\xi_{n-1}) .$$

Der so zu einem affinoiden Raum gemachte Bereich X' ist ein affinoider Teilbereich von X und heißt *rationaler Bereich* von X.

Man überlegt sich, daß die Strukturalgebra eines rationalen Bereiches X' eindeutig bestimmt ist durch die zugrundeliegende Punktmenge. Genauer: Hat X' eine weitere Darstellung durch affinoide Funktionen g_j, $0 \leqq j \leqq m$, auf X ohne gemeinsame Nullstelle, also $X' = \{x \in X\colon |\, g_j(x)\,| \leqq |\, g_m(x)\,|\}$, so ist

$$A\left\langle\frac{f_i}{f_n}\right\rangle = A\left\langle\frac{g_j}{g_m}\right\rangle .$$

Da $|\, f_i/f_n(x)\,| \leqq 1$ für $x \in X'$, genügt es, zu zeigen, daß

$$\frac{f_i}{f_n}\bigg|\, X' \in A\left\langle\frac{g_j}{g_m}\right\rangle .$$

Dies ist aber klar da f_n keine Nullstelle in X' besitzt und

$$f_i, f_n \in A\left\langle\frac{g_i}{g_m}\right\rangle$$

gilt.

Man überlegt sich sofort, daß der Durchschnitt von endlich vielen rationalen Bereichen wieder ein solcher ist. Sei etwa

$$X_1 = \{x \in X\colon |\, f_i(x)\,| \leqq |\, f_n(x)\,|,\ 0 \leqq i \leqq n\}$$
$$X_2 = \{x \in X\colon |\, g_j(x)\,| \leqq |\, g_m(x)\,|,\ 0 \leqq j \leqq m\}$$

und $\{f_i\}$ sowie $\{g_j\}$ ohne gemeinsame Nullstelle in X. Dann ist

$$X_1 \cap X_2 = \{x \in X\colon |\,(f_ig_j)(x)\,| \leqq |\,(f_ng_m)(x)\,|,\ 0 \leqq i \leqq n,\ 0 \leqq j \leqq m\}$$

und die Funktionen f_ig_j haben keine gemeinsame Nullstelle in X. Zudem gilt für die Algebra A_{12} der affinoiden Funktionen auf $X_1 \cap X_2$:

$$A_{12} = A\langle\xi_{ij}\rangle/(f_ig_j - f_ng_m\xi_{ij}) .$$

Man nennt die rationalen Bereiche der Form

$$X' = \{x \in X\colon |\, f_i(x)\,| \leqq 1\} ,$$

wobei f_i affinoide Funktionen auf X sind, *Weierstraßbereiche* von X und die Bereiche der Form $X' = \{x \in X\colon |\, f_i(x)\,| \leqq 1,\ |\, g_j(x)\,| \geqq 1\}$, wobei f_i, g_j affinoide Funktionen auf X sind, *Laurentbereiche* von X.

Beispiel. Der rationale Bereich X' von E^1 sei wie folgt gegeben:

$$X' = \{x \in E: |x| \leqq |x - a|^2\}, \qquad a \in k, 0 < |a| < 1 .$$

Es ist $X' = X_1 \cup X_2$, wobei X_1 den Weierstraßbereich

$$\{x \in E: |x| \leqq |a|^2\}$$

und X_2 den Laurentbereich $\{x \in X: |x| = 1\}$ bezeichnet. Man sieht leicht ein, daß X' kein Laurentbereich von E ist, da jeder Laurentbereich von E, der in keinem echten Weierstraßbereich von E enthalten ist, von der Form $E - \cup K_\lambda$ ist, wobei die K_λ endlich viele "Kreise" sind, vgl. [4]. Der Bereich X' ist dagegen ein Weierstraßbereich im Laurentbereich $X^* = \{x \in E: |x - a| \geqq |a|\}$. Dies zeigt insbesondere, daß der Begriff Laurentbereich nicht transitiv ist.

SATZ 1. *Es sei $X' \subset X$ ein rationaler Bereich von X und $X'' \subset X'$ ein rationaler Bereich von X', dann ist X'' sogar ein rationaler Bereich von X.*

BEWEIS. Es seien $X'' = \{x \in X': |b_j(x)| \leqq |b_m(x)|, 0 \leqq j \leqq m\}$, b_j affinoide Funktionen auf X' ohne gemeinsame Nullstelle in X'. Da die Funktionen b_j, $0 \leqq j \leqq m$, in der Algebra A' der affinoiden Funktionen auf X' die Eins aufspannen, ist

$$\varepsilon = \inf_{x \in X'} \{\max_{j=0}^{m} |b_j(x)|\} > 0 .$$

(1) Wir betrachten nun zuerst den Fall, daß X' ein Weierstraßbereich von X ist. Dann läßt sich jede affinoide Funktion auf X' approximieren durch affinoide Funktionen auf X.

Es seien b'_j affinoide Funktionen auf X, so daß $|b_j(x) - b'_j(x)| < \varepsilon$ für alle $x \in X'$. Da $|b_m(x)| = |b'_m(x)| \geqq \varepsilon$ für $x \in X''$ gilt, erhält man eine Darstellung für X'':

$$X'' = \{x \in X: |b'_j(x)| \leqq |b'_m(x)|, \varepsilon' \leqq |b'_m(x)|\} \cap X' ;$$

dabei sei $0 < \varepsilon' = |a| \leqq \varepsilon$ und $a \in k$. Also ist X'' als Durchschnitt von rationalen Bereichen wieder ein solcher Bereich.

(2) Als nächstes betrachten wir den Fall, daß X' ein Laurentbereich der Form $\{x \in X: |s(x)| \geqq 1\}$, s affinoide Funktion auf X, ist.

Es gibt dann für jede affinoide Fuktion b_j auf X' eine Darstellung

$$b_j = \sum_{\nu=0}^{\infty} a_{j\nu} s^{-\nu},\ a_{j\nu} \text{ affinoid auf } X \text{ mit } |a_{j\nu}| \to 0\ .$$

Da man für die Beschreibung von X'' die Funktionen b_j durch alle affinoiden Funktionen b'_j auf X' ersetzen kann, für welche $|b_j - b'_j| < \varepsilon$ auf X' gilt, darf man annehmen, daß fast alle $a_{j\nu} = 0$ sind. Weiter darf man dann annehmen, daß

$b_j = a_j/s$ gilt, wobei a_j affinoid auf ganz X ist, da man ja s durch eine Potenz s^n ersetzen kann.

Nunmehr gilt:

$$X'' = \{x \in X \colon |a_i(x)| \leqq |a_m(x)|,\ \varepsilon' \leqq |a_m(x)|\} \cap X'\ .$$

Dies ist der Fall, da für alle $x \in X''$ gilt: $|a_m(x)|/|s(x)| \geqq \varepsilon$, und somit, weil $|s(x)| \geqq 1$, sogar $|a_m(x)| \geqq \varepsilon \geqq \varepsilon'$ ist.

(3) Im allgemeinen Fall ist X' ein Weierstraßbereich in einem Laurentbereich X^* der Form $\{x \in X \colon |s(x)| \geqq 1\}$, s affinoid auf X. Daher folgt in diesem Fall die Behauptung sofort aus (1) und (2).

2. Es sei $u\colon Y \to X$ eine affinoide Abbildung und $\Phi\colon A \to B$ der zu u gehörende k-Homomorphismus der Strukturalgebren.

Wir nennen $u\colon Y \to X$ eine Rungesche Immersion, wenn $\Phi(A)$ in B mit der induzierten Topologie eine dichte Unteralgebra ist. Äquivalent hierzu ist, daß es ein affinoides Erzeugendensystem von B gibt, welches in $\Phi(A)$ liegt, vgl. [3, § 3].

Man zeigt leicht, daß u_0 in einer Rungeschen Immersion $u = (u_0, u_1)$ eineindeutig ist, und daß $u_1\colon H(X) \circ u_0 \to H(Y)$ surjektiv ist. Eine Rungesche Immersion ist also wieder eine Immersion im Sinne von § 1.

Man stellt weiter leicht fest, daß sich jede Rungesche Immersion u in der folgenden Weise faktorisieren läßt:

$$\begin{array}{ccc} Y & \xrightarrow{\quad u \quad} & X \\ & {\scriptstyle u'}\searrow \quad \nearrow{\scriptstyle v} & \\ & W & \end{array}$$

wobei $u'\colon Y \to W$ eine Einbettung und $W \subset X$ ein Weierstraßbereich von X, sowie $v\colon W \hookrightarrow X$ die identische Immersion bezeichnet, vgl. [3, § 3].

Falls u außerdem noch eine offene Immersion ist, kann man u' als biaffinoide Abbildung wählen. Somit ist in diesem Fall Y vermöge u ein Weierstraßbereich von X.

3. Der folgende Satz wird an entscheidender Stelle herangezogen:

SATZ 2. *Es sei u*: $Y \to X$ *eine affinoide Abbildung und* X_ε *der rationale Bereich*

$$\{x \in X : |f_i(x)| \leqq \varepsilon |f_n(x)|\}, \qquad \varepsilon \in |\bar{k}^*| .$$

Falls die von u induzierte Abbildung $Y_1 = u^{-1}(X_1) \to X_1$ *eine Rungesche Immersion ist, gibt es ein* $\varepsilon > 1$, $\varepsilon \in |\bar{k}|$, *so daß*

$$Y_\varepsilon = u^{-1}(X_\varepsilon) \to X_\varepsilon$$

ebenfalls noch eine Rungesche Immersion ist.

Wir schicken dem Beweis einige Hilfsbetrachtungen voraus.

Auf der Strukturalgebra T_n des affinoiden Raumes E^n ist auf natürliche Weise eine vollständige Norm $\| \ \|$ gegeben. Sie ist die Norm der gleichmäßigen Konvergenz und für $f = \sum f_{\nu_1 \cdots \nu_n} x_1^{\nu_1} \cdots x_n^{\nu_n}$ ist $\|f\| = \max |f_{\nu_1 \cdots \nu_n}|$. Ist der affinoide Raum X in E^n eingebettet und ist $I \subset T_n$ das definierende Ideal, so ist die Strukturalgebra $A = T_n/I$ von X ebenfalls mit einer kanonischen Norm versehen; es ist für

$$\bar{f} \in A : \|\bar{f}\| = \inf_{a \in I} \|f - a\| ,$$

wobei f eine Fortsetzung von $\bar{f}$ auf E^n bezeichnet.

Die Strukturalgebra eines affinoiden Teilbereiches $E_\varepsilon \subset E^n$ der Form

$$E_\varepsilon = \{x \in E^n : |x_i| \leqq \varepsilon_i\} , \qquad \varepsilon_i \in |k^*| ,$$

ist mit einer kanonischen Norm versehen, für welche folgendes gilt: Ist $f = \sum f_{\nu_1 \cdots \nu_n} x_n^{\nu_n} \cdots x_n^{\nu_n} \in T_n$, so ist

$$\|f \mid E_\varepsilon\| = \max_\nu |f_{\nu_1 \cdots \nu_n}| \lambda_1^{\nu_1} \cdots \lambda_n^{\nu_n} ,$$

wobei $\lambda_i = \min\{1, \varepsilon_i\}$. Insbesondere sieht man, daß die Zuordnung $\varepsilon \to \|f \mid E_\varepsilon\|$ stetig ist.

HILFSSATZ 1. *Es sei X in* E^n *eingebettet und*

$$E_\varepsilon = \{x \in E^n : |c_i x_i| \leqq \varepsilon\}$$

mit $c_i \in k$, *sowie* $X_\varepsilon = X \cap E_\varepsilon$, $\varepsilon \in |\bar{k}^*|$. *Zu jeder affinoiden Funktion f auf* E^n *mit* $\|f \mid X_\varepsilon\| < 1$ *gibt es eine affinoide Funktion g auf* E^n *mit* $f \mid X = g \mid X$ *und* $\|g \mid E_\varepsilon\| < 1$.

Weiter gibt es ein $\varepsilon' > \varepsilon$, $\varepsilon' \in |\bar{k}|$, *für welches noch* $\|g \mid E_{\varepsilon'}\| < 1$ *gilt.*

BEWEIS. Es bezeichne I (bzw. I_ε) das Ideal in der Strukturalgebra von E^n (bzw. E_ε), das die Einbettung von X (bzw. X_ε) in E^n

(bzw. E_ε) definiert. Es gibt nach Definition der Norm ein $a \in I_\varepsilon$, so daß

$$\| (f - a) \mid E_\varepsilon \| < 1 .$$

Da I über E_ε ganz I_ε erzeugt und I endlich erzeugbar ist, lassen sich die affinoiden Funktionen aus I_ε durch affinoide Funktionen aus I approximieren. Insbesondere gibt es ein $a' \in I$ mit $\| a - a' \mid E_\varepsilon \| <$ 1. Es folgt mit $g := f - a'$:

$$\| g \mid E_\varepsilon \| < 1 ,$$

und der erste Teil der Behauptung ist gezeigt. Da die Zuordnung $\varepsilon \longrightarrow \| g \mid E_\varepsilon \|$ stetig ist und $|\bar{k}|$ in $\mathbf{R}_+$ dicht liegt, folgt auch die zweite Behauptung.

HILFSSATZ 2. *Es sei B die Strukturalgebra eines affinoiden Raumes Y und $\| \ \|$ eine Norm auf B, welche durch eine Einbettung von Y in einen E^n induziert wird. Ist dann $\{b_1, \cdots, b_n\}$ das affinoide Erzeugendensystem von B, welches durch die Einbettung bestimmt wird, und $b'_i \in B$ mit $\| b_i - b'_i \| < 1$, so ist auch $\{b'_1, \cdots, b_n\}$ ein Erzeugendensystem von B.*

BEWEIS. Es sei $\Phi: T_n \longrightarrow B$ der durch die Einbettung definierte Homomorphismus. Es sei $\xi_i(x) = x_i$ und damit $\Phi(\xi_i) = b_i$. Wir wählen $h_i \in T_n$ mit $\Phi(h_i) = b'_i$ und $\| h_i - \xi_i \| < 1$. Dann erzeugen die $\{h_i\}$ die Algebra T_n und daher erzeugt das System der $\{b'_i\}$ die Algebra B, siehe [5, p. 409].

Nun zum *Beweis des Satzes* 2. (1) Wir behandeln zunächst den Fall, daß $f_n = 1$ ist. Dann ist X_ε ein Weierstraßbereich in X, und es läßt sich jede affinoide Funktion auf Y_1 approximieren durch affinoide Funktionen auf X.

Nun sei $\{b_1, \cdots, b_m\}$ ein affinoides Erzeugendensystem der Algebra B der affinoiden Funktionen auf Y. Wir wählen die Einbettung von Y in E^{m+n}, die durch das Erzeugendensystem

$$\left\{b_1, \cdots, b_m, \frac{f_j}{c_j} \circ u\right\}, \; c_j \in k^* = k - \{0\}$$

mit $|c_j| \geqq |f_j|$, gegeben wird. Weiter sei

$$E_\varepsilon = E^m \times \left\{x \in E^n \colon |x_j| \leqq \frac{\varepsilon}{|c_j|}\right\} .$$

Es ist gerade $Y_\varepsilon = Y \cap E_\varepsilon$. Es gibt nun zu b_i affinoide Funktionen a_i auf X, so daß $\| (b_i - a_i \circ u) \mid X_1 \| < 1$ gilt. Da $h_i = b_i - a_i \circ u$

eine affinoide Funktion auf Y ist, gibt es nach Hilfssatz 1 eine affinoide Funktion g_i auf E^{m+n}, so daß $g_i \mid X = h_i$ und $\| g_i \mid E_1 \| < 1$ gilt. Weiter gibt es sogar ein $\varepsilon > 1$, so daß noch $\| g_i \mid E_\varepsilon \| < 1$ ist.

Es gibt eine natürliche Zahl l, so daß $\varepsilon^l = |c|$ mit $c \in k$. Aus Hilfssatz 2 folgt dann, daß $a_i \circ u \mid Y_\varepsilon$ und $(f_j^l/c) \circ u \mid Y_\varepsilon$ Erzeugende der Strukturalgebra von Y_ε sind. Jede affinoide Funktion auf Y_ε läßt sich also durch Funktionen $p \circ u$ approximieren, wobei p ein Polynom in den a_i und den f_j^l/c ist.

(2) Der Allgemeinfall läßt sich schnell auf den eben behandelten Fall zurückführen. Es sei

$$X^* = \{x \in X\colon |f_n(x)| \geqq \zeta\}$$

und $\zeta \in |k^*|$ so klein gewählt, daß es ein $\varepsilon_0 > 1$ gibt, so daß für alle $\varepsilon \leqq \varepsilon_0$ die Inklusion $X_\varepsilon \subset X^*$ gilt. Durch Division kann man nun das f_n in X^* zu 1 machen. Man kann dann (1) auf die Abbildung $Y^* = u^{-1}(X^*) \to X^*$ anwenden und erhält das Ergebnis.

4. In [9] hat Tate gezeigt, daß jede endliche Überdeckung eines affinoiden Raumes X durch Laurentbereiche X_i von X azyklisch ist. Aus diesem Satz erhält man als Folgerung

SATZ 3. *Jede endliche Überdeckung* $\mathfrak{U} = (X_i)$ *eines affinoiden Raumes* X *durch rationale Bereiche* X_i *von* X *ist ebenfalls azyklisch.*

BEWEIS. Es seien

$$X_i = \{x \in X\colon |f_{ij}(x)| \leqq |f_{in}(x)|\}\,, \qquad 1 \leqq i \leqq m\,,$$

f_{ij} affinoide Funktionen auf X und $\{f_{ij}\}$ für jedes feste i ohne gemeinsame Nullstelle. Wir setzen

$$U_i = \{x \in X\colon |f_{in}(x)| \geqq \varepsilon\}\,, \qquad V_i = \{x \in X\colon |f_{in}| \leqq \varepsilon\}$$

und wählen $\varepsilon \in |k|$ so klein, daß jedes X_i mit V_i einen leeren Durchschnitt hat. $\mathfrak{L}$ sei die Überdeckung von X durch alle Laurentbereiche L der Form

$$L = \bigcap\nolimits_{i \in I} U_i \cap \bigcap\nolimits_{j \notin I} V_j\,, \qquad I \text{ Teilmenge von } \{1, \cdots, m\}\,.$$

Für jeden solchen Bereich L ist $L \cap X_i$ ein Weierstraßbereich in L. Somit ist $\mathfrak{U} \mid L$ für jedes $L \in \mathfrak{L}$ eine endliche Überdeckung durch Weierstraßbereiche. Daher ist $\mathfrak{U} \mid L'$ für jeden affinoiden Bereich $L' \subset L$ und somit insbesondere für alle endlichen Durchschnitte von Elementen aus $\mathfrak{L}$ azyklisch.

Da $\mathfrak{L}$ selbst azyklisch ist, muß daher nach dem Satz von Leray auch $\mathfrak{U}$ azyklisch sein.

3. Beweis des Hauptsatzes

1. Es sei X ein affinoider Raum und A die Strukturalgebra von X. $A\langle\xi_1, \cdots, \xi_n\rangle$ sei die Algebra der strikt konvergenten Potenzreihen über A in den Unbestimmten $\xi_1, \cdots, \xi_n$; sie ist die Strukturalgebra des affinoiden Raumes $X \times E^n$.

Eine Potenzreihe

$$f = \sum f_{\nu_1 \cdots \nu_n} \xi_1^{\nu_1} \cdots \xi_n^{\nu_n} \in A\langle\xi_1, \cdots, \xi_n\rangle$$

heißt ξ_n-allgemein über jedem Grundpunkt $x \in X$ von der Ordnung $\leqq s$, wenn

$$f(x) = \sum f_{\nu_1 \cdots \nu_n}(x) \xi_1^{\nu_1} \cdots \xi_n^{\nu_n} \in k(x)\langle\xi_1, \cdots, \xi_n\rangle$$

stets ξ_n-allgemein von der Ordnung $\leqq s$ ist, vgl. [5, p. 402].

SATZ 1. *Es sei*

$$f = \sum f_{\nu_1 \cdots \nu_n} \xi_1^{\nu_1} \cdots \xi_n^{\nu_n} \in A\langle\xi_1, \cdots, \xi_n\rangle$$

über jedem Grundpunkt $x \in X$ ξ_n-allgemein von der Ordnung $\leqq s$. Dann bildet die Menge Z der Grundpunkte x, über denen $f(x)$ ξ_n-allgemein von der Ordnung s ist, einen rationalen Bereich von X.

BEWEIS. Wir stellen f als Potenzreihe in ξ_n dar:

$$f = \sum_{\mu=0}^{\infty} h_\mu \xi_n^\mu, \; h_\mu \in A\langle\xi_1, \cdots, \xi_{n-1}\rangle .$$

Der konstante Term der Potenzreihe h_μ sei a_μ; somit ist $a_\mu = f_{0\cdots 0\mu}$.

Bemerkung. Es sei

$$X_\nu := \left\{ x \in X : \begin{array}{ll} |a_\mu(x)| \leqq |a_\nu(x)| & \text{für } \mu \leqq \nu \\ |a_\lambda(x)| < |a_\nu(x)| & \text{für } \nu < \lambda \leqq s \end{array} \right\}$$

Dann besteht X_ν gerade aus den Punkten, über denen f ξ_n-allgemein von der Ordnung ν ist.

In der Tat! Ist $f(x) = \sum h_\mu(x)\xi_n^\mu$ ξ_n-allgemein von der Ordnung ν, so ist $\|h_\mu(x)\| \leqq \|h_\nu(x)\|$ für $\mu \leqq \nu$ und $\|h_\lambda(x)\| < \|h_\nu(x)\|$ für $\nu < \lambda \leqq s$. Außerdem gilt für den konstanten Term $a_\nu(x)$ von $h_\nu(x)$:

$$|a_\nu(x)| = \|h_\nu(x)\| .$$

Da trivialerweise $|a_\mu(x)| \leqq \|h_\mu(x)\|$ für alle Indizes μ, gilt ersichtlich $x \in X_\nu$. Da zudem $X_\nu \cap X_\mu = \emptyset$ für $\nu \neq \mu$, ist die Bemerkung bewiesen.

Satz 1 ist vollständig bewiesen, sobald gezeigt ist, daß die affinoiden Funktionen a_μ, $0 \leqq \mu \leqq s$, keine gemeinsame Nullstelle x besitzen. Dies ist aber klar, da in einer solchen Stelle x die Potenzreihen $h_\mu(x)$ für $\mu \leqq s$ keinen konstanten Term besitzen können.

SATZ 2. *Es sei*

$$f = \sum f_{\nu_1 \cdots \nu_n} \xi_1^{\nu_1} \cdots \xi_n^{\nu_n} \in A\langle \xi_1, \cdots, \xi_n \rangle ;$$

die affinoiden Funktionen $f_{\nu_1 \cdots \nu_n}$ mögen keine gemeinsame Nullstelle in X haben. Dann gibt es eine A-Algebra-Scherung

$$\sigma \colon A\langle \xi \rangle \to A\langle \xi \rangle ,$$

so daß die Potenzreihe $\sigma(f)$ ξ_n-allgemein über jedem Grundpunkt $x \in X$ von beschänkter Ordnung ist.

BEWEIS. Es sei v_x der größte Index an einem Koeffizienten $f_{\nu_1 \cdots \nu_n}(x)$ mit $\| f(x) \| = | f_{\nu_1 \cdots \nu_n}(x) |$. Wir zeigen, daß

$$v \colon = \sup_{x \in X} v_x < \infty .$$

Da die Funktionen f_ν mit $\nu = (\nu_1, \cdots, \nu_n)$ keine gemeinsame Nullstelle besitzen, gibt es affinoide Funktionen r_ν auf X, $| r_\nu | \leqq 1$, und eine endliche Summe $\sum r_\mu f_\mu = a \in k$, $a \neq 0$. Daher ist

$$\inf_{x \in X} (\max_{\mu \in \mathbf{N}^n} | f_\mu(x) |) \geqq | a | > 0 .$$

Da fast alle Koeffizienten f_μ dem Betrag nach $< | a |$ sind, muß $v < \infty$ gelten.

Der Automorphismus σ wird nun wie in [5, p. 405], definiert:

$$\begin{aligned} \sigma(\xi_1) &= \xi_1 + \xi_n^{c_1} \\ &\vdots \\ \sigma(\xi_{n-1}) &= \xi_{n-1} + \xi_n^{c_{n-1}} \\ \sigma(\xi_n) &= \xi_n \end{aligned}$$

mit $c_{n-1} > v$, $c_{n-2} > v(c_{n-1} + 1)$, $\cdots$, $c_1 > v(c_2 + \cdots + c_{n-1} + 1)$. σ induziert $k(x)$-Scherungen $\sigma(x)$ auf $k(x)\langle \xi_1, \cdots, \xi_n \rangle$. Aus [5, p. 405], folgt, daß $\sigma(f)(x) = \sigma(x)(f(x))$ stets ξ_n-allgemein von einer Ordnung $\leqq (\sum_{\nu=1}^{n-1} c_\nu + 1)v$ ist.

2. Im folgenden sei $u \colon Y \to X$ eine fest gegebene affinoide Abbildung. Wir betrachten Einbettungen $v \colon Y \to X \times E^n$ über X; das sind solche Einbettungen v, für die $u = p \circ v$ gilt, wobei

$$p \colon X \times E^n \to X$$

die kanonische Projektion auf X bezeichnet.

SATZ 3. *Es sei* $v\colon Y \to X \times E^n$ *eine Einbettung über* X. *Weiter sei vorausgesetzt, daß die Dimension jeder Faser* $u^{-1}(x)$ *für jedes* $x \in X$ *kleiner als* n *ist. Dann gibt es eine affinoide Funktion* $f = \sum f_\nu \xi^\nu$ *auf* $X \times E^n$, *welche auf* $v(Y)$ *verschwindet und deren Koeffizienten* f_ν *keine gemeinsame Nullstelle besitzen. Weiter gibt es eine biaffinoide Transformation* $w\colon X \times E^n \to X \times E^n$ *über* X, *so daß danach* f *über jedem Grundpunkt* ξ_n*-allgemein von beschränkter Ordnung ist.*

BEWEIS. Es sei $f = \sum f_\nu \xi^\nu$ eine affinoide Funktion auf $X \times E^n$, welche auf $v(Y)$ verschwindet und N die gemeinsame Nullstellenmenge in X der Funktionen f_ν. Nun ist N bereits die gemeinsame Nullstellenmenge von endlich vielen Funktionen $f_{\nu_1}, \ldots, f_{\nu_t}$. Es sei m die größte Zahl, die in den Indextupeln $\nu_1, \cdots, \nu_t$ vorkommt. Falls $N \neq \emptyset$, wählen wir ein $x \in N$. Wegen der Bedingung über die Dimension der Fasern $u^{-1}(x)$ gibt es eine affinoide Funktion $g = \sum g_\nu \xi^\nu$ auf $X \times E^n$, die auf $v(Y)$ verschwindet und deren Koeffizienten g_ν nicht sämtlich auf x verschwinden. Wir setzen

$$h = f + \xi_1^m \cdot \cdots \cdot \xi_n^m \cdot g$$

und sehen, daß h auf $v(Y)$ verschwindet und die gemeinsame Nullstellenmenge der Koeffizienten h_ν von h in $N - \{x\}$ enthalten ist. Da eine echt absteigende Kette von abgeschlossenen Unterräumen schließlich bei der leeren Menge anlangt, gibt es eine affinoide Funktion f, wie im Satz 3 behauptet. Die weitere Behauptung folgt unmittelbar aus Satz 2.

Wir sagen, die Einbettungsdimension $\langle Y\colon X \rangle$ von Y über X ist $\leqq n$, falls es eine abgeschlossene Einbettung $v\colon Y \to X \times E^n$ über X gibt. Insbesondere gilt:

KOROLLAR. *Es sei* $\langle Y\colon X \rangle \leqq n$ *und die Dimension jeder Faser* $u^{-1}(x) < n$, $n \geqq 1$. *Dann gibt es eine abgeschlossene Einbettung* $v\colon Y \to X \times E^n$ *über* X *und eine affinoide Funktion* f *auf* $X \times E^n$, *welche auf* $v(Y)$ *verschwindet und* ξ_n*-allgemein von beschränkter Ordnung über jedem Grundpunkt* $x \in X$ *ist.*

3. Es sei $u\colon Y \to X$ wie in Abschnitt 2 eine fest vorgegebene affinoide Abbildung.

SATZ 4. *Ist* $v\colon Y \to X \times E^1$ *eine Einbettung über* X *und* f *eine affinoide Funktion auf* $X \times E^1$, *die auf* $v(Y)$ *verschwindet* *und die über jedem Grundpunkt $x \in X$ ξ-allgemein von der Ordnung s ist, dann ist u endlich.*

Beweis. Es sei A die Algebra der affinoiden Funktionen auf X. Für die Potenzreihe $f = \sum_{\nu=0}^{\infty} f_\nu \xi^\nu$ gilt unter den Voraussetzungen des Satzes:

$$f_s \text{ ist Einheit in } A\,,$$

$$\left|\frac{f_\nu}{f_s}\right| \leqq 1 \qquad \text{für } \nu \leqq s\,,$$

$$\left|\frac{f_\nu}{f_s}\right| < 1 \qquad \text{für } \nu > s\,.$$

Hierbei bezeichnet $|\ |$ die Seminorm der gleichmäßigen Konvergenz auf X. Da die Seminorm der gleichmäßigen Konvergenz auf A vollständig ist, [2], gilt für die Potenzreihe $f \cdot f_s^{-1}$ die Weierstraßsche Formel. Das bedeutet, daß $A\langle\xi\rangle/(f\cdot f_s^{-1}) = A\langle\xi\rangle/(f)$ ein endlicher A-Modul ist. Da die Algebra B der affinoiden Funktionen auf Y ein epimorphes Bild von $A\langle\xi\rangle/(f)$ ist, muß u endlich sein.

Aus § 1 folgt:

Korollar 1. *Ist in Satz 4 zusätzlich u als eineindeutige Immersion vorausgesetzt, so ist u eine Einbettung.*

Korollar 2. *Es sei $u\colon Y \to X$ eine eineindeutige Immersion mit $\langle Y\colon X\rangle \leqq n$, $n \geqq 1$, und $v\colon Y \to X \times E^n$ eine Einbettung von Y über X, sowie f eine affinoide Funktion auf $X \times E^n$, die auf $v(Y)$ verschwindet und über jedem Grundpunkt $x \in X$ ξ_n-allgemein von der Ordnung $\leqq s$ ist. Weiter sei Z der rationale Bereich von X der Grundpunkte, über denen f ξ_n-allgemein von der Ordnung s ist. Dann gilt für die von u induzierte Abbildung $u^{-1}(Z) \to Z$:*

$$\langle u^{-1}(Z)\colon Z\rangle \leqq n - 1\,.$$

Beweis. Man wendet Satz 4 an auf $Z \times E^{n-1}(\xi_1, \cdots \xi_{n-1})$ anstelle X und auf $u^{-1}(Z)$ anstelle Y. Man erhält, daß $u^{-1}(Z) \to Z \times E^n$ eine Einbettung ist.

4. Wir schicken dem unten zu beweisenden Hauptsatz dieses Abschnitts einen Hilfssatz voraus.

Hilfssatz 1. *Es sei Z ein rationaler Bereich des affinoiden Raumes X; es sei etwa*

$$Z = \{x \in X\colon |\,a_j(x)\,| \leqq |\,a_n(x)\,|\}$$

mit affinoiden Funktionen a_j, $0 \leqq j \leqq n$, ohne gemeinsame Nullstelle in X. Weiter sei

$$\mathring{Z} = \{x \in X: |\, a_j(x)\,| < |\, a_n(x)\,|\} \subset Z\,.$$

Dann läßt sich $X - \mathring{Z}$ als endliche Vereinigung von rationalen Bereichen von X erhalten.

BEWEIS. Wir setzen

$$X_i = \{x \in X: |\, a_j(x)\,| \leqq |\, a_i(x)\,| \text{ für alle } j\}\,.$$

Dann ist $Z = X_n$ und das System der X_i überdeckt ganz X. Da

$$X_i \cap X_j = \{x \in X: |\, a_k(x)\,| \leqq |\, a_i(x)\,| = |\, a_j(x)\,| \text{ für alle } k\}\,,$$

ist $X_i \cap \mathring{Z} = \emptyset$ für $i \neq n$. Ist $x \in Z$, $x \notin \mathring{Z}$, so gibt es ein i, so daß $|\, a_i(x)\,| = |\, a_n(x)\,|$. Somit ist $x \in X_i \cap X_n \subset X_i$. Also ist

$$X - \mathring{Z} = \bigcup_{i=0}^{n-1} X_i\,.$$

SATZ 5. *Es sei $u: Y \longrightarrow X$ eine eineindeutige Immersion. Dann gibt es eine endliche Überdeckung $\mathfrak{U} = (X_i)$ von X durch rationale Bereiche X_i von X, so daß die von u induzierten Abbildungen $u_i: Y_i = u^{-1}(X_i) \longrightarrow X_i$ für alle i Rungesche Immersionen sind.*

BEWEIS. Es sei $n = \langle Y: X \rangle$. Falls $n = 0$, ist die Aussage des Satzes trivial; sei daher $n \geqq 1$. Nach Satz 3 wählen wir eine Einbettung $v: Y \longrightarrow X \times E^n$ über X und eine affinoide Funktion f auf $X \times E^n$, die auf $v(Y)$ verschwindet und die in jedem Grundpunkt $x \in X$ ξ_n-allgemein von der Ordung $\leqq s$ ist. Nach Korollar 2 zu Satz 4 ist die Menge der Punkte $x \in X$, in denen $f(x)$ ξ_n-allgemein von der Ordnung s ist, ein rationaler Bereich Z von X und für die von u induzierte Abbildung $u^{-1}(Z) \longrightarrow Z$ gilt:

$$\langle u^{-1}(Z): Z \rangle \leqq n - 1\,.$$

Führt man daher Induktion über n, kann man auf Z die Induktionsannahme anwenden. Somit gibt es eine endliche Überdeckung $\mathfrak{B} = \{Z_j\}$ von Z durch rationale Bereiche Z_j von Z, so daß die von u induzierten Abbildungen $u^{-1}(Z_j) \longrightarrow Z_j$ Rungesche Immersionen sind. Da Z ein rationaler Bereich von X ist, sind die Z_j jedoch auch rationale Bereiche von X.

Seien etwa

$$Z_j = \{x \in X: |\, a_{jk}(x)\,| \leqq |\, a_j(x)\,|\}$$

und a_{jk}, a_j affinoide Funktionen auf X. Dann gibt es nach Satz 2.2 ein $\delta_j > 1$, $\delta_j \in |\bar{k}|$, derart daß für

$$Z'_j := \{x \in X\colon |a_{jk}(x)| \leqq \delta_j |a_j(x)|$$

gilt:

Die von u induzierte Abbildung $u^{-1}(Z'_j) \to Z'_j$ ist eine Rungesche Immersion.

Wir definieren

$$\mathring{Z}'_j := \{x \in X\colon |a_{jk}(x)| < \delta_j |a_j(x)|\} .$$

Die $\mathring{Z}'_j$ enthalten Z_j und überdecken deshalb Z. Wir stellen sodann $X - \mathring{Z}'_j$ als endliche Vereinigung $\mathfrak{W}_j$ von rationalen Bereichen in X dar. Wir setzen $\mathfrak{W} = \Pi \mathfrak{W}_j$; d.h. $\mathfrak{W}$ enthält gerade alle Durchschnitte $\bigcap_j W_{jk_j}$ mit $W_{jk_j} \in \mathfrak{W}_j$. Wir betrachten dann die Bereiche W_k aus $\mathfrak{W}$. Es gilt ersichtlich $W_k \cap Z = \varnothing$, da $W_k \cap Z_i = \varnothing$ für alle i ist. Die von u induzierte Abbildung $u^{-1}(W_k) \to W_k$ ist wieder eine eineindeutige Immersion. Ist daher f' die Beschränkung der affinoiden Funktion f auf $W_k \times E^n \subset X \times E^n$, so gilt für jeden Grundpunkt $x \in W_k$: $f'(x) = f(x)$ ist ξ_n-allgemein von einer Ordnung $\leqq s - 1$, falls nicht der triviale Fall $s = 0$ vorlag, aus dem $Y \cap u^{-1}(x) = \varnothing$ folgt. Führt man nun Induktion über s (für festes n), so darf man annehmen, da $\langle u^{-1}(W_k)\colon W_k\rangle \leqq n$, daß es eine endliche Überdeckung $\mathfrak{V}_k = (X_{km})$ von W_k durch rationale Bereiche X_{km} von W_k gibt, so daß die von u induzierten Abbildungen $u^{-1}(X_{km}) \to X_{km}$ Rungesche Immersionen sind. Faßt man dann die Überdeckungen $\mathfrak{V}$ und $\mathfrak{V}_k$ zu einer Überdeckung $\mathfrak{U} = (X_i)$ von X zusammen, so gilt wegen der Transitivitätseigenschaft der rationalen Bereiche der Satz für die Überdeckung $\mathfrak{U}$.

5. Wir ziehen eine einfache

Folgerung. Es sei u: $Y \to X$ eine eineindeutige offene Immersion. Es gibt eine endliche Überdeckung $\mathfrak{U} = (X_i)$ von X durch rationale Bereiche X_i von X, so daß $u^{-1}(X_i)$ vermöge der durch u induzierten Immersion

$$u^{-1}(X_i) \to X_i$$

ein Weierstraßbereich von X_i ist.

Beweis. Nach § 2.3 sind die eineindeutigen offenen Immersionen, welche gleichzeitig Rungesch sind, kanonische Immersionen von Weierstraßbereichen.

SATZ 6. *Es sei $Y \subset X$ ein affinoider Bereich von X. Dann ist Y die Vereinigung von endlich vielen rationalen Bereichen von X und es folgt Satz* 1.1.

BEWEIS. Es gibt nach obiger Folgerung endlich viele rationale Bereiche Y_i von X mit $Y = \bigcup Y_i$. Faßt man die Y_i zu einer Überdeckung $\mathfrak{Y}$ von Y zusammen, so gilt: $H^\circ(\mathfrak{Y}, H)$ ist kanonisch isomorph zur Strukturalgebra von Y. Da die Strukturalgebren der rationalen Bereiche Y_i eindeutig durch die ihnen zugrundeliegenden Punktmengen bestimmt sind, gilt dasselbe auch für Y.

Folgerung. Eine bianalytische affinoide Abbildung (=bijektive offene Immersion) ist biaffinoid.

6. Der Beweis des Satzes 1.2, d.h. der Azyklizität der endlichen affinoiden Überdeckungen, ergibt sich nunmehr, wenn man nochmals den Satz von Leray anwendet.

BEWEIS. Zu der endlichen affinoiden Überdeckung $\mathfrak{B} = \{Y_j\}$ wählen wir gemäß der Folgerung zu Satz 5 Überdeckungen $\mathfrak{U}_j = \{X_{ji}\}$ durch rationale Bereiche X_{ji} von X, so daß $\mathfrak{U}_j \mid Y_i$ eine endliche Überdeckung von Y_j durch Weierstraßbereiche von Y_j ist. Es sei weiter $\mathfrak{U} = \Pi\mathfrak{U}_i$. Dann ist auch für jeden affinoiden Bereich $Z \subset Y_j$ und somit insbesondere für endliche Durchschnitte von Bereichen Y_j die Beschränkung $\mathfrak{U} \mid Z$ eine Überdeckung von Z durch Weierstraßbereiche von Y_j, und daher azyklisch. Mit dem Satz von Leray folgt wieder die Azyklizität von $\mathfrak{U}$.

Als weitere Anwendung sei der Cartansche Verheftungssatz für endliche affinoide Überdeckungen angeführt, vgl. [8].

SATZ 7. *Es sei S eine analytische Garbe über einem affinoiden Raum X und $\mathfrak{U} = (X_i)$ eine endliche affinoide Überdeckung von X. Es seien Moduln $M_i \subset \Gamma(X_i, S)$ gegeben, so daß $(S \mid X_i, M_i)$ stets eine kohärente affinoide Garbe über X_i ist. Weiter gelte*

$$M_i \mid X_{ij} = M_j \mid X_{ij} .$$

Dann gibt es genau einen Modul $M \subset \Gamma(X, S)$ mit $M \mid X_i = M_i$, so daß (S, M) über X kohärent ist.

BEWEIS. Aus Satz 6 folgt insbesondere, daß jede endliche affinoide Überdeckung zulässig ist im Sinne von [8, p. 258]. Daher folgt der Satz aus [8, Th. 1.2].

4. Holomorphe Räume

Da jetzt die Azyklizität der affinoiden Überdeckungen affinoider Räume nachgewiesen ist, kann man die Definition des "rigiden analytischen Raumes", wie sie in [9] und [7] durchgeführt wurde, vereinfachen. Wir nennen "rigide analytische Räume" fortan *holomorphe Räume*. Unser Begriff dürfte nicht ganz mit dem Tateschen übereinstimmen.

1. Es sei X ein topologischer Raum und H eine Garbe von lokalen $\bar{k}$-Algebren auf X. Unter einem *affinoiden Bereich in* X versteht man eine offene Teilmenge $X_1 \subset X$ zusammen mit einer Unteralgebra $A_1 \subset \Gamma(X_1, H_1)$, wobei $H_1 = H \mid X_1$, so daß (X_1, H_1, A_1) ein affinoider Raum ist. Zwei affinoide Bereiche (X_1, H_1, A_1) und (X_2, H_2, A_2) in X heißen miteinander *verträglich*, wenn $X_{12} = X_1 \cap X_2$ leer ist oder wenn X_{12} affinoider Teilbereich von X_1 und X_2 ist. Es ist gefordert, daß X_{12} als affinoider Teilbereich von X_1 und als affinoider Teilbereich von X_2 die gleiche Strukturalgebra A trägt und daß A von $\mathring{A}_1 \mid X_{12}$, $\mathring{A}_2 \mid X_{12}$ erzeugt wird. Dabei bezeichnet $\mathring{A}_\nu$ die Menge

$$\{f \in A_\nu : |f(x)| \leqq 1, x \in X_\nu\} .$$

Daß A erzeugt wird, bedeutet, daß sich jedes $f \in A$ durch eine strikt konvergente Reihe $f = \sum a_\nu f_1^{\nu_1} \cdots f_m^{\nu_m}$ mit $a_\nu \in k$ und

$$f_\lambda \in \mathring{A}_1 \mid X_{12} \cup \mathring{A}_2 \mid X_{12}$$

darstellen läßt.

Unter einem *affinoiden Atlas* in X versteht man schließlich eine Menge $\mathfrak{A} = \{X_i\}$ von affinoiden Bereichen, so daß folgendes gilt:

(1) $\cup X_i = X$,

(2) X_i, X_j sind stets miteinander verträglich,

(3) Jedes X_i hat höchstens mit endlich vielen anderen X_j einen nichtleeren Durchschnitt.

Ein Atlas $\mathfrak{A}^* = \{X_j^*\}$ heißt eine *zulässige Verfeinerung* eines Atlas $\mathfrak{A} = \{X_i\}$, wenn die Überdeckung $\mathfrak{A}^*$ eine Verfeinerung von $\mathfrak{A}$ ist, jedes X_i höchstens mit endlich vielen X_j^* einen nichtleeren Durchschnitt hat, und $X_j^* \subset X_{\tau(j)}$ stets affinoider Teilbereich ist. Dabei bezeichnet τ die Verfeinerungsabbilung der Indexmengen. Zwei Atlanten $\mathfrak{A}_1$ und $\mathfrak{A}_2$ in X heißen nun *äquivalent*, wenn sie eine gemeinsame zulässige Verfeinerung besitzen. Nach dieser Definition ist jede zulässige Verfeinerung von $\mathfrak{A}$ mit $\mathfrak{A}$ äquivalent. Man zeigt leicht, daß unser Begriff "äquivalent" den Axiomen der Äquivalenzrelation genügt. Eine Äquivalenzklasse von äquivalenten Atlanten auf X nennt man sodann eine *holomorphe Struktur*, und ein $\bar{k}$-beringter Raum zusammen mit einer holomorphen Struktur $\mathfrak{S}$ heißt ein *holomorpher Raum*.

Es sei $X' \subset X$ eine offene Teilmenge. Wir setzen $H' = H \mid X'$ und nennen einen Atlas $\mathfrak{A}'$ auf X' einen von X *induzierten* Atlas, wenn es zu endlich vielen $X_1', \cdots, X_l' \in \mathfrak{A}'$ stets einen Atlas $\mathfrak{A}$ in der holomorphen Struktur $\mathfrak{S}$ von X gibt, so daß $X_1', \cdots, X_l' \in \mathfrak{A}$. Das Paar (X', H') heißt ein *holomorpher Teilbereich* von X, wenn eine holomorphe Struktur $\mathfrak{S}'$ auf X' gegeben ist, die beliebig feine von X induzierte Atlanten enthält. Natürlich ist S' durch die Menge X' keineswegs eindeutig bestimmt.

Auf holomorphen Räumen kennt man den Begriff der *holomorphen Funktion*. Das sind Schnittflächen $f \in \Gamma(X, H)$, deren Beschränkung auf jedes $X_1 \in \mathfrak{A} \in \mathfrak{S}$ eine affinoide Funktion ist. Ist H torsionsfrei, so hat man auf X auch den Begriff der *meromorphen Funktion*. Zunächst bildet man die Quotientengarbe Q von H nach dem multiplikativen System der kompletten Elemente[2]. Eine Schnittfläche $f \in \Gamma(X, Q)$ heißt dann eine meromorphe Funktion, wenn ihre Beschränkung auf jedes $X_1 \in \mathfrak{A} \in \mathfrak{S}$ als Quotient zweier affinoider Funktionen darstellbar ist.

2. Es sei (X, H, A) ein affinoider Raum. Wir sagen, daß eine Teilmenge D ganz im Innern von X liegt (in Zeichen $D \Subset X$), wenn für jedes $f \in \mathring{A}$ die Bildmenge $f(D)$ in der Vereinigung von endlichen vielen Kreisen $\{|x - x_0| \leqq q < 1\}$ liegt. Ist $\mathfrak{A} = \{X_i\} \in \mathfrak{S}$ ein Atlas, $\mathfrak{A}^* = \{X_j^*\} \in \mathfrak{S}$ eine zulässige Verfeinerung von $\mathfrak{A}$, so heißt $\mathfrak{A}^*$ eine *echte Schrumpfung* von $\mathfrak{A}$, wenn es eine Verfeinerungsabbildung τ gibt, so daß stets $X_j^* \Subset X_{\tau(j)}$ ist. Im allgemeinen gibt es zu einem Atlas keine echte Schrumpfung. Wir nennen einen holomorphen Raum *offen*, wenn ein Atlas $\mathfrak{A} \in \mathfrak{S}$ existiert, der eine echte Schrumpfung gestattet. Besitzt X sogar einen echt schrumpfbaren endlichen Atlas, so heißt X ein *geschlossener holomorpher Raum*.

Man kann auf holomorphen Räumen X die Theorie der kohärenten analytischen Garben durchführen (Man vgl. [8]). Es sei S eine analytische Garbe über X. Die Garbe S ist also eine Garbe von H-Moduln. Ist $X_1 \in \mathfrak{A} \in \mathfrak{S}$ ein affinoider Bereich, so sei stets ein $A(X_1)$-Untermodul $M_1 \subset \Gamma(X_1, S)$ gegeben. Gilt

$$X_1 \in \mathfrak{A} \in \mathfrak{S},\ X_2 \in \mathfrak{A} \in \mathfrak{S} \quad \text{und} \quad X_{12} = X_1 \cap X_2 \neq \emptyset\ ,$$

---

[2] Ein Element aus H_x heißt komplett, wenn sein Bild in der Reduktion von H_x Nichtnullteiler ist.

so sei stets $M_1 \mid X_{12} = M_2 \mid X_{12}$. Ist $\mathfrak{A}^*$ eine zulässige Verfeinerung von $\mathfrak{A}$, so sei $M_j^* = M_{\tau(j)} \mid X_j^*$. Wir nennen dann S eine *holomorphe Garbe* auf X. Wir können auf übliche Weise die Koketten $C^l(\mathfrak{A}, S)$ für jedes $\mathfrak{A} \in \mathfrak{S}$ einführen, ebenso die Čechschen Kohomologiegruppen $H^l(\mathfrak{A}, S)$. Durch Übergang zum induktiven Limes erhält man schließlich $H^1(X, S)$.

Die Garbe S heißt *kohärent*, wenn für jedes $X_1 \in \mathfrak{A} \in \mathfrak{S}$ die affinoide Garbe $(S \mid X_1, M_1)$ kohärent ist. R. Kiehl hat gezeigt: Es sei

$$\mathfrak{A} = \{X_i\} \in \mathfrak{S} \quad \text{und} \quad M_i \subset \Gamma(X_i, S \mid X_i)$$

ein A_i-Untermodul, so daß $(S \mid X_i, M_i)$ kohärente affinoide Garben sind. Es gelte stets $M_i \mid X_{ij} = M_j \mid X_{ij}$. Dann gibt es eine eindeutig bestimmte kohärente Garbe über X, die für jedes $X_i \in \mathfrak{A}$ den Schnittmodul M_i besitzt. Man vgl. Kiehl [8, Th. 1.2 und § 3, Satz 9] und den Satz 7 aus § 3.

Für geschlossene Räume gilt der Endlichkeitssatz, vgl. [7].

SATZ 1. *Ist X ein geschlossener holomorpher Raum und S eine kohärente Garbe über X, so gilt*

$$d_l = \dim_k H^l(X, S) < \infty$$

und $d_l = 0$ für $l \geqq l_0$.

3. Es sei X ein algebraischer Raum über k (=variété algébrique im Sinne von A. Weil). Wir beschränken uns auf diejenigen Punkte von X, die über $\bar{k}$ definiert sind. X besitzt eine affine Überdeckung mit endlich vielen affinen Teilräumen X_i. Jedes X_i kann in einen $\bar{k}^{n_i}$ eingebettet werden. Der Unterraum $X_i \subset \bar{k}^{n_i}$ ist über k definiert. Es sei $a_i \in k^*$ und

$$E_{a_i}^{n_i} = \{(x_1, \cdots, x_{n_i}) \colon |x_\nu| \leqq |a_i|\} .$$

$E_{a_i}^{n_i}$ ist dann (wie E^{n_i}) ein affinoider Raum und $X_i(a_i) = X_i \cap E_{a_i}^{n_i}$ ein in $E_{a_i}^{n_i}$ eingebetteter affinoider Raum. Ist nun X vollständig und sind die $|a_i|$ hinreichend groß, so bilden—wie man zeigen kann—die $X_i(a_i)$ einen (endlichen, echt schrumpfbaren) Atlas in X, der eine holomorphe Struktur bestimmt. Diese ist unabhängig von der Wahl der Überdeckung $\{X_i\}$. Der Raum X zusammen mit dieser Struktur ist ein (geschlossener) holomorpher Raum. Er sei mit X_a bezeichnet. Jede rationale Funktion auf X wird zu einer meromorphen Funktion auf X_a (Es muß dabei natürlich vorausgesetzt werden, daß die Strukturgarbe H von X_a torsionsfrei ist).

Ist X nicht vollständig, so kann man X als offene Teilmenge eines vollständigen algebraischen Raumes $\hat{X}$ erhalten. Die Differenz $\hat{X} - X$ ist eine niederdimensionale algebraische Teilmenge von $\hat{X}$. Nun trägt, wie vorhin gezeigt, $\hat{X}$ eine holomorphe Struktur, die $\hat{X}$ zu einem holomorphen Raum $\hat{X}_a$ macht. Man kann leicht zeigen, daß $X \subset \hat{X}_a$ ein holomorpher Teilbereich ist und daß die holomorphe Struktur $\mathfrak{S}$ von X unabhängig von $\hat{X}$ und eindeutig bestimmt ist, wenn man verlangt, daß jedes $X_1 \in \hat{\mathfrak{A}} \in \mathfrak{S}(\hat{X})$ mit $X_1 \subset X$ mit nur endlich vielen Elementen eines jeden $\mathfrak{A} \in \mathfrak{S}$ einen nichtleeren Durchschnitt hat.

Wir nennen einnen holomorphen Raum *algebraisch* (*projektiv algebraisch*), wenn er von einem algebraischen (projektiv algebraischen) Raum herkommt.

Aus dem Endlichkeitssatz folgt wie bei den komplexen Räumen:

SATZ 2. *Es sei X ein eindimensionaler geschlossener holomorpher Raum. Dann ist X projektiv algebraisch.*

Es soll jetzt das Beispiel eines 2-dimensionalen holomorphen Raumes X gegeben werden, der nur eine analytisch unabhängige meromorphe Funktion besitzt und deshalb nicht algebraisch sein kann. Im Anschluß an die komplexe Analysis nennen wir X eine *Hopfsche Mannigfaltigkeit.*

Es sei $a \in k$ und $0 < |a| < 1$. Wir betrachten die Ähnlichkeitstransformationen $x \longrightarrow a^s \cdot x$ des $\bar{k}^2$ in sich für $s \in \mathbf{Z}$. Diese bilden eine Gruppe Γ. Es sei $X = k^2 - \{0\}/\Gamma$. Die Menge X, versehen mit der Quotiententopologie, ist ein topologischer Raum. Die Quotientenabbildung $\mathfrak{q}\colon \bar{k}^2 - \{0\} \longrightarrow X$ ist lokaltopologisch. Liftet man die Strukturgarbe des $\bar{k}^2 - \{0\}$ nach X, so wird X zu einem $\bar{k}$-geringten Raum. Man kann $\mathfrak{q}$ dann als analytische Abbildung auffassen.

Wir definieren auf X einen Atlas, der aus 4 affinoiden Bereichen in X besteht. Der Einfachheit halber setzen wir voraus, daß die Wertegruppe $|k^*|$ dicht in

$$\mathbf{R}_+ = \{r \in \mathbf{R}\colon r > 0\}$$

liegt. Das ist z.B. der Fall, wenn k algebraisch abgeschlossen ist. Es sei $\varepsilon \in |k^*|$, $1 < \varepsilon < 1/\sqrt[4]{|a|}$. Es gilt dann

$$|a|/\varepsilon < \sqrt{|a|}/\varepsilon < \varepsilon\sqrt{|a|} < \varepsilon$$

und

$$\frac{|a|/\varepsilon}{\varepsilon\sqrt{|a|}} > |a| \quad \text{und} \quad \frac{\sqrt{a}}{\varepsilon} > |a|$$

Es sei

$$U_1 = \{(x_1, x_2) : |a|/\varepsilon \leqq |x_1| \leqq \varepsilon\sqrt{|a|}\} ;$$
$$U_2 = \{(x_1, x_2) : \sqrt{|a|}/\varepsilon \leqq |x_1| \leqq \varepsilon\} ,$$

U_2, U_3 mögen nach Vertauschung von x_1, x_2 entsprechend gebildet sein. Wir setzen $X_\nu = \mathfrak{q}(U_\nu) \subset X$. Offenbar bildet $\mathfrak{q}$ den affinoiden Raum U_ν bianalytisch auf X_ν ab. Das Bild der Strukturalgebra von U_ν in X_ν macht X_ν zu einem affinoiden Bereich in X. Man zeigt leicht, daß $\{X_1, \cdots, X_4\}$ ein Atlas $\mathfrak{A}$ in X ist. Die Menge aller zu $\mathfrak{A}$ äquivalenten Atlanten bildet eine holomorphe Struktur in X. Damit ist X zu einem holomorphen Raum gemacht.

Ist f eine meromorphe Funktion in X, so ist $\hat{f} = f \circ \mathfrak{q}$ eine meromorphe Funktion in $\bar{k}^2 - \{0\}$ (in Bezug auf die natürliche holomorphe Struktur des $\bar{k}^2$). $\hat{f}$ ist aber periodisch: $\hat{f}(a^s x_1, a^s x_2) = \hat{f}(x_1, x_2)$. Nun kann man nach [1] jede meromorphe Funktion $\hat{f}$ zu einer meromorphen Funktion $\bar{f}$ in den ganzen $\bar{k}^2$ fortsetzen. Es sei P die Polstellenmenge von $\bar{f}$ und $G \subset \bar{k}^2$ eine holomorphe Gerade $\{cx_1 + dx_2 = 0\}$, die P in 0 isoliert schneidet. $\bar{f} \mid G$ ist eine meromorphe Funktion auf G, die in der Nähe von 0 holomorph ist. Es muß dann $\bar{f} \mid G$ auf ganz G holomorph sein. Ebenso ist

$$f \mid \mathfrak{q}(G) = (\hat{f} \mid G) \circ \mathfrak{q}^{-1}$$

holomorph. Da $\mathfrak{q}(G)$ ein geschlossener holomorpher Raum ist, ist also $f \mid \mathfrak{q}(G)$ konstant. Schneidet G nicht isoliert, so gilt $G \subset P$. Mithin umfassen die Fasern von f gerade die holomorphen Mengen $\mathfrak{q}(G) \subset X$. Es folgt, daß zwei meromorphe Funktionen immer analytisch abhängig sind, q.e.d.

MATHEMATISCHES INSTITUT DER UNIVERSITÄT, MÜNSTER
MATHEMATISCHES INSTITUT DER UNIVERSITÄT, GÖTTINGEN

LITERATUR

[1] W. BARTENWERFER, Elementare Theorie der k-affinoiden Räume und ein Satz über meromorphe Fortsetzung, Dissertation, Göttingen, 1968.

[2] L. GERRITZEN, Die Norm der gleichmäßigen Konvergenz auf reduzierten affinoiden Algebren, J. reine angew. Math. 231, (1968) 114-120.

[3] ———, *On one-dimensional affinoid domains and open immersions*, Invent. Math. **5**, (1968) 106-119.

[4] H. GRAUERT, Affinoide Überdeckungen eindimensionaler affinoider Räume, Publ. Math. IHES, 34 (1968).

[5] ——— und R. REMMERT, Nichtarchimedische Funktionentheorie. Arbeitsgemeinschaft für Forschung des Landes Nordrhein-Westfalen, Bd. 33, p. 393-476, Opladen: Westdeutscher Verlag 1966.

[6] ———, On non-Archimedean analysis. Abstracts of reports. International Congress of Mathematicians 1966, Moscow, 49-51.

[7] R. KIEHL, *Der Endichkeitssatz für eigentliche Abbildungen in der nichtarchimedischen Funktionentheorie*, Invent. Math. **2**, (1967) 119-214.

[8] ———, *Theorem A und Theorem B in der nichtarchimedischen Funktionentheorie*, Invent. Math. **2** (1967), 256-273.

[9] J. TATE, Rigid analytic spaces. Private notes of J. Tate reproduced by IHES, 1962.

(Received August 9, 1968)

Hermitian differential geometry, Chern classes, and positive vector bundles

By PHILLIP A. GRIFFITHS*

0. Introduction, statement of results, and open questions

(a) *Statement of results.* Our purpose is to discuss the notion of *positivity* for holomorphic vector bundles. This paper is partly expository, and in so doing we hope to clarify, simplify, and unify, some of the existing material on the subject. There are several new results, mostly relating the analytic notion of positivity to the topological properties of the bundle. We also correct two errors in our previous paper [10].

We now give the main results to be proved in this paper.

Let V be a compact, complex manifold and $\mathbf{E} \to V$ a holomorphic vector bundle; we denote by $\Gamma(\mathbf{E})$ the space of holomorphic cross-sections of $\mathbf{E} \to V$. The relevant definitions are the following:

(0.1) $\mathbf{E} \to V$ is *positive* if there exists an hermitian metric in $\mathbf{E}$ whose *curvature tensor* $\Theta = \{\Theta^{\rho}_{\sigma i \bar{j}}\}$ has the property that the hermitian quadratic form

$$\Theta(\xi, \eta) = \sum_{\rho,\sigma,i,j} \Theta^{\rho}_{\sigma i \bar{j}} \xi^{\sigma} \bar{\xi}^{\rho} \eta^{i} \bar{\eta}^{j}$$

is positive definite in the two variables ξ, η;

(0.2) $\mathbf{E} \to V$ is *ample* if

(a) the *global sections generate each fibre*, so that we have $0 \to \mathbf{F}_z \to \Gamma(\mathbf{E}) \to \mathbf{E}_z \to 0$ (for all $z \in V$), and

(b) the natural mapping $\mathbf{F}_z \to \mathbf{E}_z \otimes \mathbf{T}^*_z$ is onto ($\mathbf{F}_z$ = sections of $\mathbf{E}$ vanishing at z);

(0.3) $\mathbf{E} \to V$ is *cohomologically positive* if, given any coherent sheaf $S \to V$, there is a $\mu_0 = \mu_0(S)$ such that $H^q(V, \mathcal{O}(E^{(\mu)}) \otimes S) = 0$ for $q > 0$, $\mu \geqq \mu_0$; and

(0.4) $\mathbf{E} \to V$ is *numerically positive* if, for any complex analytic subvariety $W \subset V$ and any quotient bundle $\mathbf{Q}$ of $\mathbf{E} \mid W$, we

* Work supported in part by National Science Foundation Grant GP7952X.

have $\int_W P(c_1, \cdots, c_s) > 0$ where $P(c_1, \cdots, c_s)$ is a *positive polynomial* in the *Chern classes* $c_1, \cdots, c_s$ of $\mathbf{Q} \to W$ (cf. § 5 (b)).

Our main general results are:

THEOREM A: *ample* $\Rightarrow$ *positive;*

THEOREM B: *positive* $\Rightarrow$ *cohomologically positive;*

THEOREM C: *positive* $\Rightarrow \mathbf{E}^{(\mu)}$ *ample for all* $\mu \geqq \mu_0$;

THEOREM D: *ample* $\Rightarrow$ *numerically positive;*

THEOREM E: *numerically positive* $\Rightarrow$ *cohomologically positive.*

THEOREM F: *cohomologically positive* $\Rightarrow \mathbf{E}^{(\mu)}$ *ample for* $\mu \geqq \mu_0$.

In summary, if we take the sequence of bundles $\mathcal{E} = (\mathbf{E}, \mathbf{E}^{(2)}, \cdots, \mathbf{E}^{(\mu)}, \cdots)$, then, modulo finitely many bundles, positive = ample = cohomologically positive.

Our principal specific results generalize the *Kodaira vanishing theorem* and the *Lefschetz theorem.* For example, we prove

THEOREM G. *If* $\mathbf{E} \to V$ *is generated by its sections, and if* $\mathbf{F} \to V$ *is a line bundle such that* $\mathbf{F}^* \otimes \mathbf{K} \otimes \det \mathbf{E} < 0$ *where* $\mathbf{K} \to V$ *is the canonical bundle, then:*

$$H^q(V, \mathcal{O}(\mathbf{E}^{(\mu)} \otimes \mathbf{F})) = 0 \qquad \textit{for } q > 0, \mu \geqq 0 .$$

The Kodaira theorem is the case $\mathbf{E}$ = trivial line bundle in Theorem G. In § 3 (b) (cf. (3.25)) we shall give a precise vanishing theorem which has Theorem G as a consequence.

As a generalization of the Lefschetz theorem, we assume that $\mathbf{E} \to V$ is a positive bundle with fibre $\mathbf{C}^r$ and where $\dim V = n$. Let $\xi \in H^0(V, \mathcal{O}(\mathbf{E}))$ be a holomorphic section whose zero locus $S \subset V$ is a smooth subvariety of codimension r.

THEOREM H. *We have* $H_q(S, \mathbf{Z}) \to H_q(V, \mathbf{Z}) \to 0$ *for* $q \leqq n - r$ *and* $0 \to H_q(S, \mathbf{Z}) \to H_q(V, \mathbf{Z}) \to 0$ *for* $q \leqq n - r - 1$.

The Lefschetz theorem is the case $r = 1$.

As an application of Theorem H, we prove in § 3 (d) (cf. (3.51)) that the cup product

$$(0.5) \qquad H^{p,q}(V) \xrightarrow{\omega} H^{p+r,q+r}(V), p + q = n - r ,$$

is an isomorphism where $\omega \in H^{r,r}(V)$ is the rth *Chern class* of $\mathbf{E} \to V$.

This result is the analogue of a well-known fact in Kähler

geometry [26].

We now give another generalization of the (*coarse*) *Kodaira vanishing theorem*. Let $\mathbf{E} \to V$ be a positive vector bundle with a non-singular section $\xi \in H^0(V, \mathcal{O}(\mathbf{E}))$; denote by I the *ideal sheaf* of the zero-locus S of ξ. We introduce the curvature form

$$(0.1)' \qquad \Theta^{\sharp}_{\mathbf{E}}(\xi, \eta) = (r+1) \sum_{\left\{\substack{\rho,\sigma \\ i,j}\right.} \Theta^{\rho}_{\sigma i} \xi_{\bar{j}}{}^{\sigma} \bar{\xi}^{\rho} \eta^{i} \bar{\eta}^{j} - \sum_{\left\{\substack{\rho,\sigma \\ i,j}\right.} \Theta^{\rho}_{\rho i \bar{j}} \xi^{\sigma} \bar{\xi}^{\sigma} \eta^{i} \bar{\eta}^{j} .$$

For the significance of this form, we refer to (3.25), Theorem G′ where it is proved that, if $\Theta^{\sharp}_{\mathbf{E}} > 0$, then $H^q(V, \mathcal{O}(\mathbf{E}^*)) = 0$ for $q \leqq n-1$. The Kodaira theorem is the case $r = 1$. Note that, if $\mathbf{L} \to V$ is a positive line bundle, then $\Theta^{\sharp}_{\mathbf{E}\otimes\mathbf{L}^{\mu}} > 0$ for μ sufficiently large; this is because of the $r+1$ factor in front of the first term.

THEOREM I. *Let* $\mathbf{F} \to V$ *be a holomorphic vector bundle. Then there exists a constant* $e = c(\mathbf{F}, V)$ *such that, if*

$$\Theta^{\sharp}_{\mathbf{E}}(\xi, \eta) > c\,|\,\xi\,|^2\,|\,\eta\,|^2 ,$$

then

$$H^q(V, I \otimes \mathcal{O}(\mathbf{F})) = 0 \qquad \textit{for } q \leqq n - r .$$

Another analogon of Kodaira's (coarse) vanishing theorem is

THEOREM J. *With the same assumption as in Theorem* I, *we have*

$$H^q(V, I^{\mu} \otimes \mathcal{O}(\mathbf{F})) = 0 \qquad \textit{for } \mu \geqq \mu_0(\mathbf{F}),\ q \leqq n - r .$$

As an unsolved problem, we would like to mention the following possible generalization of the *precise Kodaira vanishing theorem*.

(0.6) *Conjecture*. If $\mathbf{E} \to V$ is positive, then

$$H^q(V, \mathcal{O}(\mathbf{E}^*)) = 0 \qquad \textit{for } q \leqq n - r .$$

Remark. For $\mathbf{E} \to V$ a line bundle, (0.6) is just the Kodaira theorem. Taking $V = \mathbf{P}_2$ and $\mathbf{E} = \mathbf{T}(\mathbf{P}_2)$ the (positive) tangent bundle, $\mathcal{O}(\mathbf{E}^*) = \Omega^1$ and $H^1(\mathbf{P}_2, \Omega^1) \neq 0$, so that (0.6) is the best possible. We will prove the conjecture (cf. § 5 (e)) when $r = 2$ and $\mathbf{E} \to V$ has a non-singular section.

Another problem is

(0.7) *Conjecture*. If $\mathbf{E} \to V$ is positive, then $\mathbf{E} \to V$ is numerically positive.

Remark. We will prove (0.7) in case V is a surface ($n = 2$) and $\mathbf{E}$ has fibre dimension 2. This proof, given in the Appendix to

§ 5 (b), involves a Schwarz inequality for differential forms. In the context of algebraic geometry (characteristic zero), the assertion "$\mathbf{E} \to V$ cohomologically positive $\Rightarrow \mathbf{E} \to V$ numerically positive" has been proved for $r = 1$ (Nakai [21]), $n = 1$ and $r = 2$ (Hartshorne [14]), and $n = 2$ (Kleiman [17]). The first step in proving (0.7) would be to show that the *Chern classes* $c_q(\mathbf{E})$ of a positive bundle $\mathbf{E} \to V$ are positive. Still another problem we mention is

(0.8) *Problem.* Find a better definition of the cone of positive polynomials and prove that *cohomologically positive* $\Leftrightarrow$ *numerically positive.*

Remark. For $r = 1$, we have the theorem of Nakai [21] (cf. § 5 (c) below).

Another question is

(0.9) *Problem.* If $\mathbf{E} \to V$ is cohomologically positive, is $\mathbf{E} \to V$ positive?

If true, this would give the nicest general result, as we would have

(0.10) Positive (differential-geometric) $\Leftrightarrow$ cohomologically positive (algebro-geometric) $\Leftrightarrow$ numerically positive (topological) $\Leftrightarrow$ $\mathbf{E}^{\mu}$ ample for $\mu \geqq \mu_0$.

If $\mathbf{E} \to V$ is cohomologically positive, then there is a *non-linear positive metric* in $\mathbf{E}$ as follows: By sending ξ into $\underbrace{\xi \otimes \cdots \otimes \xi}_{\mu}$ (diagonal mapping), we have an embedding $\mathbf{E} \subset \mathbf{E}^{\mu}$, and $\mathbf{E}^{\mu}$ is ample for $\mu \geqq \mu_0$. Using this metric, a tubular neighborhood of the zero cross-section of $\mathbf{E}^* \to V$ is *strongly pseudo-convex* (cf. § 3 (a) below).

The reason that we use the differential-geometric notion of positivity (0.1) rather than the function-theoretic definition (cf. Grauert [9]) just mentioned is that the curvature is relevant for *precise vanishing theorems* and for the Lefschetz theorem, whereas pseudo-convexity will yield only coarse results. These precise vanishing theorems will have several uses in geometric problems; for example, Theorem G has been used by W. Schmid [23] to give a generalization of the *Borel-Weil theorem* to real, semi-simple Lie groups.

(b) *Complements and corrections to* [10]. The difficulty in [10] is that there were several definitions of positivity and ampleness given and these did not leave a clear picture of what positive and ample bundles should be. If we are not worried about the bundle $\mathbf{E} \to V$ itself but are content to take symmetric powers, then the various notions coincide, as indicated below Theorem F above. This is the approach taken by Hartshorne [14], whose definition of an ample bundle coincides with our cohomologically positive (cf. [10, Prop. (3.3)]).

It now seems that (0.1) is the best differential-geometric generalization of positivity for line bundles; other definitions are discussed in [10, § 3] and still another definition is given in [22]. Our positive here is the same as weakly positive in [10], and is a condition which turns up most naturally in geometric situations.

Our definition of ampleness (0.2) expresses the geometric assumption that $\mathbf{E} \to V$ should have "sufficiently many sections". In case $\mathbf{E}$ is a line bundle, "sufficiently many sections" means that the mapping into projective space is an immersion. However, in general the universal bundle over the grassmannian is *not* positive or ample (in any sense), and "sufficiently many sections" means that the immersion of V in a grassmannian is twisted.

The notion of sufficiently ample in [10] is not a good definition, nor is the definition of negative given above Proposition 4.1 in [10].

The definitions (0.3) and (0.4) of cohomologically positive and numerically positive are seemingly good notions.

The main error in [10] is Proposition 7.2. An application of this, Proposition 10.2 of [10], is incorrect, as the following example shows: Let $X = \mathbf{P}_2$ be projective 2-space and $\mathbf{T} \to X$ the tangent bundle. Then $\mathbf{T}$ is generated by its sections, as is $\Lambda^2\mathbf{T} = K^*$. Thus we have $0 \to \mathbf{F}^* \to \mathbf{E}^* \to \mathbf{K}^* \to 0$ where $\mathbf{E}$ is a trivial bundle; this dualizes to $0 \to \mathbf{K} \to \mathbf{E} \to \mathbf{F} \to 0$. Now $\mathbf{E}$ and $\mathbf{F}$ are generated by their sections and $\mathbf{K}$ is negative; if Proposition 10.2 were true, then $H^1(X, \mathcal{O}(\mathbf{F})) = 0 = H^2(X, \mathcal{O}(\mathbf{E}))$ and so $H^2(X, \mathcal{O}(\mathbf{K})) = H^2(X, \Omega^2) = 0$, which is absurd.

The mistake in the proof of Proposition 7.2 arises in equation (7.7), in which the row and column indices for the curvature matrix are interchanged. This error can be traced to just below equation (1.3), where the connexion form should be $\omega = \partial h \cdot h^{-1}$. With the correct formula $\Theta^{\rho}_{\sigma i \bar{j}} = -\sum A^{\alpha}_{\sigma i} \bar{A}^{\alpha}_{\rho j}$ (cf. equation (2.24) below), the curvature operator $\Theta(\xi, \xi)$, given by (4.2) in [10], on an $\mathbf{E}$-valued (0.1) form $\xi = \{\xi^{\rho}_{j} \omega^{j}\}$ becomes

$$-\sum A^\alpha_{\sigma i}\bar{A}^\alpha_{\rho i}\xi^\sigma_{\bar{j}}\xi^\rho_{\bar{j}} + \sum A^\alpha_{\sigma i}\bar{A}^\alpha_{\rho j}\xi^\sigma_{\bar{i}}\bar{\xi}^\rho_{\bar{j}}$$
$$= \sum_\alpha \{\text{Trace } (A^\alpha(\xi)^t \overline{A^\alpha(\xi)}) - |\text{ Trace } A^\alpha(\xi)\,|^2\}$$

where $A^\alpha(\xi)_{i\bar{j}} = \sum_\sigma A^\alpha_{\sigma i}\xi^\sigma_{\bar{j}}$, which is neither positive nor negative. With the incorrect formula, $\Theta(\xi, \xi) \leqq 0$ and this is the mistake.

The other error in [10] is Lemma 9.2, which is corrected in formula (2.38) below.

The corrected version of Proposition 10.2 in [10] is Theorem G above; and the corrected form of Proposition 7.2 is Theorem A above. The remaining results in [10] are presumably correct.

Referring again to Hartshorne's paper [14], he works on the associated projective bundle **P**(**E***), as was done in of [10, § 9] and is done here. Many of the general results of [10] on positive bundles, such as the fact that a quotient of a positive (ample) bundle is positive (ample), also appear in [14]. The Theorems D and E on numerical positivity, which are proved for ample bundles in [11], are given in [14, Prop. 7.5] for the case of V a curve and **E** a bundle with fibre $\mathbf{C}^2$. Interestingly, the proofs are quite similar; both use the numerical criterion of Nakai [21] on the associated projective bundle. Finally, as mentioned above, Kleiman has proved (0.7) in case $\dim V = 2$ [17].

1. Discussion of methods and calculations

Suppose that $\mathbf{E} \to V$ is ample and let $\mathbf{\Gamma}(\mathbf{E})$ be the trivial bundle $V \times \Gamma(\mathbf{E})$. Then there are exact sequences

$$0 \longrightarrow \mathbf{F} \longrightarrow \mathbf{\Gamma}(\mathbf{E}) \longrightarrow \mathbf{E} \longrightarrow 0\,, \tag{1.1}$$

and

$$\mathbf{F} \longrightarrow \mathbf{E} \otimes \mathbf{T}^* \longrightarrow 0\,, \tag{1.2}$$

where the fibre $\mathbf{F}_z = \{s \in \Gamma(\mathbf{E}) \colon s(z) = 0\}$. The flat metric in the trivial bundle $\mathbf{\Gamma}(\mathbf{E})$ induces a metric in **F** and, by orthogonality, a metric in **E**. A computation shows that the curvature matrix Θ of this metric in **E** has the local form

$$\Theta^\rho_\sigma = \sum_\alpha A^\alpha_\rho \wedge \bar{A}^\alpha_\sigma \tag{1.3}$$

where $A^\alpha_\sigma = \sum_i A^\alpha_{\sigma i}dz^i$ is essentially the differential of the mapping of V into a grassmannian. The quadratic form $\Theta(\xi, \eta)$ in (0.1) is then $\sum_\alpha |\, A^\alpha(\xi, \eta)\,|^2$ where $A^\alpha(\xi, \eta) = \sum_{\rho, i} A^\alpha_{\rho i}\bar{\xi}^\rho\eta^i$, and from (1.2) it will follow that $\Theta(\xi, \eta)$ is positive definite, which proves Theorem A.

The formula (1.3) for the curvature is a special case of studying the hermitian geometry of an exact bundle sequence

$$0 \longrightarrow \mathbf{S} \longrightarrow \mathbf{E} \longrightarrow \mathbf{Q} \longrightarrow 0 . \tag{1.4}$$

A metric in **E** induces metrics in **S** and **Q**, and the deviation of the induced connexion on **S** from the connexion of the induced metric is measured by an important invariant, the *second fundamental form of* **S** *in* **E**, introduced in [11, § VI. 3], and discussed in some detail below. This tensor has both geometric and cohomological significance.

To prove Theorem D, we use the representation of Chern classes by differential forms [5] and [12], together with the form (1.3) of the curvature. Then, as was outlined in [11, § IV], it will follow that a positive polynomial

$$P(c_1, \cdots, c_s) \geqq \left(\frac{i}{2}\right)^q (-1)^{q(q-1)/2}\{\textstyle\sum_\lambda B_\lambda \wedge \bar{B}_\lambda\} \tag{1.5}$$

where B_λ is a $(q, 0)$ form ($q = \dim W$). Using (1.2) we find

$$\left(\frac{i}{2}\right)^q (-1)^{q(q-1)/2} \int_W \{\textstyle\sum_\lambda B_\lambda \wedge \bar{B}_\lambda\} > 0 ,$$

which proves Theorem D.

It is an open question whether or not a positive bundle, or perhaps a cohomologically positive bundle, is numerically positive.

The more difficult assertions are Theorems B, G, and, later on, E. A proof of Theorem B directly, by differential geometric methods, involves (for $q = 1$) the quadratic form (cf. [10, Prop. 5.2 and [22])

$$\Theta(\varphi, \varphi) = \textstyle\sum_{\rho,\sigma,i,j} \Theta^\rho_{\sigma i \bar{j}} \varphi^\sigma_i \bar{\varphi}^\rho_j , \tag{1.6}$$

and we need to show that $\Theta(\varphi, \varphi) \geqq 0$. Suppose that **E** has a lot of sections so that Θ^ρ_σ has the form (1.3). Using the identity

$$\begin{aligned} &A^\alpha_{\sigma i}\varphi^\rho_i \overline{A^\alpha_{\rho j}\varphi^\sigma_j} + A^\alpha_{\rho i}\varphi^\sigma_i \overline{A^\alpha_{\sigma j}\varphi^\rho_j} \\ &\quad = A^\alpha_{\rho i}\varphi^\sigma_i \overline{A^\alpha_{\rho j}\varphi^\sigma_j} + A^\alpha_{\sigma i}\varphi^\rho_i \overline{A^\alpha_{\sigma j}\varphi_j} \\ &\qquad\qquad - (A^\alpha_{\rho i}\varphi^\sigma_i - A^\alpha_{\sigma i}\varphi^\sigma_i)(\overline{A^\alpha_{\rho j}\varphi^\sigma_j - A^\alpha_{\sigma j}\varphi^\rho_j}) \end{aligned}$$

(no summation), we get

$$\Theta(\varphi, \varphi) = \textstyle\sum_{\alpha,\rho,\sigma} | A(\varphi)^\alpha_{\rho\sigma} |^2 - \frac{1}{2} \sum_{\sigma,\rho,\alpha} | \hat{A}(\varphi)^\alpha_{\rho\sigma} |^2 , \tag{1.7}$$

where $A(\varphi)^{\alpha}_{\rho\sigma} = \sum_i A^{\alpha}_{\sigma i}\varphi^{\rho}_i$ and $\hat{A}(\varphi)^{\alpha}_{\rho\sigma} = \sum_i (A^{\alpha}_{\rho i}\varphi^{\sigma}_i - A^{\alpha}_{\sigma i}\varphi^{\rho}_i)$. If $\mathbf{E}$ is a line bundle, $\hat{A}(\varphi) = 0$ and, by (1.7), $\Theta(\varphi, \varphi) \geqq 0$. In general, however, the quadratic form (1.7) does not have a sign.

The conclusion is that a geometric assumption (ampleness) gives an inequality on the quadratic form $\Theta(\varphi, \varphi)$ where $\varphi = \xi \otimes \eta$ is a *decomposable tensor*, whereas to prove directly a vanishing theorem we need information on $\Theta(\varphi, \varphi)$ for *all* tensors.

A means around this trouble is suggested in [10, § 9]. Let $\mathbf{E}^* \to V$ be the dual bundle, $\mathbf{P}(\mathbf{E}^*) = \mathbf{P}$ the associated projective bundle, and $\mathbf{L} \to \mathbf{P}$ the standard line bundle whose restriction to each fibre of $\mathbf{P}(\mathbf{E}^*) \to V$ is the positive hyperplane bundle. There are two basic facts

(1.8) $\mathbf{E} \to V$ positive $\Rightarrow \mathbf{L} \to$ positive, (cf. [10, Prop. (9.1)] and § 3. (b) below); and

$$H^q(V, \mathcal{O}(\mathbf{E}^{(\mu)}) \otimes S) \cong H^q(\mathbf{P}, \mathcal{O}(\mathbf{L}^{\mu}) \otimes \pi^* S) \tag{1.9}$$

for any coherent sheaf S over V. These two facts, together with Theorem B for line bundles give the assertion for general vector bundles.

The proof of Theorem G follows from the usual Kodaira vanishing theorem on $\mathbf{P}(\mathbf{E}^*)$, coupled with a precise curvature computation. Passing from $\mathbf{E} \to V$ to $\mathbf{L} \to \mathbf{P}(\mathbf{E}^*)$ has the analytic effect of splitting all tensors, which in turn leads to the desired inequalities.

The proof of Theorem E, as outlined in [11, § A. 1], uses the result for line bundles (cf. Nakai [21]) and (1.9)), so that it will suffice to prove that $\mathbf{L} \to \mathbf{P}$ is numerically positive if $\mathbf{E} \to V$ is. This involves relating the algebraic homology ring of $\mathbf{P}$ with that of V and use of integration over the fibre in $\mathbf{P} \to V$.

The proof of Theorem H is done by suitably generalizing Bott's Morse-theoretic argument [2] for line bundles to the case of vector bundles. This result substantiates the definition of positivity. Along with the proof of Theorem H we show that $V - S$ is *r-convex* (i.e., there is an exhaustion function for $V - S$ whose *E. E. Levi form* has $n - r + 1$ positive eigenvalues).

The proof of Theorem I uses Theorem G′ and a standard locally free resolution of the ideal sheaf I of $S \subset V$. The more interesting Theorem J is proved by first blowing up V along S to obtain a codimension one situation of $\tilde{S} \subset \tilde{V}$ where $\tilde{S}$ is given by the zeros of a holomorphic section of a line bundle $\mathbf{L} \to \tilde{V}$. The metric on $\mathbf{E} \to V$ induces a metric on $\mathbf{L} \to \tilde{V}$ whose curvature we compute using the second fundamental forms. It follows that the curvature $\Theta_{\mathbf{L}}$ has everywhere $n - r + 1$ positive eigenvalues, and then a suitable vanishing theorem gives Theorem J.

In § 4 we discuss Chern classes. A direct geometric definition, involving an algebraic-geometric obstruction theory, and the definition using differential forms are given. Using the theorem of Weil, proved in § 4 (b) below, these definitions are proved to be the same. By putting a little more effort into this argument, we give in § 4 (c) another proof of the theorem of Bott-Chern [3].

In § 5 we discuss positive cohomology classes and give the proof of Theorem E. Remarks on the problem (0.7) are also given (cf. below the proof of Theorem D in § 5 (a)).

2. Hermitian differential geometry

(a) *The frame bundle.* Let V be a complex manifold and $\mathbf{E} \to V$ a *holomorphic vector bundle* with fibre $\mathbf{C}^r$. We think of $\mathbf{C}^r$ as column vectors

$$\xi = \begin{pmatrix} \xi^1 \\ \vdots \\ \xi^r \end{pmatrix}$$

and, for a matrix $g = (g^\rho_\sigma) \in G = GL(r, \mathbf{C})$, we set

$$g\xi = \begin{pmatrix} \vdots \\ \sum_\sigma g^\rho_\sigma \xi^\sigma \\ \vdots \end{pmatrix}.$$

Let $\mathbf{P} \xrightarrow{\pi} V$ be the principal bundle, with group G, of all *holomorphic frames* $f = (e_1, \cdots, e_r)$ for $\mathbf{E} \to V$. Then G acts on $\mathbf{P}$ by $fg = (\cdots, \sum_\rho g^\rho_\sigma e_\rho, \cdots)$, and a section ξ of $\mathbf{E} \to V$ is given on $\mathbf{P}$ by $\xi = \sum_{\rho=1}^r \xi^\rho(f) e_\rho$ with

$$\xi^\rho(fg) = \sum_\sigma (g^{-1})^\rho_\sigma \xi^\sigma(f) .$$

Similarly, a differential form on V with values in $\mathbf{E}$ is given on $\mathbf{P}$ as $\varphi = \sum_\rho \varphi^\rho e_\rho$ where φ^ρ is a horizontal form on $\mathbf{P}$ satisfying equivariance conditions.

As an example, consider the *Grassmann manifold* $\mathbf{G} = \mathbf{G}(r, m)$ of r-planes in $\mathbf{C}^m$. We let $\mathbf{P}$ be the r-frames $f = (e_1, \cdots, e_r)$ in $\mathbf{C}^m$, where $e_\rho = (\xi^\alpha_\rho)$ is a column vector and $e_1 \wedge \cdots \wedge e_r \neq 0$. Then $\pi\colon \mathbf{P} \to \mathbf{G}$ is given by $\pi(f) =$ subspace spanned by $e_1, \cdots, e_r$. Observe that

$$fg = (\textstyle\sum_\rho \xi^\alpha_\rho g^\rho_\sigma) = (\cdots, \sum_\rho g^\rho_\sigma e_\rho, \cdots)$$

so that our notation is consistent. The vector bundle $\mathbf{E} \to \mathbf{G}$ is the *universal bundle* whose fibre $\mathbf{E}_S$ at a subspace $S \in \mathbf{G}$ is the vector space S itself.

(b) *Metrics, connexions, and curvatures.* A *hermitian metric* in $\mathbf{E} \to V$ gives a matrix function h on $\mathbf{P}$ by the rule $h(f)_{\rho\sigma} = (e_\sigma, e_\rho)$. Then

$$(\textstyle\sum_\rho \xi^\rho e_\rho, \sum_\sigma \eta^\sigma e_\sigma) = \sum_{\rho,\sigma} \bar{\eta}^\sigma h_{\sigma\rho}\xi^\rho = {}^t\bar{\eta} h \xi \,.$$

We have $h = {}^t\bar{h}$, $h > 0$, and $h(fg) = {}^t\bar{g}h(f)g$. From this last equation, we see that the (1.0) form θ on $\mathbf{P}$ defined by $\theta = h^{-1}\partial h$; i.e., $\theta^\rho_\sigma = \sum_\tau (h^{-1})_{\rho\tau}\partial h_{\tau\sigma}$, satisfies $\theta(fg) = g^{-1}\theta(f)g$ and gives a *connexion* in $\mathbf{P} \to V$. For a section $\xi = \sum_\rho \xi^\rho e_\rho$ of $\mathbf{E} \to V$, we define the *covariant differential* $D\xi = \sum_\rho d\xi^\rho e_\rho + \sum \xi^\rho De_\rho$ where $De_\rho = \sum_\sigma \theta^\sigma_\rho e_\sigma$; i.e., $(D\xi)^\rho = d\xi^\rho + \sum_\sigma \theta^\rho_\sigma \xi^\sigma$. Then $D\xi$ is an $\mathbf{E}$-valued 1-form and $d(\xi, \eta) = (D\xi, \eta) + (\xi, D\eta)$ for sections ξ, η of $\mathbf{E}$. Writing $D = D' + D''$ where D' is type (1, 0) and D'' is of type (0, 1), we have that $D' = \partial + \theta$ and $D'' = \bar{\partial}$.

More generally, for an $\mathbf{E}$-valued q-form φ on V, we set $D\varphi = \sum_\rho (d\varphi^\rho e_\rho + (-1)^q \varphi^\rho \wedge De_\rho)$; $D\varphi$ is an $\mathbf{E}$-valued $q+1$-form on V.

The *curvature form* $\Theta = (\Theta^\rho_\sigma)$ is given by $\Theta = d\theta + \theta \wedge \theta$; i.e., $\Theta^\rho_\sigma = d\theta^\rho_\sigma + \sum \theta^\rho_\tau \wedge \theta^\tau_\sigma$. Since $\partial(h^{-1}\partial h) + h^{-1}\partial h \wedge h^{-1}\partial h = 0$, it follows that

$$\Theta = \bar{\partial}\theta = -h^{-1}\partial\bar{\partial}h - h^{-1}\bar{\partial}h \wedge h^{-1}\partial h \,, \tag{2.1}$$

and Θ is of type (1, 1).

For an $\mathbf{E}$-valued form φ, we have the *Bianchi identity*:

$$D^2\varphi = \Theta \wedge \varphi = \textstyle\sum_{\sigma,\rho} \Theta^\rho_\sigma \wedge \varphi^\sigma e_\rho \,. \tag{2.2}$$

Returning to our example of the principal bundle $\mathbf{P} \to \mathbf{G}$ over the grassmannian, we define a metric in the universal bundle by

$$h(f) = {}^t\bar{f}f \,, \tag{2.3}$$

where $f = (\xi^\alpha_\rho)$ is the $m \times r$ matrix whose columns give the frame f. Then $h(fg) = {}^t\bar{g}{}^t\bar{f}fg = {}^t\bar{g}h(f)g$ and the curvature:

$$\Theta = h^{-1}\,{}^t\overline{df} \wedge df - h^{-1}({}^t\overline{df}f)h^{-1}({}^t\bar{f}df) \,. \tag{2.4}$$

In case $r = 1$, $\mathbf{P} = \mathbf{C}^m - \{0\}$ and, by (2.4),

(2.5) $$\Theta = \frac{(df, f)(f, df) - (f, f)(df, df)}{(f, f)^2} .$$

Thus Θ is the negative of the standard Kähler form on $\mathbf{P}_{m-1}$, which checks our signs since $\mathbf{E}$ is the dual of the positive line bundle given by the divisor $\mathbf{P}_{m-1} \subset \mathbf{P}_m$.

(c) *Calculations in local coordinates.* Let $z = (z^1, \cdots, z^n)$ be local holomorphic coordinates in V and $f(z) = (e_1(z), \cdots, e_r(z))$ a local holomorphic section of $\mathbf{P} \to V$. Then $h_{\rho\sigma}(z) = (e_\sigma(z), e_\rho(z))$ is a function of z and

$$\theta^\rho_\sigma(z) = \sum\nolimits_{\tau, j} (h^{-1}(z))_{\rho\tau} \frac{\partial h_{\tau\sigma}(z)}{dz^j} dz^j .$$

The curvature $\Theta(z)$ is given by (2.4), where $h = h(z)$ is a function of z.

If $g(z) = (g^\rho_\sigma(z))$ is a holomorphic matrix, then $f(z)g(z)$ is another holomorphic frame. Taking $g(z)$ to be a suitable constant matrix, we may assume that $h_{\rho\sigma}(0) = \delta^\rho_\sigma$; i.e., $h(0) = I$. Let $A(z) = (\sum_j A^\rho_{\sigma j} z^j)$ be a linear matrix with

$$A^\rho_{\sigma j} = -\frac{\partial h_{\rho\sigma}(0)}{\partial z^j} .$$

Then $dA(0) = -\partial h(0)$ and so $d({}^t\overline{g(z)} h(z) g(z))_{z=0} = 0$ with $g(z) = I + A(z)$. In summary, we may choose our frame $f(z) = (e_1(z), \cdots, e_r(z))$ such that, at $z = 0, h = I$ and $dh = 0$. By (2.1), at the origin,

(2.6) $$\Theta = -\partial\bar{\partial} h .$$

For example, letting $Z = (\xi^\lambda_\rho)$ $(1 \leqq \lambda \leqq m - r, 1 \leqq \rho \leqq r)$ be an $(m - r) \times r$ matrix, the mapping $f(Z) = \begin{pmatrix} I_r \\ Z \end{pmatrix}$ gives a local cross-section of $\mathbf{P} \to \mathbf{G}$, the bundle over the grassmannian considered above. The metric $h(Z) = I + {}^t\bar{Z}Z$ and so $h(0) = I, dh(0) = 0$. The curvature is given, by (2.6), as

(2.7) $$\Theta = {}^t\overline{dZ} \wedge dZ ,$$

or

(2.8) $$\Theta^\rho_\sigma = -\sum\nolimits_\lambda d\xi^\lambda_\sigma \wedge \overline{d\xi^\lambda_\rho} .$$

Observe that the general linear group $GL(m, \mathbf{C})$ acts transitively on $\mathbf{P}$ by $Af = (A^\alpha_\beta \xi^\beta_\rho)$; this action preserves the fibering $\mathbf{P} \to \mathbf{G}$. The unitary group $U(m)$ acts on $\mathbf{P}$ and preserves the metric, since $h(Af) = {}^t\overline{Af}Af = {}^t\bar{f}\,{}^t\bar{A}Af = {}^t\bar{f}f = h(f)$. The action of $U(m)$ on $\mathbf{G}$

is transitive, and so the curvature in $\mathbf{E} \to V$ is determined by (2.8), which is Θ at a point.

Let $e_1(z), \cdots, e_r(z)$ be a local frame for $\mathbf{E}$ and $e_1^*(z), \cdots, e_r^*(z)$ the dual frame for the *dual bundle* $\mathbf{E}^* \to X$. Then there is defined a connexion D^* in $\mathbf{E}^*$ by the requirement

$$0 = d\langle e_\rho, e_\sigma^* \rangle = \langle De_\rho, e_\sigma^* \rangle + \langle e_\rho, D^* e_\sigma^* \rangle .$$

This gives $\theta_\rho^\sigma + \theta^{*\rho}_\sigma = 0$ or $\theta^{*\rho}_\sigma = -\theta_\rho^\sigma$. The curvature

$$\Theta^{*\rho}_\sigma = -\Theta_\rho^\sigma . \tag{2.9}$$

It is easy to verify that θ^* is the metric connexion of the induced metric in $\mathbf{E}^*$. In fact, θ^* is obviously of type (1, 0) and preserves the metric in $\mathbf{E}^*$; by uniqueness, θ^* is the metric connexion.

If $\mathbf{E}$ and $\mathbf{F}$ are bundles with frames $e_1, \cdots, e_r$; $f_1, \cdots, f_s$, then $e_\rho \otimes f_\alpha$ is a frame for $\mathbf{E} \otimes \mathbf{F}$ and we may define a connexion in $\mathbf{E} \otimes \mathbf{F}$ by

$$\begin{aligned} D(e_\rho \otimes f_\alpha) &= De_\rho \otimes f_\alpha + e_\rho \otimes Df_\alpha \\ &= \textstyle\sum_\sigma \theta_\rho^\sigma e_\sigma \otimes f_\alpha + \sum_\beta \theta_\alpha^\beta e_\rho \otimes f_\beta ; \end{aligned}$$

i.e., the connexion $\theta_{\mathbf{E}\otimes\mathbf{F}} = \theta_{\mathbf{E}} \otimes 1 + 1 \otimes \theta_{\mathbf{F}}$ and the curvature

$$\Theta_{\mathbf{E}\otimes\mathbf{F}} = \Theta_{\mathbf{E}} \otimes I_{\mathbf{F}} + I_{\mathbf{E}} \otimes \Theta_{\mathbf{F}} . \tag{2.10}$$

We may consider the bundle $\mathbf{B} \subset \mathbf{P}$ of orthonormal frames. On $\mathbf{B}$, $h_{\rho\sigma} = \delta_\sigma^\rho$ and the connexion form θ satisfies $\theta_\sigma^\rho + \bar{\theta}_\rho^\sigma = 0$. For the curvature then we have the symmetry

$$\Theta_\sigma^\rho + \bar{\Theta}_\rho^\sigma = 0 , \tag{2.11}$$

or

$$\Theta_{\sigma i \bar{j}}^\rho + \bar{\Theta}_{\rho j \bar{i}}^\sigma = 0 . \tag{2.12}$$

The holomorphic frames $f(z)$ constructed above pass through a point $f(0) \in \mathbf{B}$ and are tangent to $\mathbf{B}$ at $f(0)$.

(d) *The second fundamental form of a sub-bundle.* Suppose that we have an exact sequence

$$0 \longrightarrow \mathbf{S} \longrightarrow \mathbf{E} \longrightarrow \mathbf{Q} \longrightarrow 0 , \tag{2.13}$$

of holomorphic vector bundles over V. In this case, we let $\mathbf{P}$ be the bundle of all frames $f = (e_1, \cdots, e_r)$ for $\mathbf{E}$ where $e_1, \cdots, e_s$ is a frame for $\mathbf{S}$. The group G of $\mathbf{P}$ is now the group $GL(s, r-s)$ of all $r \times r$ matrices $g = \begin{pmatrix} A & B \\ 0 & C \end{pmatrix}$ with A an $s \times s$ matrix. We agree on the range of indices $1 \leqq \rho, \sigma \leqq r$; $1 \leqq \lambda, \mu \leqq s$; and $s + 1 \leqq \alpha$,

$\beta \leqq r$.

Suppose now that we have in $\mathbf{E}$ a hermitian metric. Then there is a connexion $D_{\mathrm{E}}\colon A^0(\mathrm{E}) \to A^1(\mathrm{E})$, where $A^q(\mathrm{E})$ are the C^∞ q-forms with values in $\mathbf{E}$. This gives in particular $D_{\mathrm{E}}\colon A^0(\mathrm{S}) \to A^1(\mathrm{E})$.

Since $\mathrm{S} \subset \mathbf{E}$ is a sub-bundle, there is an induced hermitian metric which has its own metric connexion $D_{\mathrm{S}}\colon A^0(\mathrm{S}) \to A^1(\mathrm{S})$. The difference $D_{\mathrm{E}} - D_{\mathrm{S}}\colon A^0(\mathrm{S}) \to A^1(\mathrm{E})$ is then linear over the C^∞ functions and so is given by a Hom (S, E)-valued 1-form $b \in A^1(\mathrm{Hom}(\mathrm{S}, \mathrm{E}))$. What we claim is that $b \in A^{1,0}(\mathrm{Hom}\, \mathrm{S}, \mathrm{Q}))$; i.e., b is a (1, 0) form satisfying $(b(\mathrm{S}), \mathrm{S}) = 0$.

To see this, we let $h = \begin{pmatrix} h_1 & h_2 \\ h_3 & h_4 \end{pmatrix}$ be the metric function on $\mathbf{P}$ where h_1 is the induced metric on $\mathbf{S}$. We first choose a holomorphic frame

$$f(z) = (e_1(z), \cdots, e_s(z); \hat{e}_{s+1}(z), \cdots, \hat{e}_r(z))$$

such that $h_1(0) = I_s$, $dh_1(0) = 0$. This is done by varying the A part of $g = \begin{pmatrix} A & B \\ 0 & C \end{pmatrix}$. Letting $e'_\alpha(z) = \hat{e}_\alpha(z) - \sum_\lambda h_{\alpha\lambda}(0) e_\lambda(z)$, we have a new holomorphic frame in $\mathbf{P}$ for which $h(0) = \begin{pmatrix} I & 0 \\ 0 & h_4(0) \end{pmatrix}$, $dh_1(0) = 0$. By using the C part of g, we may assume that $h(0) = \begin{pmatrix} I & 0 \\ 0 & I \end{pmatrix}$, $dh_1(0) = 0$, $dh_4(0) = 0$.

We let φ^ρ_σ be the connexion for $\mathbf{E}$ and θ^λ_μ the connexion for S; $\varphi = h^{-1}\partial h$ and $\theta = h_1^{-1}\partial h_1$. Then

$$b(e_\lambda) = (D_{\mathrm{E}} - D_{\mathrm{S}})e_\lambda = \sum_\rho \varphi^\rho_\lambda e_\rho - \sum_\mu \theta^\mu_\lambda e_\mu \,.$$

Since, at $z = 0$, $\varphi^\mu_\lambda(0) = 0$, $\theta^\mu_\lambda(0) = 0$, we have that $b(e_\lambda) = \sum_\alpha \varphi^\alpha_\lambda e_\alpha$, $\varphi^\alpha_\lambda = \partial h_{\alpha\lambda}$. This proves that b is of type (1, 0) and $(b(e_\lambda), \mathrm{S}) = 0$ as required.

By definition, $b \in A^{1,0}(\mathrm{Hom}(\mathrm{S}, \mathrm{Q}))$ is the *second fundamental form* of S in Q. This b has been used in [11, § VI. 3], where the terminology is justified.

We now compute the curvatures at $z = 0$ using (2.1). This gives $(\Theta_{\mathrm{S}})^\lambda_\mu = -\partial\bar{\partial} h_{\lambda\mu}$ and

$$\begin{aligned} (\Theta_{\mathrm{E}})^\lambda_\mu &= -\partial\bar{\partial} h_{\lambda\mu} - \textstyle\sum_\alpha \bar{\partial} h_{\lambda\alpha} \wedge \partial h_{\alpha\mu} \\ &= -\partial\bar{\partial} h_{\lambda\mu} - \textstyle\sum_\alpha \overline{\partial h_{\alpha\lambda}} \wedge \partial h_{\alpha\mu} \,. \end{aligned}$$

Combining, we have

$$(\Theta_{\mathrm{S}})^\lambda_\mu = (\Theta_{\mathrm{E}})^\lambda_\mu + \sum_\alpha \overline{\partial h_{\alpha\lambda}} \wedge \partial h_{\alpha\mu} \,. \tag{2.14}$$

In invariant terms, this gives:

$$\Theta_{S} = \Theta_{E} \mid S + {}^t\bar{b} \wedge b\,, \tag{2.15}$$

where b is the 2nd fundamental form.

Let us check our signs by computing an example. If $\mathbf{G} = \mathbf{G}(r, m)$ is the Grassmann variety and $\mathbf{S} \to \mathbf{G}$ is the universal bundle, $\mathbf{E} = \mathbf{G} \times \mathbf{C}^m$ the trivial bundle, then we have $0 \to \mathbf{S} \to \mathbf{E} \to \mathbf{Q} \to 0$. The metric in $\mathbf{S}$ is induced from the flat euclidean metric in $\mathbf{E} = \mathbf{G} \times \mathbf{C}^m$. As above, we choose the frame $f(Z) = \begin{pmatrix} I \\ Z \end{pmatrix}$ for $\mathbf{S}$ over an open set in $\mathbf{G}$. Writing $f(Z) = (e_1, \cdots, e_r)$ we complete this to the frame

$$(e_1, \cdots, e_r; e_{r+1}, \cdots, e_m) = \begin{pmatrix} I & 0 \\ Z & I \end{pmatrix} \qquad \text{for } \mathbf{E}\,.$$

The metric function for $\mathbf{E}$ is

$$h = \begin{pmatrix} h_1 & h_2 \\ h_3 & h_4 \end{pmatrix} = \begin{pmatrix} I & {}^t\bar{Z} \\ 0 & I \end{pmatrix}\begin{pmatrix} I & 0 \\ Z & I \end{pmatrix} = \begin{pmatrix} I + {}^t\bar{Z}Z & {}^t\bar{Z} \\ Z & I \end{pmatrix}.$$

Then $h_1(0) = I$, $dh_1(0) = 0$, $h_4(0) = I$, $dh_4(0) = 0$. Furthermore, $b = dZ$ and, by (2.15), we have $\Theta_S = {}^t\overline{dZ} \wedge dZ$, which agrees with (2.7).

(e) *Properties of the second fundamental form.* We want to discuss further the second fundamental form $b \in A^{1,0}(\mathrm{Hom}(\mathbf{S}, \mathbf{Q}))$ given above. The basic facts are

$$D'b = 0\,; \tag{2.16}$$

$$\bar{\partial} b = 0 \qquad \text{if } \Theta_E = 0\,. \tag{2.17}$$

To check these, we use orthonormal frames. Thus let $\mathbf{B} \subset \mathbf{P}$ be all unitary frames $f = (e_1, \cdots, e_r)$ for $\mathbf{E}$ where $e_1, \cdots, e_s$ is a frame for $\mathbf{S}$. Given $f \in \mathbf{B}$, we can find a *holomorphic* frame $f(z)$ for $\mathbf{E}$ with $f(0) = f$, but, in general, this cross-section will *not* be tangent to $\mathbf{B}$ at f. The obstruction is essentially the second fundamental form b, as was seen above.

On $\mathbf{B}$ we write $D_{\mathbf{E}} e_\rho = \sum_\sigma \varphi_\rho^\sigma e_\sigma$ and $D_{\mathbf{S}} e_\lambda = \sum_\mu \theta_\lambda^\mu e_\mu$. Then $\varphi_\rho^\sigma + \bar{\varphi}_\sigma^\rho = 0$, $\theta_\mu^\lambda + \bar{\theta}_\lambda^\mu = 0$. We claim that $\varphi_\lambda^\mu = \theta_\lambda^\mu$ and that $b = \sum_{\alpha,\lambda} \varphi_\lambda^\alpha e_\alpha \otimes e_\lambda^*$.

Let $De_\lambda = \sum_\mu \varphi_\lambda^\mu e_\mu$; this gives a connexion in $\mathbf{S}$ which preserves the metric, and to show that $D = D_{\mathbf{S}}$, we need to prove that $D'' = \bar{\partial}$. Choose a C^∞ frame $f(z) = (e_1(z), \cdots, e_r(z))$ for $\mathbf{B}$. We may then find *holomorphic* sections $\xi_\mu(z) = \sum_\lambda \xi_\mu^\lambda(z) e_\lambda(z)$ of $\mathbf{S}$ with $\xi_\mu(0) = e_\mu(0)$. Then

$$0 = D_{\mathbf{E}}''\xi_\mu(0) = \sum_\lambda \bar{\partial}\xi_\mu^\lambda(0) e_\lambda(0) + \sum_{\lambda,\rho} \xi_\mu^\lambda(0)\varphi_\lambda^\rho(0)'' e_\rho(0)$$

which gives $\varphi_\lambda^{\alpha''} = 0$, $(D_E - D)'' = 0$, and $D'' = D_E'' = \bar\partial$. Furthermore $b(e_\lambda) = (D_E - D_S)e_\lambda = \sum_\alpha \varphi_\lambda^\alpha e_\alpha$ where φ_λ^α is of type (1, 0).

We now compute

$$\begin{aligned} Db &= \textstyle\sum_{\alpha,\lambda} d\varphi_\lambda^\alpha e_\alpha \otimes e_\lambda^* - \sum_{\alpha,\beta,\lambda} \varphi_\lambda^\alpha \wedge \varphi_\alpha^\beta e_\beta \otimes e_\lambda^* \\ &\quad + \textstyle\sum \varphi_\lambda^\alpha \wedge \varphi_\mu^\lambda e_\alpha \otimes e_\mu^* \\ &= \textstyle\sum_{\alpha,\lambda} \left(d\varphi_\lambda^\alpha + \sum_\beta \varphi_\beta^\alpha \wedge \varphi_\lambda^\beta + \sum_\mu \varphi_\mu^\alpha \wedge \varphi_\lambda^\mu\right) e_\alpha \otimes e_\lambda^* \ , \end{aligned}$$

which gives:

$$Db = \textstyle\sum_{\alpha,\lambda} \Theta_\lambda^\alpha e_\alpha \otimes e_\lambda^* \ . \tag{2.18}$$

Since b is of type (1, 0) and Θ_E of type (1, 1), we get $D'b = 0$, and clearly $\bar\partial b = 0$ if $\Theta_E = 0$.

There are two applications of this. If we let $c = \sum_{\alpha,\lambda} \varphi_\alpha^\lambda e_\lambda \otimes e_\alpha^*$, then, since $\varphi_\lambda^{\alpha''} = 0$ and $\varphi_\lambda^\alpha + \bar\varphi_\alpha^\lambda = 0$, $\varphi_\alpha^{\lambda'} = 0$ and $c \in A^{0,1}(\mathrm{Hom}(\mathbf{Q},\mathbf{S}))$. From (2.16) we get $\bar\partial c = 0$, and it is proved in [11, § VI. 3] that the Dolbeault cohomology class of $c \in H^1(V, \mathrm{Hom}(\mathbf{Q}, \mathbf{S}))$ is the obstruction to splitting the sequence $0 \to \mathbf{S} \to \mathbf{E} \to \mathbf{Q} \to 0$ holomorphically.

Secondly, if we take V to be the grassmannian $\mathbf{G} = \mathbf{G}(r, m)$ of r-planes in $\mathbf{C}^m$ and $\mathbf{S} \to \mathbf{G}$ the universal bundle, then we have $0 \to \mathbf{S} \to \mathbf{E} \to \mathbf{Q} \to 0$ where $\mathbf{E} = \mathbf{G} \times \mathbf{C}^m$ is a trivial bundle. Taking the flat metric in $\mathbf{E}$, $\Theta_E = 0$ and $\bar\partial b = 0$. But then b is a holomorphic section of $\mathrm{Hom}(\mathbf{T}, \mathrm{Hom}(\mathbf{S}, \mathbf{Q}))$ and from the local coordinate description of b above, we see that the second fundamental form b gives an isomorphism:

$$b\colon \mathbf{T} \longrightarrow \mathrm{Hom}(\mathbf{S}, \mathbf{Q}) \ . \tag{2.19}$$

(f) *Curvature of ample bundles and proof of Theorem* A. As an application of the second fundamental form, we let $\mathbf{E} \to V$ be a holomorphic vector bundle such that the global sections $\Gamma(\mathbf{E}^*)$ generate $\mathbf{E}^*$. Then we have $\mathbf{\Gamma}(\mathbf{E}^*) \to \mathbf{E}^* \to 0$ and, by duality,

$$0 \longrightarrow \mathbf{E} \longrightarrow \mathbf{\Gamma}(\mathbf{E}^*)^* \longrightarrow \mathbf{F} \longrightarrow 0 \ , \tag{2.19}$$

where $\mathbf{\Gamma}(\mathbf{E}^*)^* = V \times \Gamma(\mathbf{E}^*)^*$.

Let $\mathbf{G}$ be the Grassmann manifold of r-planes in $\Gamma(\mathbf{E}^*)^*$; for each $z \in V$, $\mathbf{E}_z$ gives a subspace of $\Gamma(\mathbf{E}^*)^*$ and $\varphi\colon V \to \mathbf{G}$ given by $\varphi(z) = \mathbf{E}_z \subset \Gamma(\mathbf{E}^*)^*$ has the property that $\varphi^*(\mathbf{S}) = \mathbf{E}$ where $\mathbf{S} \to \mathbf{G}$ is the universal bundle. In fact, φ^* lifts the exact sequence

$$0 \longrightarrow \mathbf{S} \longrightarrow \mathbf{G} \times \Gamma(\mathbf{E}^*)^* \longrightarrow \mathbf{Q} \longrightarrow 0 \tag{2.20}$$

back to the exact sequence (2.19).

To find the induced metric in $\mathbf{E} \to V$, we choose a unitary basis $s^1, \cdots, s^m$ for $\Gamma(\mathbf{E}^*)$ and set

$$(2.21) \qquad (e_\rho, e_\sigma) = \sum_{j=1}^m \langle s,^j e_\rho \rangle \langle \overline{s^j, e_\sigma} \rangle ,$$

where $f = (e_1, \cdots, e_r)$ is a frame for $\mathbf{E}$. Thus

$$(2.22) \qquad h_{\sigma\rho} = \sum_{j=1}^m A^j_\rho \bar{A}^j_\sigma = ({}^t\bar{A}(f)A(f))_{\sigma\rho}$$

where $A^j_\rho = \langle s^j, e_\rho \rangle$.

If we let $\mathbf{P} \to V$ be the frame bundle for $\mathbf{E} \to V$ and $\mathbf{B} \to \mathbf{G}$ the frame bundle for $\mathbf{S} \to \mathbf{G}$, then the mapping $\varphi\colon \mathbf{P} \to \mathbf{B}$ given by $\varphi(f) = A(f)$ satisfies $\varphi(fg) = \varphi(f)g$ for $g \in G = GL(r, \mathbf{C})$; the induced mapping $\varphi\colon V \to \mathbf{G}$ ($V = \mathbf{P}/G$, $\mathbf{G} = \mathbf{B}/G$) is the same as φ above.

We now give $\varphi\colon V \to \mathbf{G}$ locally. Let $z^1, \cdots, z^n$ be local coordinates on V and assume that $s^1, \cdots, s^r \in \Gamma(\mathbf{E}^*)$ are linearly independent near $z = 0$. We let $f(z) = (e_1(z), \cdots, e_r(z))$ be the frame for $\mathbf{E}$ with $\langle s^\rho, e_\sigma \rangle = \delta^\rho_\sigma$ ($1 \leqq \rho, \sigma \leqq r$). Our range of indices is to be $1 \leqq \rho, \sigma \leqq r$; $r + 1 \leqq \alpha, \beta \leqq m$. Then $\varphi(z) = \begin{pmatrix} I \\ B(z) \end{pmatrix}$ where $B(z) = (b^\alpha_\rho(z)$ is an $(m - r) \times r$ matrix with $b^\alpha_\rho(z) = \langle s^\alpha(z), e_\rho(z) \rangle$. By (2.22),

$$h(z) = (I\,{}^t\overline{B(z)}) \begin{pmatrix} I \\ B(z) \end{pmatrix} = I + {}^t\overline{B(z)}B(z) .$$

Making a unitary change of $s^1, \cdots, s^m$, we may assume that $B(0) = 0$. Then $h(0) = I$, $dh(0) = 0$, and, by (2.6), the curvature in $\mathbf{E}$ at $z = 0$ is

$$(2.23) \qquad \Theta_{\mathbf{E}} = -\partial\bar{\partial}h = {}^t\overline{dB} \wedge dB ;$$

that is

$$(2.24) \qquad \Theta^\rho_\sigma = -\sum_{\alpha=r+1}^m db^\alpha_\sigma \wedge \overline{db^\alpha_\rho} .$$

We want to relate these formulas to the second fundamental forms. If $b_{\mathbf{E}} \in \operatorname{Hom}(\mathbf{T}(V), \operatorname{Hom}(\mathbf{E}, \mathbf{F}))$ is the 2nd fundamental form of $\mathbf{E}$ in $\Gamma(\mathbf{E}^*)^*$, then at $z = 0$, $b_{\mathbf{E}}$ is just db^α_σ. The formula (2.23) for the curvature is then the same as (2.15). Furthermore, the following diagram commutes.

$$(2.25) \qquad \begin{array}{ccc} \mathbf{T}(V) & \xrightarrow{\varphi_*} & \mathbf{T}(\mathbf{G}) \\ \downarrow{\scriptstyle b_{\mathbf{E}}} & & \downarrow{\scriptstyle b_{\mathbf{S}}} \\ \operatorname{Hom}(\mathbf{E}, \mathbf{F}) & \xrightarrow{\varphi} & \operatorname{Hom}(\mathbf{S}, \mathbf{Q}) . \end{array}$$

Now, as discussed in (e) above, b_S is an isomorphism (cf. (2.19)) and φ is an algebraic isomorphism. With these identifications, we conclude that φ_*: $\mathbf{T}(V) \to \mathbf{T}(\mathbf{G})$ is the same as

$$b_E: \mathbf{T}(V) \longrightarrow \mathrm{Hom}(\mathbf{E}, \mathbf{F}) . \tag{2.26}$$

In terms of the local coordinates above $\varphi(z) = B(z)$ and (2.26) is clear.

PROOF OF THEOREM A. It will suffice to show that if $\mathbf{E}^* \to V$ is ample, then $\mathbf{E} \to V$ is negative in the sense that the quadratic form

$$\sum_{\rho,\sigma,i,j} (\Theta_E)^{\rho}_{\sigma i \bar{j}} \xi^{\sigma} \bar{\xi}^{\rho} \eta^i \bar{\eta}^j = \Theta_E(\xi, \eta) \tag{2.27}$$

is negative definite. Indeed, $(\Theta_{E^*})^{\rho}_{\sigma} = -(\Theta_E)^{\sigma}_{\rho}$ by (2.4) and so $-\Theta_E(\xi, \eta) = \Theta_{E^*}(\bar{\xi}, \eta)$.

At $z = 0$, write $db^{\alpha}_{\rho} = \sum_{i=1}^{n} b^{\alpha}_{\rho i} dz^i$. Then, by (2.24), $\Theta^{\rho}_{\sigma i \bar{j}} = -\sum_{\alpha} b^{\alpha}_{\sigma i} \bar{b}^{\alpha}_{\rho j}$ and so

$$\Theta_E(\xi, \eta) = \sum_{\alpha} \left| \sum_{\sigma,i} b^{\alpha}_{\sigma i} \xi^{\sigma} \eta^i \right|^2 . \tag{2.28}$$

Since $\mathbf{E}^*$ is ample, we have

$$\begin{cases} 0 \longrightarrow \mathbf{F}^* \longrightarrow \Gamma(\mathbf{E}^*) \longrightarrow \mathbf{E}^* \longrightarrow 0 \\ \mathbf{F}^* \longrightarrow \mathbf{E}^* \otimes \mathbf{T}^* \longrightarrow 0 . \end{cases} \tag{2.29}$$

Taking the second exact sequence at $z = 0$ and dualizing, we get

$$0 \longrightarrow \mathbf{E}_z \otimes \mathbf{T}_z \xrightarrow{\psi} \mathbf{F}_z \longrightarrow 0 . \tag{2.30}$$

In terms of the frames $(e_1, \cdots, e_r)$ for $\mathbf{E}$, $(f_1, \cdots, f_{m-r})$ for $\mathbf{F}$, and coordinates above, we have that

$$\psi\left(e_{\sigma} \otimes \frac{\partial}{\partial z^i}\right) = \sum_{\alpha} b^{\alpha}_{\sigma i} f_{\alpha} . \tag{2.31}$$

Combining (2.31) and (2.28), we have that

$$\Theta_E(\xi, \eta) = -\left| \psi(\xi \otimes \eta) \right|^2 , \tag{2.32}$$

which is negative by (2.30). This proves Theorem A.

Remark. If $r > 1$, the universal bundle $\mathbf{S} \to \mathbf{G}$ is *not* ample, and is in fact *not* positive, as follows easily from (2.8).

(g) *Curvatures in the associated projective bundle*. We now compute an example discussed in [10, Lemma 9.1]. Let $\mathbf{E} \to V$ be a holomorphic vector bundle in which we have a hermitian metric. On $\mathbf{E} - V$ we define a positive real function h by $h(z, \xi) = (\xi, \xi)_z =$

${}^t\xi h(z)\xi$, where $\xi \in \mathbf{E}_z$, the fibre of $\mathbf{E}$ at $z \in V$. For $\lambda \in \mathbf{C}^*$, $h(z, \lambda\xi) = |\lambda|^2 h(z, \xi)$.

The quotient space $\mathbf{E} - V/\mathbf{C}^* = P(\mathbf{E})$ is a bundle $P(\mathbf{E}) \to V$, whose fibre $P(\mathbf{E})_z$ is the projective space $P(\mathbf{E}_z)$ of lines in $\mathbf{E}_z$. Clearly $\mathbf{E} - V \to P(\mathbf{E})$ is a principal bundle and $h(z, \xi)$ is a metric in the associated line bundle $\mathbf{L}(\mathbf{E})$. By (2.1), the curvature is given by $\Theta_{\mathbf{L}(\mathbf{E})} = -\partial\bar{\partial} \log h(z, \xi)$.

To calculate $\Theta_{\mathbf{L}(\mathbf{E})}$, we choose local coordinates $z^1, \cdots, z^n$ on V and a frame $f(z) = (e_1(z), \cdots, e_r(z))$ for $\mathbf{E} \to V$ such that $h(0) = I$, $dh(0) = 0$. Evaluated at $z = 0$, we have (cf. the proof of Lemma 9.1 in [10]).

$$\Theta_{\mathbf{L}(\mathbf{E})} = -\frac{{}^t\xi\partial\bar{\partial}h\xi}{|\xi|^2} - \left\{\frac{(d\xi, d\xi)(\xi, \xi) - (d\xi, \xi)(\xi, d\xi)}{|\xi|^2}\right\}. \tag{2.33}$$

To interpret this, we consider the dual projective bundle $P(\mathbf{E}^*) \to V$ and the line bundle $\mathbf{L} \to P(\mathbf{E}^*)$ which, on each fiber of $P(\mathbf{E}^*) \to V$, is positive; thus $\mathbf{L} = \mathbf{L}(\mathbf{E}^*)^*$. By (2.33), the curvature of $\mathbf{L}$ at $(\xi, 0)$ is

$$\Theta_{\mathbf{L}} = -\frac{{}^t\xi\partial\bar{\partial}h\xi}{|\xi|^2} + \left\{\frac{(d\xi, d\xi)(\xi, \xi) - (d\xi, \xi)(\xi, d\xi)}{|\xi|^2}\right\}. \tag{2.34}$$

If we set

$$\Theta_{\mathbf{E}}(\xi) = \sum_{\rho,\sigma,i,j} \Theta^{\rho}_{\sigma i \bar{j}} \xi^{\sigma} \bar{\xi}^{\rho} dz^i \wedge d\bar{z}^j , \tag{2.35}$$

and

$$\omega(\xi) = \frac{(d\xi, d\xi)(\xi, \xi) - (d\xi, \xi)(\xi, d\xi)}{|\xi|^2} ,$$

then we have:

$$\Theta_{\mathbf{L}} = \frac{\Theta_{\mathbf{E}}(\xi)}{|\xi|^2} + \omega(\xi) . \tag{2.36}$$

Comparing (2.35) with the definition (0.1) of the introduction, we find (cf. [10, Prop. 9.1]):

(2.37) $\mathbf{L} \to P(\mathbf{E}^*)$ is *positive* if $\mathbf{E} \to V$ is.

We now want to compute the *canonical bundle* $\mathbf{K}_{\mathbf{P}}$ of $\mathbf{P}(\mathbf{E}^*)$. The formula to be verified is

$$\mathbf{K}_{\mathbf{P}} = \mathbf{L}^{-r} \otimes \det(\mathbf{E}) \otimes \mathbf{K}_V . \tag{2.38}$$

In (2.38), $\det(\mathbf{E}) \otimes \mathbf{K}_V$ is a line bundle on V which has been lifted in the fibering $\mathbf{P}(\mathbf{E}^*) \to V$.

In order to keep our signs straight, we need a few preliminary remarks. Let E be a vector space, E^* the dual space, and $\mathbf{P}(E^*)$ the projective space. If $\mathbf{L}^* \to \mathbf{P}(E^*)$ is the tautological line bundle whose fibre $\mathbf{L}^*_\lambda$ over a line $\lambda \subset E^*$ is just the 1-dimensional vector space λ, then E is the space $H^0(\mathbf{P}(E^*), \mathcal{O}(\mathbf{L}))$ of holomorphic cross-sections of the dual bundle $\mathbf{L} \to \mathbf{P}(E^*)$. In particular, each vector $e \in E$ gives a holomorphic function on $\mathbf{L}^*$; note that $\mathbf{L}^* - \mathbf{P}(E^*) = E^* - \{0\}$ and then it is clear how to think of e as a function on $\mathbf{L}^*$.

Now let $e_1, \cdots, e_r$ be a basis for E and let ξ_ρ be the function on $E^* - \{0\}$ defined by e_ρ. Then

$$\eta = \sum^r (-1)^{\sigma-1}\xi_\sigma d\xi_1 \wedge \cdots \wedge d\hat{\xi}_\sigma \wedge \cdots \wedge d\xi_r$$

is an $(r-1)$-form on $\mathbf{P}(E^*)$ with values in $\mathbf{L}^{*r} = \mathbf{L}^{-r}$. If we have a linear substitution $\hat{e}_\rho = \sum_\sigma g^\sigma_\rho e_\sigma$, then $\hat{\xi}_\rho = \sum g^\sigma_\rho \xi_\sigma$ and $d\hat{\xi}_\rho = \sum_\sigma g^\sigma_\rho d\xi_\sigma$. Clearly then $\hat{\eta} = \det(g^\sigma_\rho)\eta$.

If now $z^1, \cdots, z^n$ are local coordinates on V and $e_1(z), \cdots, e_r(z)$ is a frame for $\mathbf{E}$, we let $\xi_\rho = \xi_\rho(z)$ be the corresponding functions on $\mathbf{E}^* - \{0\}$ and set

$$\varphi(z, \xi) = dz^1 \wedge \cdots \wedge dz^n \{\textstyle\sum_\sigma (-1)^{\sigma-1}\xi_\sigma d\xi_1 \wedge \cdots \wedge d\hat{\xi}_\sigma \wedge \cdots \wedge d\xi_r\} .$$

Then $\varphi(z, \xi)$ is an $(n + r - 1)$-form on $\mathbf{P}(\mathbf{E}^*)$ with values in $\mathbf{L}^{-r}$. If $\hat{e}_1(z), \cdots, \hat{e}_r(z)$ is a new frame for $\mathbf{E}$, then $\hat{e}_\rho(z) = \sum_\sigma g^\sigma_\rho(z) e_\sigma(z)$ and $d\hat{\xi}_\rho(z) = \sum_\sigma g^\sigma_\rho(z) d\xi_\sigma(z)$ modulo $dz^1, \cdots, dz^n$. Thus $\hat{\varphi} = (\det g)\varphi$. From this it follows that $\mathbf{K}_P = \mathbf{L}^{-r} \otimes \det(\mathbf{E}) \otimes \mathbf{K}_V$ as desired.

To close this section, let us prove (1.9) for S a locally free sheaf (this is the case to be used below). Thus, given $\mathbf{F} \to V$, we must show

$$H^q(V, \mathcal{O}(\mathbf{E}^{(\mu)} \otimes \mathbf{F})) \cong H^q(P, \mathcal{O}(\mathbf{L}^\mu \otimes \pi^*\mathbf{F})) . \tag{2.40}$$

For this, we use the *Leray spectral sequence* [8, p. 201]. The pth derived sheaf $R^p_\pi(\mathbf{L}^\mu \otimes \pi^*\mathbf{F})$ for $\mathcal{O}(\mathbf{L}^\mu \otimes \pi^*\mathbf{F})$; $\mathbf{P}(\mathbf{E}^*) \xrightarrow{\pi} V$ is the sheaf arising from the pre-sheaf

$$U \longrightarrow H^p(\pi^{-1}(U), \mathcal{O}(\mathbf{L}^\mu \otimes \pi^*\mathbf{F}) \mid \pi^{-1}(U))$$

for $U \subset V$ an open set. Taking U for which $\mathbf{F} \mid U$ is trivial, we see that $R^p_\pi(\mathbf{L}^\mu \otimes \pi^*\mathbf{F}) \cong R^p_\pi(\mathbf{L}^\mu) \otimes \mathcal{O}_V(\mathbf{F})$. Taking U for which $\mathbf{E}^* \mid U$ is trivial, $\pi^{-1}(U) \cong U \times \mathbf{P}_{r-1}$ and $\mathbf{L}^\mu \mid \pi^{-1}(U) \cong \mathcal{O}_U \otimes \mathcal{O}_{\mathbf{P}_{r-1}}(\mathbf{H}^\mu)$ where $\mathbf{H} \to \mathbf{P}_{r-1}$ is the hyperplane bundle. If $\mathbf{P}_{r-1} = \mathbf{P}(E^*)$ for a vector space E, then $H^p(\mathbf{P}_{r-1}, \mathcal{O}(\mathbf{H}^\mu)) = 0$ for $p > 0, \mu \geqq 0$ and $H^0(\mathbf{P}_{r-1}, \mathcal{O}(\mathbf{H}^\mu)) \cong E^{(\mu)}$. It follows that $R^p_\pi(\mathbf{L}^\mu) = 0$ for $p > 0$ and $R^0_\pi(\mathbf{L}^\mu) = \mathcal{O}(\mathbf{E}^\mu)$. The assertion (2.40) now follows from the spectral sequence.

(h) PROOF OF THEOREM H. We let $\mathbf{E} \to V$ be a positive bundle with fibre $\mathbf{C}^r$, and we consider a section $\xi \in H^0(V, \mathcal{O}(\mathbf{E}))$ whose divisor $S = \{z \in V: \xi(z) = 0\}$ is a non-singular subvariety of dimension $n - r$, where $n = \dim V$. We define a non-negative function φ on V by $\varphi(z) = |\xi(z)|^2 = {}^t\overline{\xi(z)}h(z)\xi(z)$, where $h(z)$ is the metric in $\mathbf{E} \to V$. We want to check first that S is a *non-degenerate critical manifold* of φ, which means that we must show:

$$(2.40) \quad \begin{cases} d\varphi = 0 \text{ along } S\,; \\ \text{if } z \in S, \text{ the null-space of the } \textit{hessian } H(\varphi) \\ \text{at } z \text{ is the tangent space } \mathbf{T}_z(S)\,. \end{cases}$$

We may choose local coordinates $z^1, \cdots, z^n$ on V and a frame $e_1(z), \cdots, e_r(z)$ for $\mathbf{E}$ such that

$$\xi(z) = \begin{pmatrix} z^1 \\ \vdots \\ z_r \end{pmatrix}.$$

We may also assume that $h(0) = I$. Then $\varphi(z) = \sum_{\rho,\sigma} h_{\rho\sigma}(z)\bar{z}^\rho z^\sigma$ and the hessian $H(\varphi)$ of φ at the origin is

$$H(\varphi) = \begin{pmatrix} I_r & 0 \\ 0 & 0 \end{pmatrix}.$$

From this, (2.40) is clear.

Let now z_0 be a *critical point* of φ on $V - S$; i.e., $d\varphi(z_0) = 0$. In local coordinates $z^1, \cdots, z^n$, we may assume that z_0 is the origin, and we may compute the hessian $H(\varphi)$ of φ at z_0. The *index* of φ at z_0 is the dimension of the subspace of $\mathbf{T}_{z_0}(V)$ on which $H(\varphi)$ is negative definite. We want to show

(2.41) The index of φ at z_0 is no less than $n - r + 1$.

We assume that $h(0) = I$ and $dh(0) = 0$. Then $\varphi(z) = {}^t\overline{\xi(z)}h(z)\xi(z)$ and, at $z = 0$, we have ${}^t\overline{d\xi}\xi + {}^t\bar{\xi}d\xi = 0$ since z_0 is a critical point, and $\partial\bar{\partial}\varphi = {}^t\bar{\xi}\partial\bar{\partial}h\xi - {}^t\overline{d\xi} \wedge d\xi$. Thus, by (2.6),

$$(2.42) \qquad H(\varphi)_{i\bar{j}} = -\sum_{\rho,\sigma} \Theta^\rho_{\sigma i\bar{j}}\xi^\sigma\bar{\xi}^\rho + \sum_\rho \frac{\overline{\partial\xi^\rho}}{\partial z^j}\frac{\partial\xi^\rho}{\partial z^i}\,.$$

Let A be the $r \times n$ matrix $A^\rho_i = \partial\xi^\rho/\partial z^i$. For

$$\eta = \begin{pmatrix} \eta^1 \\ \vdots \\ \eta^n \end{pmatrix},$$

we have

$$\sum_{i,j} H(\varphi)_{i\bar{j}}\eta^i\bar{\eta}^j = -\Theta(\xi, \eta) + ({}^t\overline{A\eta})(A\eta) , \tag{2.43}$$

where $-\Theta(\xi, \eta)$ is negative definite since $\mathbf{E} \to V$ is positive. If we set $W = \{\eta \in \mathbf{C}^n : A\eta = 0\}$, from (2.43), to prove (2.41) it will suffice to show that $\dim W \geqq n - r + 1$.

Now $\dim W = n - \operatorname{rank}(A)$ where $\operatorname{rank}(A)$ is the number of independent row vectors $A^\rho = (A_1^\rho, \cdots, A_n^\rho)$ in A. From ${}^t\bar{\xi}d\xi = 0$ we get

$$\sum_\rho \bar{\xi}^\rho \frac{\partial \xi^\rho}{\partial z^i} = 0 \qquad (i = 1, \cdots, n) ,$$

so that $\sum_\rho \bar{\xi}^\rho A^\rho = 0$. Thus $\operatorname{rank}(A) \leqq r - 1$ and $\dim W \geqq n - r + 1$ as required.

The same argument as used in [2, Prop. 4.1 and Th. II] shows that

$$V = S \cup e_1 \cup \cdots \cup e_s \qquad \text{where } \dim e_k \geqq n - r + 1 .$$

This notation means that, up to homotopy type, V is obtained by attaching cells $e_1, \cdots, e_s$ to S. From this it follows that $H_q(S, \mathbf{Z})$ maps onto $H_q(V, \mathbf{Z})$ for $q \leqq n - r$ and $H_q(S, Z)$ maps into $H_q(V, Z)$ for $q \leqq n - r - 1$, which proves Theorem H.

Remark. We let $W = V - S$ be the open manifold obtained by removing S from V. Then $\psi(z) = -\log \varphi(z)$ is an *exhaustion function* on W, and we let $L(\psi) = \partial\bar{\partial}\psi$ be the *E. E. Levi form* of ψ.

(2.44) $L(\psi)$ has everywhere at least $n - r$ positive eigenvalues

PROOF. $L(\psi) = -\partial\bar{\partial} \log \varphi = -\partial\left(\dfrac{\bar{\partial}\varphi}{\varphi}\right) = \dfrac{-\partial\bar{\partial}\varphi}{\varphi} + \dfrac{\partial\varphi\bar{\partial}\varphi}{\varphi^2}$.

Now $\partial\varphi = \partial(\xi, \xi) = (D'\xi, \xi)$ (since $\bar{\partial}\xi = 0$); $\bar{\partial}\varphi = \bar{\partial}(\xi, \xi) = (\xi, D'\xi)$; and $\partial\bar{\partial}\varphi = \partial(\xi, D'\xi) = (D'\xi, D'\xi) + (\xi, \Theta\xi)$ (since $\bar{\partial}D'\xi = (\bar{\partial}D' + D'\bar{\partial})\xi = D^2\xi = \Theta\xi$). This gives that

$$L(\psi) = \frac{-(D'\xi, D'\xi)(\xi, \xi) + (D'\xi, \xi)(\xi, D'\xi)}{\varphi^2} + \frac{(\Theta\xi, \xi)}{(\xi, \xi)} . \tag{2.44$'$}$$

If $\eta = \{\eta^i\}$ is a vector and $\xi = \sum_{\rho=1}^r \xi^\rho e_\rho$ our section of $\mathbf{E}$, then

$$(\Theta\xi, \xi)\eta \wedge \bar{\eta} = \sum_{\rho,\sigma,i,j} \Theta_{\sigma i\bar{j}}^\rho \xi^\sigma \bar{\xi}^\rho \eta^i \bar{\eta}^j = \Theta(\xi, \eta) > 0$$

so that $L(\psi)$ is positive on the space of vectors η such that

$\langle D'\xi, \eta\rangle = 0$. But the dimension of this space is less than or equal r (since $D'\xi = \sum_{\rho=1}^{r} D'\xi^\rho e_\rho$), so that $L(\psi)$ has everywhere at least $n - r$ positive eigenvalues.

We observe that the calculation just used shows that $d\varphi = (D'\xi, \xi)$ and $H(\varphi) = (D'\xi, D'\xi) - (\Theta\xi, \xi)$ where $\varphi = (\xi, \xi)$ is the function used in the proof of Theorem H.

3. Positive, ample, and cohomologically positive bundles

(a) *General properties.* We want to give some functorial properties of positive, ample, and cohomologically positive bundles. The first are

$$(3.1)\quad \begin{cases} \text{If } \mathbf{E} \to V \text{ is positive, ample, or cohomologically} \\ \text{positive, then so is } \mathbf{L} \to \mathbf{P}(\mathbf{E}^*)\ ; \end{cases}$$

$$(3.2)\quad \begin{cases} \text{If } \mathbf{L} \to \mathbf{P}(\mathbf{E}^*) \text{ is ample or cohomologically posi-} \\ \text{tive, then } \mathbf{E} \to V \text{ is also.} \end{cases}$$

We have already proved (cf. (2.37)) that

$$\mathbf{E} \text{ positive} \Longrightarrow \mathbf{L} \text{ positive}\ .$$

We now shall show that $\mathbf{E}$ ample $\Leftrightarrow$ $\mathbf{L}$ ample. Let $z_0 \in V$ and $\mathbf{E}_{z_0}$ be the fibre of $\mathbf{E}$ at z_0. If $(z_0, \xi) \in \mathbf{P}(\mathbf{E}^*)$ is a point lying over z_0, then ξ is a line in $\mathbf{E}^*_{z_0}$ and we let

$$\mathbf{F}_{(z_0,\xi)} = \{e \in \mathbf{E}_{z_0} : \langle e, \xi\rangle = 0\}\ .$$

As in §2 (g), we see that $\mathbf{L}_{(z_0,\xi)} = \mathbf{E}_{z_0}/\mathbf{F}_{(z_0,\xi)}$ and, in the exact sequence $0 \to \mathbf{F} \to \pi^*(\mathbf{E}) \to \mathbf{L} \to 0$ over $\mathbf{P} = \mathbf{P}(\mathbf{E}^*)$, $\mathbf{F}_{(z_0,\xi)}$ is the fibre of $\mathbf{F}$ at (z_0, ξ). Using the isomorphisms

$$H^0(V, \mathcal{O}(\mathbf{E})) \cong H^0(\mathbf{P}, \mathcal{O}(\pi^*\mathbf{E})) \cong H^0(\mathbf{P}, \mathcal{O}(\mathbf{L}))\ ,$$

we see that $\mathbf{E}$ is generated by its sections if, and only if, $\mathbf{L}$ is. This checks the first condition in the definition (0.2) of ampleness. To verify the second condition, we choose local coordinates $z^1, \cdots, z^n$ on V such that z_0 is the origin and a frame for $\mathbf{E}$ so that $\mathbf{E}_{z_0} = \mathbf{C}^r$ is column vectors

$$\zeta = \begin{pmatrix} \zeta^1 \\ \vdots \\ \zeta^r \end{pmatrix}$$

and $\mathbf{F}_{(z_0,\xi)}$ is given by $\zeta^1 = 0$. Suppose that $\mathbf{F}_{z_0} = \{s \in \Gamma(\mathbf{E}) : s(z_0) = 0\}$ and that $\mathbf{F}_{z_0} \to \mathbf{E}_{z_0} \otimes \mathbf{T}^*_{z_0} \to 0$. Then we may find $s^j \in \mathbf{F}_{z_0}$ with

$$s^j(z) = \begin{pmatrix} z^j \\ 0 \\ \vdots \\ 0 \end{pmatrix} + (\text{higher powers}) \ ,$$

and $\mathbf{s}^\rho \in \Gamma(\mathbf{E})$ with

$$\mathbf{s}^\rho(z) = \begin{pmatrix} 0 \\ \vdots \\ 1 \\ \vdots \\ 0 \end{pmatrix} + (\text{higher powers}) \ .$$

Then $s^1, \cdots, s^n; \mathbf{s}^2, \cdots, \mathbf{s}^r$ lie in $\mathbf{F}_{(z_0,\xi)}$ and the differentials $ds^1, \cdots, ds^n; d\mathbf{s}^2, \cdots, d\mathbf{s}^r$ span $\mathbf{L}_{(z_0,\xi)} \otimes \mathbf{T}_{(z_0,\xi)}(\mathbf{P})^*$, so that $\mathbf{L}$ is ample. By reversing this argument, we see that $\mathbf{L}$ ample $\Rightarrow$ $\mathbf{E}$ ample, which proves our assertion.

To prove cohomological positivity (cf. definition (0.3)), it will suffice to take S to be locally free sheaf $\mathcal{O}(\mathbf{F})$ where $\mathbf{F}$ is a holomorphic vector bundle (cf. §3 (b), the proof of Theorem B). From (2.40) it follows that $\mathbf{L}$ cohomologically positive $\Rightarrow$ $\mathbf{E}$ cohomologically positive. To prove the converse, by examining the proof of (2.40) above, it will suffice to show that

(a) there is $\nu = \nu(\mathbf{F})$ such that $R^p_\pi(\mathbf{F} \otimes \mathbf{L}^\nu) = 0$ for $p > 0$, and

(b) $R^0_\pi(\mathbf{F} \otimes \mathbf{L}^\nu \otimes \mathbf{L}^\mu) \cong R^0_\pi(\mathbf{F} \otimes \mathbf{L}^\nu) \otimes R^0_\pi(\mathbf{L}^\mu)$.

In fact, given (a) and (b),

$$\begin{aligned} H^q(\mathbf{P}(\mathbf{E}^*), \mathcal{O}(\mathbf{F} \otimes \mathbf{L}^{\nu+\mu})) &\cong H^q(V, R^0_\pi(\mathbf{F} \otimes \mathbf{L}^{\nu+\mu})) \\ &\cong H^p(V, R^0_\pi(\mathbf{F} \otimes \mathbf{L}^\nu) \otimes \mathcal{O}(\mathbf{E}^{(\mu)})) = 0 \\ &\qquad \text{for } \mu > \mu_0(\mathbf{F}), q > 0 \ , \end{aligned}$$

since $R^0_\pi(\mathbf{F} \otimes \mathbf{L}^\nu)$ is locally free. Now both (a) and (b) are well-known.

We cannot prove that $\mathbf{L} \to \mathbf{P}(\mathbf{E}^*)$ positive $\Rightarrow$ $\mathbf{E} \to V$ positive. However, for the notion of *weakly positive*, due to Grauert [9], it is true that

(3.3) $\mathbf{E}$ weakly positive $\Leftrightarrow$ $\mathbf{L}$ weakly positive.

In fact, Grauert says that $\mathbf{E} \to V$ is weakly positive if, and only if, a tubular neighborhood of the zero section in $\mathbf{E}^* \to V$ is *strongly pseudo-convex* (cf. [13]). Since $\mathbf{E}^* - V = \mathbf{L}^* - \mathbf{P}(\mathbf{E}^*)$ (cf. 2(g)), it is clear that (3.3) is verified.

On the other hand, it is true that

(3.4) E positive $\Rightarrow$ E weakly positive.

The proof goes as follows. On $\mathbf{E} \to V$, we define a positive function φ by $\varphi(z, \xi) = |\xi|_z^2 = {}^t\bar{\xi}h(z)\xi$. The *tubular neighborhood* T of V in E is given by $\varphi < \varepsilon$. We must calculate the *E. E. Levi form* $L(\varphi) = \partial\bar{\partial}\varphi$ evaluated on the tangent space to the boundary ∂T_ε of T_ε. Choose coordinates $z^1, \cdots, z^n$ and a frame for E such that $h(0) = I, dh(0) = 0$. Then, at $(0, \xi)$, $d\varphi = \sum_\rho (d\xi^\rho\bar{\xi}^\rho + \xi^\rho d\bar{\xi}^\rho)$ and the tangent space to the boundary is all vectors

$$\sum_j \varphi^j \frac{\partial}{\partial z^j} + \sum \eta^\rho \frac{\partial}{\partial \xi^\rho}$$

with $\sum_\rho \bar{\xi}^\rho\eta^\rho = 0$. The Levi form is given by

$$(3.5) \qquad L(\varphi) = -\sum_{\rho,\sigma,i,j} \Theta^\rho_{\sigma i\bar{j}}\xi^\sigma\bar{\xi}^\rho dz^i d\bar{z}^j = \sum_\gamma d\xi^\gamma d\bar{\xi}^\gamma ,$$

since $\partial\bar{\partial}h = -\Theta_{\mathrm{E}}$ at $z = 0$ (cf. 2.6)). From this, (3.4) follows.

Grauert's weakly positive is a better function-theoretic notion than our positive. However, the differential-geometric methods lead to Theorems D, G, and H, and we know of no function-theoretic argument which gives these results.

Suppose now that $0 \to \mathbf{S} \to \mathbf{E} \to \mathbf{Q} \to 0$ is an exact sequence of holomorphic bundles. Then

(3.6) If E is positive, ample, cohomologically positive, or numerically positive, then so is Q .

Assume that E is positive and let $f = (e_1, \cdots, e_r)$ be an orthonormal frame for E such that $e_1, \cdots, e_s$ is a frame for S (cf. 2(e); as done there, we let $1 \leqq \mu, \lambda \leqq s$; $s + 1 \leqq \alpha, \beta \leqq r$; $1 \leqq \rho, \sigma \leqq r$). From (2.15) we have that

$$(3.7) \qquad (\Theta_{\mathrm{Q}})^\alpha_\beta = (\Theta_{\mathrm{E}})^\alpha_\beta + \sum_\lambda b^\alpha_\lambda \wedge \bar{b}^\beta_\lambda$$

where $b = (b^\alpha_\lambda) \in A^{1,0}(\mathrm{Hom}\,(\mathbf{S}, \mathbf{Q}))$ is the 2nd fundamental form of S in E. Thus

$$\begin{aligned}\sum_{\alpha,\beta,i,j} (\Theta_{\mathrm{Q}})^\alpha_{\beta i\bar{j}}\xi^\beta\bar{\xi}^\alpha\eta^i\bar{\eta}^j &= \sum_{\alpha,\beta,i,j} (\Theta_{\mathrm{E}})^\alpha_{\beta i\bar{j}}\, \xi^\beta\bar{\xi}^\alpha\eta^i\bar{\eta}^j \\ &\quad + \sum_{\alpha,\beta,\lambda,i,j} b^\alpha_{\lambda i}\bar{\xi}^\alpha\eta^i\overline{b^\beta_{\lambda j}\bar{\xi}^\beta\eta^j} \\ &\geqq \sum (\Theta_{\mathrm{E}})^\alpha_{\beta i j}\xi^\beta\bar{\xi}^\alpha\eta^i\bar{\eta} > 0\end{aligned}$$

since E is positive. This proves that E positive $\Rightarrow$ Q positive.

It is clear that: E ample $\Rightarrow$ Q ample.

Equally trivially, we note that: E weakly positive $\Rightarrow$ Q weakly positive.

Suppose now that **E** is cohomologically positive; we want to show that, for $\mathbf{F} \to V$ a bundle, there is $\mu = \mu_0(\mathbf{F})$ such that $H^q(V, \mathcal{O}(\mathbf{Q})^{(\mu)} \otimes \mathbf{F})) = 0$ for $\mu > \mu_0$, $q > 0$. An algebraic proof begins by observing the exact sequences

$$0 \longrightarrow \mathbf{\Sigma}_\mu \longrightarrow \mathbf{E}^{(\mu)} \longrightarrow \mathbf{Q}^{(\mu)} \longrightarrow 0 , \tag{3.8}$$

$$0 \longrightarrow \Lambda^2\mathbf{S} \otimes \mathbf{E}^{(\mu-2)} \longrightarrow \mathbf{S} \otimes \mathbf{E}^{(\mu-1)} \longrightarrow \mathbf{\Sigma}_\mu \longrightarrow 0 . \tag{3.9}$$

From (3.8), it will suffice to check that $\mathbf{\Sigma}_\mu$ is cohomologically positive; this follows from (3.9) and the fact that **E** is cohomologically positive. This proves that **E** cohomologically positive $\Rightarrow$ **Q** cohomologically positive.

A more instructive geometric proof can be given using the fact that $\mathbf{L}(\mathbf{E}) \to \mathbf{P}(\mathbf{E}^*)$ is cohomologically positive (cf. (3.1) above). For $z \in V$, let $\mathbf{P}(\mathbf{Q}^*)_z \subset \mathbf{P}(\mathbf{E}^*)_z = \mathbf{P}(\mathbf{E}_z^*)$ be those lines ξ in $\mathbf{P}(\mathbf{E}_z^*)$ with $\langle \xi, \mathbf{S}_z \rangle = 0$. Then $\bigcup_{z \in V} \mathbf{P}(\mathbf{Q}^*)_z = \mathbf{P}(\mathbf{Q}^*)$ is a sub-bundle of $\mathbf{P}(\mathbf{E}^*) \to V$. As the notation suggests, $\mathbf{P}(\mathbf{Q}^*)$ is the projective bundle associated to $\mathbf{Q}^* \to V$. Clearly $\mathbf{L}(\mathbf{E}) \mid \mathbf{P}(\mathbf{Q}^*) = \mathbf{L}(\mathbf{Q})$, and so we have the exact sheaf sequences

$$\begin{aligned} 0 \longrightarrow I \otimes \mathcal{O}(\mathbf{L}(\mathbf{E})^\mu \otimes \pi^*\mathbf{F}) \longrightarrow \mathcal{O}(\mathbf{L}(\mathbf{E})^\mu \otimes \pi^*\mathbf{F}) \\ \longrightarrow \mathcal{O}(\mathbf{L}(\mathbf{Q})^\mu \otimes \pi^*\mathbf{F}) \longrightarrow 0 , \end{aligned} \tag{3.10}$$

where $I \subset \mathcal{O}(\mathbf{P}(\mathbf{E}^*))$ is the ideal sheaf of $\mathbf{P}(\mathbf{Q}^*)$. Using the cohomological positivity of $\mathbf{L}(\mathbf{E}) \to \mathbf{P}(\mathbf{E}^*)$, we find that

$$H^q(\mathbf{P}(\mathbf{Q}^*), \mathcal{O}(\mathbf{L}(\mathbf{Q})^\mu \otimes \pi^*\mathbf{F})) = 0 \qquad \text{for } \mu \geqq 0, q > 0 .$$

From (2.40) it follows that **Q** is cohomologically positive.

The final general property is:

(3.11) $\begin{cases} \text{If } \mathbf{E} \text{ and } \mathbf{F} \text{ are positive, ample, or cohomo-} \\ \text{logically positive, then so is } \mathbf{E} \otimes \mathbf{F} . \end{cases}$

Letting $(e_1, \cdots, e_r) = (\cdots, e_\rho, \cdots)$ be a frame for **E** and $(f_1, \cdots, f_s) = (\cdots, f_\alpha, \cdots)$ a frame for **F**, by (2.10) we have

$$(\Theta_{\mathbf{E}\otimes\mathbf{F}})^{\rho\alpha}_{\sigma\beta i\bar{j}} = \delta^\alpha_\beta(\Theta_{\mathbf{E}})^\rho_{\sigma i\bar{j}} + \delta^\rho_\sigma(\Theta_{\mathbf{F}})^\alpha_{\beta i\bar{j}} . \tag{3.12}$$

Then

$$\begin{aligned} \sum (\Theta_{\mathbf{E}\otimes\mathbf{F}})^{\sigma\alpha}_{\sigma\beta ij}\xi^{\sigma\beta}\bar{\xi}^{\rho\alpha}\eta^i\bar{\eta}^j &= \sum (\Theta_{\mathbf{E}})^\rho_{\sigma ij}\xi^{\sigma\alpha}\bar{\xi}^{\rho\alpha}\eta^i\eta^j \\ &\quad + \sum (\Theta_{\mathbf{F}})^\alpha_{\beta ij}\xi^{\rho\beta}\bar{\xi}^{\rho\alpha}\eta^i\bar{\eta}^j > 0 . \end{aligned}$$

Note in fact that $\mathbf{E} \otimes \mathbf{F} > 0$ if $\mathbf{E} > 0$, $\mathbf{F} \geqq 0$.

It is trivial that, if **E** and **F** are ample, then $\mathbf{E} \otimes \mathbf{F}$ is ample. In fact, if **E** is ample and **F** is generated by its sections, then $\mathbf{E} \otimes \mathbf{F}$ is ample.

For the proof of **E**, **F** cohomologically positive $\Rightarrow \mathbf{E} \otimes \mathbf{F}$ cohomologically positive, we refer to Hartshorne's paper [14]. This paper contains a through account of the relationship between cohomological positivity and the algebraic operations $\oplus$, $\otimes$, etc. on vector bundles.

(b) *The Nakano inequalities and proofs of Theorems* B *and* G. The vanishing theorems we shall use are based on representing the cohomology groups $H^q(V, \mathcal{O}(\mathbf{E}))$ by *harmonic forms* [19], [22], and [15]. Suppose that we have a hermitian metric in $\mathbf{E} \to V$ and *Kähler metric* on V. We choose locally an orthonormal frame $f = (e_1, \cdots, e_r)$ for **E** and an orthonormal co-frame $\omega^1, \cdots, \omega^n$ for V; the Kähler form is then

$$\gamma = \frac{i}{2} \sum_{j=1}^{n} \omega^j \wedge \bar{\omega}^j .$$

An **E**-valued (p, q)-form φ is written

$$\varphi = \frac{1}{p!q!} \sum_{\rho, I, J} \varphi^{\rho}_{I\bar{J}} e_\rho \otimes \omega^I \otimes \bar{\omega}^J$$

where $I = (i_1, \cdots, i_p)$ and $J = (j_1, \cdots, j_q)$, $\omega^I = \omega^{i_1} \wedge \cdots \wedge \omega^{i_q}$, etc. The point inner product is

$$\langle \varphi, \psi \rangle = \frac{1}{p!q!} \left(\sum_{\rho, I, J} \varphi^{\rho}_{I\bar{J}} \bar{\psi}^{\rho}_{I\bar{J}}\right) .$$

Setting

$$(\varphi, \psi) = \int_V \langle \varphi, \psi \rangle \left\{\left(\frac{i}{2}\right)^n \omega^1 \wedge \bar{\omega}^1 \wedge \cdots \wedge \omega^n \wedge \bar{\omega}^n\right\} ,$$

we obtain a global inner product and we let $\bar{\partial}^*$ be the adjoint of $\bar{\partial}$; thus $(\bar{\partial}^* \varphi, \psi) = (\varphi, \bar{\partial} \psi)$ for all ψ, and this equation defines $\bar{\partial}^* \varphi$. The *laplacian* $\square$ is defined by $\square = \bar{\partial}^* \bar{\partial} + \bar{\partial} \bar{\partial}^*$, and the space of *harmonic forms* is

$$\mathbf{H}^{p,q}(\mathbf{E}) = \{\varphi : \square \varphi = 0\} = \{\varphi : \bar{\partial} \varphi = 0 = \bar{\partial}^* \varphi\} . \tag{3.13}$$

For φ an **E**-valued (p, q) form, we now write $\varphi = \sum_\rho \varphi^\rho \otimes e_\rho$ where φ^ρ is a (p, q) form and we set

$$\Theta \wedge \varphi = \sum_{\rho, \sigma} \Theta^\rho_\sigma \wedge \varphi^\sigma \otimes e_\rho , \tag{3.14}$$

(3.15) $$L\varphi = \sum_\rho \gamma \wedge \varphi^\rho \otimes e_\rho \ ,$$

and we define the adjoint L^* of L by

(3.16) $$\langle L^*\varphi, \psi\rangle = \langle \varphi, L\psi\rangle \ ,$$

for all ψ. We remark that usually L^* is denoted by Λ and $\bar{\partial}^*$ by $\mathfrak{D}$. Comparing (3.14) with (2.2), we see that

$$\Theta \wedge \varphi = D^2\varphi = (D'\bar{\partial} + \bar{\partial}D')\varphi \ .$$

The following basic inequalities are due to S. Nakano [22]. For $\varphi \in H^{p,q}(\mathbf{E})$,

(3.17) $$\frac{i}{2}(L^*\Theta \wedge \varphi, \varphi) \geqq 0 \ ,$$

(3.18) $$\frac{i}{2}(\Theta \wedge L^*\varphi, \varphi) \leqq 0 \ .$$

For the proofs, which use only one basic Kähler identity and no tensor calculations, we refer to [22], [10], or [4]. As hinted above the operator $\Theta \wedge \varphi$ in (3.14) arises from $\Theta \wedge \varphi = D^2\varphi$.

We shall use the Nakano inequalities primarily in case $\mathbf{E} \rightarrow V$ is a *negative line bundle*. Then we take as the Kähler metric $\gamma = (-i/2)\Theta$ and subtract (3.18) from (3.17) to get: For $\varphi \in \mathbf{H}^{p,q}(\mathbf{E})$,

(3.19) $$(\{L^*L - LL^*\}\varphi, \varphi) \leqq 0 \ .$$

Combining this with the elementary identity $(L^*L - LL^*)\varphi = (n - p - q)\varphi$ [26, p. 21] gives

(3.20) $$(n - p - q)(\varphi, \varphi) \leqq 0 \qquad \text{for } \varphi \in \mathbf{H}^{p,q}(\mathbf{E}) \ .$$

From (3.20) we obtain our vanishing theorem

(3.21) $$\begin{cases} H^q(V, \Omega^p(\mathbf{E})) = 0 \text{ for } p + q < n \text{ and } \mathbf{E} \rightarrow V \text{ a} \\ \text{negative line bundle} \ . \end{cases}$$

For $p = 0$, (3.21) is the original Kodaira theorem [18], which may be dualized to read

(3.22) $$H^q(V, \mathcal{O}(\mathbf{E})) = 0 \qquad \text{for } q > 0 \text{ if } \mathbf{E} \otimes \mathbf{K}^* \text{ is positive} \ .$$

Here $\mathbf{K} \rightarrow V$ is the *canonical bundle* and we have used the *duality theorem*: $H^q(V, \mathcal{O}(\mathbf{E})) \cong H^{n-q}(V, \mathcal{O}(\mathbf{K} \otimes \mathbf{E}^*))^*$ (cf. [15]).

PROOF OF THEOREM G. Let now $\mathbf{E} \rightarrow V$ be a general holomorphic vector bundle with fibre $\mathbf{C}^r$, $\mathbf{P}(\mathbf{E}^*) \xrightarrow{\pi} V$ and $\mathbf{L} \rightarrow \mathbf{P}(\mathbf{E}^*)$ the tautological bundles discussed in §2 (g), and $\mathbf{F} \to V$ a holomorphic vector bundle. We shall use the isomorphism (1.9) (cf. (2.40)).

$$H^q(V, \mathcal{O}(\mathbf{E}^{(\mu)} \otimes \mathbf{F})) \cong H^q(\mathbf{P}, \mathcal{O}(\mathbf{L}^\mu \otimes \pi^*\mathbf{F})) , \tag{3.23}$$

and the curvature calculations in §2 (g) to prove a vanishing theorem for the groups $H^q(V, \mathcal{O}(\mathbf{E}^\mu \otimes \mathbf{F}))$, $\mathbf{F} \to V$ being a line bundle. To state this result, suppose that we have hermitian metrics in $\mathbf{E} \to V$, $\mathbf{F} \to V$, and a Kähler metric on V. We denote the curvature forms for $\mathbf{E}$, $\mathbf{F}$, $\mathbf{K}_V$ by $\sum_{i,j} \Theta^\rho_{\sigma i\bar{j}} dz^i \wedge d\bar{z}^j$, $\sum_{i,j} \varphi_{i\bar{j}} dz^i \wedge d\bar{z}^j$, and $\sum_{i,j} k_{i\bar{j}} dz^i \wedge d\bar{z}^j$ respectively. We introduce the quadratic form:

$$\begin{aligned} Q_\mu(\xi, \eta) = (\mu + r) \textstyle\sum_{\rho,\sigma,i,j} \Theta^\rho_{\sigma i\bar{j}} \xi^\sigma \bar{\xi}^\rho \eta^i \bar{\eta}^j \\ - \textstyle\sum_{i,j} \{k_{i\bar{j}} - \varphi_{i\bar{j}} + \textstyle\sum_\rho \Theta^\rho_{\rho i\bar{j}}\} \eta^i \bar{\eta}^j |\xi|^2 . \end{aligned} \tag{3.24}$$

(3.25) THEOREM G′. *If $Q_\mu(\xi, \eta)$ is positive definite, then*

$$H^q(V, \mathcal{O}(\mathbf{E}^{(\mu)} \otimes \mathbf{F})) = 0 \qquad \text{for } q > 0 .$$

To prove Theorem G′, by (3.23) it will suffice to show that $H^q(\mathbf{P}, \mathcal{O}(\mathbf{L}^\mu \otimes \mathbf{F})) = 0$ for $q > 0$. This will be done using (3.22) on $\mathbf{P}$; thus, in (3.22), replace V by $\mathbf{P}$, $\mathbf{E}$ by $\mathbf{L}^\mu \otimes \mathbf{F}$, and $\mathbf{K}$ by $\mathbf{K}_P$. Using the formula (2.38) for $\mathbf{K}_P$, we must show

(3.26) $\mathbf{L}^{\mu+r} \otimes \mathbf{K}_V^* \otimes \det(\mathbf{E})^* \otimes \mathbf{F}$ is positive if Q_μ in (3.24) is.

But (3.24) follows immediately from (2.36), and this proves Theorem G′.

To prove Theorem G, we need to show that $Q_\mu(\xi, \eta)$ in (3.24) is positive definite for all $\mu \geqq 0$ if $\mathbf{F}^* \otimes \mathbf{K}_V \otimes \det(\mathbf{E}) < 0$, provided that $\mathbf{E}$ is generated by its sections. By (2.24), at a given point, $\Theta^\rho_{\sigma i\bar{j}} = \sum_\alpha A^\alpha_{\rho i} \bar{A}^\alpha_{\sigma j}$ and so

$$\textstyle\sum_{\rho,\sigma,i,j} \Theta^\rho_{\sigma i\bar{j}} \xi^\sigma \bar{\xi}^\rho \eta^i \bar{\eta}^j = \textstyle\sum_{\alpha,\rho,\sigma i,j} A^\alpha_{\rho i} \bar{\xi}^\rho \eta^i \overline{A^\alpha_{\sigma j} \bar{\xi}^\sigma \eta^j} \geqq 0 .$$

Then $\Theta_\mu(\xi, \eta) \geqq$ (curvature form for $\mathbf{K}_V^* \otimes \mathbf{F} \otimes \det(\mathbf{E})^* |\xi|^2) > 0$. This completes the proof of Theorem G.

PROOF OF THEOREM B. Suppose we can prove that, for any *vector bundle* $\mathbf{F} \to V$,

$$H^q(V, \mathcal{O}(\mathbf{E}^{(\mu)} \otimes \mathbf{F})) = 0 \qquad \text{for } q > 0,\ \mu > \mu_0(\mathbf{F}) . \tag{3.27}$$

Then, taking as *definition* of a coherent sheaf $S \to V$ a sheaf having a global resolution

$$0 \longrightarrow F_k \longrightarrow F_{k-1} \longrightarrow \cdots \longrightarrow F_1 \longrightarrow S \longrightarrow 0 \tag{3.28}$$

by locally free sheaves, using exact cohomology sequences we will find that $H^q(V, \mathcal{O}(\mathbf{E}^{(\mu)} \otimes S)) = 0$ for $q > 0$, $\mu \geqq \mu_0(S)$.

Now that usual definition of a coherent sheaf is that there are *locally* resolutions by free sheaves; then as a conseqence of cohomological positivity for some particular line bundle, we can find a global resolution (3.28). Taking a projective embedding $V \subset \mathbf{P}_N$, one of Serre's basic theorems [24] and [25] is the cohomological positivity of the standard positive line bundle. The conclusion then is that, to prove Theorem B, it will suffice to show (3.27).

We shall prove (3.27) by showing that (3.27) holds in case $\mathbf{E} \rightarrow V$ is a positive line bundle. Then, in the general case, $\mathbf{L} \rightarrow \mathbf{P}(\mathbf{E}^*)$ is a positive line bundle (cf. (2.37)) and so $H^q(\mathbf{P}, \mathcal{O}(\mathbf{L}^\mu \otimes \mathbf{F})) = 0$ for $q > 0$, $\mu \geqq \mu_0(\mathbf{F})$. Using the isomorphism (3.23), we get (3.27).

To prove (3.27) for $\mathbf{E} \rightarrow V$ a positive line bundle, it will suffice to show that

$$H^q(V, \mathcal{O}(\mathbf{E}^{-\mu} \otimes \mathbf{F})) = 0 \qquad \text{for } q < n,\ \mu \geqq \mu_0(\mathbf{F})\ . \tag{3.29}$$

Indeed, one may pass from (3.27) to (3.29) and back by using the duality theorem. We shall prove (3.29) by using the first Nakano inequality (3.17). As our Kähler metric, we take $\gamma = (i/2)\Theta_{\mathbf{E}}$ where $\Theta_{\mathbf{E}}$ is the curvature in $\mathbf{E}$. By (2.10), we have

$$\Theta_{\mathbf{E}^{-\mu}\otimes\mathbf{F}} = -\mu\Theta_{\mathbf{E}} \otimes 1 + 1 \otimes \Theta_{\mathbf{F}}\ .$$

For φ a $(0, q)$ form with values in $\mathbf{E}^{-\mu} \otimes F$,

$$\begin{aligned}\frac{i}{2}L^*\Theta_{\mathbf{E}^{-\mu}\otimes\mathbf{F}} \wedge \varphi &= -\mu L^*L\varphi + \frac{i}{2}L^*\Theta_{\mathbf{F}} \wedge \varphi \\ &= -\mu(n-q)\varphi + \frac{i}{2}L^*\Theta_{\mathbf{F}} \wedge \varphi\ .\end{aligned}$$

From (3.17) we get:

$$\frac{i}{2}(L^*\Theta_{\mathbf{F}} \wedge \varphi, \varphi) \geqq \mu(n-q)(\varphi, \varphi)\ . \tag{3.30}$$

Taking μ large in (3.30), we get $\varphi = 0$, which implies that

$$H^q(V, \mathcal{O}(\mathbf{E}^{-\mu} \otimes \mathbf{F})) = 0 \qquad \text{for } \mu \geqq \mu_0(\mathbf{F}),\ q < n\ .$$

(c) PROOF OF THEOREMS C AND F. We shall prove Theorem C using Theorem G′ above (cf. (3.25)) and using *quadratic transformations* [20]. Let $z_0 \in V$ be a fixed point and $W \xrightarrow{\pi} V$ the quadratic transform of V at z_0. This may be described as follows. Let

$z^1, \cdots, z^n$ be coordinates in an open set $U \subset V$ with $z_0 = (0, \cdots, 0)$. We consider the principal bundle $\mathbf{C}^n - \{0\} \to \mathbf{P}_{n-1} = \mathbf{P}$ and let $\mathbf{H}^* \to \mathbf{P}$ be the corresponding line bundle. Then $\mathbf{H}^* - \mathbf{P} = \mathbf{C}^n - \{0\}$ and so U corresponds to a tubular neighborhood $T(U) \subset \mathbf{H}^*$ of the zero section, and W is obtained by replacing U with $T(U)$. Geometrically, we have replaced z_0 with the lines through z_0.

Letting $\mathbf{P} = \pi^{-1}(z_0)$, $\mathbf{P}$ is a $\mathbf{P}_{n-1}$ embedded in W and $W - \mathbf{P} \cong V - \{z_0\}$. If $\mathbf{L} \to W$ is a line bundle determined by the divisor $\mathbf{P} \subset W$, then $\mathbf{L} \mid P = \mathbf{H}^*$. To find a metric in $\mathbf{L} \to W$, we choose a concentric open set $U_1 \subset U$ with $\bar{U}_1 \subset U$, $z_0 \in U_1$ and let ρ be a function which is one on U_1 and zero on $W - U$. Then there is defined a metric in $\mathbf{L}$ whose curvature is [20]

$$\Theta_{\mathrm{L}} = \partial\bar{\partial}(\rho \log \{\textstyle\sum_{i=1}^{n} | z^i |^2\}) . \tag{3.31}$$

Thus $\Theta_{\mathrm{L}} = 0$ on $W - U$ and, on U_1,

$$\Theta_{\mathrm{L}} = -\left\{\frac{(dz, dz)(z, z) - (dz, z)(z, dz)}{(z, z)^2}\right\} \tag{3.32}$$

(cf. (2.5)). In particular, $\Theta_{\mathrm{L}} < 0$ on U_1.

Over W we consider the exact bundle sequences

$$\begin{cases} 0 \to \pi^*\mathbf{E}^{(\mu)} \otimes \mathbf{L}^* \to \pi^*\mathbf{E}^{(\mu)} \to \pi^*\mathbf{E}^{(\mu)} \mid \mathbf{P} \to 0 , \\ 0 \to \pi^*\mathbf{E}^{(\mu)} \otimes \mathbf{L}^{*2} \to \pi^*\mathbf{E}^{(\mu)} \otimes \mathbf{L}^* \to \pi^*\mathbf{E}^{(\mu)} \otimes \mathbf{L}^* \mid \mathbf{P} \to 0 . \end{cases} \tag{3.33}$$

We observe that

$$\begin{aligned} H^0(W, \mathcal{O}(\pi^*\mathbf{E}^{(\mu)})) &\cong H^0(V, \mathcal{O}(\mathbf{E}^{(\mu)}) , \\ H^0(\mathbf{P}, \mathcal{O}(\pi^*\mathbf{E}^{(\mu)})) &\cong \mathbf{E}_{z_0}^{(\mu)} , \\ H^0(\mathbf{P}, \mathcal{O}(\pi^*\mathbf{E}^{(\mu)} \otimes \mathbf{H})) &\cong \mathbf{E}_{z_0}^{(\mu)} \otimes \mathbf{T}_{z_0}^*(V) \end{aligned}$$

(since $H^0(\mathbf{P}, \mathcal{O}(\mathbf{H}))$ has basis $z^1, \cdots, z^n$). Thus, if

$$H^1(W, \mathcal{O}(\pi^*\mathbf{E}^{(\mu)} \otimes \mathbf{L}^*)) = 0 = H^1(W, \mathcal{O}(\pi^*\mathbf{E}^{(\mu)} \otimes \mathbf{L}^{*2})) ,$$

we get from (3.33) that

$$\begin{cases} 0 \longrightarrow \mathbf{F}_{z_0}^{(\mu)} \longrightarrow \Gamma(\mathbf{E}^{(\mu)}) \longrightarrow \mathbf{E}_{z_0}^{(\mu)} \longrightarrow 0 \\ \mathbf{F}_{z_0}^{(\mu)} \longrightarrow \mathbf{F}_{z_0} \otimes \mathbf{T}_{z_0}^* \longrightarrow 0 , \end{cases} \tag{3.34}$$

where $\mathbf{F}_{z_0}^{(\mu)} = \{s \in \Gamma(\mathbf{E}^{(\mu)}) \colon s(z_0) = 0\} \cong H^0(W, \mathcal{O}(\pi^*\mathbf{E}^{(\mu)} \otimes \mathbf{L}^*))$.

Thus, to prove Theorem C, we need to show that

$$\begin{cases} \mathbf{E} \text{ positive} \Longrightarrow H^1(W, \mathcal{O}(\pi^*\mathbf{E}^{(\mu)} \otimes \mathbf{L}^*)) = 0 = \\ H^1(W, \mathcal{O}(\pi^*\mathbf{E}^{(\mu)} \otimes \mathbf{L}^{*2})) \text{ for all } \mu > \mu_0, \text{ and } all\ z_0 \in V . \end{cases} \tag{3.35}$$

We want to look at the quadratic form (3.24) on W for $\pi^*\mathbf{E}$ and where $\mathbf{F} = \mathbf{L}^*$ or $\mathbf{L}^{*2}$. Since $\mathbf{E} \to V$ is positive, the quadratic form $\Theta_{\pi^*\mathbf{E}}(\xi, \eta)$ on W is positive definite outside of $\mathbf{P}$. From (3.32), the curvature $\Theta_{\mathbf{L}^*}$ is positive definite in U_1 and zero outside U. Finally, the canonical bundle [20]

$$\mathbf{K}_W = \pi^*\mathbf{K}_V \otimes \mathbf{L}^{n-1} . \tag{3.36}$$

Thus $\Theta_{\mathbf{K}_W} = \Theta_{\pi^*\mathbf{K}_V} + (n-1)\Theta_{\mathbf{L}^*}$. From the explicit form of (3.24), we see that $Q_\mu(\xi, \eta)$ is positive definite on W for $\mu \geqq \mu_0$. In fact, $\Theta_\mu(\xi, \eta) = (\mu + r)\Theta_{\pi^*\mathbf{E}}(\xi, \eta) + \{n\Theta_{\mathbf{L}^*}(\eta) - \Theta_{\pi^*\det\mathbf{E}}(\eta) - \Theta_{\pi^*\mathbf{K}_V}(\eta)\} \,|\, \xi\,|^2$, from which our assertion is clear. Furthermore, by using obvious estimates from continuity, we see that the quadratic forms will be positive definite for $\mu \geqq \mu_0$ and *all* $z_0 \in V$. Then (3.35) follows from Theorem G′ ((3.25)).

This completes the proof of Theorem C.

Proof of Theorem F. Let $z_0 \in V$ and $I_{z_0} \subset \mathcal{O}$ be the ideal sheaf of z_0. We can choose $\mu(z_0)$ such that

$$H^q(V, I_{z_0} \otimes \mathcal{O}(\mathbf{E}^\mu)) = 0 = H^q(V, I_{z_0}^2 \otimes \mathcal{O}(\mathbf{E}^{(\mu)})) \qquad \text{for } \mu \geqq \mu(z_0),\ q > 0 .$$

From the cohomology sequences of $0 \to I_{z_0} \otimes \mathcal{O}(\mathbf{E}^{(\mu)}) \to \mathcal{O}(\mathbf{E}^{(\mu)}) \to \mathbf{E}_{z_0}^{(\mu)} \to 0$ and $0 \to I_{z_0} \otimes \mathcal{O}(\mathbf{E}^{(\mu)}) \to I_{z_0} \otimes \mathcal{O}(\mathbf{E}^{(\mu)}) \to \mathbf{E}_{z_0}^{(\mu)} \otimes \mathbf{T}_{z_0}^* \to 0$, we get

$$\begin{cases} 0 \longrightarrow \mathbf{F}_{z_0}^{(\mu)} \longrightarrow \Gamma(\mathbf{E}^{(\mu)}) \longrightarrow \mathbf{E}_{z_0}^{(\mu)} \longrightarrow 0 , \\ \mathbf{F}_{z_0}^{(\mu)} \longrightarrow \mathbf{E}_{z_0}^{(\mu)} \otimes \mathbf{T}_{z_0}^* \longrightarrow 0 \end{cases}$$

for $\mu \geqq \mu(z_0)$. In particular then, there is a neighborhood $U(z_0)$ such that, for $z \in U(z_0)$, we have

$$\begin{cases} 0 \longrightarrow \mathbf{F}_z^{(\mu(z_0))} \longrightarrow \Gamma(\mathbf{E}^{(\mu(z_0))}) \longrightarrow \mathbf{E}_z^{(\mu(z_0))} \longrightarrow 0 , \\ \mathbf{F}_z^{(\mu(z_0))} \longrightarrow \mathbf{E}_z^{(\mu(z_0))} \otimes \mathbf{T}_z^* \longrightarrow 0 . \end{cases} \tag{3.37}$$

Observe that (3.37) holds for any $\mu = \mu_1\mu(z_0)$.

Given now $z_1 \in V$, we may find $\mu(z_1)$ such that

$$\begin{cases} 0 \longrightarrow \mathbf{F}_{z_1}^{(\nu_1)} \longrightarrow \Gamma(\mathbf{E}^{(\nu_1)}) \longrightarrow \mathbf{E}_{z_1}^{(\nu_1)} \longrightarrow 0 \\ \mathbf{F}_{z_1}^{(\nu_1)} \longrightarrow \mathbf{E}_{z_1}^{(\nu_1)} \otimes \mathbf{T}_{z_1}^* \longrightarrow 0 \end{cases} \tag{3.38}$$

where $\nu_1 = \nu(z_1)\mu(z_0)$. Also, (3.38) holds in a neighborhood $U(z_1)$, as well as in $U(z_0)$. Continuing, and using the compactness of V, we may find a μ such that $\mathbf{E}^{(\mu)}$ is ample. That is

(3.39) $\mathbf{E}$ cohomomologically positive $\Rightarrow \mathbf{E}^{(\mu)}$ ample for some μ.

This is weak version of Theorem F, which asserts that $\mathbf{E}^{(\mu)}$ is ample for *all* $\mu > \mu_0$. To prove this, it will suffice to show

$$(3.40)\quad \begin{cases} H^q(V, I_{z_0} \otimes \mathcal{O}(\mathbf{E}^{(\mu)})) = 0 = H^q(V, I_{z_0}^2 \otimes \mathcal{O}(\mathbf{E}^{(\mu)})) \\ \text{for } q > 0,\ \mu \geqq \mu_0, \text{ and } all\ z_0 \in V\,. \end{cases}$$

Consider now $\mathbf{L} \to \mathbf{P}(\mathbf{E}^*)$. Since $\mathbf{E}$ is cohomologically positive, $\mathbf{L}$ is colomologically positive and, by (3.39) applied to $\mathbf{L}$, $\mathbf{L}^\mu \to \mathbf{P}$ is ample for some μ. Thus $\mathbf{L}^\mu$ is positive; i.e., there is a metric in $\mathbf{L}^\mu$ whose curvature $\Theta_{\mathbf{L}^\mu} > 0$. Since $\Theta_{\mathbf{L}} = \frac{1}{\mu}(\Theta_{\mathbf{L}^\mu})$, we have that $\Theta_{\mathbf{L}} > 0$.

In the fibering $\mathbf{P} \xrightarrow{\pi} V$, let $P(z_0) = \pi^{-1}(z_0)$ and $I(z_0) \subset \mathcal{O}_{\mathbf{P}}$ be the ideal sheaf of $P(z_0)$. Then, by (1.9),

$$H^q(V, I_{z_0}^k \otimes \mathcal{O}(\mathbf{E}^{(\mu)})) \cong H^q(\mathbf{P}, I(z_0)^k \otimes \mathcal{O}(\mathbf{L}^\mu))$$

and, using (3.40), to prove Theorem F it will suffice to show:

$$(3.41)\quad \begin{matrix} H^q(\mathbf{P}, I(z_0)^k \otimes \mathcal{O}(\mathbf{L}^\mu)) = 0 \\ k = 1, 2\,. \end{matrix} \qquad \text{for } q > 0;\ \mu \geqq \mu_0\,;$$

We shall prove (3.41) by a method similar to the proof of Theorem C above.

Let $Q \xrightarrow{\varpi} \mathbf{P}$ be the quadratic transform of $\mathbf{P}$ along $P(z_0)$. For the equations giving Q, see [11, §V]. We set $S = \tilde{\omega}^{-1}(P(z_0))$ so that $Q - S = \mathbf{P} - P(z_0)$, and we let $\mathbf{J} \to Q$ be the line bundle determined by the divisor $S \subset Q$. We recall that $R^p_{\varpi}(\mathbf{J}^{*l}) = 0$ for $p > 0$ and $R^0_{\varpi}(\mathbf{J}^{*l}) = I(z_0)^l$. This, plus Leray's theorem, gives

$$(3.42)\quad H^q(Q, \mathcal{O}(\mathbf{J}^{*l} \otimes \tilde{\omega}^{-1}\mathbf{L}^\mu)) = H^q(\mathbf{P}, I(z_0)^l \otimes \mathcal{O}(\mathbf{L}^\mu))\,.$$

We used special cases of (3.42) just below (3.33) above.

To prove (3.41), using (3.42) we need to show

$$(3.43)\quad \begin{cases} H^q(Q, \mathcal{O}(\mathbf{J}^{*k} \otimes \tilde{\omega}^{-1}\mathbf{L}^\mu)) = 0 \text{ for } q > 0;\ \mu \geqq \mu_0\,; \\ k = 1, 2; \text{ and for all } z_0 \in V\,. \end{cases}$$

As in the proof of Theorem C above, we shall prove (3.43) for fixed z_0 by a curvature estimate and then, using continuity of these estimates relative to z_0, we will get (3.43). Thus we need to compute the curvature $\Theta_{\mathbf{J}}$ on Q.

Let U be a polycylindrical neighborhood of z_0 such that $\mathbf{E} \mid U \cong U \times \mathbf{C}^r$. Then $\pi^{-1}(U) \cong U \times \mathbf{P}_{r-1}$ and $\pi^{-1}(U) \subset \mathbf{P}(\mathbf{E}^*)$ is a tubular neighborhood $N(P(z_0))$ of $P(z_0) = \{0\} \times \mathbf{P}_{r-1}$, where $z_0 = (0, \cdots, 0)$. Letting $\tilde{U}$ be the ordinary quadratic transform of U at $\{0\}$, $\tilde{\omega}^{-1}(N(P(z_0))) =$

$T(S)$ is a tubular neighborhood of S and $T(S) \cong \tilde{U} \times \mathbf{P}_{r-1}$. If $z^1, \cdots, z^n$ are coordinates on U and $\xi^0, \cdots, \xi^r$ homogeneous coordinates on $\mathbf{P}_{r-1}$, we have that $\Theta_{\mathrm{L}} = \Theta_{\mathrm{L}}(dz^i, d\xi^\rho)$ is positive definite on $N(P(z_0))$. Furthermore, there is a metric in $\mathbf{J} \mid T(S)$ such that the curvature $\Theta_{\mathrm{J}} = \partial\bar{\partial} \log (\sum |z^i|^2)$ (cf. (3.32)). Thus $\Theta_{\varpi^{-1}(\mathrm{L})} + \Theta_{\mathrm{J}^*}$ is positive on $T(S)$. Since $\Theta_{\varpi^{-1}(\mathrm{L})} \mid Q - T(S)$ is positive, we may fit the above metric on $\mathbf{J} \mid T(S)$ is to get a metric on $\mathbf{J} \to Q$ such that $\nu\Theta_{\varpi^{-1}(\mathrm{L})} + \Theta_{\mathrm{J}^*} > 0$ for $\nu \geqq \nu_0$. By proceeding now just as in the proof of Theorem C, we will get (3.43) as desired.

(d) *Positive bundles and topological properties of algebraic varieties.* Let $\mathbf{E} \to V$ be a positive vector bundle and $\xi \in H^0(V, \mathcal{O}(\mathbf{E}))$ a holomorphic section whose divisor $S \subset V$ is a nice subvariety as discussed in § 2(h). Then Theorem H on the topology of S in V yields the following vanishing theorem for sheaf cohomology

$$H^q(V, I_S) = 0 \qquad \text{for } 0 \leqq q \leqq n - r, \tag{3.44}$$

where I_S is the ideal sheaf of S in V. For the proof of (3.44), we consider the exact sheaf sequence $0 \to I_S \to \mathcal{O}_V \to \mathcal{O}_S \to 0$. Since S and V are Kähler manifolds, we have a diagram

$$\begin{array}{ccccccc} & & & 0 & & 0 & \\ & & & \downarrow & & \downarrow & \\ \cdots \longrightarrow & H^q(I_S) & \longrightarrow & H^q(\mathcal{O}_V) & \longrightarrow & H^q(\mathcal{O}_S) & \longrightarrow \cdots \\ & & & \downarrow & & \downarrow & \\ & & & (H^q(V,\mathbf{C}) & \longrightarrow & H^q(S,\mathbf{C}) . & \end{array} \tag{3.45}$$

Using that $0 \to H^q(V,\mathbf{C}) \to H^q(S,\mathbf{C})$ for $q \leqq n-r$ and $0 \to H^q(V,\mathbf{C}) \to H^q(S,\mathbf{C}) \to 0$ for $q \leqq n-r-1$, we find $H^q(I_S) = 0$ for $0 \leqq q \leqq n-r$ as required.

In case S is a hypersurface ($\mathbf{E}$ is a line bundle), we have $I_S \cong \mathcal{O}_V(\mathbf{E}^*)$ and (3.44) becomes

$$H^q(V, \mathcal{O}(\mathbf{E}^*)) = 0 \qquad \text{for } 0 \leqq q \leqq n - 1, \tag{3.46}$$

which is the original Kodaira theorem [18].

Still considering the case when $S \subset V$ is a hypersurface, we shall show that Theorem H gives

$$H^q(V, \Omega_V^p(\mathbf{E}^*)) = 0 \qquad \text{for } p + q < n, \tag{3.47}$$

which generalizes (3.46). For the proof, consider the pair of exact sheaf sequences [15, page 127]

$$(3.48)\quad \begin{cases} 0 \longrightarrow \Omega_V^p(\mathbf{E}^*) \xrightarrow{\ \xi\ } \Omega_V^p \longrightarrow \Omega_{V|S}^p \longrightarrow 0 \\ 0 \longrightarrow \Omega_S^{p-1}(\mathbf{E}^*) \xrightarrow{\ d\xi\ } \Omega_{V|S}^p \longrightarrow \Omega_S^p \longrightarrow 0 . \end{cases}$$

In cohomology, we have:

$$(3.49)\quad \begin{cases} \to H^q(V, \Omega_V^p(\mathbf{E}^*)) \to H^q(V, \Omega_V^p) \to H^q(S, \Omega_{V|S}^p) \to \\ \qquad\qquad\qquad\qquad\qquad\qquad\qquad \| \\ \cdots \to H^q(S, \Omega_S^{p-1}(\mathbf{E}^*)) \to H^q(S, \Omega_{V|S}^p) \to H^q(S, \Omega_S^p) . \end{cases}$$

Using that $0 \to H^q(V, \Omega_V^p) \to H^q(S, \Omega_S^p)$ for $p + q < n$, and that this mapping is onto for $p + q < n - 1$, we get (3.47).

Conversely, from (3.47), we obtain

$$(3.50)\quad \begin{cases} 0 \longrightarrow H^q(V, \Omega_V^p) \longrightarrow H^q(S, \Omega_S^p) \longrightarrow 0 , & p + q < n - 1 \\ 0 \longrightarrow H^q(V, \Omega_V^p) \longrightarrow H^q(S, \Omega_S^p) , & p + q = n - 1 . \end{cases}$$

This in turn gives Theorem H over **Q**.

In conclusion, for a positive line bundle $\mathbf{E} \to V$ which has a non-singular section, we see that Theorem H (over **Q**) and the vanishing theorem (3.21) are equivalent.

As another application of topology to sheaf cohomology, we let $\mathbf{E} \to V$ be a positive bundle having a non-singular holomorphic section ξ with divisor S. Let $\omega \in H^{2r}(V, \mathrm{C})$ be the rth *Chern class* [15] of $\mathbf{E} \to V$; then $\omega \in H^r(V, \Omega_V^r)$ is dual to the homology class $[S] \in H_{2n-2r}(V, \mathrm{C})$ defined by S. We shall prove

$$(3.51)\quad \begin{cases} \text{The cup product } H^q(V, \Omega_V^p) \xrightarrow{\ \omega\ } H^{q+r}(V, \Omega_V^{p+r}) \text{ is} \\ \text{an isomorphism for } p + q = n - r . \end{cases}$$

For $r = 1$, this is a well-known result in Kähler varieties [26]. We have to show that $H^{n-r}(V, \mathrm{C}) \xrightarrow{\ \omega\ } H^{n+r}(V, C)$ is an isomorphism. First we dualize $H_{n-r+1}(V, S) \to H_{n-r}(S) \to H_{n-r}(V) \to 0$ to get $H^{n+r-1}(V - S) \to H^{n-r}(S) \to H^{n+r}(V) \to 0$. Combining this with $0 \to H^{n-r}(V) \to H^{n-r}(S) \to H^{n-r-1}(V, S)$, we find a diagram

$$(3.52)\quad \begin{array}{ccccccc} & & & & H^{n+r-1}(V - S) & & \\ & & & & \downarrow \partial^* & \searrow \psi^* & \\ 0 & \longrightarrow & H^{n-r}(V) & \longrightarrow & H^{n-r}(S) & \longrightarrow & H^{n-r+1}(V, S) \\ & & & \searrow \omega & \downarrow & & \\ & & & & H^{n+r}(V) & & \\ & & & & \downarrow & & \\ & & & & 0 . & & \end{array}$$

From (3.52), to prove (3.51) it will suffice to show that kernel $\partial^* =$ kernel ψ^*. Dually, we must show that, in

$$\begin{array}{ccc} H_{n-r+1}(V, S) & & \\ \downarrow \partial & \searrow \psi & \\ H_{n-r}(S) & \xrightarrow{\tau} & H_{n+r-1}(V - S) . \end{array}$$

ψ and ∂ have the same kernel; i.e., if $\partial\sigma \neq 0$, then $\psi(\sigma) \neq 0$.

At each point $z \in S$, the normal sphere S_z^{2r-1} to S in V at z is a $(2r - 1)$-sphere. If $\gamma \in H_{n-r}(S)$ is a cycle, then $\tau(\gamma) \in H_{n+r-1}(V - S)$ is the cycle traced out by the spheres S_z^{2r-1} for $z \in \gamma$ (this follows from the Thom isomorphism [15]). Let $\sigma \in H_{n-r+1}(V, S)$ and $\partial\sigma = \delta \in H_{n-r}(S)$. Now find a cycle $\hat{\delta} \in H_{n-r}(S)$ whose intersection number with δ is ± 1. Then it is essentially clear that the intersection number $\sigma \cdot \tau(\hat{\delta}) = \pm 1$, and so $\psi(\sigma) \neq 0$. This proves (3.51).

We remark that (3.51) is *false* for the universal bundle over the grassmannian (cf. [11, § IV. 4, page 397]), of course, this bundle is *not* positive.

To close this section, let us give one of the most noteworthy "vanishing theorems" derived from topological considerations; this is the famous *regularity of the adjoint* theorem of Picard [27]. Let V be an algebraic surface, $\mathbf{E} \to V$ a positive line bundle, and $\xi \in H^0(V, \mathcal{O}(\mathbf{E}))$ a non-singular section with divisor $S \subset V$. We set $\mathbf{N} = \mathbf{E} \mid S$, $\mathbf{K}_V =$ canonical bundle of V, $\mathbf{K}_S =$ cannonical bundle of S; the *adjunction formula* gives (cf. [15]):

(3.53) $$\mathbf{K}_S = \mathbf{L} \otimes \mathbf{K}_V \mid S .$$

Using (3.53), we get the exact sheaf sequences

(3.54) $$\begin{aligned} 0 \longrightarrow \mathcal{O}_V(\mathbf{K}_V \otimes \mathbf{E}^{r-1}) &\xrightarrow{\xi} \mathcal{O}_V(\mathbf{K}_V \otimes \mathbf{E}^r) \\ &\longrightarrow \mathcal{O}_S(\mathbf{K}_S \otimes \mathbf{N}^{r-1}) \longrightarrow 0 . \end{aligned}$$

The regularity of the adjoint theorem is

(3.55) $$H^0(\mathcal{O}_V(\mathbf{K}_V \otimes \mathbf{E}^r)) \longrightarrow H^0(\mathcal{O}_S(\mathbf{K}_S \otimes \mathbf{N}^{r-1})) \longrightarrow 0 , \qquad r \geqq 2 .$$

In classical language, if $V \subset P_3$ is a surface of degree n with ordinary singularities given by an affine equation $f(x, y, z) = 0$, then the adjoint polynomials $P(x, y, z)$ of degree at least $n - 3 + r$ $(r \geqq 1)$ cut out on a generic plane section $x =$ constant of V a complete linear series.

It will suffice to prove that $H^1(\mathbf{K}_V \otimes \mathbf{E}^{r-1})) = 0$ for $r \geqq 2$. We shall do this for $r = 2$ by giving Picard's original argument. For $r = 1$, (3.54) becomes

$$0 \longrightarrow \Omega_V^2 \xrightarrow{\ \xi\ } \Omega_V^2(\mathbf{E}) \longrightarrow \Omega_S^1 \longrightarrow 0 . \tag{3.56}$$

In cohomology we get a diagram:

$$\begin{array}{ccccccccc} H^0(\Omega_S^1) & \longrightarrow & H^1(\Omega_V^2) & \longrightarrow & H^1(\Omega_V^2(\mathbf{E})) & \longrightarrow & H^1(\Omega_S^1) & \xrightarrow{\ \delta\ } & H^2(\Omega_V^2) \\ \uparrow & \nearrow_{\omega} & & & & & & & \\ H^0(\Omega_V^1) & & & & & & & & \\ \uparrow & & & & & & & & \\ 0 & & & & & & & & \end{array} \tag{3.57}$$

As discussed above (cf. (3.51) and (3.52)), it is a purely topological fact that ω is an isomorphism, and obviously δ is an injection. Thus $H^1(\Omega_V^2(\mathbf{E})) = 0 = H^1(\mathcal{O}_V(\mathbf{K}_V \otimes \mathbf{E}))$. Using this in the exact cohomology sequences of (3.54) when $r = 2$, we get

$$H^0(\mathcal{O}_V(\mathbf{K}_V \otimes \mathbf{E}^2)) \longrightarrow H^0(\mathcal{O}_S(\mathbf{K}_S \otimes \mathbf{N})) \longrightarrow 0$$

as required.

PROOF OF THEOREMS I AND J. Let $\mathbf{E} \to V$ be a positive vector bundle and $\xi \in H^0(V, \mathcal{O}(\mathbf{E}))$ a holomorphic section such that $S = \{\xi(z) = 0\}$ is non-singular. Letting $I = I_S$ be the ideal sheaf of S and $\mathbf{F} \to V$ and arbitrary holomorphic bundle, we want to show that there is a constant $c = c(\mathbf{F}, V)$ such that, if the curvature form (0.1)′ has the property $\Theta_{\mathbf{E}}^{\sharp}(\xi, \eta) \geqq c\,|\,\xi\,|^2\,|\,\eta\,|^2$, then we have

$$H^q(V, I \otimes \mathcal{O}(\mathbf{F})) = 0 \qquad \text{for } q \leqq n - r . \tag{3.58}$$

We remark that, for $\mathbf{F} = 1$, (3.58) follows from (3.44).

Now we use the *Koszul complex*

$$\begin{aligned} 0 \longrightarrow \mathcal{O}(\Lambda^r\mathbf{E}^*) \xrightarrow{\ \xi\ } \cdots \longrightarrow \mathcal{O}(\Lambda^2\mathbf{E}^*) \\ \xrightarrow{\ \xi\ } \mathcal{O}(\mathbf{E}^*) \xrightarrow{\ \xi\ } I \longrightarrow 0 , \end{aligned} \tag{3.59}$$

where $\Lambda^q\mathbf{E}^* \to \Lambda^{q-1}\mathbf{E}^*$ is contraction with $\xi \in \Gamma(\mathbf{E})$. For example, if $r = 2$ then (3.59) becomes

$$0 \longrightarrow \mathcal{O}(\Lambda^2\mathbf{E}^*) \longrightarrow \mathcal{O}(\mathbf{E}^*) \longrightarrow I \longrightarrow 0 . \tag{3.60}$$

The sequence (3.59) remains exact after tensoring with $\mathcal{O}(\mathbf{F})$. From (3.25) we see that, if for $1 \leqq q \leqq r$ we have

(3.61) $$\Theta^{\#}_{\Lambda^q \mathbf{E}}(\xi, \eta) \geqq c(q) \mid \xi \mid^2 \mid \eta \mid^2$$

where $c(q)$ becomes large as c becomes large, then we will have

(3.62) $$H^q(V, \mathcal{O}(\Lambda^q \mathbf{E}^* \otimes \mathbf{F})) = 0$$

for $0 \leqq p \leqq n - 1$. But, by using (3.59), (3.62) implies that

$$H^p(V, \mathcal{O}(\mathbf{F}) \otimes I) = 0 \qquad \text{for } q \leqq n - r$$

as required.

Suppose now that $\Theta^{\#}_{\mathbf{E}}(\xi, \eta) \geqq c \mid \xi \mid^2 \mid \eta \mid^2$ and that we have $\mathbf{E} = \mathbf{S} \oplus \mathbf{Q}$ as a direct sum of holomorphic bundles. Then, for the curvature of the induced metric in $\mathbf{S}$, we will have $\Theta^{\#}_{\mathbf{S}}(\xi, \eta) \geqq c \mid \xi \mid^2 \mid \eta \mid^2$. Since $\Lambda^q \mathbf{E}$ is such a unitary direct summand of $\bigotimes^q \mathbf{E}$, it will suffice to show that

(3.63) $$\Theta^{\#}_{\mathbf{E}}(\xi, \eta) \geqq c \mid \xi \mid^2 \mid \eta \mid^2 \Longrightarrow \Theta^{\#}_{\otimes^q \mathbf{E}}(\xi, \eta) \geqq c_q \mid \xi \mid^2 \mid \eta \mid^2 ,$$

where c_q becomes large as c does. This is easy to verify using the tensor product rule for curvatures.

Proof of (0.6) *for* $r = 2$. Here we simply use (3.60),

$$H^q(V, \mathcal{O}(\Lambda^2 \mathbf{E}^*)) = 0 \qquad \text{for } q \leqq n - 1$$

(Kodaira vanishing theorem), and $H^q(V, I) = 0$ for $q \leqq n - 2$ (by (3.58)) to conclude that $H^q(V, \mathcal{O}(\mathbf{E}^*)) = 0$ for $q \leqq n - 2$.

PROOF OF THEOREM J. Let $S \subset V$ be the zero locus of $\xi \in H^0(V, \mathcal{O}(\mathbf{E}))$ and $\tilde{S} \subset \tilde{V}$ the result of blowing V up along S. Thus, if $\mathbf{N} \rightarrow S$ is the normal bundle, $\tilde{S} = \mathbf{P}(\mathbf{N})$ and so $z \in S$ is blown up by sending z into $\mathbf{P}(\mathbf{N}_z)$. Let $\mathbf{L} \rightarrow \tilde{V}$ be the line bundle determined by $\tilde{S}$; then $\mathbf{L} \mid \tilde{S}$ is the normal bundle and $\mathbf{L} \mid \mathbf{P}(\mathbf{N}_z)$ is the negative of the hyperplane bundle. If our blowing up diagram is

$$\begin{array}{ccc} \tilde{S} & \subset & \tilde{V} \\ \downarrow{\scriptstyle \pi} & & \downarrow{\scriptstyle \pi} \\ S & \subset & V , \end{array}$$

we want to show that

(3.64) There is an exact bundle sequence $0 \rightarrow \mathbf{L} \rightarrow \pi^*(\mathbf{E}) \rightarrow \mathbf{Q} \rightarrow 0$.

(*Proof.* $\tilde{S}$ determines a section $\sigma \in H^0(\tilde{V}, \mathcal{O}(\mathbf{L}))$ and the locus $\sigma = 0$ is the same as $\pi^* \xi = 0$, where $\pi^* \xi \in H^0(\tilde{V}, \mathcal{O}(\pi^* \mathbf{E}))$. It wiil suffice to show that $\pi^* \xi / \sigma$ is a non-vanishing holomorphic section of $\mathbf{L}^* \otimes \pi^* \mathbf{E}$. Locally on V we may choose coordinates $z^1, \cdots, z^n$ such that

$$\xi(z) = \begin{pmatrix} z^1 \\ \vdots \\ z^r \end{pmatrix}.$$

If $\lambda = [\lambda_1, \cdots, \lambda_r]$ are homogeneous coordinates on $\mathbf{P}_{r-1}$, locally $\tilde{V}$ is the set of (z, λ) satisfying the quadratic relations $z^\alpha\lambda_\beta - z^\beta\lambda_\alpha = 0$ $(1 \leqq \alpha, \beta \leqq r)$. In the open set $\lambda_r \neq 0$, $\tilde{S}$ is given by $z^r = 0$, since $z^r = 0$ and $z^\alpha\lambda_r - z^r\lambda_\alpha = 0$ taken together give $z^\alpha = 0$ for $1 \leqq \alpha \leqq r$. Thus σ is locally a unit times z^r, and it will suffice to have $\pi^*\xi/z^r$ a non-vanishing holomorphic vector. But $z^\alpha = z^r\lambda_\alpha$ and so

$$\frac{\pi^*\xi}{z^r} = \begin{pmatrix} \lambda_1 \\ \vdots \\ \lambda_{r-1} \\ 1 \end{pmatrix}$$

which is as required.)

We now recall that, setting $\tilde{I} = \mathcal{O}(\mathbf{L}^*)$, the direct image sheaves $R^q_\pi(\tilde{I}^\mu)$ are given by

$$R^q_\pi(\tilde{I}^\mu) = 0 \text{ for } q > 0,\ \mu \geqq 0 \text{ and } R^0_\pi(\tilde{I}^\mu) = I^\mu . \tag{3.65}$$

Using the Leray spectral sequence, it follows that

$$H^q(V, I^\mu \otimes \mathcal{O}(\mathbf{F})) \cong H^q(\tilde{V}, \tilde{I} \otimes \mathcal{O}(\pi^*\mathbf{F})) \tag{3.66}$$

for any holomorphic bundle $\mathbf{F} \to V$. Now by assumption $\mathbf{E} \to V$ has a metric such that the curvature form $\Theta_{\mathbf{E}}$ given by (0.1) is positive. Since $\pi^*\Theta_{\mathbf{E}} = \Theta_{\pi^*\mathbf{E}}$, there is induced a metric in $\pi^*\mathbf{E}$ whose curvature is positive semi-definite. In fact, $\Theta_{\pi^*\mathbf{E}}$ is zero on the $r-1$ dimensional tangent spaces to the fibering $\tilde{S} \to S$.

By (3.64), there is induced a metric in $\mathbf{L}$ and, from § 2. (c), we have that

$$\Theta_{\mathbf{L}} = \Theta_{\pi^*\mathbf{E}} \,|\, \mathbf{L} - b \wedge {}^t\bar{b} \tag{3.67}$$

where $b \in A^{1,0}(\mathrm{Hom}(\mathbf{Q}, \mathbf{L}))$ is the *second fundamental form* of $\mathbf{L}$ in $\pi^*\mathbf{E}$. Choosing unitary frames $e_1, \cdots, e_r$ for $\pi^*\mathbf{E}$ such that e_1 is a frame for $\mathbf{L} \to \tilde{V}$, (3.67) gives

$$\begin{aligned} \Theta_{\mathbf{L}} &= \Theta^1_1 - \textstyle\sum_{\alpha=2}^r b^\alpha_1 \wedge \bar{b}^\alpha_1 , \\ b &= \textstyle\sum_{\alpha-2}^r b^\alpha_1 e_1 \otimes e^*_\alpha . \end{aligned} \tag{3.68}$$

We want to show that

(3.69) Θ_L has everywhere at least $n-r+1$ positive eigenvalues.

From (3.68), this is clearly the case on $\tilde{V} - \tilde{S}$ where Θ_1^1 is positive definite. Let $\tilde{z} \in \tilde{S}$ and $z = \pi(\tilde{z})$ so that $\tilde{z} \in \mathbf{P}(\mathbf{N}_z)$. Since $b \wedge {}^t\bar{b}$ has rank less than or equal $r-1$, if we show that $b \wedge {}^t\bar{b}$ is non-singular on the tangent space to the projective space $\mathbf{P}(\mathbf{N}_z)$ passing thru $\tilde{z}$, then $\Theta_1^1 - b \wedge {}^t\bar{b}$ will be positive definite on the $n-r+1$ dimensional null-space of b. This will prove (3.69). But $\Theta_L \mid \mathbf{P}(\mathbf{N}_z) = -b \wedge {}^t\bar{b} \mid \mathbf{P}(\mathbf{N}_z)$ is the curvature of the negative of the hyperplane bundle over $\mathbf{P}_{r-1}$, and so $-b \wedge {}^t\bar{b}$ is negative definite on the tangent space to $\mathbf{P}(\mathbf{N}_z)$.

Now we use the theorem of Andreotti-Grauert [1] to conclude

(3.70) $$H^q(\tilde{V}, \mathcal{O}(\mathbf{L}^{-\mu} \otimes \pi^*\mathbf{F})) = 0 \qquad \text{for } \mu \geqq \mu_0,\ q \leqq n-r\,.$$

Combining (3.70) with (3.66) and using $\tilde{I}^\mu \cong \mathcal{O}(\mathbf{L}^{-\mu})$ gives Theorem J.

4. Chern classes and numerically positive bundles

(a) *Chern homology classes of holomorphic bundles.* Let V be an algebraic manifold and $\mathbf{E} \to V$ an *ample* holomorphic vector bundle (cf. (0.2)). Then we have

(4.1) $$0 \longrightarrow \mathbf{F} \longrightarrow \mathbf{\Gamma} \longrightarrow \mathbf{E} \longrightarrow 0$$

where $\mathbf{\Gamma} = V \times \Gamma(\mathbf{E})$ is a trivial bundle. Dualizing (4.1) gives

(4.2) $$0 \longrightarrow \mathbf{E}^* \longrightarrow \mathbf{\Gamma}^* \longrightarrow \mathbf{F}^* \longrightarrow 0\,.$$

We assume that $\mathbf{E}^*$ has fibre $\mathbf{C}^r$ and $\mathbf{F}^*$ has fibre $\mathbf{C}^{m+r}$. Choose a sequence of linear subspaces $\Gamma_1 \subset \Gamma_2 \subset \cdots \subset \Gamma_{m+r-1} \subset \mathbf{\Gamma}^*$ where $\dim \Gamma_\alpha = \alpha$. For each r-tuple of integers $\rho = (\rho_1, \cdots, \rho_r)$ with $\rho_1 \leqq \rho_2 \leqq \cdots \leqq \rho_r \leqq m$, we define $S_\rho \subset V$ as follows

(4.3) $$S_\rho = \{z \in V \text{ such that } \dim(\mathbf{E}_z^* \cap \Gamma_{j+\rho_j}) \geqq j \text{ for } j = 1, \cdots, r\}\,.$$

Referring to Hodge-Pedoe [16, Ch. XIV], we see that S_ρ is the intersection of V with the *Schubert cycle* of symbol ρ on the grassmannian $G(r, m)$. Taking the Γ_α to be generic, we find

(4.4) $$\begin{cases} S_\rho \text{ is an irreducible variety of dimension} \\ n + (\rho_1 + \cdots + \rho_r) - mr\,. \end{cases}$$

For later use, we record the trivial formula:

(4.5) $$\dim(\mathbf{E}_z^* \cap \Gamma_\alpha) = r + \alpha - \dim(\mathbf{E}_z^* + \Gamma_\alpha)\,.$$

The dimension formula (4.4) above will be checked in the case of all S_ρ we shall use.

Let now ρ_q be the Schubert symbol

$$(4.6) \qquad (\underbrace{m-1, \cdots, m-1}_{q}, \underbrace{m, \cdots, m}_{r-q}) \, ,$$

and $S_q = S_{\rho_q}$. By (4.4) $\dim S_q = n + q(m - 1 + m(r-q) - mr = n - q$ so that S_q defines $\sigma_q \in H_{2n-2q}(V, \mathbf{Z})$.

(4.7) *Definition.* σ_q is the qth *Chern* (*homology*) *class* of $\mathbf{E} \to V$.

Remarks. (i) There are r irreducible algebraic subvarieties $S_1, \cdots, S_r$ which carry the Chern classes $\sigma_1, \cdots, \sigma_r$ of $\mathbf{E} \to V$. Since $S_\rho \subset S_\tau$ if $\rho \leqq \tau$ (i.e., $\rho_j \leqq \tau_j$ for $j = 1, \cdots, r$), we have $S_1 \supset \cdots \supset S_r$.

(ii) Let $\xi \in \Gamma(\mathbf{E})$ be a section of $\mathbf{E}$ and (ξ) the zero locus of ξ. Then (ξ) carries the homology class σ_r. In words, the rth Chern class is the divisor of a (generic) section of $\mathbf{E}$.

PROOF. We consider ξ as a linear function on Γ^* and choose the Γ_α such that Γ_{m+r-1} is the set of vectors in Γ^* annilated by ξ. From $\dim (\mathbf{E}_z^* \cap \Gamma_{r+m-1}) = 2r + m - 1 - \dim (\mathbf{E}_z^* + \Gamma_{r+m-1})$, we see that: $\dim (\mathbf{E}_z^* \cap \Gamma_{r+m-1}) \geqq r \Leftrightarrow \dim (\mathbf{E}_z^* + \Gamma_{r+m-1}) \leqq r + m - 1 \Leftrightarrow \mathbf{E}_z^* \subset \Gamma_{r+m-1} \Leftrightarrow \xi(z) = 0$; i.e.,

$$(4.8) \qquad \dim (\mathbf{E}_z^* \cap \Gamma_{r+m-1}) \geqq r \Longleftrightarrow \xi(z) = 0 \, .$$

On the other hand, if $\dim (\mathbf{E}_z^* \cap \Gamma_{r+m-1}) \geqq r$, then $\mathbf{E}_z^* \subset \Gamma_{r+m-1}$ and $\dim (\mathbf{E}_z^* + \Gamma_{j+m-1}) \leqq r + m - 1$ for $j = 1, \cdots, r$. Using this in $\dim (\mathbf{E}_z^* \cap \Gamma_{j+m-1}) = r + j + m - 1 - \dim (\mathbf{E}_z^* + \Gamma_{j+m-1})$, we find:

$$(4.9) \qquad \dim (\mathbf{E}_z^* \cap \Gamma_{r+m-1}) \geqq r \Longleftrightarrow \dim (\mathbf{E}_z^* \cap \Gamma_{j+m-1}) \geqq j \quad \text{for } j = 1, \cdots, r \, .$$

Combining (4.8) and (4.9) gives our assertion.

(iii) The general rule is: Let $\xi_1, \cdots, \xi_r$ be r generic sections of $\mathbf{E}$. Then $\xi_1 \cdots \xi_{r-q+1}$ is a section of $\Lambda^{r-q+1}\mathbf{E}$ and $S_q \subset V$ is the zero locus of this section.

Proof. We may let Γ_{j+m-1} be given by $\xi_1 = 0, \cdots, \xi_{r-j+1} = 0$ where the ξ's are elements of the dual space of Γ^*. The Schubert symbol of S_q is $(\underbrace{m-1, \cdots, m-1}_{q}, \underbrace{m, \cdots, m}_{r-q})$. Since $\dim (\mathbf{E}_z^* \cap \Gamma_{j+m}) \geqq m$, the only non-trivial conditions are

$$\dim (\mathbf{E}_z^* \cap \Gamma_{j+m-1}) \geqq j \qquad \text{for } j = 1, \cdots, q \, .$$

This condition is equivalent to

$$(4.10) \qquad \dim (\mathbf{E}_z^* + \Gamma_{j+m-1}) \leqq m + r - 1 \qquad \text{for } j = 1, \cdots, q \, .$$

Since all $\Gamma_{j+m-1} \subset \Gamma_{q+m-1}$ $(j = 1, \cdots, q)$, we see that $(4.10)_q \Rightarrow (4.10)_j$ for $j = 1, \cdots, q$. Thus

$$(4.11) \qquad z \in S_q \Longleftrightarrow \dim (\mathbf{E}_z^* \cap \Gamma_{q+m-1}) \geqq q .$$

Now $(\xi_1 \wedge \cdots \wedge \xi_{r-q+1})(z) = 0$ in $\Lambda^{r-q+1}\mathbf{E}_z \Leftrightarrow$ there exist $f_1, \cdots, f_q \in \mathbf{E}_z^*$ with $f_1 \wedge \cdots \wedge f_q \neq 0$ and $\langle \xi_j(z), f_\alpha \rangle = 0$ for $j = 1, \cdots, r - q + 1$ and $\alpha = 1, \cdots, q$. Combining this with (4.11), we have

$$(4.12) \qquad z \in S_q \Longleftrightarrow \xi_1(z) \wedge \cdots \wedge \xi_{r-q+1}(z) = 0 ,$$

as required.

(iv) A few comments about (4.12) are relevant. The section $\xi_1 \wedge \cdots \wedge \xi_{r-q+1}$ of $\Lambda^{r-q+1}\mathbf{E}$ is *not* generic unless $q = r$ or $q = 1$. The reason is, of course, that $\xi_1 \wedge \cdots \wedge \xi_{r-q+1}(z)$ is the vector of an $r - q + 1$-plane in $\mathbf{E}_z$, and so satisfies the Cayley-Grassman relations.

A better way to think of S_q is to let $\mathbf{P}(r - q + 1) \rightarrow V$ be the bundle whose fibre $\mathbf{P}(r - q + 1)_z$ is the complex *Stiefel manifold* of $(r - q + 1)$-frames in $\mathbf{E}_z$. For $q = r$, we have $\mathbf{P}(1) = \mathbf{E} - \{0\}$ is the bundle of non-zero vectors. Now $\xi_1 \wedge \cdots \wedge \xi_{r-q+1}$ gives a rational cross-section of $\mathbf{P}(r - q + 1) \rightarrow V$ and $S_q \subset V$ are precisely the points where this cross-section is not defined.

(4.13) PROPOSITION. *Let $T \subset V$ be an irreducible subvariety of dimension q and $\tau \in H_{2q}(V, \mathbf{Z})$ the homology class of T. Then the intersection number $\tau \cdot \sigma_q > 0$.*

(4.14) *Remark.* In words, we may say that the Chern classes of an ample bundle are *numerically positive.* For the universal bundle over the grassmannian, the Chern classes are non-negative, but not positive except in the projective space case.

PROOF. The proof is based on the following general fact: Let $\{S_\lambda\}_{\lambda \in P_N}$ be a rational system of $(n - q)$-dimensional subvarieties $S_\lambda \subset V$ such that the generic S_λ is irreducible. Suppose that, given $z_0 \in V$ and an $(n - q)$-plane $\prod \subset \mathbf{T}_{z_0}(V)$, there is an S_λ passing through z_0 and with $\mathbf{T}_{z_0}(S_\lambda) = \prod$. Then the homology class carried by a generic S_λ is numerically positive.

Let now z_0 be the origin in a coordinate system $z^1, \cdots, z^n$ on V. We may locally trivialize $\mathbf{E}$ so that sections of $\mathbf{E} \rightarrow V$ in a neighborhood of z_0 are $\mathbf{C}^r$-valued holomorphic functions of $z^1, \cdots, z^n$. Since $\mathbf{E} \rightarrow V$ is ample (cf. (2.29)), we may find $\xi_1, \cdots, \xi_{r-q+1} \in \Gamma(\mathbf{E})$ such that

$$\xi_1 \equiv \begin{pmatrix} 1 \\ 0 \\ \vdots \\ 0 \end{pmatrix}, \cdots, \xi_{r-q} \equiv \begin{pmatrix} 0 \\ \vdots \\ 1 \\ \vdots \\ 0 \end{pmatrix}, \xi_{r-q+1} \equiv \begin{pmatrix} 1 \\ 0 \\ \vdots \\ z^1 \\ \vdots \\ 0 \end{pmatrix},$$

where the symbol "$\equiv$" means "modulo quadratic terms in $z^1, \cdots, z^n$". Then $(\xi_1 \wedge \cdots \wedge \xi_{r-q+1})(z_0) = 0$ and the tangent space to $\xi_1 \wedge \cdots \wedge \xi_{r-q+1} = 0$ at $z_0 = 0$ is given by $z^1 = 0, \cdots, z^q = 0$. Using the general intersection number principle above, we complete the argument for Proposition (4.13).

It remains to define the Chern (homology) classes of a general holomorphic bundle $\mathbf{E} \to V$. Let $\mathbf{L} \to V$ be an ample line bundle such that $\mathbf{E} \otimes \mathbf{L}$ is ample. Then the Chern classes $d_1, \cdots, d_r$ of $\mathbf{E} \otimes \mathbf{L}$ are defined, as is the Chern class e of $\mathbf{L}$, by using (4.7). To determine the Chern classes $c_1, \cdots, c_r$ of $\mathbf{E}$, we use the formulas in [15, page 66]. Thus writing $1 + c_1 t + \cdots + c_r t^r = \prod_{i=1}^r (1 + \psi_i t)$, $1 + d_1 t + \cdots + d_r t^r = \prod_{i=1}^r (1 + \gamma_i t)$, $\gamma_i = \psi_i + e$, we can determine $c_1, \cdots, c_r$ from $d_1, \cdots, d_r$ and e. The multiplication is in the sense of intersection of homology classes. To show that this is consistent, we must prove.

(4.15) PROPOSITION. *Let* $\mathbf{E}_1, \mathbf{E}_2$ *be two ample bundles over* V *such that there is a line bundle* $\mathbf{L}$ *with* $\mathbf{E}_2 = \mathbf{E}_1 \otimes \mathbf{L}$. *Then the Chern* (*homology*) *classes of* $\mathbf{E}_1, \mathbf{E}_2, \mathbf{L}$, *as defined by* (4.7), *are consistent with the* $\otimes$ *rule given above.*

This proposition will be proved in § 4. (c) below. In particular it follows that the Chern (homology) classes are algebraic cycles [6].

(b) *Chern classes as differential forms.* We now give the definition of the Chern classes using differential forms [5] and a brief discussion of the Weil homomorphism [12] and [3].

Let $\mathbf{E} \to V$ be a vector bundle with fibre $\mathbf{C}^r$; for this discussion, we only need a C^∞ bundle. Let $\mathbf{P} \to V$ be the principle bundle of C^∞ frames $f = (e_1, \cdots, e_r)$ for $\mathbf{E} \to V$ and choose a connexion θ for $\mathbf{P}$. Thus θ is a matrix-valued 1-form on $\mathbf{P}$ satisfying $\theta(hg) = g^{-1}\theta(h)g$ and where θ is the Maurer-Cartan form on each fibre of $\mathbf{P} \to V$. If we choose a trivialization $\mathbf{E} \mid U \cong U \times \mathbf{C}^r$ for an open set $U \subset V$, then each

$$e_\rho = \begin{pmatrix} \xi_\rho^1 \\ \vdots \\ \xi_\rho^r \end{pmatrix}$$

is a column vector and $f = (e_1, \cdots, e_r)$ is a non-singular $r \times r$ matrix. Letting $z = (z^1, \cdots, z^n)$ be a coordinate system in U, we easily have that $\theta(z, f) = f^{-1}\theta(z)f + f^{-1}df$ where $\theta(z)$ is a matrix-valued 1-form on U.

The curvature Θ of θ is given by

$$\Theta = d\theta + \theta \wedge \theta . \tag{4.16}$$

In terms of the local expressions above,

$$\Theta(z, f) = f^{-1}\Theta(z)f \tag{4.17}$$

where $\Theta(z) = d\theta(z) + \theta(z) \wedge \theta(z)$.

Consider now a multilinear form $P(A_1, \cdots, A_q)$ where the A_j are $r \times r$ matrices. By linearity, if $A_j = (A_{j\sigma}^\rho)$, then,

$$P(A_1, \cdots, A_q) = \sum_{\substack{\rho=(\rho_1,\cdots,\rho_q) \\ \sigma=(\sigma_1,\cdots,\sigma_q)}} c_{\rho\sigma} A_{1\sigma_1}^{\rho_1} \cdots A_{\sigma_q}^{\rho_q} . \tag{4.18}$$

We will call P *symmetric* if $P(\cdots A_i, \cdots, A_j \cdots) = P(\cdots A_j, \cdots, A_i \cdots)$ and *invariant* if $P(g^{-1}A_1g, \cdots, g^{-1}A_qg) = P(A_1, \cdots, A_q)$ for all $g \in GL(r, \mathbf{C})$. The condition that P be invariant is, in infinitesimal form

$$\sum_{j=1}^q P(\cdots, [B, A_j], \cdots) = 0 , \tag{4.19}$$

where B is an arbitrary $r \times r$ matrix. These symmetric invariant q-linear forms will be called *invariant polynomials* and form a vector space $\mathbf{I}_q$. By an obvious multiplication, $\mathbf{I}_p \cdot \mathbf{I}_q \subset \mathbf{I}_{q+p}$ and we let $\mathbf{I} = \sum_{q \geq 0} \mathbf{I}_q$ be the graded ring of invariant polynomials.

To justify the terminology, we remark that a symmetric q-linear invariant form P gives a polynomial $P(A) = P(\underbrace{A, \cdots, A}_{q})$ in the matrix entries A_σ^ρ of A satisfying $P(g^{-1}Ag)$. Conversely, given such an invariant polynomial in the A_σ^ρ, we may recover the corresponding q-linear invariant form. For example, when $q = 2$,

$$P(A_1, A_2) = 1/2\{P(A_1 + A_2) - P(A_1) - P(A_2)\} .$$

We now define the *Weil homomorphism*

$$W: \mathbf{I} \longrightarrow H^*(V, \mathbf{C}) \tag{4.20}$$

by letting, for $P \in \mathbf{I}$,

(4.21) $$W(P) = P(\Theta) .$$

By definition, $P(\Theta) = P(\Theta, \cdots, \Theta)$ is obtained by plugging in the curvature matrix Θ for A. This makes sense, since Θ is a matrix-valued form of degree 2. From (4.17), $P(\Theta) = P(\Theta(z, f)) = P(f^{-1}\Theta(z)f) = P(\Theta)$ so that $P(\Theta)$ is a C^∞ form on V. Also, $dP(\Theta) = \sum P(\cdots, d\Theta, \cdots) = \sum P(\cdots, [\Theta, \theta], \cdots) = 0$ (by (4.19)) since $d\Theta = \Theta \wedge \theta - \theta \wedge \Theta$ by differentiating (4.16).

The main fact is:

(4.22) $W\colon \mathbf{I} \to H^*(V, \mathbf{C})$ is an algebra homomorphism which is independent of the connexion θ.

Weil's proof of (4.22) is short enough to be given here. Let θ, θ_1 be connexions in $\mathbf{P} \to V$ so that $\theta_1 - \theta = \eta$ is a Hom $(\mathbf{E}, \mathbf{E})$-valued one-form. Setting $\theta_t = \theta + t\eta$, we get connexions θ_t with curvatures $\Theta_t = d\theta + td\eta + (\theta + t\eta) \wedge (\theta + t\eta)$. Then $\dot{\Theta}_t = d\eta + [\theta + t\eta, \eta] = D_t\eta = D_t\dot{\theta}_t$ so that

$$\begin{aligned}\dot{P}(\Theta_t, \cdots, \Theta_t) &= \sum P(\Theta_t, \cdots, \dot{\Theta}_t, \cdots, \Theta_t) \\ &= \sum P(\Theta_t, \cdots, D_t\dot{\theta}_t, \cdots, \Theta_t) \\ &= d\{\sum P(\Theta_t, \cdots, \dot{\theta}_t, \cdots, \Theta_t)\} .\end{aligned}$$

Thus

$$P(\Theta_1) - P(\Theta) = d\Big\{\sum \int_0^1 P(\Theta_t, \cdots, \dot{\theta}_t, \cdots, \Theta_t)dt\Big\}$$

so that $P(\Theta_1) = P(\Theta)$ in $H^*(V, \mathbf{C})$. This proves (4.22), since it is trivial that W is an algebra homomorphism.

We also observe that, on $\mathbf{P}$,

(4.23) $$P(\Theta, \cdots, \Theta) = dP(\theta, \Theta, \cdots, \Theta) .$$

To define Chern classes, we define invariant polynomials $P_q(A)$ by setting

(4.24) $$\det\Big(\lambda I_r + \frac{1}{2\pi i}A\Big) = \sum_{q=0}^{r} (-1)^q P_q(A)\lambda^{r-q} .$$

(4.25) *Definition.* The qth *Chern class* (in cohomology) $c_q \in H^{2q}(V, \mathbf{C})$ is given by $c_q = P_q(\Theta) = W(P_q)$.

The *total Chern class* $c(\mathbf{E})$ is defined by

(4.26) $$c(\mathbf{E}) = \sum_{q=0}^{r} c_q t^q .$$

Remarks. (1) Suppose that $\mathbf{E} \to V$ is a holomorphic vector bundle in which we have a hermitian metric. We let $\mathbf{B} \subset \mathbf{P}$ be the bundle of unitary frames $f = (e_1, \cdots, e_r)$. Then the curvature form Θ of the metric connexion is of type (1, 1) (cf. (2.1)) and on $\mathbf{B}$, $\Theta^\rho_\sigma + \bar{\Theta}^\sigma_\rho = 0$. Thus

$$\overline{{}^t\left(\frac{\Theta}{2\pi i}\right)} = \frac{\Theta}{2\pi i}$$

and so

$$\overline{\det\left(\lambda I + \frac{\Theta}{2\pi i}\right)} = \det\left(\bar{\lambda} I + \frac{{}^t\Theta}{2\pi i}\right) = \det\left(\lambda I + \frac{\Theta}{2\pi i}\right)$$

(if λ is real), so that we have

(4.27) The Chern classes (4.25) of a holomorphic vector bundle are real and of type (q, q) .

(2) Suppose that $\mathbf{E}_1 \to V$, $\mathbf{E}_2 \to V$ are (C^∞) vector bundles. Then we have the *duality theorem*

$$c(\mathbf{E}_1 \oplus \mathbf{E}_2) = c(\mathbf{E}_1)c(\mathbf{E}_2) \; . \tag{4.28}$$

PROOF. If θ_1, θ_2 are connexions for $\mathbf{E}_1$, $\mathbf{E}_2$ and with curvatures Θ_1, Θ_2, then

$$\theta = \begin{pmatrix} \theta_1 & 0 \\ 0 & \theta_2 \end{pmatrix}$$

is a connexion for $\mathbf{E}_1 \oplus \mathbf{E}_2$ with curvature

$$\Theta = \begin{pmatrix} \Theta_1 & 0 \\ 0 & \Theta_2 \end{pmatrix} .$$

Letting $\lambda = 1/t$, we have by (4.24) and (4.26) that

$$\begin{aligned} c(\mathbf{E}) &= t^r \det\left(\lambda I_r + \frac{i}{2\pi}\Theta\right) \\ &= t^{r_1} \det\left(\lambda I_{r_1} + \frac{i}{2\pi}\Theta_1\right) t^{r_2} \det\left(\lambda I_{r_2} + \frac{i}{2\pi}\Theta_2\right) \\ &= c(\mathbf{E}_1)c(\mathbf{E}_2) \; , \end{aligned}$$

where r_1, r_2 are the fibre dimensions of $\mathbf{E}_1$, $\mathbf{E}_2$.

(3) Let $\mathbf{E} \to V$, $\mathbf{F} \to V$ be (C^∞) bundles and write $c(\mathbf{E}) = \prod_{\rho=1}^r (1 + \gamma_\rho t)$, $c(\mathbf{F}) = \prod_{\alpha=1}^s (1 + \gamma_\alpha t)$ where r, s are the fibre dimensions of $\mathbf{E}$, $\mathbf{F}$. Then we have

(4.29) $$c(\mathbf{E}\otimes\mathbf{F}) = \prod_{\rho,\alpha}(1+(\gamma_\rho+\gamma_\alpha)t)\,.$$

PROOF. Choosing connexions $\theta_{\mathbf{E}}, \theta_{\mathbf{F}}$ in $\mathbf{E}, \mathbf{F}$, then $\theta_{\mathbf{E}\otimes\mathbf{F}} = \theta_{\mathbf{E}}\otimes 1 + 1\otimes\theta_{\mathbf{F}}$ gives a connexion in $\mathbf{E}\otimes\mathbf{F}$ with curvature $\Theta_{\mathbf{E}\otimes\mathbf{F}} = \Theta_{\mathbf{E}}\otimes 1 + 1\otimes\Theta_{\mathbf{F}}$ (cf. (2.10)). Using this, (4.29) will follow from the algebraic facts,

(i) $\det(A\otimes B) = (\det A)^s(\det B)^r$ if A is $r\times r$ and B is $s\times s$; and

(ii) if A, B are general $u\times u$ matrices satisfying $AB - BA = 0$, if $\det(I+tA) = \sum_{q=1}^{u} P_q(A)t^q = \prod_{j=1}^{u}(1+\gamma_j(A)t)$ and $\det(I+tB) = \prod_{j=1}^{n}(1+\gamma_j(B)t)$, then

$$\det(I+t(A+B)) = \prod_{j=1}^{u}\{1+(\gamma_j(A)+\gamma_j(B))t\}\,.$$

Of course (i) is standard, and (ii) follows by simultaneously diagonalizing A and B. We apply (ii) letting

$$A = \frac{i}{2\pi}\Theta_{\mathbf{E}}\otimes 1,\ B = 1\otimes\frac{i}{2\pi}\Theta_{\mathbf{F}}$$

$\big($then $[A, B] = 0\big)$ and $u = rs$. Then

$$\begin{aligned} c(\mathbf{E}\otimes\mathbf{F}) &= \det\left(I + t\left(\frac{i}{2\pi}\Theta_{\mathbf{E}}\otimes 1 + 1\otimes\frac{i}{2\pi}\Theta_{\mathbf{F}}\right)\right) \\ &= \prod_{j=1}^{rs}\{1+t(\gamma_j(A)+\gamma_j(B))\} \\ &= \prod_{\rho\,\alpha=1}^{r,s}\left\{1+t\left(\gamma_\rho\left(\frac{i}{2\pi}\Theta_{\mathbf{E}}\right)+\gamma_\alpha\left(\frac{i}{2\pi}\Theta_{\mathbf{F}}\right)\right)\right\}, \end{aligned}$$

which proves (4.29).

(c) *Proof of the equivalence of definitions.* Let now V be an algebraic manifold and $\mathbf{E}\to V$ a holomorphic vector bundle. Then we have defined the Chern (homology) classes $\sigma_q \in H_{2n-2q}(V, \mathbf{Z})$ (cf. § 4 (a)) and the Chern (cohomology) classes $c_q \in H^{2q}(V, \mathbf{R})$.

(4.30) THEOREM. *σ_q is the Poincaré dual of c_q.*

Before giving the proof, we make some preliminary remarks. If we can prove (4.30) in case $\mathbf{E}\to V$ is an *ample* bundle, then (4.30) will be true for all bundles. This follows from the definition of σ_q for a general holomorphic bundle $\mathbf{E}\to V$, together with (4.29). At the same time, if Theorem (4.30) is true for ample bundles, then by (4.29) it follows that the definition of σ_q for general $\mathbf{E}\to V$ is consistent, which proves Proposition (4.15). So it will suffice to prove Theorem (4.30) for ample bundles, and we now give the argument in this case.

What has to be shown is this: Let $S_q \subset V$ be a general $2n-2q$ dimensional subvariety which carries the homology class $\sigma_q \in H_{2n-2q}(V, \mathbf{Z})$.

We let Γ be a $2q$ cycle which meets S_q simply at a finite number of points. Then we need to have

(4.31) $$\sigma_q \cdot \Gamma = \langle c_q, \Gamma \rangle ,$$

where the left hand side of (4.31) is the intersection number.

PROOF. We assume that Γ is an algebraic submanifold, the general case is based on the same ideas. Let $\xi_1, \cdots, \xi_{r-q+1}$ be general holomorphic sections of $\mathbf{E} \rightarrow V$. Then S_q is given by $\xi_1 \wedge \cdots \wedge \xi_{r-q+1} = 0$ and we may assume that $\xi_1 \wedge \cdots \wedge \xi_{r-q} \neq 0$ on $S_q \cdot \Gamma$ (since $\xi_1 \wedge \cdots \wedge \xi_{r-q}$ defines a $2n-2q-2$ dimensional subvariety of V). Then $\xi_1 \wedge \cdots \wedge \xi_{r-q} \neq 0$ on Γ and so, over Γ, we have

(4.32) $$0 \longrightarrow \mathbf{S} \longrightarrow \mathbf{E} \longrightarrow \mathbf{Q} \longrightarrow 0 ,$$

where $\mathbf{S}$ is the trivial sub-bundle generated by $\xi_1, \cdots, \xi_{r-q}$. From $c(\mathbf{S})c(\mathbf{Q}) = c(\mathbf{E})$ and $c(\mathbf{S}) = 1$, it follows that $\langle c_q, \Gamma \rangle = \langle \omega, \Gamma \rangle$ where ω is the qth Chern class of $\mathbf{Q} \rightarrow \Gamma$. Now ξ_{r-q+1} gives a section ξ of $\mathbf{Q} \rightarrow \Gamma$ such that $\sigma_q \cdot \Gamma$ is just the zero locus of ξ. In conclusion:

(4.33) To prove (4.31), it will suffice to take $\mathbf{E} \rightarrow V$ a holomorphic bundle with fibre $\mathbf{C}^n$ ($n = \dim V$), $q = n$, S_q the divisor of a general section ξ of $\mathbf{E}$, and ω the nth Chern class of $\mathbf{E} \rightarrow V$.

There are two steps in the proof of (4.33). One is the geometric notion of *transgression*, due to Chern [7], and the other consists of applying certain formulas in the unitary geometry of $\mathbf{E} \rightarrow V$. These formulas have an independent interest.

Suppose then that ξ vanishes at z_0 (the case of several zeroes is the same) and let $S(\mathbf{E}) \rightarrow V$ be the bundle of unit vectors in $\mathbf{E} \rightarrow V$. Then $\pi^*(\mathbf{E}) \rightarrow S(\mathbf{E})$ has a non-vanishing section and so, by (4.28), $c_n(\pi^*(\mathbf{E})) = 0$. Thus $\pi^*\omega = d\psi$ where ψ is a $2n-1$ form on $S(\mathbf{E})$. Letting $S_z \subset S(\mathbf{E})$ be the unit sphere in $\mathbf{E}_z$, we claim that $\int_{S_z} \psi$ is independent of z. (*Proof*. If γ is a curve joining z_1 and z_2 and $T = \bigcup_{z \in \gamma} S_z$, then $\int_{S_{z_1}} \psi - \int_{S_{z_2}} \psi = \int_{\partial T} \psi = \int_T \pi^*\omega = 0$ since ω is "horizontal".)

Let $\alpha = -\int_{S_z} \psi$; we claim that

(4.34) $\int_V \omega = \alpha\cdot$(number of zeroes of ξ).

For the proof we let Δ be a spherical neighborhood of z_0 such that $\mathbf{E}\,|\,\Delta \cong \Delta \times \mathbf{C}^r$. Then $\xi(z) = (z, \zeta(z))$ for $z \in \Delta$, where ζ has an isolated zero at z_0. Using $\sim$ for "approximately equal to", we have:

$$\int_V \omega \sim \int_{V-\Delta} \omega = \int_{V-\Delta} \xi^*\pi^*\omega = \int_{V-\Delta} d\xi^*\psi = -\int_{\Delta-\Delta} \xi^*\psi = -\int_{\zeta(\partial\Delta)} \psi\ ,$$

where this last integral is taken over the unit sphere S_{z_0} in $\mathbf{E}_{z_0}$ and ζ is considered as a mapping of $\partial\Delta$ into S_{z_0}. Thus

$$\int_V \omega \sim \left(-\int_{S_{z_0}} \psi\right)(\deg \zeta)$$

and, letting Δ shrink to z_0, we get (4.34).

To prove (4.33), it will thus suffice to show that:

(4.35) $-\psi\,|\,S_z$ is the normalized volume element.

We shall prove (4.35) from the structure equations in the hermitian geometry of $\mathbf{E} \to V$.

Let then $\mathbf{E} \to V$ be an hermitian vector bundle and suppose that we have an exact sequence

$$0 \longrightarrow \mathbf{S} \longrightarrow \mathbf{E} \longrightarrow \mathbf{Q} \longrightarrow 0 \tag{4.36}$$

of holomorphic bundles. Then $\mathbf{S}$ and $\mathbf{Q}$ have each induced metrics with their respective hermitian geometries; the relevant formulas are discussed in § 1 (d). We let Θ be the curvature in $\mathbf{E}$ and $\hat{\Theta}$ the curvature in $\hat{\mathbf{E}} = \mathbf{S} \oplus \mathbf{Q}$. As C^∞ bundles, $\mathbf{E} \cong \hat{\mathbf{E}}$ so that Θ and $\hat{\Theta}$ are both curvatures in the *same* C^∞ complex vector bundle. By Weil's theorem (4.22), if $P \in \mathbf{I}_q$ is any invariant polynomial (cf. §4 (b)), we have

$$P(\Theta) - P(\hat{\Theta}) = d\varphi\ , \tag{4.37}$$

where φ is given in the proof of (4.22). We shall prove Bott and Chern's refinement [3] of Weil's theorem by showing that

$$P(\Theta) - P(\hat{\Theta}) = \partial\bar{\partial}\psi\ , \tag{4.38}$$

where ψ is a form of type $(q-1, q-1)$ which may be given explicitly. Our proof is a continuation of Weil's proof of (4.37) and is somewhat different from the proof in [3].

Let then $\mathbf{B} \to V$ be the principal bundle of unitary frames $f = (e_1, \cdots, e_r)$ for $\mathbf{E} \to V$ such that $e_1, \cdots, e_s$ is a frame for $\mathbf{S} \to V$ (cf. § 1 (e)). We write the connexion form for $\mathbf{E}$ as

$$\theta = \begin{pmatrix} \theta_{11} & \theta_{12} \\ \theta_{21} & \theta_{22} \end{pmatrix}$$

where θ_{11} is an $s \times s$ matrix giving the induced connexion in $\mathbf{S}$, θ_{22} gives the induced connexion in $\mathbf{Q}$, $\theta_{21} \in A^{1,0}(\mathrm{Hom}(\mathbf{S}, \mathbf{Q}))$ gives the second fundamental form of $\mathbf{S}$ in $\mathbf{E}$, and $\theta_{12} = -{}^t\bar{\theta}_{21} \in A^{0,1}(\mathrm{Hom}(\mathbf{Q},\mathbf{S}))$. Then

$$\Theta = d\theta + \theta \wedge \theta \, , \tag{4.39}$$

and

$$\hat{\Theta} = d\hat{\theta} + \hat{\theta} \wedge \hat{\theta} \tag{4.40}$$

where

$$\hat{\theta} = \begin{pmatrix} \theta_{11} & 0 \\ 0 & \theta_{22} \end{pmatrix}$$

is the metric connexion in $\mathbf{S} \oplus \mathbf{Q}$.

Write $\hat{\theta} = \theta + \varphi$ where

$$\varphi = \begin{pmatrix} 0 & -\theta_{12} \\ -\theta_{21} & 0 \end{pmatrix}$$

and let

$$\varphi' = \begin{pmatrix} 0 & 0 \\ -\theta_{21} & 0 \end{pmatrix}$$

be the $(1, 0)$ part of φ,

$$\varphi'' = \begin{pmatrix} 0 & -\theta_{12} \\ 0 & 0 \end{pmatrix}$$

the $(0, 1)$ part. Then $\varphi = \varphi' + \varphi''$ and

$$\begin{aligned} D\varphi &= d\varphi + [\theta, \varphi] \\ &= \begin{pmatrix} 0 & -d\theta_{12} \\ -d\theta_{21} & 0 \end{pmatrix} + \begin{pmatrix} \theta_{11} & \theta_{12} \\ \theta_{21} & \theta_{22} \end{pmatrix} \begin{pmatrix} 0 & -\theta_{12} \\ -\theta_{21} & 0 \end{pmatrix} \\ &\quad + \begin{pmatrix} 0 & -\theta_{12} \\ -\theta_{21} & 0 \end{pmatrix} \begin{pmatrix} \theta_{11} & \theta_{12} \\ \theta_{21} & \theta_{22} \end{pmatrix} \\ &= \begin{pmatrix} 0 & -\Theta_{12} \\ -\Theta_{21} & 0 \end{pmatrix} + 2 \begin{pmatrix} -\theta_{12}\theta_{21} & 0 \\ 0 & -\theta_{21}\theta_{12} \end{pmatrix} . \end{aligned}$$

We collect this equation and the ones resulting from it by using decomposition into type as follows:

$$D\varphi = \begin{pmatrix} 0 & -\Theta_{12} \\ -\Theta_{21} & 0 \end{pmatrix} + 2\begin{pmatrix} -\theta_{12}\theta_{21} & 0 \\ 0 & -\theta_{21}\theta_{12} \end{pmatrix}, \tag{4.41}$$

$$\begin{cases} D'\varphi = D'\varphi'' = \begin{pmatrix} 0 & -\Theta_{12} \\ 0 & 0 \end{pmatrix} + 2\begin{pmatrix} 0 & 0 \\ 0 & -\theta_{21}\theta_{12} \end{pmatrix} \\ D''\varphi = D''\varphi' = \begin{pmatrix} 0 & 0 \\ -\Theta_{21} & 0 \end{pmatrix} + 2\begin{pmatrix} -\theta_{12}\theta_{21} & 0 \\ 0 & 0 \end{pmatrix} \\ D'\varphi' = D''\varphi'' = 0\,. \end{cases} \tag{4.42}$$

Let $\theta_t = \theta + t\varphi$ be the linear 1-parameter family of connexions with $\theta_0 = \theta$, $\theta_1 = \hat{\theta}$. Then we have

$$\begin{cases} \Theta_t = \Theta + tD\varphi + \dfrac{t^2}{2}[\varphi, \varphi] \\ \dot{\Theta}_t = D\varphi + t[\varphi, \varphi]\,. \end{cases} \tag{4.43}$$

Here, as opposed to the proof of (4.22), D is always taken with respect to the connexion θ. We claim that (cf. the proof of (4.22)

$$\dot{P}(\Theta_t, \cdots, \Theta_t) = d\{\textstyle\sum P(\Theta_t, \cdot, \varphi, \cdot, \Theta_t)\}\,, \tag{4.44}$$

where $\dot{P} = dP/dt$. In fact, $\dot{P}(\Theta_t, \cdot, \Theta_t) = \sum P(\Theta_t, \cdot, \dot{\Theta}_t, \cdots, \Theta_t) = \sum P(\Theta_t, \cdot, D\varphi + t[\varphi, \varphi], \cdot, \Theta_t)$; while

$$\begin{aligned} &P(\Theta_t, \cdot, D\varphi + t[\varphi, \varphi], \cdot, \Theta_t) \\ &\quad = -\textstyle\sum P(\cdot, t[\varphi, \Theta_t], \cdot, \varphi, \cdot) + P(\Theta_t, \cdot, D\varphi, \cdot, \Theta_t) \\ &\quad = \textstyle\sum P(\cdot, D\Theta_t, \varphi, \cdot) + P(\Theta_t, D\varphi, \cdot, \Theta_t) \\ &\quad = dP(\Theta_t, \cdot, \varphi, \cdot, \Theta_t) \end{aligned}$$

since $D\Theta_t = d\Theta_t + [\theta, \Theta_t] = d\Theta_t + [\theta_t, \Theta_t] - [\theta_t - \theta, \Theta_t] = -t[\varphi, \Theta_t]$ because $D_t\Theta_t = 0$.

Taking types in (4.44), we obtain

$$\begin{aligned} \dot{P}(\Theta_t) = \partial\,&\{\textstyle\sum P(\Theta_t, \cdot, \varphi'', \cdot, \Theta_t)\} \\ &+ \bar{\partial}\,\{\textstyle\sum P(\Theta_t, \varphi', \cdot, \Theta_t)\}\,. \end{aligned} \tag{4.45}$$

Let $\xi = \begin{pmatrix} 1 & 0 \\ 0 & 0 \end{pmatrix}$ where 1 is an $s \times s$ unit matrix (ξ is the orthogonal projection on $\mathbf{S}$). We set

$$Q_t = \frac{2}{(1-t)}\{\textstyle\sum P(\Theta_t, \cdot, \xi, \cdot, \Theta_t) - \sum P(\hat{\Theta}, \cdot, \xi, \cdot, \hat{\Theta})\} \tag{4.46}$$

and assert that

$$(4.47)\qquad \begin{cases} Q_t \text{ is a smooth family of } C^\infty \text{ forms on } V \text{ and} \\ \bar\partial Q_t = 2\{\sum P(\Theta_t, \cdot, \varphi'', \cdot, \Theta_t)\} \\ \partial Q_t = -2\{\sum P(\Theta_t, \cdot, \varphi', \cdot, \Theta_t)\} . \end{cases}$$

Combining (4.47) and (4.45) gives $\dot{P}(\Theta_t) = \partial\bar\partial Q_t$ or

$$(4.48)\qquad P(\hat\Theta) - P(\Theta) = \partial\bar\partial\left\{\int_0^1 Q_t dt\right\} .$$

Proof of (4.47). We have

$$D\xi = [\theta, \xi] = \begin{pmatrix} \theta_{11} & \theta_{12} \\ \theta_{21} & \theta_{22} \end{pmatrix}\begin{pmatrix} 1 & 0 \\ 0 & 0 \end{pmatrix} - \begin{pmatrix} 1 & 0 \\ 0 & 0 \end{pmatrix}\begin{pmatrix} \theta_{11} & \theta_{12} \\ \theta_{21} & \theta_{22} \end{pmatrix} = \begin{pmatrix} 0 & -\theta_{12} \\ \theta_{21} & 0 \end{pmatrix},$$

which gives

$$\begin{cases} D'\xi = -\varphi' \\ D''\xi = \varphi'' . \end{cases}$$

Now

$$\begin{aligned} &\bar\partial P(\Theta_t, \cdot, \xi, \cdot, \Theta_t) \\ &\quad = \sum P(\cdot, D''\Theta_t, \cdot, \xi, \cdot) + P(\Theta_t, \cdot, D''\xi, \cdot, \Theta_t) \\ &\quad = -\sum P(\cdot, t[\varphi'', \Theta_t], \cdot, \xi, \cdot) + P(\Theta_t, \cdot, \varphi'', \cdot, \Theta_t) \\ &\quad = P\big(\Theta_t, \cdot, t[\varphi'', \xi], \cdot, \Theta_t\big) + P(\Theta_t, \cdot, \varphi'', \cdot, \Theta_t) \\ &\quad = (1-t)P(\Theta_t, \cdot, \varphi'', \cdot, \Theta_t) \end{aligned}$$

since $[\varphi'', \xi] = -\varphi''$. This gives

$$(4.49)\qquad \bar\partial\frac{1}{1-t}\{P(\Theta_t, \cdot, \xi, \cdot, \Theta_t)\} = P(\Theta_t, \cdot, \varphi'', \cdot, \Theta_t) .$$

Since $P(\Theta_t, \cdot, \xi, \cdot, \Theta_t)$ is a polynomial in t whose coefficients are C^∞ forms on V, and since $P(\Theta_1, \cdot, \xi, \cdot, \Theta_1) = P(\hat\Theta, \cdot, \xi, \cdot, \hat\Theta)$ to prove (4.47) (for $\bar\partial$) from (4.49) we must show that $\bar\partial P(\hat\Theta, \cdot, \xi, \hat\Theta) = 0$. Letting $\hat D$ be covariant differentiation with respect to $\hat\theta$, we have $\hat D'' = \bar\partial$, $\hat D''\hat\Theta = 0$, $\hat D''\xi = 0$. Thus

$$\begin{aligned} &\bar\partial P(\hat\Theta, \cdot, \xi, \cdot, \hat\Theta) \\ &\quad = \sum P(\cdot, \hat D''\hat\Theta, \cdot, \xi, \cdot) + P(\hat\Theta, \cdot, \hat D''\xi, \cdot, \hat\Theta) \\ &\quad = 0 . \end{aligned}$$

This proves (4.47) for $\bar\partial$; the proof for ∂ is similar.

PROOF OF (4.35.) What we have to show is this.

(4.50) Let $\mathbf{E} \to V$ be an hermitian vector bundle with fiber $\mathbf{C}^n$

and with a non-vanishing section σ. Let ω be the form of type (n, n) on V giving $c_n(\mathbf{E})$ computed from the curvature in $\mathbf{E}$. Then $\omega = d\psi$, where ψ gives the negative volume element on the unit spheres $S_z \subset \mathbf{E}_z$.

To prove (4.35) from (4.50), we look at $\pi^*\mathbf{E} \to \hat{\mathbf{E}}$ where $\hat{\mathbf{E}} = \mathbf{E} - \{0\}$ is the bundle of non-zero vectors in $\mathbf{E}$. There is a canonical section σ of $\pi^*\mathbf{E} \to \hat{\mathbf{E}}$, and (4.50) applied to this situation gives (4.35).

PROOF OF (4.50). Let $\mathbf{S} \subset \mathbf{E}$ be the bundle generated by σ so that we have an exact sequence (4.36) with curvatures $\Theta_{\mathbf{E}}$, $\Theta_{\mathbf{S}}$, $\Theta_{\mathbf{Q}}$. For a $q \times q$ matrix Φ, we set $c(\Phi) = [(1/2\pi i)]^q \det \Phi$; then, by (4.38),

$$c(\Theta_{\mathbf{E}}) = c(\Theta_{\mathbf{S}})c(\Theta_{\mathbf{Q}}) + \partial\bar{\partial}\gamma \,, \tag{4.51}$$

and

$$c(\Theta_{\mathbf{S}}) = d\eta \tag{4.52}$$

since $\mathbf{S}$ is trivial. Combining (4.51) and (4.52) gives

$$c(\Theta_{\mathbf{E}}) = d\{\eta c(\Theta_{\mathbf{Q}}) + \bar{\partial}\gamma\} = d\psi \tag{4.53}$$

where $\psi = \eta c(\Theta_{\mathbf{Q}}) + \bar{\partial}\gamma$. We want to compute $\psi \mid S_z$, and, in so doing, we may ignore all terms $\Theta_{\rho\sigma}$, which are horizontal forms. Using the frames in the proof of (4.38), and using "$\equiv$" to mean modulo terms $\Theta_{\rho\sigma}$", we need to show that

$$\psi \equiv -\left(\frac{1}{2\pi i}\right)^n \frac{1}{(n-1)!}\{\theta_{11}\theta_{21}\theta_{12} \cdots \theta_{n1}\theta_{1n}\} \,. \tag{4.54}$$

Now, by (4.44)

$$\bar{\partial}\gamma = \int_0^1 \{\sum P(\Theta_t, \cdot, \varphi, \cdot, \Theta_t)\}dt \,, \tag{4.54}$$

where P is the invariant polynomial corresponding to $c(\Theta)$. Using (4.41) and (4.43),

$$\Theta_t \equiv tD\varphi + \frac{t^2}{2}[\varphi, \varphi] \equiv \left(-t + \frac{t^2}{2}\right)[\varphi, \varphi] \,.$$

Thus

$$\begin{aligned}\sum P(\Theta_t, \cdot, \varphi, \cdot, \Theta_t) &\equiv \left(-t + \frac{t^2}{2}\right)^{n-1} \sum P(\cdot, [\varphi, \varphi], \cdot, \varphi, \cdot) \\ &\equiv \left(t - \frac{t^2}{2}\right)^{n-1} P(\varphi, \cdot, [\varphi, [\varphi, \varphi]], \cdot, \varphi) \\ &= 0 \,.\end{aligned}$$

since $[\varphi, [\varphi, \varphi]] = 0$. Thus we obtain the equation:

$$\psi \equiv \eta c(\Theta_Q) . \tag{4.55}$$

Now, by (4.43) at $t = 1$,

$$\Theta_1 = \begin{pmatrix} \Theta_S & 0 \\ 0 & \Theta_Q \end{pmatrix} \equiv \begin{pmatrix} \theta_{12}\theta_{21} & 0 \\ 0 & \theta_{21}\theta_{12} \end{pmatrix} .$$

Also, θ_{11} is defined on V (using σ) and $d\theta_{11} \equiv -\theta_{12}\theta_{21}$ so that $\eta \equiv -(1/2\pi i)\theta_{11}$ and

$$\eta c(\Theta_Q) \equiv -\left(\frac{1}{2\pi i}\right)^n \theta_{11} \det(\theta_{21}\theta_{12}) . \tag{4.56}$$

But $\det(\theta_{21}\theta_{12}) = (n-1)!\ \theta_{21}\theta_{12}\theta_{31}\theta_{13} \cdots \theta_{n1}\theta_{1n}$ which, using (4.56) and (4.55), gives (4.54).

Remarks. (a) Let "$\sim$" denote "congruent modulo commutators" (so that, e.g., $AB \sim BA$). Referring to (4.43) we have

$$\dot{\Theta}_t \sim D\varphi , \tag{4.57}$$

$$\dot{\Theta}_t \sim D'D''\left(\frac{\xi}{2}\right) . \tag{4.58}$$

(*Proof.* $D\varphi = (D' + D'')\varphi = (D' + D'')(D'' - D')\xi = (D'D'' - D''D')\xi$. Now $D^2\xi = (D'D'' + D''D')\xi = [\Theta, \xi] \sim 0$ so that $-D''D'\xi \sim D'D''\xi$.) Equation (4.57) which holds for general vector bundles, is the basis for Weil's theorem (4.37); it says that, modulo commutators (which essentially give zero in an invariant polynomial), the variation in Θ_t is an exact form. Similarly, (4.58), which holds for hermitian vector bundles, is the basis for (4.38).

(b) Referring to (4.38), suppose that $P(\Theta)$ is an invariant polynomial of degree q. Then we claim that, in (4.38),

$$\psi = q \sum_{s+t>0} \left(\frac{1}{2}\right)^{s-1} \left(\frac{1}{2s+t}\right) \binom{q-1}{s+2} \binom{s+t}{s} \times P(\underbrace{\hat{\Theta}; \tilde{\Theta}}_{t}; \underbrace{[\varphi, \varphi]}_{s}; \xi) \tag{4.59}$$

where

$$P(\hat{\Theta}; \underbrace{\tilde{\Theta}}_{t}; \underbrace{[\varphi, \varphi]}_{s}; \xi) = P(\hat{\Theta}, \cdots, \hat{\Theta}, \underbrace{\tilde{\Theta}, \cdots, \tilde{\Theta}}_{t}, \underbrace{[\varphi, \varphi], \cdots, [\varphi, \varphi]}_{s}, \xi)$$

and

$$\widetilde{\Theta} = \begin{pmatrix} 0 & \Theta_{12} \\ \Theta_{21} & 0 \end{pmatrix}.$$

Proof. By (4.46), $Q_t = 2q/(1-t)\{P(\Theta_t, \cdots, \Theta_t, \xi) - P(\hat{\Theta}, \cdots, \hat{\Theta}, \xi)\}$. Now $\Theta_t = \hat{\Theta} + \Phi$ where $\Phi = -(1-t)D\varphi - \{(1-t^2)/2\}[\varphi, \varphi] = (1-t)\{\widetilde{\Theta} + (1-t/2)[\varphi, \varphi]\}$, by (4.41) since $D\varphi = -\widetilde{\Theta} - [\varphi, \varphi]$. Thus

$$\begin{aligned} Q_t &= \frac{2q}{1-t}\{P(\hat{\Theta} + \Phi, \cdots, \hat{\Theta} + \Phi, \xi) - P(\hat{\Theta}, \cdots, \hat{\Theta}, \xi)\} \\ &= \frac{2q}{1-t}\left\{\sum_{l>0}\binom{q-1}{l}P(\hat{\Theta}, \cdot, \hat{\Theta}, \Phi, \xi)\right\} \\ &= 2q\left\{\sum_{l>0}\left(\frac{1}{2}\right)^m\binom{q-1}{l}\binom{l}{m}(1-t)^{l+m-1}P(\underbrace{\hat{\Theta}; \widetilde{\Theta}}_{l-m}; \underbrace{[\varphi, \varphi]}_{m}; \xi)\right\} \end{aligned}$$

and

$$\begin{aligned} \int_0^1 Q_t &= 2q\left\{\sum_{l>0}\left(\frac{1}{2}\right)^m\left(\frac{1}{l+m}\right)\binom{q-1}{l}\binom{l}{m}P(\underbrace{\hat{\Theta}; \widetilde{\Theta}}_{l-m}; \underbrace{[\varphi, \varphi]}_{m}; \xi)\right\} \\ &= (4.59)\,. \end{aligned}$$

When $q = 1$, $P(\Theta)$ is a multiple of Trace Θ and $\operatorname{Tr}\Theta = \operatorname{Tr}\hat{\Theta}$ since $\operatorname{Tr}[\varphi, \varphi] = 0 = \operatorname{Tr}\widetilde{\Theta}$. When $q = 2$, there are only two terms ($t = 0, s = 1$ and $t = 1, s = 0$) in (4.59) and $\psi = P([\varphi, \varphi], \xi) + 4P(\widetilde{\Theta}, \xi)$. If $r = 2$ and $P(\Theta) = \det\Theta$, then $\psi = P([\varphi, \varphi], \xi) = \theta_{21}\theta_{12}$, which means that

$$\det\begin{pmatrix} \Theta_{11} & \Theta_{12} \\ \Theta_{21} & \Theta_{22} \end{pmatrix} - \det\begin{pmatrix} \Theta_{11} - \theta_{21}\theta_{12} & 0 \\ 0 & \Theta_{22} + \theta_{21}\theta_{12} \end{pmatrix} = \partial\bar{\partial}(\theta_{21}\theta_{12})\,, \tag{4.60}$$

an equation which may be verified directly.

(c) Suppose now that, in the exact sequence (4.36), **S** has fibre dimension one, **E** has fibre $\mathbf{C}^r$, and $P(\Theta) = \det\Theta$. Then ψ given by (4.59) may be written as

$$\psi = \psi_0 + \cdots + \psi_{r-1}$$

where ψ_q is homogeneous of degree q in terms $\theta_{\alpha 1}\theta_{1\beta}$. Recall that, using natural frames,

$$\varphi = -\begin{pmatrix} 0 & \theta_{12} \cdots \theta_{1r} \\ \theta_{21} & 0 \cdots 0 \\ \vdots & \vdots \quad\;\; \vdots \\ \theta_{r1} & 0 \cdots 0 \end{pmatrix}$$

and

$$[\varphi, \varphi] = 2\begin{pmatrix} \sum_{\alpha=2}^{r} \theta_{1\alpha}\theta_{\alpha 1} & 0 & \cdots & 0 \\ 0 & \theta_{21}\theta_{12} & \cdots & \theta_{21}\theta_{1r} \\ \vdots & \vdots & & \vdots \\ 0 & \theta_{r1}\theta_{12} & \cdots & \theta_{r1}\theta_{1r} \end{pmatrix}.$$

Now $P(A_1, \cdots, A_r)$ is the polarized determinant function and, since

$$\widetilde{\Theta} = \begin{pmatrix} 0 & \Theta_{12} & \cdots & \Theta_{1r} \\ \Theta_{21} & 0 & \cdots & 0 \\ \vdots & \vdots & & \vdots \\ \Theta_{r1} & 0 & \cdots & 0 \end{pmatrix} \quad \text{and} \quad \xi = \begin{pmatrix} 1 & 0 & \cdots & 0 \\ 0 & & & \\ \vdots & & & \vdots \\ 0 & & \cdots & 0 \end{pmatrix}$$

we see that $P(\widetilde{\Theta}; \hat{\Theta}; [\varphi, \varphi]; \xi) = 0$ if $t > 0$. Thus

$$\psi = \sum_{s>0} \binom{r}{s} \left(\frac{1}{2}\right)^s \binom{r-1}{s} P(\hat{\Theta}; \widetilde{\Theta}; \underbrace{[\varphi, \varphi]}_{s}; \xi).$$

Let

$$\gamma = \begin{pmatrix} \theta_{21}\theta_{12} & \cdots & \theta_{21}\theta_{1r} \\ \vdots & & \vdots \\ \theta_{r1}\theta_{12} & \cdots & \theta_{r1}\theta_{1r} \end{pmatrix} \quad \text{and} \quad \eta = \begin{pmatrix} \Theta_{22} & \cdots & \Theta_{2r} \\ & & \vdots \\ \Theta_{r2} & & \Theta_{rr} \end{pmatrix}$$

$(= \Theta_Q)$. Then $P(\hat{\Theta}; \underbrace{[\varphi, \varphi]}_{s}; \xi) = 2^s R(\underbrace{\eta + \gamma}_{r-s-1}, \underbrace{\gamma}_{s})$, where R is the polynomial obtained by polarizing the determinant function on $(r-1) \times (r-1)$ matrices. It is then clear that $\psi = \psi_0 + \cdots + \psi_{r-1}$ where $\psi_q = \lambda_q R(\eta, \underbrace{\gamma}_{q})$, the λ_q being suitable positive constants. This gives:

(4.61) In case S has fibre dimension one, we have

$$c_r(\Theta_E) - c_1(\Theta_S)c_{r-1}(\Theta_Q) = \partial\bar{\partial}(\psi_1 + \cdots + \psi_{r-1})$$

where

$$\psi_q = \frac{c_q}{(2\pi i)^r} \sum \operatorname{sgn} \pi\theta_{\alpha_1 1}\theta_{1\pi(\alpha_1)} \cdots \theta_{\alpha_q 1}\theta_{1\pi(\alpha_q)} \times \Theta_{\beta_1\pi(\beta_1)} \cdots \Theta_{\beta_{r-q-1}\pi(\beta_{r-q-1})} \tag{4.62}$$

and where the summation in (4.62) is over permutations π of $(2 \cdots r)$ into disjoint sets $(\alpha_1 \cdots \alpha_q)$, $(\beta_1 \cdots \beta_{r-q-1})$ of increasing indices. In particular,

$$\psi_{r-1} = \frac{c_r}{(2\pi i)^r}\theta_{21}\theta_{12} \cdots \theta_{r1}\theta_{1r}. \tag{4.63}$$

5. Numerically and arithmetically positive bundles

(a) *Positive forms and cohomology classes.* Let V be a complex manifold and ω a differential form of type (q, q) in an open set $U \subset V$. We say that ω is *positive*, written $\omega > 0$, if $\omega \neq 0$ and if there exist $(q, 0)$ forms φ_α such that

$$(5.1) \qquad \omega = (-1)^{\frac{q(q-1)}{2}}\left(\frac{\sqrt{-1}}{2}\right)^q \{\textstyle\sum_\alpha \varphi_\alpha \wedge \bar{\varphi}_\alpha\} .$$

The signs are such that, in $\mathbf{C}^q$,

$$(-1)^{\frac{q(q-1)}{2}}\left(\frac{\sqrt{-1}}{2}\right)^q (dz^1 \wedge \cdots \wedge dz^q \wedge d\bar{z}^1 \wedge \cdots \wedge d\bar{z}^q)$$
$$= dx^1 \wedge dy^1 \wedge \cdots \wedge dx^q \wedge dy^q .$$

The symbol $\omega \geqq 0$ has the obvious meaning. If $\omega \geqq 0$, $\varphi \geqq 0$, then $\omega \wedge \varphi \geqq 0$ and $\omega + \varphi \geqq 0$ if $\deg \omega = \varphi$.

We let $\mathbf{A}^{q,q}$ be the vector space of C^∞ (q, q) forms on V and $\mathbf{A}^* = \sum \mathbf{A}^{q,q}$. A form $\omega \in \mathbf{A}^{q,q}$ is positive if locally $\omega > 0$. The space $\mathbf{P}^q \subset \mathbf{A}^{q,q}$ of positive (q, q) forms is a convex cone.

If $\omega \geqq 0$ in an open set $U \subset V$ and $S \subset U$ is a q-dimensional analytic set, then $\int_S \omega \geqq 0$. If $Z \subset U$ is a subvariety, then $\omega \mid Z \geqq 0$.

Consider now $H^{q,q}(V)$ and let ω be a cohomology class which is real; $\omega = \bar{\omega}$. Then we write $\omega > 0$ if $\langle \omega, \sigma \rangle > 0$ for all $\sigma \in H_{2q}(V, \mathbf{Z})$ where σ is the cycle carried by an irreducible subvariety of dimension q lying in V. Obviously we have:

(5.2) Let $\omega \in H^{q,q}(V)$ and suppose $\omega \in \mathbf{A}^{q,q}$ is a closed (q, q) form representing the cohomology class ω. Then $\omega > 0$ if $\omega > 0$ and $\omega \geqq 0$ if $\omega \geqq 0$.

(b) *The cone of positive polynomials and proof of Theorem* D. Let V be a compact, complex manifold and $\mathbf{E} \rightarrow V$ a holomorphic vector bundle with Chern classes $c_1, \cdots, c_r$. For a q-tuple $I = (i_1, \cdots, i_q)$, we set $|I| = i_1 + \cdots + i_q$ and $c_I = c_{i_1} \cdots c_{i_1} \in H^{2|I|}(V, \mathbf{Z})$. We let $\mathbf{R} = \bigoplus_{q \geqq 0} \mathbf{R}_q$ be the graded ring of polynomials $P = \sum p_I c_I$ in $c_1, \cdots, c_r$ with rational coefficients; clearly $\mathbf{R}_q \cdot \mathbf{R}_s \subset \mathbf{R}_{q+s}$. We want to define the *cone of positive polynomials* $\mathbf{\Pi} = \bigoplus_{q \geqq 0} \mathbf{\Pi}_q$; $\mathbf{\Pi}$ will be a convex graded cone (over $\mathbf{Q}$) with $\mathbf{\Pi}_q \mathbf{\Pi}_s \subset \mathbf{\Pi}_{q+s}$. Then we will prove

$$(5.3) \qquad \begin{cases} \text{If } \mathbf{E} \rightarrow V \text{ is ample and } P \in \mathbf{\Pi}_q \ (q \leqq \dim V), \text{ then} \\ \mathbf{P}(c_1, \cdots, c_r) > 0 \text{ in } H^{2q}(V, \mathbf{Q}) . \end{cases}$$

This is Theorem D. It should remain true when $\mathbf{E} \to V$ is positive, but we can only prove that certain $P \in \mathbf{\Pi}$ give positive cohomology classes (cf. the appendix to 5(b) below).

To describe $\mathbf{\Pi}$, we follow Hirzebruch [15] and write formally $1 + c_1 t + \cdots + c_r t^r = (1 + \gamma_1 t) \cdots (1 + \gamma_r t)$.

(5.4) $\begin{cases} \text{Then } \mathbf{R} \cong \mathbf{R}^* \text{ where } \mathbf{R}^* \text{ is the ring of polynomials} \\ \text{in } \gamma_1, \cdots, \gamma_r \text{ which are invariant under the permu-} \\ \text{tation group .} \end{cases}$

We remark that $\gamma_I = \gamma_{i_1} \cdots \gamma_{i_q}$ now has weight q, so that $\mathbf{R}_q^*$ consists of all invariant polynomials $p(\gamma) = \sum_{I=(i_1,\cdots,i_q)} p_I \gamma_I$.

We now let $B = (B_{\rho\sigma})$ be a variable $r \times r$ matrix $(1 \leqq \rho, \sigma \leqq r)$ and $\gamma_\rho = B_{\rho\rho}$. Then

(5.5) $\begin{cases} \text{The ring } \mathbf{R}^* \text{ of polynomials in } \gamma_1, \cdots, \gamma_r, \text{ invariant} \\ \text{under the permutation group, is isomorphic to the} \\ \text{ring } \mathbf{I} \text{ of polynomials in } B_{\rho\sigma} \text{ invariant under } B \to \\ MBM^{-1} (M \in GL(r)) \text{ .} \end{cases}$

PROOF. $\mathbf{I}$ is the ring of polynomials $P(B)$ satisfying $P(MBM^{-1}) = P(B) (M \in GL(r))$. The mapping $\mathbf{I} \to \mathbf{R}^*$ is given by $P(B) \to P(\gamma)$ where

$$\gamma = \begin{pmatrix} \gamma_1 & & 0 \\ & \ddots & \\ 0 & & \gamma_r \end{pmatrix}$$

(i.e., $\gamma_\rho = B_{\rho\rho}, B_{\rho\sigma} = 0$ for $\rho \neq \sigma$). To see that this makes sense, we let $\mathbf{h}$ be the vector space of diagonal matrices and $N = \{M \in GL(r) \colon M\mathbf{h}\mathbf{M}^{-1} \subset \mathbf{h}\}$. Then $N \supset H$ where H is the group of non-singular diagonal matrices; H acts trivially on $\mathbf{h}$ (i.e., $M\gamma M^{-1} = \gamma$ for $M \in H$) and $N/H = W$ is the permutation group acting on $\mathbf{h}$. Consequently, if $P \in \mathbf{I}$, then $P(\gamma)$ is a polynomial in $\gamma_1, \cdots, \gamma_r$ invariant under $\gamma_\rho \to \gamma_{\pi(\rho)}$ where $\pi \in S(r)$, $S(r)$ being the permutation group on r symbols. Thus the ring homomorphism $\mathbf{I} \to \mathbf{R}^*$ is well-defined.

If $P(B) \in \mathbf{I}$ and $P(\gamma) \equiv 0$, then $P(B) = 0$ for any matrix B which can be diagonalized. This implies that $P = 0$ in $\mathbf{I}$, and so $\mathbf{I} \to \mathbf{R}^*$ is injective.

Similarly, if $P(\gamma) \in \mathbf{R}^*$ and B is diagonalizable, then we may set $P(B) = P(MBM^{-1})$ where $MBM^{-1} \in \mathbf{h}$. This is well-defined since

$P(\gamma)$ is invariant under N. By continuity, we may then define $P(B)$ for all B and so $\mathbf{I} \to \mathbf{R}^*$ is onto, which proves (5.5).

The gist of (5.4) and (5.5) is that the graded ring $\mathbf{I} = \bigoplus_{q \geqq 0} \mathbf{I}_q$ of invariant polynomials (cf. 4(b)) gives isomorphically the ring $\mathbf{R}$ of polynomials in the Chern classes. A direct mapping $W: \mathbf{I} \to \mathbf{R}$ is the *Weil homomorphism* (4.20).

We now describe those polynomials $P(B) \in \mathbf{I}$ which will be positive in $\mathbf{R}$. As motivation for this, we first observe that any $P(B) \in \mathbf{I}_q$ can be written as

$$(5.6) \qquad P(B) = \sum\nolimits_{\left\{\substack{\rho=(\rho_1, \cdots \rho_q) \\ \pi, \tau \in S(q)}\right.} p_{\rho,\pi,\tau} B_{\rho_{\pi(1)}\rho_{\tau(1)}} \cdots B_{\rho_{\pi(q)}\rho_{\tau(q)}} .$$

PROOF. If we let $I_0(B) = 1$ and, for $q \geqq 1$,

$$(5.7) \qquad I_q(B) = \sum\nolimits_{\substack{\rho_1 < \cdots < \rho_q \\ \pi \in S(q)}} \operatorname{sgn} \pi B_{\rho_1 \rho_{\pi(1)}} \cdots B_{\rho_q \rho_{\pi(q)}} ,$$

then $\mathbf{I}$ is just the ring of polynomials in $I_0(B), \cdots, I_r(B)$. This is because of (5.5) and the fact that $I_q(\gamma) = \sum_{\rho_1 < \cdots < \rho_q} \gamma_{\rho_1} \cdots \gamma_{\rho_q}$ is the qth elementary symmetric function of $\gamma_1, \cdots, \gamma_r$. Note that $I_q(B) \in \mathbf{I}_q$ because

$$(5.8) \qquad \det(B + tI) = \sum\nolimits_{q=0}^{r} I_q(B) t^{r-q} .$$

Since

$$I_q(B) = \left(\frac{1}{q!}\right)^2 \sum\nolimits_{\substack{\rho=(\rho_1, \cdots, \rho_q) \\ \pi, \tau \in S(q)}} \operatorname{sgn} \pi \operatorname{sgn} \tau B_{\rho_{\pi(1)}\rho_{\tau(1)}} \cdots B_{\rho_{\pi(q)}\rho_{\tau(q)}} ,$$

it is clear that any $P(B) \in \mathbf{I}_q$ is of the form (5.6). To define $\mathbf{\Pi}_q$, we only need to say which polynomials (5.6) are to be positive.

5.9 *Definition*. $P(B) \in \mathbf{I}_q$ is positive if

$$(5.10) \qquad P(B) = \sum\nolimits_{\left\{\substack{\rho=(\rho_1, \cdots, \rho_q) \\ \pi, \tau \in S(q) \\ j}\right.} \lambda_{\rho,j} q_{\rho,j,\pi} \bar{q}_{\rho,j,\tau} B_{\rho_{\pi(1)}\rho_{\tau(1)}} \cdots B_{\rho_{\pi(q)}\rho_{\tau(q)}}$$

where $\lambda_{\rho,j} \geqq 0$.

Remarks. Using (5.6), $P(B) > 0$ if

$$(5.11) \qquad p_{\rho,\pi,\tau} = \sum\nolimits_j \lambda_{\rho,j} q_{\rho,j,\pi} \bar{q}_{\rho,j,\tau} .$$

Observe that if $P(B) \in \mathbf{\Pi}_q$, then

$$\begin{aligned} P(\gamma) &= \sum\nolimits_{\left\{\substack{\rho=(\rho_1, \cdots, \rho_q) \\ \pi \in S(q) \\ j}\right.} \lambda_{\rho,j} \,|\, q_{\rho,j,\pi} \,|^2 \gamma_{\rho_{\pi(1)}} \cdots \gamma_{\rho_{\pi(q)}} \\ &= q! \sum\nolimits_\rho \left\{\sum\nolimits_{j,\pi} \lambda_{\rho,j} \,|\, q_{\rho,j,\pi} \,|^2\right\} \gamma_{\rho 1} \cdots \gamma_{\rho q} \end{aligned}$$

so that, numerically, $P(\gamma) > 0$ if γ is real and positive (i.e., $\gamma_\rho > 0$).

More generally, we have

(5.12) If $B = {}^t\bar{B}$ and $B \geqq 0$, then $P(B) \geqq 0$ if $P \in \mathbf{\Pi}_q$.

PROOF. We may write $B = A^t\bar{A}$ for some matrix $A = (A_\rho^\alpha)$. Then $B_{\rho\sigma} = \sum_\alpha A_\rho^\alpha \bar{A}_\sigma^\alpha$ and

$$\begin{aligned} P(B) &= P(A^t\bar{A}) \\ &= \sum_{\left\{\begin{smallmatrix}\rho, j \\ \pi, \tau \in S(q) \\ \alpha = (\alpha_1, \cdots, \alpha_q)\end{smallmatrix}\right.} \lambda_{\rho,j} q_{\rho,\pi,j} \bar{q}_{\rho,\tau j} A^{\alpha_1}_{\rho_{\pi(1)}} \bar{A}^{\alpha_1}_{\rho_{\tau(1)}} \cdots A^{\alpha_q}_{\rho_{\pi(q)}} \bar{A}^{\alpha_q}_{\rho_{\tau(q)}} \\ &= \sum_{\left\{\begin{smallmatrix}\rho, j, \alpha \\ \pi, \tau\end{smallmatrix}\right.} \lambda_{\rho,j} q_{\rho,\pi^{-1},j} \bar{q}_{\rho,\tau^{-1} j} A^{\alpha_{\pi(1)}}_{\rho_1} \cdots A^{\alpha_{\pi(q)}}_{\rho_q} \bar{A}^{\alpha_{\tau(1)}}_{\rho_1} \cdots \bar{A}^{\alpha_{\tau(q)}}_{\rho_q} \\ &= \sum_{\rho,j} \lambda_{\rho,j} \,|\, Q_{\rho,\alpha,j}(A) \,|^2 \end{aligned}$$

where

$$Q_{\rho,\alpha,j} = \sum_\pi q_{\rho,\pi^{-1},j} A^{\alpha_{\pi(1)}}_{\rho_1} \cdots A^{\alpha_{\pi(q)}}_{\rho_q} . \tag{5.13}$$

From the definition it is clear that

(5.14) $\mathbf{\Pi}_q$ is a convex cone and $\mathbf{\Pi}_q \mathbf{\Pi}_s \subset \mathbf{\Pi}_{q+s}$.

Roughly speaking, $P(B) \in \mathbf{I}_q$ is positive if, upon writing $B = A^t\bar{A}$,

$$P(B) = \sum_\lambda |\, Q_\lambda(A) \,|^2$$

where $Q_\lambda(A)$ is a polynomial of degree q in A.

Examples of Positive Polynomials. (i) The qth Chern class $c_q = \sum_{\rho_1 < \cdots < \rho_q} \gamma_{\rho_1} \cdots \gamma_{\rho_q}$ corresponds to the polynomial

$$I_q(B) = \left(\frac{1}{q!}\right)^2 \sum_{\rho;\pi,\tau} \operatorname{sgn} \pi \operatorname{sgn} \tau B_{\rho_{\pi(1)}\rho_{\tau(1)}} \cdots B_{\rho_{\pi(q)}\rho_{\tau(q)}} .$$

In (5.10) we then take $\lambda_{\rho,j} = (1/q!)^2$ and $q_{\rho,j,\pi} = \operatorname{sgn} \pi$. Thus $c_q > 0$ and, in fact, $c_I = c_{i_1} \cdots c_{i_q} > 0$. This gives

(5.15) Any polynomial $\sum_I p_I c_I$ with $p_I \geqq 0$ is positive .

(ii) If $\mathbf{E} \to V$ is a *line bundle* with Chern class ω, then $\mathbf{\Pi}_q$ are the classes $\lambda\omega^q$ with $\lambda > 0$.

(iii) Consider the polynomial $P_q(\gamma) = \sum_{\alpha_1 + \cdots + \alpha_r = q} \gamma_1^{\alpha_1} \cdots \gamma_r^{\alpha_r}$. Obviously $P_q(\gamma)$ is invariant under $S(q)$ and so $P_q(\gamma) \in \mathbf{R}_q^*$. We claim

$$P_q(\gamma) = \sum_{\alpha_1 + \cdots + \alpha_r = q} \gamma_1^{\alpha_1} \cdots \gamma_r^{\alpha_r} \tag{5.16}$$

is positive.

PROOF. Let $\rho = (\rho_1, \cdots, \rho_q)$ be a q-tuple where the ρ_i need *not* be distinct. We let $\xi(\rho)$ be the number of permutations $\pi \in S(q)$

which leave ρ invariant; i.e. which satisfy $\rho_1 = \rho_{\pi(1)}, \cdots, \rho_q = p_{\pi(q)}$. If the ρ_i are distinct, then $\xi(\rho) = 1$; if the ρ_i are all equal, then $\xi(\rho) = q!$. It is clear that $q!/\xi(\rho)$ is the number of *distinct* q-tuples $\sigma = (\sigma_1, \cdots, \sigma_q)$ which are rearrangements of ρ. Thus we find

$$\sum_{\alpha_1+\cdots+\alpha_r=q} \gamma_1^{\alpha_1} \cdots \gamma_r^{\alpha_q} = \frac{1}{q!} \sum_{\rho=(\rho_1,\cdots,\rho_q)} \xi(\rho)\gamma_{\rho_1} \cdots \gamma_{\rho_q} ,$$

which gives us that

$$P_q(\gamma) = \left(\frac{1}{q!}\right)^2 \sum_{\substack{\rho=(\rho_1,\cdots\rho_q) \\ \pi\in S(q)}} \xi(\rho)\gamma_{\rho_{\pi(1)}} \cdots \gamma_{\rho_{\pi(q)}} . \tag{5.17}$$

Consider now

$$P_q(B) = \left(\frac{1}{q!}\right)^2 \sum_{\rho,\pi,\tau} B_{\rho_{\pi(1)}\rho_{\tau(1)}} \cdots B_{\rho_{\pi(q)}\rho_{\tau(q)}} . \tag{5.18}$$

If

$$B = \begin{pmatrix} B_{11} & & 0 \\ & \ddots & \\ 0 & & B_{rr} \end{pmatrix}$$

is diagonal, then

$$\begin{aligned} P_q(B) &= \left(\frac{1}{q!}\right)^2 \sum_{\rho,\pi} \left\{ \sum_{\substack{\tau(1)=\pi(1) \\ \vdots \\ \tau(q)=\pi(q)}} B_{\rho_{\pi(1)}\rho_{\tau(1)}} \cdots B_{\rho_{\pi(q)}\rho_{\tau(q)}} \right\} \\ &= \left(\frac{1}{q!}\right)^2 \sum_{\rho,\pi} \xi(\rho) B_{\rho_{\pi(1)}\rho_{\tau(1)}} \cdots B_{\rho_{\pi(q)}\rho_{\tau(q)}} = P_q(\gamma) , \end{aligned}$$

where $\gamma_\rho = B_{\rho\rho}$ and $P_q(\gamma)$ is given by (5.17) above. Thus we need to show that $P_q(B)$ is invariant and positive. For the latter assertion, we simply take $\lambda_\rho = (1/q!)^2$ and $q_{\rho,\pi} = 1$ in (5.10).

To see that $P_q(B)$ is invariant, we set $B = ACA^{-1}$ so that

$$\begin{aligned} &P_q(ACA^{-1}) \\ &= \left(\frac{1}{q!}\right)^2 \sum_{\alpha,\gamma,\rho,\pi,\tau} A_{\rho_{\pi(1)}\alpha_1} C_{\alpha_1\gamma_1} (A^{-1})_{\gamma_1\rho_{\tau(1)}} \cdots A_{\rho_{\pi(q)}\alpha_q} C_{\alpha_q\gamma_q} (A^{-1})_{\gamma_q\rho_{\tau(q)}} \\ &= \left(\frac{1}{q!}\right)^2 \sum_{\alpha,\gamma,\pi,\tau} \{ \sum_\rho (A^{-1})_{\gamma_{\tau^{-1}(1)}\rho_1} A_{\rho_1\alpha_{\pi^{-1}(1)}} \cdots \\ &\qquad\qquad (A^{-1})_{\gamma_{\tau^{-1}(q)}\rho_q} \cdot A_{\rho_q\alpha_{\pi^{-1}(q)}} C_{\alpha_1\gamma_1} \cdots C_{\alpha_q\gamma_q} \} \\ &= \left(\frac{1}{q!}\right) \sum_{\alpha,\gamma,\pi,\tau} \delta^{\gamma_{\tau^{-1}(1)}}_{\alpha_{\pi^{-1}(1)}} \cdots \delta^{\gamma_{\tau^{-1}(q)}}_{\alpha_{\pi^{-1}(q)}} C_{\alpha_1\gamma_1} \cdots C_{\alpha_q\gamma_q} \\ &= \left(\frac{1}{q!}\right)^2 \sum_{\alpha,\pi,\tau} C_{\alpha_1\alpha_{\pi^{-1}\tau(1)}} \cdots C_{\alpha_q\alpha_{\pi^{-1}\tau(q)}} \\ &= P_q(C) . \end{aligned}$$

This proves that $P_q(B)$ is invariant and completes the proof of (5.16).

We list here the first few polynomials $P_q(\gamma)$.

$$
(5.19) \qquad \begin{cases} P_1(\gamma) = c_1 \\ P_2(\gamma) = c_1^2 - c_2 \\ P_3(\gamma) = c_1^3 - 2c_1c_2 + c_3 \end{cases}
$$

Remark. For later interpretation, we record here a fact proved in [11, Lem. A. 1, p. 405]. Let $\mathbf{E} \to V$ be a holomorphic bundle with fibre $\mathbf{C}^r$, $P(\mathbf{E}^*) \xrightarrow{\pi} V$ the associated projective bundle, $\mathbf{L} \to P(\mathbf{E}^*)$ the standard line bundle, and $\omega \in H^2(P(\mathbf{E}^*), \mathbf{Z})$ the characteristic class of $\mathbf{L}$. Recall that there is defined the *integration over the fibre* ([3]):

$$
(5.20) \qquad \pi_*\colon H^{2(q+r-1)}(P(\mathbf{E}^*)) \longrightarrow H^{2q}(V) ,
$$

which satisfies

$$
(5.21) \qquad \pi_*(\xi \cup \pi^*\eta) = (\pi_*\xi) \cup \eta \qquad (\xi \in H^*(P(\mathbf{E}^*)), \eta \in H^*(V)) .
$$

(5.22) PROPOSITION. $\pi_*(\omega^{q+r-1}) = P_q(\gamma) \in H^{2q}(V, \mathbf{Z})$.

(iv) Let $P(B) = \sum_{\rho,\pi,\tau} q_\pi \bar{q}_\tau B_{\rho_{\pi(1)}\rho_{\tau(1)}} \cdots B_{\rho_{\pi(q)}\rho_{\tau(q)}}$ where q_π is independent of ρ. Then

$$
\begin{aligned}
P(ACA^{-1}) &= \sum_{\alpha,\gamma,\rho,\pi,\tau} q_\pi \bar{q}_\tau A_{\rho_{\pi(1)}\alpha_1} \cdots A_{\rho_{\pi(q)}\alpha_q} (A^{-1})_{\gamma_1\rho_{\tau(1)}} \cdots \\
&\qquad\qquad (A^{-1})_{\gamma_q\rho_{\tau(q)}} \cdot C_{\alpha_1\gamma_1} \cdots C_{\alpha_q\gamma_q} \\
&= \sum_{\alpha,\gamma,\rho,\pi,\tau} q_{\pi^{-1}} \bar{q}_{\tau^{-1}} (A^{-1})_{\gamma_{\tau(1)}\rho_1} A_{\rho_1\alpha_{\pi(1)}} \cdots \\
&\qquad\qquad (A^{-1})_{\gamma_{\tau(q)}\rho_q} A_{\rho_q\alpha_{\pi(q)}} \cdot C_{\alpha_1\gamma_1} \cdots C_{\alpha_q\gamma_q} \\
&= \sum_{\alpha,\gamma,\pi,\tau} q_{\pi^{-1}} \bar{q}_{\tau^{-1}} \delta^{\gamma_{\tau(1)}}_{\sigma_{\pi(1)}} \delta^{\gamma_{\tau(q)}}_{\alpha_{\pi(q)}} C_{\alpha_1\gamma_1} \cdots C_{\alpha_q\gamma_q} \\
&= \sum_{\alpha,\pi,\tau} q_\pi \bar{q}_\tau C_{\alpha_1\alpha_{\pi^{-1}\tau(1)}} \cdots C_{\alpha_q\alpha_{\pi^{-1}\tau(q)}} \\
&= \sum_{\rho,\pi,\tau} q_\pi \bar{q}_\tau C_{\alpha_{\pi(1)}\alpha_{\pi^{-1}\tau\pi(1)}} \cdots C_{\alpha_{\pi(q)}\alpha_{\pi^{-1}\tau\pi(q)}} .
\end{aligned}
$$

Thus we will have $P(ACA^{-1}) = P(C)$ if $q_\tau = q_{\pi^{-1}\tau\pi}$ for $\pi, \tau \in S(q)$. In other words:

$$
(5.22) \quad \begin{cases} P(B) = \sum_{\rho,\pi,\tau} q_\pi \bar{q}_\tau B_{\rho_{\pi(1)}\rho_{\tau(1)}} \cdots B_{\rho_{\pi(q)}\rho_{\tau(q)}} \text{ is an in-} \\ \text{variant polynomial} \Leftrightarrow q_\tau = q_{\pi\tau\pi^{-1}} \text{ for all } \tau, \pi \in S(q) . \end{cases}
$$

This allows us essentially to determine the positive polynomials of low degree. For example, when $q = 2 = r$, $S(q)$ has two elements e, f (e is $(1, 2) \to (1, 2)$ and f is $(1, 2) \to (2, 1)$). We let $q_e = \alpha$ and $q_f = \beta$ and, supposing that α, β are real (and rational), we have $P(B)=(\alpha+\beta)^2\{B_{11}B_{11}+B_{22}B_{22}\}+2(\alpha^2+\beta^2)B_{11}B_{22}+4\alpha\beta B_{12}B_{21}$. Thus $P(\gamma) = (\alpha+\beta)^2\{\gamma_1^2+2\gamma_1\gamma_2+\gamma_2^2\} + 2\{\alpha^2+\beta^2-(\alpha+\beta)^2\}\gamma_1\gamma_2 = (\alpha+\beta)^2c_1^2 - 4\alpha\beta c_2$; i.e.,

$$P(\gamma) = (\alpha+\beta)^2c_1^2 - 4\alpha\beta c_2 \tag{5.23}$$

is the general positive polynomial when $q = 2 = r$.

If $\alpha+\beta=0$ (this is essentially the case $q_\pi = \operatorname{sgn}\pi$ in example (i) above), then $P(\gamma) = \mu c_2$ where $\mu > 0$. If $\alpha = \beta$, then $P(\gamma) = \mu(c_1^2 - c_2)$ where $\mu > 0$, and this is essentially (iii) above.

Assume now that $\alpha+\beta \neq 0$. Then $P(\gamma) = c_1^2 - \{4\alpha\beta/(\alpha+\beta)^2\}c_2$. Now $\{4\alpha\beta/(\alpha+\beta)^2\} \leqq 1$ so that $P(\gamma) = (c_1^2 - c_2) + \mu c_2$ where $\mu \geqq 0$. Thus $\mathbf{\Pi}_2$ is generated by c_2 and $c_1^2 - c_2$.

PROOF OF THEOREM D. Suppose that $\mathbf{E} \to V$ is ample and $P(B) \in \mathbf{\Pi}_q$. It will suffice to find a metric in $\mathbf{E} \to V$ with curvature Θ such that

$$\left(\frac{1}{2\pi i}\right)^q P(\Theta) > 0$$

in the sense of 5(a). Of course we take in $\mathbf{E}$ the metric given by the global sections, which is the same as the metric induced in $\mathbf{E}$ from the universal bundle over the grassmannian (cf. § 1(f)).

Given $z_0 \in V$, we can find a local holomorphic frame $f(z) = (e_1(z), \cdots, e_r(z))$ for $\mathbf{E} \to V$ and a matrix $A = (A_\rho^\alpha(z))$ of (1, 0) forms such that $f(z_0)$ is unitary and

$$\Theta_{\rho\sigma}(z_0) = \textstyle\sum_\alpha A_\rho^\alpha(z_0)\bar{A}_\sigma^\alpha(z_0) . \tag{5.24}$$

This follows from (2.24).

Let $P(B)$ be given by (5.10). Then

$$\begin{aligned} P(\Theta) &= \textstyle\sum_{\left\{\substack{\rho,\pi,\tau,j \\ \alpha_1,\cdots,\alpha_q}\right.} \lambda_{\rho,j}q_{\pi,\rho,j}\bar{q}_{\tau,\rho,j}A_{\rho_{\pi(1)}}^{\alpha_1}\bar{A}_{\rho_{\tau(1)}}^{\alpha_1} \cdots A_{\rho_{\pi(q)}}^{\alpha_q}\bar{A}_{\rho_{\tau(q)}}^{\alpha_q} \\ &= (-1)^{\{q(q-1)/2\}} \textstyle\sum \lambda_{\rho,j}q_{\pi^{-1},\rho,j}\bar{q}_{\tau^{-1},\rho,j}A_{\rho_1}^{\alpha_{\pi(1)}} \cdots \\ &\qquad\qquad A_{\rho_q}^{\alpha_{\pi(q)}}\bar{A}_{\rho_1}^{\alpha_{\tau(1)}} \cdots \bar{A}_{\rho_q}^{\alpha_{\tau(q)}} \\ &= (-1)^{\{q(q-1)/2\}} \textstyle\sum_{\alpha,\rho,j} \lambda_{\rho,j}\theta_{\rho,j,\alpha} \wedge \bar{\theta}_{\rho,j,\alpha} \end{aligned}$$

where $\theta_{\rho,j,\alpha} = \sum_\pi q_{\pi^{-1},\rho,j}A_{\rho_1}^{\alpha_{\pi(1)}} \cdots A_{\rho_q}^{\alpha_{\pi(q)}}$ is a form of type $(q, 0)$. This proves that $(1/2\pi i)^q P(\Theta) \geqq 0$; we need only show that some $\theta_{\rho,j,\alpha}(z_0) \neq 0$.

Changing notation slightly, let $\theta_{\rho,\alpha} = \sum_\pi q_{\pi,\rho}A_{\rho_1}^{\alpha_{\pi(1)}} \cdots A_{\rho_q}^{\alpha_{\pi(q)}}$. If all $\theta_{\rho,\alpha} = 0$, then we have

$$\textstyle\sum_\pi q_{\pi,\rho}A_{\rho_1}^{\alpha_{\pi(1)}} \cdots A_{\rho_q j_1}^{\alpha_{\pi(q)}} = 0 \tag{5.25}$$

for all ρ, α, and $j_1 < \cdots < j_q$ where $A^\alpha_\rho = \sum_j A^\alpha_{\rho j} dz^j$.

We need now to interpret the matrices A^α_ρ. This is given in Section 2 (f) where it is shown that: In terms of the holomorphic frame $f(z)$ above (with $f(z_0)$ unitary), we may choose a basis $s^1(z)$, $\cdots$, $s^m(z) = (\cdots s^\alpha(z) \cdots)$ for the sections of $\mathbf{E} \to V$ which vanish at z_0 such that

$$s^\alpha(z) = \sum\nolimits_{\rho,j} A^\alpha_{\rho j} z^j e_\rho + (\text{terms of order } 2) . \tag{5.26}$$

Since $\mathbf{E} \to V$ is ample, the forms $\sum_{\rho,j} A^\alpha_{\rho j} e_\rho \otimes dz^j$ span $\mathbf{E}_{z_0} \otimes \mathbf{T}^*_{z_0}$. We may choose the sections s^α such that the matrix $(A^\alpha_{\rho j})_{1 \leqq \alpha \leqq rn}$ is non-singular and $A^\alpha_{\rho j} = 0$ for $\alpha > rn$. Relabeling, we write $s^{\sigma k}(z) = \sum_{\rho,j} A^{\sigma k}_{\rho j} z^j e_\rho + (\cdots)$ where $(A^{\sigma k}_{\rho j})$ is non-singular and $ds^\alpha(z_0) = 0$ for $\alpha > rn$. Then (5.25) becomes:

$$\sum\nolimits_\pi q_{\pi,\rho} A^{\sigma_{\pi(1)} k_{\pi(1)}}_{\rho_1 j_1} \cdots A^{\sigma_{\pi(q)} k_{\pi(q)}}_{\rho_q j_1} = 0 . \tag{5.27}$$

Multiplying (5.27) on the right by $(A^{-1})^{\tau_1 i_1}_{\sigma_1 k_1} \cdots (A^{-1})^{\tau_q i_q}_{\sigma_q k_q}$ and summing on $\sigma_1, \cdots, \sigma_q$; $k_1, \cdots, k_q$, we get

$$\sum\nolimits_\pi q_{\pi,\rho} \delta^{\tau_{\pi(1)}}_{\rho_1} \cdots \delta^{\tau_{\pi(q)}}_{\rho_q} \delta^{i_{\pi(1)}}_{j_1} \cdots \delta^{i_{\pi(q)}}_{j_q} = 0 \tag{5.28}$$

for all ρ, τ, $\mathbf{i} = (i_1, \cdots, i_q)$, and $j_1 < \cdots < j_q$. From (5.28) we get $q_{\pi,\rho} = 0$, a contradiction which completes the proof of Theorem D.

Remarks. As mentioned below (5.3), it should be the case that $P(c_1, \cdots, c_r) > 0$ if $\mathbf{E} \to V$ is *positive* and $P \in \mathbf{\Pi}_q$. In particular, we should be able to prove:

$$\begin{cases} c_q > 0 \\ P_q(c_1, \cdots, c_r) > 0 , & \text{where } P_q \text{ is given by (5.16)} . \end{cases} \tag{5.29}$$

Now in the Appendix to 5(b), we show that $c_2 > 0$ by proving that $(1/2\pi i)^2 I_2(\Theta) > 0$ if $\mathbf{E} \to V$ is positive. It is probably true that $(1/2\pi i)^q I_q(\Theta) > 0$, but this will require a better understanding of the algebraic properties of the curvature form Θ.

If $\mathbf{E} \to V$ is spanned by its sections, then $c_q \geqq 0$. Using (3.51), let us prove:

(5.30) If $\mathbf{E} \to V$ is spanned by its sections, if $\mathbf{E} \to V$ is positive, and if $Z_q \subset V$ is an algebraic *submanifold*, then $\langle c_q, Z \rangle > 0$.

PROOF. Using a standard result, we have over Z an exact sequence

$$0 \longrightarrow \mathbf{S} \longrightarrow \mathbf{E} \,|\, Z \longrightarrow \mathbf{Q} \longrightarrow 0$$

where S is a trivial bundle with fibre $\mathbf{C}^{r-q}$. Now $\mathbf{E} \mid Z$ is positive and spanned by its sections; by (3.6) the same is true for $\mathbf{Q}$. Since $c_q(\mathbf{E} \mid Z) = c_q(\mathbf{Q})$ and $c_q\colon H^0(Z) \to H^{2q}(Z)$ is an isomorphism (by (3.51)), it follows that $\langle c_q, Z\rangle > 0$ as required.

A possible alternative approach to (5.29) is to split $\mathbf{E}$ into a sum of line bundles (certainly, if Θ is diagonal, then (5.29) holds). We shall show that this method gives $P_q(c_1, \cdots, c_r) > 0$, but not $c_q > 0$.

For simplicity, suppose that $\mathbf{E} \to V$ has fibre $\mathbf{C}^2$ and consider the exact sequence

$$(5.31) \qquad 0 \longrightarrow \mathbf{F} \longrightarrow \pi^*(\mathbf{E}) \longrightarrow \mathbf{L} \longrightarrow 0$$

over $P(\mathbf{E}^*)$. Here $\mathbf{F}_{(z,\xi)} = \{\lambda \in \mathbf{E}_z\colon \langle \xi, \lambda\rangle = 0\}$ where $z \in V, \xi \in P(\mathbf{E}_z^*)$. We choose unitary frames $f = (e_1, e_2)$ for $\pi^*(\mathbf{E})$ such that $e_1 \in \mathbf{F}$ and $e_2 \in \mathbf{L}$ (cf. 2(e)). Then, if

$$\Theta = \begin{pmatrix} \Theta_{11} & \Theta_{12} \\ \Theta_{21} & \Theta_{22} \end{pmatrix}$$

is the curvature for $\mathbf{E}$ (and also $\pi^*(\mathbf{E})$), the curvature in $\mathbf{F}$ is $\Theta_{\mathbf{F}} = \Theta_{11} - \theta \wedge \bar{\theta}$ and the curvature in $\mathbf{L}$ is $\Theta_{\mathbf{L}} = \Theta_{22} + \theta \wedge \bar{\theta}$. Here $\theta \in A^{1,0}(\mathrm{Hom}\,(\mathbf{F}, \mathbf{L}))$ is the 2nd fundamental form of $\mathbf{F}$ in $\pi^*(\mathbf{E})$ (cf. 2 (d)).

It is easy to check that $\theta \mid P(\mathbf{E}_z^*)$ is non-zero, so that $\Theta_{22} + \theta \wedge \bar{\theta} > 0$ on $P(\mathbf{E}^*)$ and $\mathbf{L} \to P(\mathbf{E}^*)$ is positive (cf. 2(g)).

We now prove that $P_q(\gamma) > 0$ in $H^{2q}(V, \mathbf{Z})$, where $P_q(\gamma)$ is given by (5.16). Let $Z \subset V$ be a q-dimensional algebraic subvariety and $Z_\pi = \pi^{-1}(Z) \subset P(\mathbf{E}^*)$. Then, by (5.22),

$$\langle P_q(c_1, \cdots, c_r), Z\rangle = \langle \pi_* \omega^{q+s-1}, Z\rangle = \langle \omega^{r+q-1}, Z_\pi\rangle > 0$$

since $\omega > 0$ on $P(\mathbf{E}^*)$.

If we try to use this argument to show that, e.g., $c_2 > 0$, we have $\langle c_2, Z\rangle = \langle \pi^* c_2 \omega, Z_\pi\rangle$ so that we need $\pi^* c_2 \omega > 0$. By (5.31), $\pi^* c_2 = c_1(\mathbf{F}) c_1(\mathbf{L})$, so we want

$$\Theta_{\mathbf{F}} \Theta_{\mathbf{L}} \Theta_{\mathbf{L}} = (\Theta_{22} - \theta\bar{\theta})(\Theta_{11} + \theta\bar{\theta})(\Theta_{11} + \theta\bar{\theta})$$

to be positive. The relevant term in this product is $(2\Theta_{11}\Theta_{22} - \Theta_{11}^2)\theta\bar{\theta}$, which however need *not* be positive.

Here we use $\pi^* c_2 \omega$ because $\pi_*(\pi^* c_2 \omega) = c_2$ (by (5.21)). Any $\eta \in H^6(P(\mathbf{E}^*))$ with $\pi_* \eta = c_2$ would work equally well (e.g., $\pi_* \omega^3 = P_2(\gamma) = \pi_*(\pi^*(c_1^2 - c_2)\omega)$); to prove $c_2 > 0$ by this method we need to choose the class η correctly and then make a judicious choice of a differential form representing η. This we are so far unable to do.

Appendix to § 5. (b). We shall prove.

$$(5.32) \quad \begin{cases} \text{Let } \mathbf{E} \to V \text{ be a positive holomorphic bundle} \\ \text{with fibre } \mathbf{C}^2. \text{ Then } c_2(\mathbf{E}) > 0 . \end{cases}$$

PROOF. It will suffice to assume that $\dim V = 2$ because, if $\mathbf{E} \to V$ is a positive bundle according to (0.1) and if $Z \subset V$ is an algebraic subvariety, then $\mathbf{E} \mid Z$ is positive.

By assumption, there is a metric in $\mathbf{E} \to V$ with curvature Θ such that the form

$$\Theta(\xi, \eta) = \sum_{\rho,\sigma,i,j} \Theta^{\rho}_{\sigma ij}\, \xi^{\sigma}\bar{\xi}^{\rho}\eta^{i}\bar{\eta}^{j}$$

$(1 \leqq \rho, \sigma, i, j \leqq 2)$ is positive. Let $\Phi = (1/2\pi i)^2 \det \Theta$ be the 2nd Chern class of $\mathbf{E} \to V$ according to (4.26). We shall prove that $\Phi > 0$.

Write

$$\Theta = \begin{pmatrix} \Theta_{11} & \Theta_{12} \\ \Theta_{21} & \Theta_{22} \end{pmatrix} .$$

Then, using at a point a unitary frame for $\mathbf{E} \to V$, we have $\Theta + {}^t\bar{\Theta} = 0$. This gives $\Theta_{11} + \bar{\Theta}_{11} = 0$, $\Theta_{22} + \bar{\Theta}_{22} = 0$, and $\Theta_{12} + \bar{\Theta}_{21} = 0$. Since $\Theta_{11} > 0$, we may choose a co-frame ω^1, ω^2 for V such that $\Theta_{11} = \omega^1 \wedge \bar{\omega}^1 + \omega^2 \wedge \bar{\omega}^2$. Since $\Theta_{22} > 0$, by a unitary change of ω^1, ω^2 we may assume that $\Theta_{22} = \alpha\omega^1\bar{\omega}^1 + \beta\omega^2\bar{\omega}^2$ where $\alpha, \beta > 0$ (we omit the "$\wedge$" symbol). Thus

$$\Theta = \begin{pmatrix} \omega^1\bar{\omega}^1 + \omega^2\bar{\omega}^2 & \theta \\ -\bar{\theta} & \alpha\omega^1\bar{\omega}^1 + \beta\omega^2\bar{\omega}^2 \end{pmatrix}$$

where $\theta = \sum_{i,j} h_{ij}\omega^i\bar{\omega}^j$. Letting $\omega = \omega^1\bar{\omega}^1\omega^2\bar{\omega}^2$,

$$\det \Theta = (\alpha + \beta - h_{11}\bar{h}_{22} - h_{22}\bar{h}_{11} + h_{12}\bar{h}_{12} + h_{21}\bar{h}_{21})\omega .$$

Now, by the Schwarz inequality, $h_{11}\bar{h}_{22} + h_{22}\bar{h}_{11} \leqq (|h_{11}|^2 + |h_{22}^2)$, so that:

$$(5.33) \qquad \det \Theta \geqq (\alpha - |h_{11}|^2 + \beta - |h_{22}|^2 + |h_{12}|^2 + |h_{21}|^2)\omega .$$

From (5.33), it will suffice to prove $\alpha > |h_{11}|^2$, $\beta > |h_{22}|^2$.

For $\xi = \begin{pmatrix}\xi^1 \\ \xi^2\end{pmatrix}$, we let $\Theta(\xi) = \sum_{\rho,\sigma} \Theta^{\rho}_{\sigma ij}\xi^{\sigma}\bar{\xi}^{\rho}$. We have used that $\Theta\begin{pmatrix}1 \\ 0\end{pmatrix} = \Theta_{11} > 0$, $\Theta\begin{pmatrix}0 \\ 1\end{pmatrix} = \Theta_{22} > 0$. Now

$$0 < \Theta\begin{pmatrix}\lambda \\ 1\end{pmatrix} = \Theta_1^1 |\lambda|^2 + \Theta_2^1\lambda + \Theta_1\bar{\lambda} + \Theta_2^2$$
$$= \begin{pmatrix} |\lambda|^2 + \lambda h_{11} + \bar{\lambda}\bar{h}_{11} + \alpha & h_{12}\lambda + \bar{h}_{21}\bar{\lambda} \\ h_{21}\lambda + \bar{h}_{12}\bar{\lambda} & |\lambda|^2 + \lambda h_{22} + \bar{\lambda}\bar{h}_{22} + \beta \end{pmatrix}.$$

Write $\lambda = x + iy$ and $f(x, y) = |\lambda|^2 + \lambda h_{11} + \bar{\lambda}\bar{h}_{11} + \alpha$; then $f(x, y) > 0$ for all x, y. Seeking a minimum for $f(x,y)$, we set $\partial f/\partial x(x,y) = 0$, $\partial f/\partial y(x, y) = 0$ and find $x = -(h_{11} - \bar{h}_{11}/2)$, $y = i\{(-h_{11} + \bar{h}_{11})/2\}$, and $\lambda = -\bar{h}_{11}$. At this point, $0 < f(\lambda) = \alpha - |h_{11}|^2$ so that $|h_{11}|^2 < \alpha$, $|h_{22}|^2 < \beta$ as desired.

PRINCETON UNIVERSITY

REFERENCES

[1] A. ANDREOTTI and H. GRAUERT, *Théorèmes de finitude pour la cohomologie des espaces complexes,* Bull. Soc. Math. France, **90** (1962), 193-259.

[2] R. BOTT, *On a theorem of Lefschetz,* Mich. Math. J. **6** (1959), 211-216.

[3] ——— and S. S. CHERN, *Hermitian vector bundles and the equidistribution of the zeroes of their holomorphic sections,* Acta Math. **114** (1966), 71-112.

[4] E. CALABI and E. VESENTINI, *On compact, locally symmetric Kähler manifolds,* Ann. of Math. **71** (1960), 472-507.

[5] S. S. Chern, *Characteristic classes of hermitian manifolds,* Ann. of Math. **57** (1946), 85-121.

[6] ———, *On the characteristic classes of complex sphere bundles and algebraic varieties,* Amer. J. Math. **75** (1953), 565-597.

[7] ———, *A simple intrinsic proof of the Gauss-Bonnet formula for closed riemannian manifolds,* Ann. of Math., **45** (1944), 747-752.

[8] R. GODEMENT, Topologie algébrique et théorie des faisceaux, Hermann Paris, 1958.

[9] H. GRAUERT, *Über modifikationen und exzeptionelle analytische mengen,* Math. Ann. **146** (1962), 331-368.

[10] P. GRIFFITHS, *Hermitian differential geometry and the theory of positive and ample holomorphic vector bundles,* J. Math. Mech. **14** (1965), 117-140.

[11] ———, *The extension problem in complex analysis*: II, Amer. J. Math **88** (1966), 366-446.

[12] ———, *On a theorem of Chern,* III. J. Math. **6** (1962), 468-479.

[13] R. GUNNING and H. ROSSI, Analytic Functions of Several Complex Variables, Prentice-Hall, 1965.

[14] R. HARTSHORNE, *Ample vector bundles,* Publ. I. H. E. S. **29** (1966), 319-350.

[15] F. HIRZEBRUCH, Neue topologische methoden in der algebraischen geometrie, Ergebnisse der Mathematik, vol. 9 Springer, 1956.

[16] W. V. D. HODGE and D. PEDOE, Methods of Algebraic Geometry, vol. II. Cambridge Univ. Press, 1952.

[17] S. KLEIMAN, to appear in Proc. Nat. Acad. Sci. U.S.A.

[18] K. Kodaira, *On a differential-geometric method in the theory of analytic stacks*, Proc. Nat. Acad. Sci. U.S.A., **39** (1953), 1268-1273.

[19] ———, *On cohomology groups of compact analytic varieties with coefficients in some analytic faisceaux*, Proc. Nat. Acad. Sci. U.S.A., **39** (1953), 66-74.

[20] ———, *On Kähler varieties of restricted type*, Ann. of Math. **60** (1954), 38-48.

[21] Y. Nakai, *A criterion of an ample sheaf on a projective scheme*, Amer. J. Math., **85** (1963), 14-26.

[22] S. Nakano, *On complex analytic vector bundles*, J. Math. Soc. Japan **7** (1955), 1-12.

[23] W. Schmid, *Homogeneous complex manifolds and representations of semi-simple Lie groups*, Proc. Nat. Acad. Sci. U.S.A., **59** (1968), 56-59.

[24] J-P. Serre, *Faisceaux algébriques cohérents*. Ann. of Math. **61** (1955), 197-278.

[25] ———, *Géométrie algébrique et géométrie analytique*, Ann. Inst. Fourier, Grenoble **6** (1955-1956), 1-42

[26] A. Weil, Varieties kähleriennes, Hermann (6), Paris.

[27] O. Zariski, Algebraic surfaces, Ergebnisse der Mathematik, vol. 5. Springer, 1935.

(Received December 16, 1968)

The signature of ramified coverings*

By F. HIRZEBRUCH

1.

Let X be a compact oriented differentiable manifold of dimension m without boundary on which the cyclic group G_n of order n acts by orientation preserving diffeomorphisms. The set Y of fixed points of this action is a differentiable submanifold of X not necessarily connected. The various connectedness components of Y can have different dimensions, they are not necessarily orientable.

We assume that all components of Y have codimension 2 and G_n operates freely on $X - Y$. Then X/G_n is an oriented manifold. The natural projection

$$\pi\colon X \longrightarrow X/G_n$$

maps Y bijectively onto a submanifold Y' of X/G_n. For any point $x \in Y$ we can introduce local coordinates (ξ, z) of X centered at x with

$$\xi \in \mathbf{R}^{m-2},\ z \in \mathbf{C}\ , \qquad\qquad Y \text{ given locally by } z = 0\ ,$$

and local coordinates (ξ', z') of X/G_n centered at $\pi(x)$ with

$$\xi' \in \mathbf{R}^{m-2},\ z' \in \mathbf{C}\ , \qquad\qquad Y' \text{ given locally by } z' = 0$$

such that $\pi\colon X \to X/G_n$ has the local description

$$z' = z^n\ , \qquad \xi' = \xi\ .$$

Under these circumstances we call X an n-fold ramified covering of X/G_n with branching locus Y'.

From now on suppose X has dimension $4k$. The signature of X is defined as follows. Consider over $H^{2k}(X, \mathbf{R})$ the quadratic form

$$(1) \qquad Q(\alpha, \beta) = (\alpha \cup \beta)[X] \qquad \text{for } \alpha, \beta \in H^{2k}(X, \mathbf{R})\ .$$

Q is a bilinear symmetric form over a real vector space and by

---

* Lecture given at the Summer Institute on Global Analysis, AMS, Berkeley, July 1968.

definition

$$\operatorname{sign}(X) = p^{+} - p^{-} ,$$

where p^{+} is the number of positive entries, p^{-} the number of negative entries in a diagonalisation of Q. If, as before, G_n operates on X, then $H^{2k}(X, \mathbf{R})$ is a G_n-module. The action of G_n on $H^{2k}(X, \mathbf{R})$ preserves Q. We can decompose $H^{2k}(X, \mathbf{R})$ as follows

$$H^{2k}(X, \mathbf{R}) = H_{+} \oplus H_{-} , \tag{2}$$

where H_{+} and H_{-} are Q-orthogonal, and where Q is positive-definite on H_{+} and negative-definite on H_{-} and

$$T(H_{+}) = H_{+} , \quad T(H_{-}) = H_{-} \qquad \text{for } T \in G_n .$$

For any $T \in G_n$ we define

$$\operatorname{sign}(X, T) = \operatorname{tr}(T \mid H_{+}) - \operatorname{tr}(T \mid H_{-}) \tag{3}$$

where tr denotes the trace. It is easy to show that sign (X, T) does not depend on the choice of H_{+} and H_{-} (compare [4, p. 578]). For $T = 1$ we have

$$\operatorname{sign}(X, T) = \operatorname{sign}(X) .$$

We wish to relate sign (X) and sign (X/G_n). For this we need information on sign (X, T). This is furnished by the fixed point theorem of Atiyah-Bott-Singer [2], [3], [4], or rather by a special case of it. There is an elliptic operator on X of order 1 whose index is sign (X) (see [4, § 6]). The Atiyah-Singer index theorem applied to this operator gives sign (X) in terms of Pontrjagin numbers [7, Th. 8.2.2], the Atiyah-Bott-Singer fixed point theorem gives a formula for sign (X, T) involving the Pontrjagin classes of X and the normal bundle of Y as a vector bundle on which G_n acts. We shall give the precise formula for sign (X, T) later.

2.

The map $\pi \colon X \to X/G_n$ induces a ring homomorphism

$$\pi^{*} \colon H^{*}(X/G_n, \mathbf{R}) \longrightarrow H^{*}(X, \mathbf{R}) .$$

LEMMA. *π^{*} maps $H^{*}(X/G_n, \mathbf{R})$ bijectively on $H^{*}(X, \mathbf{R})^{G_n}$, the ring of elements of $H^{*}(X, \mathbf{R})$ invariant under the operations of G_n.*

This is a well-known fact true under much more general circumstances. (Compare A. Grothendieck, Tôhoku Math. J. 9 (1957),

119–221; Chap. V.)

We have for $\alpha, \beta \in H^{2k}(X/G_n, \mathbf{R})$

$$(\pi^*\alpha \cup \pi^*\beta)[X] = n(\alpha \cup \beta)[X/G_n] .$$

Therefore sign (X/G_n) equals the signature of the form Q of X (see (1)) when restricted to

$$H^{2k}(X, \mathbf{R})^{G_n} = H_+^{G_n} \oplus H_-^{G_n} \qquad \text{(see (2))} .$$

Thus

(4) $$\operatorname{sign}(X/G_n) = \dim H_+^{G_n} - \dim H_-^{G_n} .$$

Using a well-known formula for the dimension of the invariant part of a representation of a finite group, we obtain from formulas (3) and (4)

(5) $$\operatorname{sign}(X/G_n) = \frac{1}{n} \sum_{T \in G_n} \operatorname{sign}(X, T)$$

3.

Let us first consider the case $n = 2$. Then $G_2 = \{1, T\}$ where T is an orientation preserving involution of X. We have

(6) $$\operatorname{sign}(X, T) = \operatorname{sign}(Y \circ Y)$$

where $Y \circ Y$ is the oriented self-intersection cobordism class (see [8], [9], [4, Prop. 6.15]). Thus (5) becomes

(7) $$2 \operatorname{sign}(X/G_2) = \operatorname{sign}(X) + \operatorname{sign}(Y \circ Y) .$$

Remarks. (i) The formula (6) is a consequence of the Atiyah-Bott-Singer fixed point theorem. Formula (6) holds for any orientation preserving differentiable involution, the fixed point set Y need not have codimension 2. The fixed theorem gives sign (X, T) in terms of characteristic numbers. Applying the signature theorem to $Y \circ Y$ eliminates the characteristic numbers to give a theorem which is trivial for $T =$ Identity and therefore is weaker than the version coming from the Atiyah-Bott-Singer fixed point theorem. Jänich and Ossa [10] have proved (6), for arbitrary codimension of Y, by elementary methods. It is not true for manifolds with boundary, even if one assumes $Y \cap \partial X = \emptyset$. The mistake gives rise to an interesting invariant for free involutions on $(4k-1)$-manifolds as studied in [4, § 7], [8], [9]. Also (7) is not true for manifolds with boundary.

(ii) In (6) and (7) the fixed point set Y need not be orientable.

Example. If X is the complex projective plane canonically oriented and T complex conjugation with respect to homogeneous coordinates (z_0, z_1, z_2), then Y is the real projective plane and

$$\operatorname{sign}(Y \circ Y) = -1 .$$

X/G_2 is a rational homology 4-sphere which checks with (7).

(iii) If $\dim X = 4$ and Y is orientable, then $\operatorname{sign}(Y \circ Y)$ is the self-intersection number $Y \circ Y$ in X which equals $(Y' \circ Y')/2$ where $Y' \circ Y'$ is the self-intersection number of Y' in X/G_2 (compare (13)). Thus (7) can be written in this case as

$$(8) \qquad \operatorname{sign}(X) = 2 \operatorname{sign}(X/G_2) - \frac{1}{2}(Y' \circ Y') .$$

Formula (8) is true also if Y is non-orientable. Then $Y' \circ Y'$ has to be considered as oriented self-intersection cobordism class in X/G_2.

4.

We make the assumptions of § 1 and wish to give a formula for $\operatorname{sign}(X, T)$ with $T \in G_n$, but T different from the identity. Observe that Y is orientable if $n \geqq 3$. For $n = 2$ we assume Y orientable, the non-orientable case having been settled in § 3. We orient Y and the normal bundle ν of Y such that these orientations span the given orientation of X. Then ν has $\mathbf{SO}(2)$ as structural group and may therefore be regarded as a complex line bundle with $\mathbf{U}(1)$ as structural group. This can be done equivariantly with respect to G_n. Then the operation of T in ν determines a complex eigenvalue t with $t^n = 1$ and $t \neq 1$.

THEOREM. *Under the assumptions of* § 1 (*with* $\dim X = 4k$) *we have*

$$(9) \qquad \operatorname{sign}(X, T) = \operatorname{sign} \frac{(t+1) + (t-1)Y}{(t-1) + (t+1)Y} Y .$$

This formula is to be interpreted as follows: Develop

$$\frac{(t+1) + (t-1)y}{(t-1) + (t+1)y} y ,$$

where $t \in \mathbf{C}$ $(t \neq 1)$ and y an indeterminate, as a formal power series in y.

$$(10)\qquad \frac{(t+1)+(t-1)y}{(t-1)+(t+1)y}y = \frac{t+1}{t-1}y - \frac{4t}{(t-1)^2}y^2 + \cdots .$$

We construct a sequence of oriented submanifolds of X

$$\cdots \subset Y_3 \subset Y_2 \subset Y_1 = Y \subset X .$$

If Y_r is already constructed, then we make the embedding $i\colon Y_r \to X$ transversal to Y. Let j be a transversal map approximating i, then $Y_{r+1} = j^{-1}(Y)$. The orientations of Y_{r+1} and $j^*\nu$ span the orientation of Y_r. The oriented cobordism classes of the Y_r are independent of all other choices involved. The cobordism class of Y_r is denoted by

$$Y^r = Y \circ \cdots \circ Y \in \Omega^{4k-2r} .$$

If we replace in (10) the power y^r by Y^r, then we get an element of the cobordism algebra $\Omega^* \otimes \mathbf{C}$, where of course $Y^r = Y \circ \cdots \circ Y$ does not represent a power with respect to the multiplication in Ω^*. Recall that the signature is a ring homomorphism $\Omega^* \to \mathbf{Z}$ which vanishes by definition on Ω^m for $m \not\equiv 0 \pmod 4$. The right side of (9) is the signature of the element of $\Omega^* \otimes \mathbf{C}$ obtained by replacing y^r by Y^r. Thus (9) means (for $\dim X = 4k$)

$$\operatorname{sign}(X, T) = -\frac{4t}{(t-1)^2}\operatorname{sign}(Y \circ Y) + \cdots .$$

If Y is not connected, then the eigenvalue t has to be taken separately for each connectedness component of Y, and the right side of (9) represents a sum over the connectedness components. Changing the orientation of Y and simultaneously of the normal bundle ν has the effect that $Y^r \in \Omega^{4k-2r}$ is replaced by $(-1)^r Y^r$ and t by t^{-1}. Since

$$\frac{(t+1)+(t-1)y}{(t-1)+(t+1)y}y$$

remains unchanged under the substitution $t \to t^{-1}$, $y \to -y$ it does not matter which orientations we take as long as the orientations of Y and ν span the given orientation of X. Actually we have even more freedom with the orientations, since in (9) we have $\operatorname{sign}(Y^r) \neq 0$ only for r even. Formula (9) is a consequence of the Atiyah-Bott-Singer fixed point theorem, more precisely of the G-signature theorem [4, p. 582]. When applying it the invariance under the substitution $t \to t^{-1}$, $y \to -y$ has to observed because in [4] *loc cit.* the eigenvalue t is supposed to have a positive imaginary part.

When deducing (9) from the G-signature theorem we may assume $t \neq -1$. If $t = -1$, then T is an involution (which can happen only if n is even). But (9) reduces to (6) for $t = -1$. Without further explanation we write down the G-signature theorem of [4, p. 582] for our case and $t \neq -1$. We use precisely the notation of [4]. This gives for $t = e^{i\theta}$ $(0 < \theta < \pi)$

$$\operatorname{sign}(X, T) = 2^{2k-1}\left\{\mathfrak{L}(Y)\frac{1}{\tanh\dfrac{x+i\theta}{2}}\right\}[Y]$$

where $x \in H^2(X, \mathbf{Z})$ is the Poincaré dual of Y. Substitute in the expression in $\{\quad\}$ each $(2r)$-dimensional class α by $2^r\alpha$, then we get

$$\operatorname{sign}(X, T) = \left\{\widetilde{\mathfrak{L}}(Y)\frac{1}{\tanh\left(x + i\dfrac{\theta}{2}\right)}\right\}[Y],$$

where $\widetilde{\mathfrak{L}}(Y) = \sum_{j=0}^{\infty} L_j(Y)$ is the total L-class of Y introduced in [7]. Since for $t = e^{i\theta}$

$$\frac{1}{\tanh\left(x + i\dfrac{\theta}{2}\right)} = \frac{(t+1) + (t-1)\tanh x}{(t-1) + (t+1)\tanh x},$$

we get (9) by using the virtual indices or signatures of [7, § 9].

5.

Still making the assumptions of § 1 we calculate $\operatorname{sign}(X/G_n)$ using (5) and (9). Observe that (9) remains correct if T is the identity and $t = 1$, because then the right side of (9) reduces to $\operatorname{sign}(Y^\circ)$ and $Y^\circ = X$. There is the following identity between rational functions.

$$\frac{1}{n}\sum_{t^n=1}\frac{(t+1)+(t-1)y}{(t-1)+(t+1)y} = \frac{(1+y)^n + (1-y)^n}{(1+y)^n - (1-y)^n}. \tag{11}$$

The sum is over all n^{th} roots of unity. By virtue of (11), the formulas (5) and (9) imply

THEOREM. *Let X be a compact oriented differentiable manifold of dimension $4k$ without boundary on which the cyclic group G_n of order n acts by orientation preserving diffeomorphisms. If all components of the fixed point set Y of this action have* *codimension* 2 *and if* G_n *acts freely on* $X - Y$, *then* X/G_n *is a compact oriented differentiable manifold with*

$$(12)\quad \begin{cases} \operatorname{sign}(X/G_n) = \operatorname{sign} \dfrac{(1+Y)^n + (1-Y)^n}{(1+Y)^n - (1-Y)^n} Y \\ \qquad = \dfrac{1}{n} \operatorname{sign}(X) + \dfrac{n^2-1}{3n} \operatorname{sign}(Y \circ Y) + \cdots . \end{cases}$$

Suppose Y is orientable, consider the submanifold Y' of X/G_n ($Y = Y'$ under the projection $\pi: X \to X/G_n$). Orient Y and Y' in the same way and regard their normal bundles ν in X and ν' in X/G_n as complex line bundles. Then it follows easily that

$$(13)\qquad \nu' = \nu^n = \nu \otimes_{\mathbf{C}} \cdots \otimes_{\mathbf{C}} \nu .$$

As we shall see later ν' and ν (as bundles over Y') can be extended to complex line bundles E' and E over X/G_n with

$$E' = E^n \qquad (n\text{-fold tensor product})$$

and such that the first Chern class

$$c_1(E') = x' \in H^2(X/G_n, \mathbf{Z})$$

is the Poincaré dual of Y' in X/G_n. Therefore $x' = nc_1(E)$. Represent $c_1(E)$ by an oriented submanifold U of X/G_n. Then we can deduce from (12) and (13) a formula for sign (X) in terms of the signatures of the oriented self-intersection cobordism classes $U^r = U \circ \cdots \circ U$ where the "self-intersection" takes place in X/G_n. Leaving the details to the reader we obtain

$$(14)\quad \begin{aligned} \operatorname{sign} X &= \operatorname{sign} \frac{(1+U)^n - (1-U)^n}{(1+U)^n + (1-U)^n} \cdot \frac{1}{U} \\ &= n \operatorname{sign}(X/G_n) - \frac{n(n^2-1)}{3} \operatorname{sign}(U \circ U) + \cdots . \end{aligned}$$

If dim $X = 4$, then (14) can be written as

$$(15)\qquad \operatorname{sign} X = n \operatorname{sign}(X/G_n) - \frac{n^2-1}{3n}(Y' \circ Y')$$

where $Y' \circ Y'$ is the self-intersection number of the branching locus in X/G_n.

Remarks. (i) If Y is empty, then we have an unramified covering and (12) gives the well-known formula

$$\operatorname{sign}(X) = n \operatorname{sign}(X/G_n) .$$

This is not known in the *topological case*. Though formula (5) remains correct, the preceding formula cannot be deduced from it. Namely, for topological manifolds it is not known whether sign (X, T) vanishes for T without fixed points.

(ii) In the formulas (12), (14), and (15) the orientations of Y, U, and Y' respectively, do not play any role. Observe that in (12) and (14) the rational functions on the right side of the equations are even. Therefore only self-intersections Y^r etc. with r even occur. The orientation of Y^r does not change if one changes the orientation of Y.

6.

PROPOSITION. *Let M be a compact oriented differentiable manifold without boundary. Let Y' be an oriented differentiable submanifold of codimension 2, and $x' \in H^2(M, \mathbf{Z})$ the Poincaré dual of Y'. Suppose that x' is divisible by n in $H^2(M, \mathbf{Z})$. Then there exists a G_n-manifold X with $X/G_n = M$ and with branching locus Y' such that all the assumptions of* § 1 *are satisfied.*

For the proof we consider the complex line bundle E' with $c_1(E') = x'$. It has a differentiable section $s: M \to E'$ which vanishes on Y', is different from zero on $M - Y'$, and is transversal to $s_0(M)$ where s_0 is the zero section of E'. Since x' is divisible by n we can find a complex line bundle E with

$$E^n = E \otimes \cdots \otimes E = E' .$$

If $\rho: E \to E'$ denotes the map

$$\rho: v \longmapsto v \otimes \cdots \otimes v ,$$

then $X = \rho^{-1}s(M)$ satisfies all the assumptions of § 1.

Of course, X in the above proposition is in general not uniquely determined. Its signature is given by (14) where U is a submanifold representing $c_1(E)$.

If the assumptions of § 1 are satisfied for a given G_n-manifold X, then X is obtainable by the method of the preceding proposition. This can be seen as follows. Over $X/G_n - Y'$ we have a principal G_n-bundle, and over a tubular neighborhood $\mathcal{V}$ of Y' a principal $\mathbf{C}^*$-bundle coming from the normal bundle ν of Y in X. If we extend the structural group of the principal G_n-bundle to $\mathbf{C}^*$ we get a principal $\mathbf{C}^*$-bundle over $X/G_n - Y'$ which can be identified on $\mathfrak{V} - Y'$ with the principal $\mathbf{C}^*$-bundle of ν. Thus we get a principal $\mathbf{C}^*$-bundle over X/G_n. We denote its associated complex line bundle by E. Then E^n is a complex line bundle E' with $c_1(E') = x'$, the Poincaré dual of Y'. As can be checked, X is obtained from E and E' as in the above proposition.

7.

The construction of **6** can be done in the complex analytic case. We need a minor modification. Let M be a compact complex manifold, and D a divisor on M given by meromorphic functions f_i defined in open sets U_i with $\bigcup U_i = M$ such that, on $U_i \cap U_j$, the function f_i/f_j has neither zeros nor poles (see [7, § 15.2]). We say, D has simple zeros and poles if and only if for a suitable open covering each f_i is a coordinate function z_1 in a local coordinate system $(z_1, z_2, \cdots, z_n)$ defined in U_i or is the inverse z_1^{-1} of such a coordinate function or is constant and different from zero and ∞. A divisor D has simple zero and poles if and only if $D = D_1 - D_2$ where D_1, D_2 are non-singular divisors [7, § 15.2, p. 115] with no common zeros, i.e., the complex submanifolds Y_1' and Y_2' of complex codimension 1 determined by D_1 and D_2 do not intersect.

We assume that

$$\{D\} = E \otimes \cdots \otimes E = E^n . \tag{16}$$

Here $\{D\}$ is the holomorphic complex line bundle defined by D and E is a holomorphic complex line bundle whose n^{th} power as a holomorphic line bundle is $\{D\}$. Any holomorphic line bundle L has an associated holomorphic bundle $\hat{L}$ with the complex projective line as fibre obtained by adding a point at infinity to each fibre of L.

The line bundle $\{D\}$ has transition functions

$$f_{ij} = f_i/f_j \colon U_i \cap U_j \longrightarrow \mathbf{C}^*$$

and a meromorphic section given by the functions f_i. This meromorphic section defines a holomorphic section

$$s \colon M \longrightarrow \widehat{\{D\}} .$$

Because of (16) we have the "n^{th} power maps"

$$\rho \colon E \longrightarrow \{D\} , \qquad \hat{\rho} \colon \hat{E} \longrightarrow \widehat{\{D\}} .$$

Then $X = \hat{\rho}^{-1}s(M)$ is a compact complex manifold which is an n-fold holomorphic ramified covering of M with $Y' = Y_1' \cup Y_1'$. Here

Y' is determined by the zeros and poles of the divisor D. Since we are interested in the signature of X, the orientations given to the components of Y' do not matter (§ 5, *Remark* (ii)).

Formula (16) implies that the Chern class $c_1\{D\}$ is divisible by n. This divisibility is also sufficient for the existence of a holomorphic line bundle E with $E^n = \{D\}$. For the proof we use the exact sequence

$$H^1(M, \Omega) \xrightarrow{\exp} H^1(M, \mathbf{C}^*_\omega) \xrightarrow{c_1} H^2(M, \mathbf{Z}) \xrightarrow{j} H^2(M, \Omega)$$

where Ω is the sheaf of germs of holomorphic functions and $H^1(M, \mathbf{C}^*_\omega)$ the group of holomorphic line bundles under the tensor product [7, § 15.9]. If $c_1\{D\} = nx$, then x is in the kernel of j, therefore in the image of c_1. Thus $c_1\{D\} = c_1(E_1^n)$, where E_1 is a holomorphic line bundle, and $\{D\}E_1^{-n}$ is in the image of exp. In $H^1(M, \Omega)$ every element is divisible by n. Therefore $\{D\}E_1^{-n} = L^n$ where L is in the image of exp. q.e.d.

If M is an algebraic surface, then the ramified covering X is again algebraic. We have the following theorem.

THEOREM. *Let M be a non-singular algebraic surface and $D = D_1 - D_2$ a divisor on M where D_1, D_2 are non-singular curves which do not intersect. Suppose that the integral homology class of D is divisible by n. Then there exists an algebraic surface X which is a ramified covering of M along D (we have $M = X/G_n$ where the cyclic group G_n acts on X by holomorphic maps and freely outside the set of fixed points). For any such X we have*

$$\text{sign}(X) = n\,\text{sign}(M) - \frac{n^2-1}{3n} D \circ D\,, \qquad (\text{see } (15))\,. \tag{17}$$

Here $D \circ D$ is the self-intersection number of the homology class of the divisor D.

Remark. If in the preceding theorem $\text{sign}(M) = 0$ and $D \circ D < 0$ with $n \geqq 2$, we get examples of algebraic surfaces with $\text{sign}(X) \geqq 2$. In fact, Atiyah [1] and Kodaira [11] have in this way constructed algebraic surfaces with arbitrary large signatures, as we shall recall in § 8. The existence of algebraic surfaces with $\text{sign}(X) \geqq 2$ contradicts an earlier conjecture in algebraic geometry. Borel [5] has proved the existence of discontinuous groups operating freely on bounded homogeneous symmetric domains and which have a compact orbit space. Also his result led to examples of algebraic surfaces with arbitrary large signatures (compare [7, § 22.3]).

8.

The formula for the signature of ramified coverings was motivated by papers of Atiyah [1] and Kodaira [11] who studied the signature of ramified coverings in some special cases which are of particular interest because they show that the signature of the total space of a differentiable fibre bundle need not be equal to the product of the signatures of base and fibre. This multiplicative property holds however if the fundamental group of the base operates trivially on the cohomology of the fibre [6]. The construction in §§ 6 and 7 occurs essentially in [1] except that Atiyah studies only double coverings. We report briefly on the family of algebraic surfaces studied by Kodaira [11]. The calculation of Pontrjagin classes occuring in [1] and [11] can be replaced by formula (17).

Let C_1 be a Riemann surface (algebraic curve) of genus $g_1 \geqq 1$. Since $H_1(C_1, \mathbf{Z}_2) \neq 0$ there exists a Riemann surface C_2 which is an unramified double covering of C_1. Let $\tau\colon C_2 \to C_2$ be the covering translation; τ is a free involution. The genus g_2 of C_2 is given by

$$2(2 - 2g_1) = 2 - 2g_2\ , \qquad g_2 = 2g_1 - 1\ .$$

The group $H_1(C_2, \mathbf{Z}_n)$ is a homomorphic image of $\pi_1(C_2)$. Let C_3 be the Riemann surface (algebraic curve) which is associated to the universal unramified covering of C_2 with $H_1(C_2, \mathbf{Z}_n)$ as fibre. The degree of this covering map $f\colon C_3 \to C_2$ is n^{2g_2}. The genus g_3 of C_3 is

$$g_3 = n^{2g_2}(g_2 - 1) + 1\ .$$

In the cartesian product $C_3 \times C_2$ we consider the graph Γ_f of f. The self-intersection number of Γ_f in $C_3 \times C_2$ equals

$$\Gamma_f \circ \Gamma_f = \deg(f)\cdot(2 - 2g_2) = 4n^{4g_1-2}\cdot(1 - g_1)\ . \tag{18}$$

The class of Γ_f in $H_2(C_3 \times C_2, Z_n)$ is determined by $f^*\colon H^*(C_2, \mathbf{Z}_n) \to H^*(C_3, \mathbf{Z}_n)$. By the very construction f^* is 0 in dimensions 1 and 2. The same holds for $(\tau f)^* = f^*\tau^*$. In dimension 0, both f^* and $(\tau f)^*$ are the identity. Therefore Γ_f and $\Gamma_{\tau f}$ have the same homology class in $H_2(C_3 \times C_2, \mathbf{Z}_n)$. We can now apply the theorem of § 7 with $M = C_3 \times C_2$ and $D = \Gamma_f - \Gamma_{\tau f}$. Since

$$\operatorname{sign}(M) = \operatorname{sign}(C_3)\operatorname{sign}(C_2) = 0$$

and

$$D \circ D = 2\Gamma_f \circ \Gamma_f = -8n^{4g_1-2}(g_1 - 1) ,$$

we have

THEOREM. *For any $g_1 \geqq 1$ and $n \geqq 2$ there exists an algebraic surface $X(g_1, n)$ with*

$$\text{sign}\, X(n, g_1) = 8\,\frac{n^2 - 1}{3}(g_1 - 1)n^{4g_1-3} .$$

The Kodaira algebraic surface $X(g_1, n)$ is fibered differentiably over C_3 with algebraic curves $C_2'(x)$ as fibres which are n-fold ramified coverings of C_2 with the two branching points $f(x)$ and $\tau f(x)$ for $x \in C_3$. The fibres therefore have genus $ng_2 = n(2g_1 - 1)$. Thus $X(g_1, n)$ is a 4-dimensional manifold fibered differentiable over a 2-dimensional manifold C_3 of genus $n^{4g_1-2}(2g_1 - 2) + 1$ with a 2-dimensional manifold C_2' of genus $n(2g_1-1)$ as fibre. For $g_1 \geqq 2$ we have sign $X(g_1, n) \neq 0$ whereas sign (C_2') sign $(C_3) = 0\cdot 0 = 0$. Thus the signature does not behave multiplicatively in this differentiable fibre bundle. Observe that the signature is defined to be 0 for manifolds of a dimension not divisible by 4. A differentiable bundle with total space, base, and fibre of dimensions divisible by 4 is given by the cartesian product of $X(g_1, n) = X$ with itself. Then

$$\text{sign}\, X^2 = (\text{sign}\, X)^2 \neq 0 ,$$

whereas the base C_3^2 and the fibre $(C_2')^2$ have vanishing signatures.

$X(g_1, n)$ gives for $g_1 \geqq 2$ a family of Riemann surfaces $C_2'(x)$ $(x \in C_3)$ which is locally not trivial. The complex structure of $C_2'(x)$ varies with x (see [1] and [11]). As Atiyah [1] shows, this phenomenon is closely related to the non-multiplicativity of the signature.

UNIVERSITY OF BONN

REFERENCES

[1] M. F. ATIYAH, The signature of fibre bundles (in this volume).

[2] M. F. ATIYAH and I. M. SINGER, *The index of elliptic operators*: I, Ann. of Math. **87** (1968), 484-530.

[3] M. F. ATIYAH and G. B. SEGAL, *The index of elliptic operators*: II, Ann. of Math. **87** (1968), 531-545.

[4] M. F. ATIYAH and I. M. SINGER, *The index of ellipic operators*: III, Ann. of Math. **87** (1968), 546-604.

[5] A. BOREL, *Compact Clifford-Klein forms of symmetric spaces*, Topology **2** (1963), 111-122.

[6] S. S. CHERN, F. HIRZEBRUCH, and J-P. SERRE, *On the index of a fibered*

manifold, Proc. Amer. Math. Soc. **8** (1957), 587-596.

[7] F. Hirzebruch, Topological Methods in Algebraic Geometry, Third enlarged edition, Springer Verlag, Berlin-Heidelberg-New York, 1966.

[8] ———, "Involutionen auf Mannigfaltigkeiten," Proceedings of the Conference on Transformation Groups, Tulane University, New Orleans, 1967. Springer Verlag, Berlin-Heidelberg-New, York 1968, pp. 148-166.

[9] ——— and K. Jänich, "Involutions and singularities," to appear in Proceedings of the Conference on Algebraic Geometry, Tata Institute, Bombay, 1968.

[10] K. Jänich, and E. Ossa, *On the signature of an involution*, to appear in Topology.

[11] K. Kodaira, *A certain type of irregular algebraic surfaces*, J. Anal. Math. **19** (1967), 207-215.

(Received August 13, 1968)

Deformations of compact complex surfaces

By Shigeru Iitaka

By a surface we shall mean a compact connected complex manifold of complex dimension 2. For notation and terminology we follow that of Kodaira [4]. Thus we fix our notation as follows.

S	: a surface,
$\mathcal{O}(D)$	: an invertible sheaf on S associated to a divisor D,

$l(D) = \dim H^0(S, \mathcal{O}(D))$,

$s(D) = \dim H^1(S, \mathcal{O}(D))$,

$i(D) = \dim H^2(S, \mathcal{O}(D))$,

$\mathbf{C}$	: a complex plane,
$\mathbf{D}$	: a complex upper half plane,
$\sim$	: linear equivalence of divisors,
K	: a canonical divisor of S,
$P_m = l(mK)$	: the m-genus of S,
$p_g = l(K)$	: the geometric genus of S,
$q = \dim H^1(S, \mathcal{O})$	: the irregularity of S,
c_1^2, c_2	: the Chern numbers of S,
$b_\nu (\nu = 0, 1, 2, 3, 4)$	: the Betti numbers of S,
π_1	: the fundamental group of S.

If necessary, we denote the m-genus of S by $P_m(S)$ instead of $P_m \cdots$.

Some important relations among them are

$$l(D) - s(D) + i(D) = \frac{1}{2} D(D - K) + \frac{1}{12} (c_1^2 + c_2)$$

(Theorem of Riemann-Roch by Atiyah-Singer)

$$i(D) = l(K - D) \qquad \text{(Serre Duality)}$$

$$12(p_g - q + 1) = c_1^2 + c_2 \qquad \text{(Noether Formula).}$$

We call a surface of geometric genus zero having an abelian variety as an unramified finite covering manifold, hyperelliptic, following the usage in Enriques-Severi [1].

In this paper we shall prove the following two theorems.

THEOREM I. *Any deformation of a rational surface is also rational.*

THEOREM II. *The following Table* I *is a classification of surfaces with $c_1^2 = c_2 = 0$, each surface having the fundamental group π_1 containing an abelian subgroup with finite index.*

Table I

b_1	π_1	structure of surfaces
2	$\mathbf{Z} \oplus \mathbf{Z}$	ruled surfaces
4	$\mathbf{Z} \oplus \mathbf{Z} \oplus \mathbf{Z} \oplus \mathbf{Z}$	complex tori
2	non abelian	hyperelliptic surfaces
1	$\mathbf{Z}$ is contained as a subgroup of finite index	Hopf surfaces

1. Proof of Theorem I

Consider a complex analytic family $\{S_t \mid t \in B\}$ of surfaces S_t, $t \in B$, such that for a point 0 of B, S_0 is a rational surface. It is obvious that $p_g(S_t) = p_g(S_0) = 0$, $q(S_t) = q(S_0) = 0$, $c_1^2(S_t) = c_1^2(S_0)$ for any point t of B. Let Σ be the set of points t such that S_t is rational. By the theorem of Castelnuovo to the effect that S is rational if and only if $P_2(S) = q(S) = 0$ (see Kodaira [4, IV], for its short proof), and the theorem of upper semi-continuity, the set Σ is open. We shall show that $B - \Sigma$ is open. $P_2(S_1)$ is positive for any point 1 in $B - \Sigma$. Hence, if $P_{-2}(S_1)$ is positive, then the bicanonical system of S_1 is trivial, while the canonical system is non-trivial. This contradicts $\pi_1(S_1) \simeq \pi_1(S_0) = 1$. Therefore $P_{-2}(S_1) = 0$. By the theorem of upper semi-continuity, we can choose a small neighborhood U_1 of 1 such that $P_{-2}(S_t) = 0$ for any $t \in U_1$.

(i) If S_1 is minimal, then $c_1^2(S_1) \geqq 0$. By the theorem of Riemann-Roch,

$$P_{-2}(S_t) = l(-2K) \geqq -P_3(S_t) + 3c_1^2(S_t) + 1 ,$$

while $P_3(S_t) = 0$ for any t in Σ.
Hence $P_{-2}(S_t) \geqq 1$ for $t \in \Sigma$. Therefore U_1 is contained in $B - \Sigma$.

(ii) S_1 is not minimal. By means of the stability theorem of an (irreducible) exceptional curve (of the first kind) of Kodaira [3], we can choose a small neighborhood V_1 of 1 such that the exceptional curve of S_1 extends to that of S_t for any point t in V_1. In a more precise formulation, we can say as follows. Let $\varpi: \mathfrak{M} \to B$ be a fibre space in the category of connected complex manifolds such that $\varpi^{-1}(t)$ is the given S_t for any $t \in B$. We can obtain a submanifold $\mathcal{E} \subset \varpi^{-1}(V_1)$ such that, for any $t \in V_1$, $E_t = \mathcal{E} \cap \varpi^{-1}(t)$ is an exceptional curve of S_t, and E_1 is the given exceptional curve of S_1. Since $p_g(S_t) = 0$, S_t is algebraic. Replacing B by a smaller neighborhood of 1, if necessary, we may assume ϖ to be projective in the sense of Grothendieck [2]. Then we infer that $\mathcal{E}$ can be contractible, i.e., E_t can be contractible to a point uniformly for every $t \in V_1$. By means of a finite number of contractions as described above, we can reduce the case (ii) to the case (i).

Remark A. It seems very difficult to obtain all the complex structures on topological models of rational surfaces or of Enriques surfaces. For example, we can construct non-rational surfaces which have the same homotopy type as a rational elliptic surface, as follows. Let L be a linear system of cubic curves having assigned base points $p_1, p_2, \cdots, p_8$ which lie generically on a projective plane $\mathbf{P}^2$. Then L has nine base points $p_1, p_2, \cdots, p_9$. We obtain a rational elliptic surface $\Psi: S \to \Delta$ by means of quadric transforms with the centers $p_1, p_2, \cdots, p_9$.

$$S = Q_{p_9} Q_{p_8} \cdots Q_{p_1}(\mathbf{P}^2) .$$

We obtain the desired surface $L_{p_2}(m_2)L_{p_1}(m_1)S$ by means of logarithmic transforms $L_{p_1}(m_1)$ and $L_{p_2}(m_2)$, where p_1 and p_2 are distinct points of Δ, m_1 is an odd number, m_2 is an even number, and m_1 and m_2 are relatively prime. Furthermore, for any finite abelian group $A = \mathbf{Z}_{m_1} \oplus \cdots \oplus \mathbf{Z}_{m_r}$, we obtain a surface S_A of which the numerical characters p_g, q, c_1^2 vanish and the 1-homology group $H_1(S, \mathbf{Z})$ is isomorphic to A by means of logarithmic transforms,

$$S_A = L_{p_{r+1}}(m_1 \cdots m_r) L_{p_r}(m_r) \cdots L_{p_1}(m_1) S ,$$

where $p_1, p_2, \cdots, p_r$ are arbitrary distinct points of Δ. Note that Gaeta has obtained the surfaces with $p_g = q = c_1^2 = 0$, $\sigma = 2^r$ in Roth [5, p. 36]. (σ denotes the order of Tor $H_1(S, \mathbf{Z})$.)

2. Proof of Theorem II

First we shall consider the case

(A) π_1 is abelian.

LEMMA 1. *The universal covering manifold of S is not* $\mathbf{C} \times \mathbf{D}$.

Let Γ be a properly discontinuous abelian subgroup of the biholomorphic automorphism group of $\mathbf{C} \times \mathbf{D}$ such that $\mathbf{C} \times \mathbf{D}/\Gamma$ is biholomorphically equivalent to S. We shall describe a holomorphic map γ in Γ in the form

$$(z, \zeta) \longrightarrow (f_\gamma(z, \zeta), g_\gamma(z, \zeta)) .$$

Then, by the theorem of Liouville, $g_\gamma(z, \zeta)$ depends only on ζ. Hence we get a homomorphism $\Gamma \to PSL_2(R)$. By Γ' we denote its homomorphic image. Then, D/Γ' is compact, but this is impossible, for Γ' is abelian.

LEMMA 2. *If $E \times \mathbf{C}$ is an unramified covering manifold of S, then S is a complex torus, where E denotes an elliptic curve.*

PROOF. In the same way as in the proof of Lemma 1, we have the description $(z, \zeta) \to (f_\gamma(z, \zeta), g_\gamma(\zeta))$ for any biholomorphic automorphism γ of $E \times \mathbf{C}$. If we consider $f_\gamma(z, \zeta)$ as a holomorphic map $E \to E$ for a fixed ζ, we can express $f_\gamma(z, \zeta)$ as $A_\gamma(\zeta)z + B_\gamma(\zeta)$ on the universal covering manifold of E: $\mathbf{C}$ while $A_\gamma(\zeta) = \alpha + \beta \cdot \omega$, where α and β are rational integers and $[1, \omega]$ denote periods of E. Hence the holomorphic map $A_\gamma(\zeta)$ reduces to a constant A_γ. Note that $A_\gamma^{12} = 1$, because γ is an automorphism. Obviously homomorphic image Γ' in the biholomorphic automorphism group of $\mathbf{C}$, of Γ operates properly discontinuously on $\mathbf{C}$. Hence the absolute value of $\lambda_\gamma = 1$, where $g_\gamma(\zeta) = \lambda_\gamma \zeta + \mu_\gamma$. Therefore we have

$$\text{abs det} \begin{pmatrix} \dfrac{\partial f_\gamma}{\partial z} & \dfrac{\partial f_\gamma}{\partial \zeta} \\ \dfrac{\partial g_\gamma}{\partial z} & \dfrac{\partial g_\gamma}{\partial \zeta} \end{pmatrix} = \text{abs}\,(A_\gamma \cdot \lambda_\gamma) = 1 .$$

This means that S has a volume preserving complex structure [4, II]. Hence Γ is a subgroup of the complex affine transformation group of $\mathbf{C} \times \mathbf{C}$ by Kodaira [4, II]. Consequently, S is a complex torus.

I. The case in which S is elliptic with the projection $\Psi: S \to \Delta$. Its singular fibre is only a multiple of a connected elliptic curve, and any general fibre is biholomorphically equivalent to a fixed elliptic curve E, because of vanishing of c_2. Let

$$m_1C_1, m_2C_2, \cdots, m_sC_s \; (C_i = \Psi^{-1}(p_i),\; m_iC_i = \Psi^*(p_i))$$

be all the singular fibres of $S \to \Delta$, and let α be the rational number $2\pi - 2 + \sum_{i=1}^{s} \{1 - (1/m_i)\}$, where π denotes the genus of Δ.

(i) If $\alpha < 0$, then S is a Hopf surface or a ruled surface of genus 1 by Kodaira [4, II] (cf. Suwa [6]).

(ii) If $\alpha = 0$, then the universal normal covering manifold with the ramification number m_i at each point p_i $(i = 1, 2, \cdots, s)$ is **C**. Therefore, $E \times \mathbf{C}$ is an unramified covering manifold of S. From Lemma 2, we conclude that S is a complex torus.

(iii) If $\alpha > 0$, then we derive a contradiction from Lemma 1 by the same reasoning as in case (ii).

II. The case in which S is general. (i) If $p_g = 0$, $b_1 \neq 1$, then S is algebraic by Kodaira [4, I]. By $\Psi: S \to \Delta$ we denote the Albanese variety Δ of S with the Albanese map Ψ. Its general fibre is connected algebraic curve of genus g. If $g = 0$, S is ruled and if $g \geqq 1$, S is elliptic by Kodaira [4, IV].

(ii) If $p_g = 0$, $b_1 = 1$, then S is a Hopf surface by Kodaira [4, III].

(iii) If $p_g \geqq 1$, $q = 1 + p_g \geqq 2$, then S is a complex torus or an elliptic surface by the table of classification of surfaces [4, I]. Next we must consider the case

(B) π_1 has an abelian subgroup Γ of finite index.

From the purely algebraic consideration, we may assume Γ to be a normal subgroup of π_1. Therefore, S has an unramified finite covering manifold $\tilde{S}$ whose fundamental group is Γ such that the degree of the mapping $\tilde{S} \to S$ is $[\pi_1: \Gamma]$, which we denote by r. From the obvious relations: $c_1^2(\tilde{S}) = r \cdot c_1^2(S) = 0$, $c_2(\tilde{S}) = r \cdot c_2(S) = 0$, we can conclude that S can be classified into the following four classes

(i) ruled surfaces of genus 1,

(ii) complex tori,

(iii) hyperelliptic surfaces,

(iv) Hopf surfaces.

In fact, if $\tilde{S}$ is ruled, then $P_{12}(S) \leqq P_{12}(\tilde{S}) = 0$, and hence S is ruled.

Remark B. Hyperelliptic surfaces can be classified into the following four types.

(I) $2K \sim 0$ $(K \not\sim 0)$,

(II) $3K \sim 0$ $(K \not\sim 0)$,

(III) $4K \sim 0$ $(2K \not\sim 0)$,

(IV) $6K \sim 0$ $(2K, 3K \not\sim 0)$.

T. Suwa and the author calculated the fundamental group G of hyperelliptic surfaces and obtained the following Table II.

Table II

type	Tor $H_1(S, \mathbf{Z})$	abelianized group of (G/center of G)
I	$\mathbf{Z}_2 \oplus \mathbf{Z}_2$ $\mathbf{Z}_2$	$\mathbf{Z}_2 \oplus \mathbf{Z}_2 \oplus \mathbf{Z}_2$
II	$\mathbf{Z}_3$ 0	$\mathbf{Z}_3 \oplus \mathbf{Z}_3$
III	$\mathbf{Z}_2$ 0	$\mathbf{Z}_2 \oplus \mathbf{Z}_4$
IV	0	$\mathbf{Z}_2 \oplus \mathbf{Z}_3$

UNIVERSITY OF TOKYO

REFERENCES

[1] F. ENRIQUES and F. SEVERI, *Memoire sur les surfaces hyperelliptiques,* Acta Math. **32** (1909), 283-392.

[2] A. GROTHENDIECK, *Technique de construction en géométrie analytique complexe,* I~X. Seminaire Cartan, t. **13**, 1960/61.

[3] K. KODAIRA, *On stability of compact submanifold of complex manifolds,* Amer. J. Math. **85** (1963), 79-94.

[4] ———, *On the structure of compact complex analytic surfaces,* I, II, III, IV, Amer. J. Math. **86** (1964), 751-798, **88** (1966), 687-721, **90** (1968), 55-83, **90** (1968), 1048-1066.

[5] L. ROTH, Algebraic Threefolds, Berlin, 1955.

[6] T. SUWA, *On ruled surfaces of genus* 1, to appear in J. Math. Soc. Japan.

(Received October 7, 1968)

On the volume of polyhedra

By S. IYANAGA

1. Definition of polyhedra and topological preliminaries

Let n be a positive integer and $\mathbf{R}$ be the field of real numbers as usual. $\mathbf{R}^n$ is then the n-dimensional euclidean space with the well-known metric. Elements of $\mathbf{R}^n$ will be denoted as $\mathbf{a} = (a_1, \cdots, a_n)$, $\mathbf{b} = (b_1, \cdots, b_n)$, $\mathbf{x} = (x_1, \cdots, x_n)$, $\mathbf{y} = (y_1, \cdots, y_n)$, etc. The inner product of $\mathbf{a}$ and $\mathbf{x}$ will be denoted by $(\mathbf{a}, \mathbf{x}) = \sum_{i=1}^n a_i x_i$. If $\mathbf{a} \neq 0$ and $b \in \mathbf{R}$,

$$\pi_{\mathbf{a},b} = \{\mathbf{x}; (\mathbf{a}, \mathbf{x}) = b\}$$

is a *hyperplane*, and

$$H_{\mathbf{a},b} = \{\mathbf{x}; (\mathbf{a}, \mathbf{x}) \geq b\}$$

is a *half-space* of $\mathbf{R}^n$. $H_{\mathbf{a},b}$ is a closed, convex and unbounded subset of $\mathbf{R}^n$, whose boundary is $\pi_{\mathbf{a},b}$, i.e., $\partial H_{\mathbf{a},b} = \pi_{\mathbf{a},b}$.

The subsets $A, B, \cdots$ of $\mathbf{R}^n$ form a boolean lattice $\mathcal{B}_n$ with respect to the set operations $A \cup B$ (union), $A \cap B$ (intersection) and $A^c = \mathbf{R}^n - A$ (complement). Let $\mathcal{A}_n$ be the subboolean-lattice of $\mathcal{B}_n$ generated by the half-spaces of $\mathbf{R}^n$. We call a bounded element of $\mathcal{A}_n$ a *polyhedron* of $\mathbf{R}^n$, and denote with $\mathcal{P}_n$ the set of all polyhedra of $\mathbf{R}^n$. Then the following is obvious.

1.1. *If $A, B \in \mathcal{P}_n$, then $A \cup B \in \mathcal{P}_n$, $A \cap B \in \mathcal{P}_n$ and $A - B = A \cap B^c \in \mathcal{P}_n$. The boundary ∂A of A, the closure and the open kernel of A are also $\in \mathcal{P}_n$.*

An element of $\mathcal{P}_n$ may or may not be closed; it may or may not have an inner point. When it has no inner point, it will be called *degenerate*.

The following two propositions can be proved by induction on n.

1.2. *The boundary of any element of $\mathcal{P}_n$ is degenerate.*

1.3. *A degenerate element of $\mathcal{P}_n$ can be covered by a finite number of hyperplanes.*

Let r be an integer $\geqq 2$ and $A_1, \cdots, A_r$ be r elements of $\mathcal{P}_n$. If $A_1 \cap A_2$, $(A_1 \cup A_2) \cap A_3, \cdots, (A_1 \cup \cdots \cup A_{r-1}) \cap A_r$ are degenerate, we write $\sum_{i=1}^{r} A_i = A_1 + \cdots + A_r$ instead of $\bigcup_{i=1}^{r} A_i = A_1 \cup \cdots \cup A_r$ and call it the *direct sum* of $A_1, \cdots, A_r$. (For $r = 1$, one puts $\sum_{i=1}^{r} A_i = A_1$.) We can easily prove

1.4. *Any non-degenerate element of $\mathcal{P}_n$ can be represented as the direct sum of a finite number of convex, non-degenerate elements of $\mathcal{P}_n$.*

2. Simplexes and parallelipipeds

Let k be any integer $\geqq 0$. $k + 1$ points $\mathbf{x}_0, \mathbf{x}_1, \cdots, \mathbf{x}_k$ of $\mathbf{R}^n$ are independent, when k vectors $\mathbf{x}_1 - \mathbf{x}_0, \cdots, \mathbf{x}_k - \mathbf{x}_0$ are linearly independent (over $\mathbf{R}$). Let $\mathbf{x}_0, \mathbf{x}_1, \cdots, \mathbf{x}_k$ be k independent points. Then the set

$$S(\mathbf{x}_0, \mathbf{x}_1, \cdots, \mathbf{x}_k) = \{\textstyle\sum_{j=0}^{k} \lambda_j \mathbf{x}_j;\ \lambda_j \in \mathbf{R},\ \sum_{j=0}^{k} \lambda_j = 1,\ \lambda_j \geqq 0\}$$

is a *k-simplex* with $\mathbf{x}_0, \mathbf{x}_1, \cdots, \mathbf{x}_k$ as *vertices*. It is a closed convex polyhedron (in fact, the convex closure of vertices) and is non-degenerate if and only if $k = n$.

Furthermore, let $\mathbf{a}_1, \cdots, \mathbf{a}_k$ be k linearly independent vectors and $\mathbf{x}$ a point of $\mathbf{R}^n$. The k-dimensional *parallelepiped* $P(\mathbf{x}; \mathbf{a}_1, \cdots, \mathbf{a}_k)$ with $\mathbf{x}$ as origin and with $\mathbf{a}_1, \cdots, \mathbf{a}_k$ as edges is defined as

$$P(\mathbf{x}; \mathbf{a}_1, \cdots, \mathbf{a}_k) = \{\mathbf{x} + \textstyle\sum_{i=1}^{k} \lambda_i \mathbf{a}_i;\ \lambda_i \in \mathbf{R},\ 0 \leqq \lambda_i \leqq 1\}\ .$$

This is again a closed convex polyhedron, which is non-degenerate if and only if $k = n$. In the following, we deal only with n-dimensional parallelepipeds, so that our parallelepipeds are always non-degenerate. In particular, the parallelepiped $P(0; \mathbf{e}_1, \cdots, \mathbf{e}_n)$ will be called the *unit parallelepiped* and denoted with E, where 0 is the origin $(0, \cdots, 0)$ and $\mathbf{e}_1 = (1, 0, \cdots, 0), \cdots, \mathbf{e}_n = (0, \cdots, 0, 1)$ are unit vectors.

A *regular affinity* of $\mathbf{R}^n$ is a mapping of $\mathbf{R}^n$ into itself given by

$$(*) \qquad \mathbf{y} = T\mathbf{x} + \mathbf{b}\ ,$$

where T is an $n \times n$ regular matrix and $\mathbf{b}$ any element of $\mathbf{R}^n$. It is a bijective mapping $\mathbf{R}^n \to \mathbf{R}^n$, and the set of all these mappings forms a well known group of transformations of $\mathbf{R}^n$, the *affine group*. Two sub-sets A, B of $\mathbf{R}^n$ are *affine equivalent*, if there exists an element φ of this group such that $\varphi(A) = B$. The following is a well-known elementary result:

2.1. *Two k-simplexes are affine equivalent. Two parallelepi-*

peds (more generally two parallelepipeds of the same dimension) are affine equivalent.

The affinities of the form (*) with $|T| = \pm 1$ form a subgroup of the affine group, which we call the *special-affine group*. If $\varphi(A) = B$ with φ in this group, then A, B will be called *special-affine equivalent*. We write $A \sim B$ to mean that A and B are special-affine equivalent.

Now we have

2.2. *A parallelepiped $P(\mathbf{x}; \mathbf{a}_1, \cdots, \mathbf{a}_n)$ can be represented as the direct sum of $n!$ n-simplexes which are all special-affine equivalent to the simplex with vertices $\mathbf{x}, \mathbf{x} + \mathbf{a}_1, \cdots, \mathbf{x} + \mathbf{a}_n$.*

PROOF. By virtue of 2.1, we have only to show that E can be divided into $n!$ simplexes, which are special-affine equivalent to the simplex

$$S_0 = S(0, \mathbf{e}_1, \cdots, \mathbf{e}_n) = \{(\lambda_1, \cdots, \lambda_n); \textstyle\sum_{i=1}^n \lambda_i \leqq 1, 0 \leqq \lambda_i\} .$$

Now any point of E has the coordinates $(\lambda_1, \cdots, \lambda_n)$, $0 \leqq \lambda_i \leqq 1$. Let σ be an element of symmetric group $\mathfrak{S}$ of degree n, so that $\sigma(1), \cdots, \sigma(n)$ is a permutation of $1, 2, \cdots, n$. Denote with S_σ the subset:

$$\{(\lambda_1, \cdots, \lambda_n); 0 \leqq \lambda_{\sigma(1)} \leqq \cdots = \lambda_{\sigma(n)} \leqq 1\} .$$

of E. It is easy to see that $S_\sigma \cap S_{\sigma'}$ is degenerate, if $\sigma \neq \sigma'$, and $\bigcup_{\sigma \in \mathfrak{S}} S_\sigma = E$.

We shall show that S_σ; $\sigma \in \mathfrak{S}$ are all special-affine equivalent to S_0. Firstly, we have

$$S_0 \sim S_1 = \{(\lambda_1, \cdots, \lambda_n); 0 \leqq \lambda_1 \leqq \cdots \leqq \lambda_n \leqq 1\} ,$$

because one obtains S_1 from S_0 by the special affinity

$$\mathbf{x} \longmapsto \begin{pmatrix} 1 & & & & \\ 1 & 1 & & & 0 \\ 1 & 1 & 1 & & \\ \cdots & \cdots & & & \\ 1 & \cdots & \cdots & & 1 \end{pmatrix} \mathbf{x}$$

and secondly $S_1 \sim S_\sigma$, as S_σ is obtained from S_1 by the special affinity: $\mathbf{x} \mapsto T\mathbf{x}$, where T is the "permutation matrix" whose ij-entry is $\delta_{\sigma^{-1}(i), j}$, q.e.d.

A closed non-degenerate convex polyhedron can be obviously divided into a direct sum of simplexes by projection from any one of its inner point. From 1.4 follows therefore

2.3. *Any closed non-degenerate polyhedron can be represented as direct sum of n-simplexes.*

3. Definition of volume and formulation of the main theorem

The special-affine group has a subgroup, constituted by the transformations of the form: $\mathbf{x} \mapsto T\mathbf{x} + \mathbf{b}$, where T is a (proper or non-proper) orthogonal matrix. It is the *group of euclidean motions*. When there exists an element φ of this group such that $\varphi(A) = B$, then A, B will be said to be *congruent*. We shall write $A \approx B$ in this case.

This group has again a subgroup, constituted by the transformations $\mathbf{x} \mapsto \mathbf{x} + \mathbf{b}$, called *translations*. If B is obtained from A by a translation, A, B are *translation-equivalent*. Then we write $A \cong B$.

Now we define (n-dimensional) volume as follows.

Definition. The volume is as functional $v: \mathcal{P}_n \to \mathbf{R}$ with the following properties:

(1) $v(A) \geqq 0$.

(2) $v(A) + v(B) = v(A \cup B) + v(A \cap B)$.

(3) $v(\emptyset) = 0$.

(4) $v(E) = 1$.

(5) $A \cong B$ implies $v(A) = v(B)$.

(A, B denote any elements of $\mathcal{P}_n$; E is the unit parallelepiped.)

Our *Main Theorem* states

The volume v exists and it is unique. It has moreover the property

(5′) $A \sim B$ *implies* $v(A) = v(B)$.

Remark. From (5′) follows, of course,

(5″) $A \approx B$ *implies* $v(A) = v(B)$.

Thus, congruent polyhedra have the same volume.

This result is usually taken for granted, most often implicitly, in courses in mathematics. Once E. Schmidt [2] noticed this point and gave an ingenious proof. In the following I shall reproduce a proof, which I used in my course in geometry at the University of Tokyo. It proceeds by induction on n, using an idea found in Hilbert [3], for the case $n = 2$.

4. Addition formula

We begin by proving the unicity. So we shall first assume the existence of v satisfying (1) – (5), and see what other properties it should have.

Thus, let v be, in this and subsequent sections (§§ 4 – 7), a functional on $\mathcal{P}_n$ with the properties (1) – (5). $A, B, C, \cdots$ will denote elements of $\mathcal{P}_n$.

4.1. *$A \supset B$ implies $v(A) \geqq v(B)$.*

PROOF. Put $C = A - B = A \cap B^c$. Then we have $A = B \cup C$, $B \cap C = \varnothing$ (so that $A = B + C$). From (1), (2), (3), we obtain

$$v(A) = v(B \cup C) = v(B) + v(C) \geqq v(B) , \qquad \text{q.e.d.}$$

4.2. *If A is degenerate, then $v(A) = 0$.*

PROOF. By 1.3, A is covered by a finite number of hyperplanes. So it is sufficient to show that $v(A) = 0$, for A contained in a hyperplane π.

Suppose therefore $A \subset \pi$. Let $\mathbf{x}$ be any point in π and $\mathbf{a}_1, \cdots, \mathbf{a}_{n-1}$ independent vectors lying on π and $\mathbf{a}_n$ a vector independent from $\mathbf{a}_1, \cdots, \mathbf{a}_{n-1}$. Then it is clear that

$$(\ast) \qquad A \subset P(\mathbf{x}; \lambda_1\mathbf{a}_1, \cdots, \lambda_{n-1}\mathbf{a}_{n-1}, \lambda_n\mathbf{a}_n)$$

for a suitably chosen $\mathbf{x}, \lambda_1, \cdots, \lambda_{n-1}$ and for arbitrarily small λ_n. Now we have, for any positive integer k,

$$v\left(P\left(\mathbf{x}; \lambda_1\mathbf{a}_1, \cdots, \lambda_{n-1}\mathbf{a}_{n-1}, \frac{1}{2k}\,\mathbf{a}_n\right)\right)$$
$$\leqq \frac{1}{k}\, v\big(P(\mathbf{x}; \lambda_1\mathbf{a}_1, \cdots, \lambda_{n-1}\mathbf{a}_{n-1}, \mathbf{a}_n)\big) ,$$

because k parallelepipeds

$$P\left(\mathbf{x} + \frac{j}{k}\,\mathbf{a}_n; \lambda_1\mathbf{a}_1, \cdots, \lambda_{n-1}\mathbf{a}_{n-1}, \frac{1}{2k}\,\mathbf{a}_n\right) , \qquad j = 0, 1, \cdots, k-1 ,$$

which are disjoint and translation-equivalent to each other are contained in $P(\mathbf{x}; \lambda_1\mathbf{a}_1, \cdots, \lambda_{n-1}\mathbf{a}_{n-1}, \mathbf{a}_n)$.

From ($\ast$) and 4.1 follows therefore $v(A) = 0$, q.e.d.

From (2), 4.2 and the definition of direct sum, we obtain the Addition formula

$$(4.3) \qquad v(A + B) = v(A) + v(B) ,$$

or more generally, $v(\sum_{i=1}^{r} A_i) = \sum_{i=1}^{r} v(A_i)$.

Furthermore, we can "neglect the boundary" in calculating the volume by virtue of 4.2. Consequently, we shall limit our consideration in the sequel to closed polyhedra, to simplify the matter.

5. Volume of parallelepipeds

The volume of parallelepiped $P(\mathbf{x}; \mathbf{a}_1, \cdots, \mathbf{a}_n)$ does not depend on the origin $\mathbf{x}$ because of (5). So we may put

$$v(P(\mathbf{x}; \mathbf{a}_1, \cdots, \mathbf{a}_n)) = f(\mathbf{a}_1, \cdots, \mathbf{a}_n) .$$

We put also, in view of 4.2,

$$f(\mathbf{a}_1, \cdots, \mathbf{a}_n) = 0 , \qquad \text{if } \mathbf{a}_1, \cdots, \mathbf{a}_n \text{ are dependent} .$$

If k is any positive integer, $P(\mathbf{x}; \mathbf{a}_1, \cdots, \mathbf{a}_{n-1}, \mathbf{a}_n)$ is the direct sum of k translation-equivalent parallelepipeds:

$$P\left(\mathbf{x} + \frac{j}{k}\mathbf{a}_n; \mathbf{a}_1, \cdots, \mathbf{a}_{n-1}, \frac{\mathbf{a}_n}{k}\right), \qquad j = 0, 1, \cdots, k-1$$

so that 4.3, implies

$$f\left(\mathbf{a}_1, \cdots, \mathbf{a}_{n-1}, \frac{\mathbf{a}_n}{k}\right) = \frac{1}{k} f(\mathbf{a}_1, \cdots, \mathbf{a}_n) .$$

From this type of reasoning, we obtain

$$(5.1) \qquad f(\lambda_1\mathbf{a}_1, \cdots, \lambda_n\mathbf{a}_n) = |\lambda_1 \cdots \lambda_n| f(\mathbf{a}_1, \cdots, \mathbf{a}_n) ,$$

first for positive rational numbers $\lambda_1, \cdots, \lambda_n$, then for any real numbers $\lambda_1, \cdots, \lambda_n$.

Now f has another property:

$$(5.2) \qquad f(\mathbf{a}_1 + \mu_2\mathbf{a}_2 + \cdots + \mu_n\mathbf{a}_n, \mathbf{a}_2, \cdots, \mathbf{a}_n) = f(\mathbf{a}_1, \mathbf{a}_2, \cdots, \mathbf{a}_n) .$$

PROOF. We shall prove the invariance

$$f(\mathbf{a}_1 + \mu\mathbf{a}_2, \mathbf{a}_2, \cdots, \mathbf{a}_n) = f(\mathbf{a}_1, \mathbf{a}_2, \cdots, \mathbf{a}_n)$$

for the case $0 < \mu < 1$. The more general result as formulated above follows in the same way.

Now the parallelepiped $P = P(\mathbf{x}; \mathbf{a}_1, \mathbf{a}_2, \cdots, \mathbf{a}_n)$ can be decomposed into the direct sum of two "prisms"; the one P_1 has the triangle with vertices $\mathbf{x}$, $\mathbf{x} + \mathbf{a}_1$ and $\mathbf{x} + \mathbf{a}_1 + \mu\mathbf{a}_2$ as its base; the other P_2 the quadrangle with $\mathbf{x}$, $\mathbf{x} + \mathbf{a}_2$, $\mathbf{x} + \mathbf{a}_1 + \mu\mathbf{a}_2$ and $\mathbf{x} + \mathbf{a}_1 + \mathbf{a}_2$ as its base; both are generated by vectors $\mathbf{a}_3, \cdots, \mathbf{a}_n$; more exactly, we have $P = P_1 + P_2$, where

$$P_i = \{\mathbf{y}; \mathbf{y} = \mathbf{y}_0 + \lambda_3\mathbf{a}_3 + \cdots + \lambda_n\mathbf{a}_n; 0 \leqq \lambda_j \leqq 1,$$
$$j = 3, \cdots, n, \mathbf{y}_0 \in \text{Base of } P_i\}\ , \qquad i = 1, 2\ .$$

Now by the translation $\mathbf{x} \mapsto \mathbf{x} + \mathbf{a}_2$, the prism P_1 is transformed to another prism P_1' with the triangle with vertices $\mathbf{x}+\mathbf{a}_2$, $\mathbf{x}+\mathbf{a}_1+\mathbf{a}_2$, $\mathbf{x} + \mathbf{a}_1 + (1 + \mu)\mathbf{a}_2$ as its base and generated as before by $\mathbf{a}_3, \cdots, \mathbf{a}_n$. And one sees easily $P(\mathbf{x}; \mathbf{a}_1 + \mu\mathbf{a}_2, \mathbf{a}_2, \cdots, \mathbf{a}_n) = P_1' + P_2$.

Thus it follows

$$\begin{aligned} f(\mathbf{a}_1, \mathbf{a}_2, \cdots, \mathbf{a}_n) &= v(P_1) + v(P_2) \\ &= v(P_1') + v(P_2) \\ &= f(\mathbf{a}_1 + \mu\mathbf{a}_2, \mathbf{a}_2, \cdots, \mathbf{a}_n)\ . \end{aligned}$$

Put now

$$F(\mathbf{a}_1, \mathbf{a}_2, \cdots, \mathbf{a}_n) = f(\mathbf{a}_1, \mathbf{a}_2, \cdots, \mathbf{a}_n) \operatorname{sgn} \det (\mathbf{a}_1, \cdots, \mathbf{a}_n)$$

where $\det (\mathbf{a}_1, \cdots, \mathbf{a}_n)$ means the determinant of $n \times n$ matrix with n column vectors $\mathbf{a}_1, \cdots, \mathbf{a}_n$, and $\operatorname{sgn} x$ for $x \in \mathbf{R}$ means the sign of x. (i.e., $\operatorname{sgn} x = +1, 0$ or, -1, according as $x > 0, = 0$, or < 0.) Then we have

5.3. $F(\mathbf{a}_1, \mathbf{a}_2, \cdots, \mathbf{a}_n)$ *is an alternating multi-linear functional of* $\mathbf{a}_1, \cdots, \mathbf{a}_n$.

PROOF. From the definition it follows that $F(\mathbf{a}_1, \cdots, \mathbf{a}_n) = 0$ if $\mathbf{a}_1, \cdots, \mathbf{a}_n$ are dependent. Therefore F is alternating. The relation

(i) $F(\lambda\mathbf{a}_1, \mathbf{a}_2, \cdots, \mathbf{a}_n) = \lambda F(\mathbf{a}_1, \cdots, \mathbf{a}_n)$

follows from 5.1 and the definition of F.

The additivity

(ii) $F(\mathbf{a}_1 + \mathbf{b}_1, \mathbf{a}_2, \cdots, \mathbf{a}_n) = F(\mathbf{a}_1, \mathbf{a}_2, \cdots, \mathbf{a}_n) + F(\mathbf{b}_1, \mathbf{a}_2, \cdots, \mathbf{a}_n)$

is shown as follows. We may suppose that $\mathbf{a}_2, \cdots, \mathbf{a}_n$ are independent. If both $\mathbf{a}_1$ and $\mathbf{b}_1$ depend on $\mathbf{a}_2, \cdots, \mathbf{a}_n$, then both sides of (ii) are zero. Suppose therefore that $\mathbf{a}_1, \mathbf{a}_2, \cdots, \mathbf{a}_n$ are independent. Then we can put

$$\mathbf{b}_1 = \lambda_1\mathbf{a}_1 + \lambda_2\mathbf{a}_2 + \cdots + \lambda_n\mathbf{a}_n\ .$$

Substitute this in the both sides of the above formula. Then it is reduced to (i) by virtue of 5.2., q.e.d.

As the property (4) of v implies $F(\mathbf{e}_1, \cdots, \mathbf{e}_n) = 1$, it follows by the well-known characterization of determinant that

$$F(\mathbf{a}_1, \cdots, \mathbf{a}_n) = \det (\mathbf{a}_1, \cdots, \mathbf{a}_n)\ .$$

Thus we obtain:

5.4. $f(\mathbf{a}_1, \cdots, \mathbf{a}_n) = \text{abs det}(\mathbf{a}_1, \cdots, \mathbf{a}_n)$ *where* $\text{abs}\, x$ *for* $x \in \mathbf{R}$ *means the absolute value* $|x|$.

6. An approximation theorem

Given a point $\mathbf{x}_0$ of $\mathbf{R}^n$ and n independent vectors $\mathbf{a}_1, \cdots, \mathbf{a}_n$, we call a *lattice point* of the *lattice* $\mathfrak{L} = \mathfrak{L}(\mathbf{x}_0; \mathbf{a}_1, \cdots, \mathbf{a}_n)$, any point of the form

$$\mathbf{x}_0 + m_1\mathbf{a}_1 + \cdots + m_n\mathbf{a}_n , \qquad m_i \in \mathbf{Z} .$$

If $\mathbf{x}$ is a lattice point of $\mathfrak{L}$, the parallelepiped $P(\mathbf{x}; \mathbf{a}_1, \cdots, \mathbf{a}_n)$ is called a *mesh* of $\mathfrak{L}$. (Note that $\mathfrak{L}(\mathbf{x}; \mathbf{a}_0, \cdots, \mathbf{a}_n) = \mathfrak{L}(\mathbf{x}_0; \mathbf{a}_0, \cdots, \mathbf{a}_n)$, if $\mathbf{x}$ is any lattice point of $\mathfrak{L}(\mathbf{x}_0; \mathbf{a}_0, \cdots, \mathbf{a}_n)$.) Let k be any positive integer. We call the lattice

$$\mathfrak{L}_k = \mathfrak{L}\left(\mathbf{x}_0; \frac{\mathbf{a}_1}{2^k} \cdots, \frac{\mathbf{a}_n}{2^k}\right)$$

the kth *refined lattice* of $\mathfrak{L}$. (We have $\mathfrak{L} = \mathfrak{L}_0$.) Every mesh of the kth refined lattice has the volume.

$$v_k = \frac{1}{2^{kn}} \text{abs det}(\mathbf{a}_1, \cdots, \mathbf{a}_n) = \frac{1}{2^{kn}} v_0$$

according to 5.4.

Now let A be any polyhedron, and $\mathfrak{L}$ a fixed lattice. Denote with $\bar{A}_k$ the union of meshes of the kth refined lattice $\mathfrak{L}_k$ of $\mathfrak{L}$, which have non-empty intersections with A and with $\underline{A}_k$ the union of meshes of $\mathfrak{L}_k$ which are entirely contained in A. Then we have, of course, $\bar{A}_k \supset \bar{A}_{k+1} \supset A \supset \underline{A}_{k+1} \supset \underline{A}_k$ and therefore

$$v(\bar{A}_k) \geqq v(\bar{A}_{k+1}) \geqq v(A) \geqq v(\underline{A}_{k+1}) \geqq v(\underline{A}_k) .$$

We have an approximation theorem.

6.1. *For* $k \to \infty$, *we have*

$$\lim v(\bar{A}_k) = \lim v(\underline{A}_k) = v(A) .$$

For the proof, we have only to show $v(\bar{A}_k - \underline{A}_k) \to 0$, as $k \to \infty$.

Now it is clear that each mesh of $\mathfrak{L}_k$ contained in $\bar{A}_k - \underline{A}_k$ has a non-empty intersection with the boundary ∂A of A. ∂A is degenerate and is covered by a finite number of hyperplanes. So it is sufficient to prove

6.2. *Let* B *be a (degenerate) polyhedron contained in a hyperplane* π, *and* B_k^* *the union of meshes of* $\mathfrak{L}_k$ *with non-empty intersections with* B. *Then* $v(B_k^*) \to 0$, *as* $k \to \infty$.

To clarify the proof of this proposition, we introduce the following definitions.

Let $\mathfrak{L}$ be a lattice $\mathfrak{L}(\mathbf{x}; \mathbf{a}_1, \cdots, \mathbf{a}_n)$. The parallelepiped $P = P(\mathbf{x}; \mathbf{a}_1, \cdots, \mathbf{a}_n)$ is a mesh of this lattice; the point $\mathbf{x} + \frac{1}{2}(\mathbf{a}_1 + \cdots + \mathbf{a}_n)$ is the "center" of this mesh; the distance of this center from the boundary ∂P of this mesh is called the *breadth* of $\mathfrak{L}$, denoted by $\beta(\mathfrak{L})$; the diameter $d(P)$ of the mesh P will be denoted by $\delta(\mathfrak{L})$. Then we have the following lemma.

6.3. *Let d be any positive real number and $\mathfrak{L}$ a lattice. Then there exists a constant $c > 0$ depending only on d, $\beta(\mathfrak{L})$, and $\delta(\mathfrak{L})$ with the following property:*

If M is any bounded subset with the diameter $d(M) \leqq d$, then the number of meshes of $\mathfrak{L}$ with non-empty intersections with M does not exceed c. One may put $c = (2l + 1)^n$, where

$$l = \left[\frac{d + \beta(\mathfrak{L}) + \delta(\mathfrak{L})}{2\beta(\mathfrak{L})}\right].$$

PROOF. Let P be a mesh of $\mathfrak{L}$ with a non-empty intersection with M, and $\mathbf{p}$ the center of P. Let $\mathbf{q}$ be a point in $P \cap M$. If $\mathbf{m}$ is any point in M, we have $\operatorname{dis}(\mathbf{q}, \mathbf{m}) \leqq d$, and so $\operatorname{dis}(\mathbf{p}, \mathbf{m}) \leqq \delta(\mathfrak{L}) + d < (2l + 1)\beta(\mathfrak{L})$. Thus M is contained in the ball of radius $(2l + 1)\beta(\mathfrak{L})$ with the center $\mathbf{p}$, so *a fortiori* in the parallelepiped $\tilde{P}$ obtained from P by enlarging it to the ratio $1 : (2l + 1)$ around the center $\mathbf{p}$. And $\tilde{P}$ contains just $(2l + 1)^n$ meshes of $\mathfrak{L}$.

PROOF OF 6.2. Let $\mathbf{y}$ be a point on π and $\mathbf{b}_1, \cdots, \mathbf{b}_{n-1}$ be $(n - 1)$ independent vectors lying on π. They determine an $(n - 1)$-dimensional lattice $\mathfrak{M} = \mathfrak{M}(\mathbf{y}; \mathbf{b}_1, \cdots, \mathbf{b}_{n-1})$ on π. B will be covered by a finite number of meshes of $\mathfrak{M}$. So it suffices to show that the conclusion holds for a mesh $Q = P(\mathbf{y}; \mathbf{b}_1, \cdots, \mathbf{b}_{n-1})$ of $\mathfrak{M}$. In forming k^{th} refinements of $\mathfrak{L}$, $\mathfrak{M}$ at the same time, one obtains $\mathfrak{L}_k$, $\mathfrak{M}_k$ for which we have obviously $\beta(\mathfrak{L}_k) = \frac{1}{2^k}\beta(\mathfrak{L})$, $\delta(\mathfrak{L}_k) = \frac{1}{2^k}\delta(\mathfrak{L})$, $\beta(\mathfrak{M}_k) = \frac{1}{2^k}\beta(\mathfrak{M})$, $\delta(\mathfrak{M}_k) = \frac{1}{2^k}\delta(\mathfrak{M})$. By 6.3, the number of meshes of $\mathfrak{L}_k$ intersecting with one mesh of $\mathfrak{M}_k$ is never larger than $c = (2l + 1)^n$, where

$$l = \left[\frac{\delta(\mathfrak{M}_k) + \beta(\mathfrak{L}_k) + \delta(\mathfrak{L}_k)}{2\beta(\mathfrak{L}_k)}\right]$$

which is independent of k. Now Q is covered by $2^{k(n-1)}$ meshes of $\mathfrak{M}_k$. So it is covered by $c \cdot 2^{k(n-1)}$ meshes of $\mathfrak{L}_k$. But each mesh of $\mathfrak{L}_k$ has the volume $(1/2^{kn})$ abs det $(\mathbf{a}_1, \cdots, \mathbf{a}_n)$. So

$$v(Q_k^*) \leqq \frac{c \cdot 2^{k(n-1)}}{2^{kn}} \text{abs det}\,(\mathbf{a}_1, \cdots, \mathbf{a}_n) \to 0\ , \quad \text{when } k \to \infty\ ,$$

q.e.d.

7. Unicity of volume

We can now prove (supposing always the existence of volume)

7.1. *Let φ be a regular affine transformation $x \mapsto T\mathbf{x} + \mathbf{b}$. If $\varphi(A) = B$ for polyhedra A, B, then $v(B) = v(A)$* abs det (T).

PROOF. From a lattice $\mathfrak{L} = \mathfrak{L}(\mathbf{x}; \mathbf{a}_1, \cdots, \mathbf{a}_n)$, one obtains another lattice $\mathfrak{M} = \mathfrak{L}(\mathbf{y}; \mathbf{b}_1, \cdots, \mathbf{b}_n)$ by the affine transformation φ, in setting $\mathbf{y} = \varphi(\mathbf{x})$, $\mathbf{b}_1 = T\mathbf{a}_1, \cdots, \mathbf{b}_n = T\mathbf{a}_n$.

By 6.3, one can calculate $v(A)$ as follows. Let $\bar{a}_k$ be the number of meshes of $\mathfrak{L}_k$, with non-empty intersections with A. Then $v(A)$ is obtained as lim $(\bar{a}_k/2^{kn})$ abs det $(\mathbf{a}_1, \cdots, \mathbf{a}_n)$, as $k \to \infty$.

In the same way, we have

$$v(B) = \lim \frac{\bar{b}_k}{2^{kn}} \text{abs det}\,(\mathbf{b}_1, \cdots, \mathbf{b}_n)$$

where $\bar{b}$ has the corresponding meaning; but we have here $\bar{a}_k = \bar{b}_k$ as φ is a bijection, sending A to B and $\mathfrak{L}_k$ to $\mathfrak{M}_k$.

We have on the other hand

$$\det\,(\mathbf{b}_1, \cdots, \mathbf{b}_n) = \det\,(\mathbf{a}_1, \cdots, \mathbf{a}_n) \det\,(T)$$

whence 7.1 follows.

This is of course a stronger results than (5′).

From 2.2, 4.3, 5.4, and 7.1 now follows easily

7.2. *The volume of the n-simplex with vertices $\mathbf{x}_0, \mathbf{x}_1, \cdots, \mathbf{x}_n$ is given by*

$$\frac{1}{n!} \text{abs det} \begin{pmatrix} 1 & 1 \cdots & 1 \\ \mathbf{x}_0 & \mathbf{x}_1 \cdots & \mathbf{x}_n \end{pmatrix} .$$

As mentioned in 2.3, any closed non-degenerate polyhedron A can be represented as a direct sum of simplexes. $v(A)$ must be equal to the sum of the volumes of these simplexes. Thus the unicity of volume is proved.

8. Subdivisions

A finite set of n-simplexes $\{S^{(1)}, \cdots, S^{(r)}\} = \mathfrak{S}$ is said to be a *subdivision* of A, if $A = \sum_{i=1}^{r} S^{(i)}$. From the result of the previous paragraphs we conclude

8.1. *Let v be a functional on $\mathcal{P}_n$ satisfying* (1) – (5) *in* § 3. *Then $v(A)$ must be obtained as follows.*

(a) *If A is degenerate, put $v(A) = 0$,*

(b) *If A is non-degenerate, then take a subdivision $\mathfrak{S} = \{S^{(1)}, \cdots, S^{(r)}\}$ of A and put*

$$(\dagger) \qquad v(A) = \frac{1}{n!} \sum_{i=1}^{r} \text{abs det} \begin{pmatrix} 1 & 1 & \cdots 1 \\ \mathbf{x}_0^{(i)} & \mathbf{x}_1^{(i)} & \cdots \mathbf{x}_n^{(i)} \end{pmatrix}$$

where $\mathbf{x}_0^{(i)}, \mathbf{x}_1^{(i)}, \cdots, \mathbf{x}_n^{(i)}$ are vertices of $S^{(i)}$.

To prove the existence of the functional in question, we are going to show that $v(A)$ thus obtained is well-defined and satisfies (1) – (5).

Now the right hand side of (†) depends on the subdivision $\mathfrak{S}$. It will be denoted with $v_{\mathfrak{S}}(A)$ to make clear this dependence. The central point is

8.2. *$v_{\mathfrak{S}}(A)$ does not depend on $\mathfrak{S}$,* i.e., *$v(A)$ is well-defined.*

Once this is proved, it is easy to see that v satisfies (1) – (5); e.g. (2) is shown as follows.

To limit ourselves to the interesting case, suppose that all $A \cap B$, $A - (A \cap B)$, $B - (A \cap B)$ are non-degenerate. Let $\mathfrak{S}_1$, $\mathfrak{S}_2$, $\mathfrak{S}_3$ be subdivisions of $A \cap B$, and of the closures of $A - (A \cap B)$, $B - (A \cap B)$ respectively. Then $\mathfrak{S}_1 \cup \mathfrak{S}_2$, $\mathfrak{S}_1 \cup \mathfrak{S}_3$ and $\mathfrak{S}_1 \cup \mathfrak{S}_2 \cup \mathfrak{S}_3$ are obviously subdivisions of A, B, and $A \cup B$, whence (2) follows.

Thus the essential point of our proof is reduced to that of 8.2.

9. Further reductions

Some definitions and lemmas. Let $\mathfrak{S}$, $\mathfrak{S}'$ be two subdivisions of A. When each element of $\mathfrak{S}'$ is contained in an element of $\mathfrak{S}$, then $\mathfrak{S}'$ is said to be a *refinement* of $\mathfrak{S}$. A set $\mathfrak{S}$ of simplexes is said to form a *c-set,* if elements of $\mathfrak{S}$ intersect only in their sides. (Then the elements of $\mathfrak{S}$ together with their sides form a simplicial complex.) For any two subdivisions $\mathfrak{S}_1$, $\mathfrak{S}_2$ of A, it is easy to construct a subdivision $\mathfrak{S}$ of A, which is a c-set and is a refinement of $\mathfrak{S}_1$ and of $\mathfrak{S}_2$ at the same time. So the proof of 8.2 is reduced to that of

9.1. *$v_{\mathfrak{S}}(A) = v_{\mathfrak{S}'}(A)$ when $\mathfrak{S}'$ is a c-set and is a refinement of $\mathfrak{S}$.*

For an n-simplex S with vertices $\mathbf{x}_0, \mathbf{x}_1, \cdots, \mathbf{x}_n$ put:

$$u(S) = \text{abs det} \begin{pmatrix} 1 & 1 & \cdots 1 \\ \mathbf{x}_0 & \mathbf{x}_1 & \cdots \mathbf{x}_n \end{pmatrix}$$

The proof of 9.1 is again clearly reduced to that of the following special case.

9.2. *Let* $\mathfrak{S} = \{S^{(1)}, \cdots, S^{(r)}\}$ *be a c-subdivision of an n-simplex* S (i.e., *a c-set which is a subdivision of* S.). *Then*

$$u(S) = \sum_{i=1}^{r} u(S^{(i)}) .$$

Let us now give an orientation to the n-simplex S. That means that we determine an ordering of vertices $\mathbf{x}_0, \mathbf{x}_1, \cdots, \mathbf{x}_n$ of S up to even permutations. Then the value of $\det\begin{pmatrix}1 & 1 & \cdots & 1\\ \mathbf{x}_0 & \mathbf{x}_1 & \cdots & \mathbf{x}_n\end{pmatrix}$ is determined (inclusive of sign which is not altered by even permutations of columns.) Denote with $\tilde{S}$ the *oriented simplex* and with $d(\tilde{S})$ the value of this determinant.

Let $\mathfrak{S} = \{S^{(1)}, \cdots, S^{(r)}\}$ be a c-subdivision of S. When S is oriented, $S^{(i)}$ can be oriented "coherently", i.e., the vertices $\mathbf{x}_0^{(i)}$, $\mathbf{x}_1^{(i)}, \cdots, \mathbf{x}_n^{(i)}$ of $S^{(i)}$ can be so arranged that

$$\det\begin{pmatrix}1 & 1 & \cdots & 1\\ \mathbf{x}_0^{(i)} & \mathbf{x}_1^{(i)} & \cdots & \mathbf{x}_n^{(i)}\end{pmatrix}$$

may have the same sign as

$$\det\begin{pmatrix}1 & 1 & \cdots & 1\\ \mathbf{x}_0 & \mathbf{x}_1 & \cdots & \mathbf{x}_n\end{pmatrix};$$

and this determines the orientation of $S^{(i)}$. The so oriented $S^{(i)}$ will be denoted with $\tilde{S}^{(i)}$ and the set of oriented simplexes $\{\tilde{S}^{(1)}, \cdots, \tilde{S}^{(r)}\}$ will be called thd subdivision of $\tilde{S}$ given by $\mathfrak{S}$.

Consider now an n-complex K consisting of a c-set $\{S_1, \cdots, S_k\}$ and of the sides of its elements. Then $C = \sum_{i=1}^{r} m_i\tilde{S}_i$, $m_i \in \mathbf{Z}$ is an *n-chain* on K. We put $d(C) = \sum_{i=1}^{k} m_i d(\tilde{S}_i)$. Then d is a linear functional defined on the module of n-chains on K.

Suppose now that a c-subdivision $\mathfrak{S}_i = \{S_i^{(1)}, \cdots, S_i^{(r_i)}\}$ is given for each $S_i \in K$. Then $\mathfrak{S} = \bigcup_{i=1}^{k} S_i$ is a set of n-simplexes. Suppose that $\mathfrak{S}$ forms again a c-set. Then this set together with sides of its elements forms another n-complex K'. The *subdivision operator* $Sd_{\mathfrak{S}}$ given by $\mathfrak{S}$ is the linear map from the module of n-chains on K to that of n-chains on K' defined by

$$Sd_{\mathfrak{S}}\left(\sum_{i=1}^{k} m_i\tilde{S}_i\right) = \sum_{i=1}^{k} m_i \sum_{j=1}^{ri} \tilde{S}_i^{(j)} ,$$

where $\{\tilde{S}_i^{(1)}, \cdots, \tilde{S}_i^{(r_i)}\}$ is the subdivision of $\tilde{S}_i$ given by $\mathfrak{S}_i$.

In these notations, 9.2 can be reformulated as

9.3. *$d(\tilde{S}) = d(Sd_{\mathfrak{S}}\tilde{S})$ for any c-subdivision $\mathfrak{S}$ of S.*

Now there is the well-known *boundary operator* ∂ in topology, mapping linearly the module of n-chains on K to that of $(n-1)$-chains on K. It is defined by

$$\partial\left(\sum_{i=1}^{k} m_i\tilde{S}_i\right) = \sum_{i=1}^{k} m_i\partial(S_i) ,$$

where $\partial(\tilde{S}_i) = \partial(\mathbf{x}_{i0}, \mathbf{x}_{10}, \cdots, \mathbf{x}_{in}) = \sum_{j=0}^{n} (-1)^j(\mathbf{x}_{i0}, \cdots, \hat{\mathbf{x}}_{ij}, \cdots, \mathbf{x}_{in})$.

Now $\mathfrak{S}_i$ induces a subdivision of the sides of S_i in natural manner, and consequently a subdivision of oriented $(n-1)$-simplexes $(\mathbf{x}_{i0}, \cdots, \hat{\mathbf{x}}_{ij}, \cdots, \mathbf{x}_{in})$. The module of $(n-1)$-chains on K' is generated by these $(n-1)$-simplexes. In taking the boundary of an n-chain on K', one obtains a linear combination of them. This boundary operator (although no more on K, but on another complex K') is also denoted with ∂ as it is customary.

On the other hand, the subdivision $\mathfrak{S}$ of complex K gives rise to a linear map of the module of $(n-1)$-chains on K to that of $(n-1)$-chains on K'. This map will be also denoted with $Sd_{\mathfrak{S}}$.

Then it is known in topology, that the following holds (cf. Alexandroff-Hopf [1]).

(9.4) $$Sd_{\mathfrak{S}}\cdot\partial = \partial\cdot Sd_{\mathfrak{S}} ,$$

where $Sd_{\mathfrak{S}}$ in the left hand side and in the right hand side means the maps for the module of $(n-1)$-chains and of n-chains respectively, and ∂ means boundary operators on K in the left hand side and on K' in the right hand side.

This relation is essential for our proof.

We have to use another simple concept of algebraic topology, that of *cone* $(\mathbf{p}, C)$ with a vertex $\mathbf{p}$ and a base C, which is an l-chain $(0 \leqq l \leqq n-1)$:

$$C = \sum_{i=1}^{k} m_i\tilde{S}_i^l , \qquad m_i \in Z ,$$

where $\tilde{S}_i^l = (\mathbf{x}_{i0}, \mathbf{x}_{i1}, \cdots, \mathbf{x}_{il})$, $i = 1, \cdots, k$ are oriented l-simplexes. It is assumed that $\mathbf{p}$ does not lie on any of l-planes spanned by $\mathbf{x}_{i0}, \mathbf{x}_{i1}, \cdots, \mathbf{x}_{il}$, $i = 1, \cdots, k$. Then $(\mathbf{p}, C)$ is defined as the $(l+1)$-chain

$$(\mathbf{p}, C) = \sum_{i=1}^{k} m_i(\mathbf{p}, \tilde{S}_i^l) ,$$

where $(\mathbf{p}, \tilde{S}_i^l) = (\mathbf{p}, \mathbf{x}_{i0}, \mathbf{x}_{i1}, \cdots, \mathbf{x}_{il})$.

If C is an n-chain, ∂C is an $(n-1)$-chain and $(\mathbf{p}, \partial C)$ is again an n-chain. We have

(9.5) $$d(C) = d\big((\mathbf{p}, \partial C)\big) \qquad \textit{for any } n\textit{-chain } C .$$

PROOF. If suffices to prove this for $C = \tilde{S} = (\mathbf{x}_0, \mathbf{x}_1, \cdots, \mathbf{x}_n)$. This amounts to the identity:

$$\det\begin{pmatrix} 1 & 1 \cdots 1 \\ \mathbf{x}_0 & \mathbf{x}_1 \cdots \mathbf{x}_n \end{pmatrix} = \det\begin{pmatrix} 1 & 1 \cdots 1 \\ \mathbf{p} & \mathbf{x}_1 \cdots \mathbf{x}_n \end{pmatrix} - \det\begin{pmatrix} 1 & 1 & 1 \cdots 1 \\ \mathbf{p} & \mathbf{x}_0 & \mathbf{x}_2 \cdots \mathbf{x}_n \end{pmatrix}$$
$$+ \cdots + (-1)^n \det\begin{pmatrix} 1 & 1 \cdots 1 \\ \mathbf{p} & \mathbf{x}_0 \cdots \mathbf{x}_{n-1} \end{pmatrix}.$$

Now as $\mathbf{x}_1 - \mathbf{x}_0, \cdots, \mathbf{x}_n - \mathbf{x}_0$ are independent vectors of $\mathbf{R}^n$, and $\mathbf{p} - \mathbf{x}_0$ is a vector of $\mathbf{R}^n$, $\mathbf{p}$ can be expressed in the form

$$\mathbf{p} = \lambda_0\mathbf{x}_0 + \cdots + \lambda_n\mathbf{x}_n , \qquad \textstyle\sum_{i=1}^{n} \lambda_i = 1 .$$

Substituting this in the right hand side, one obtains immediately the above identity, q.e.d.

We need another lemma.

9.6. *Let π be a hyperplane in $\mathbf{R}^n$, and C' an $(n-1)$-chain such that all $(n-1)$-simplexes appearing in C' lie on π. Let a coordinate system $\mathfrak{S}'$ on π, consisting of n independent points on π, be fixed, and a functional $d'(C')$ be defined as follows. Let $\tilde{S}' = (\mathbf{x}_0, \mathbf{x}_1, \cdots, \mathbf{x}_{n-1})$ be an oriented $(n-1)$-simplex on π. Then the vertices $\mathbf{x}_0, \mathbf{x}_1, \cdots, \mathbf{x}_{n-1}$ of S' will be represented by elements $\mathbf{x}'_0$, $\mathbf{x}'_1, \cdots, \mathbf{x}'_{n-1}$ of $\mathbf{R}^{n-1}$ according to $\mathfrak{S}'$. We put*

$$d'(\tilde{S}') = \det\begin{pmatrix} 1 & 1 \cdots 1 \\ \mathbf{x}'_0 & \mathbf{x}'_1 \cdots \mathbf{x}'_{n-1} \end{pmatrix}$$

(the matrix being of $n \times n$ type) and write $d'(C')$ for the sum of d'-values of $(n-1)$-simplexes appearing in C' multiplied with respective coefficients. Let $\mathbf{p}$ be a point outside π. Then we have

$$d\big((\mathbf{p}, C')\big) = k \cdot d'(C') ,$$

where k is a constant independent of C'.

PROOF. Let $\varphi(\mathbf{x}) = T\mathbf{x} + \mathbf{b}$ be an affinity. Then we have clearly

$$d\big(\varphi(\mathbf{x}_0), \cdots, \varphi(\mathbf{x}_n)\big) = \det T \cdot d(\mathbf{x}_0, \cdots, \mathbf{x}_n) .$$

So in computing $d((\mathbf{p}, C'))$, we may use a coordinate system obtained by adding $\mathbf{p}$ to $\mathfrak{S}'$ as the nth independent point. Then our conclusion is obvious.

As a corollary we have

9.7. *Let C'_1, C'_2 be two $(n-1)$-chains on π. Then $d'(C'_1) = d'(C'_2)$ implies $d((\mathbf{p}, C'_1)) = d((\mathbf{p}, C'_2))$.*

10. Existence of volume

The proof of existence of volume was reduced to that of 9.3. Now we have all the necessary tools for this proof, which will proceed by induction on n. We have to prove:

$$d(\tilde{S}) = d(Sd_{\mathfrak{S}}(\tilde{S}))\ .$$

Take any point $\mathbf{p}$ outside the hyperplanes spanned by the vertices of $(n-1)$-simplexes appearing in ∂S. Then we have by 9.5 and 9.4,

$$d(\tilde{S}) = d((\mathbf{p}, \partial\tilde{S}))\ ,$$

and

$$\begin{aligned} d(Sd_{\mathfrak{S}}(\tilde{S})) &= d((\mathbf{p}, \partial Sd_{\mathfrak{S}}(\tilde{S}))) \\ &= d((\mathbf{p}, Sd_{\mathfrak{S}}\partial(\tilde{S})))\ . \end{aligned}$$

Here we have

$$\partial\tilde{S} = \tilde{S}_0^{n-1} + \cdots + \tilde{S}_n^{n-1}$$

where $\tilde{S}_i^{n-1}$, $i = 0, 1, \cdots, n$ are (oriented) sides of $\tilde{S}$, and

$$Sd_{\mathfrak{S}}(\partial S) = Sd_{\mathfrak{S}}\tilde{S}_0^{n-1} + \cdots + Sd_{\mathfrak{S}}\tilde{S}_n^{n-1}\ .$$

Let π_i be the hyperplane of $(n-1)$-simplex $\tilde{S}_i^{n-1}$, $i = 0, 1, \cdots, n$. Then by the hypothesis of induction, we may assume that

$$d'(\tilde{S}_i^{n-1}) = d'(Sd_{\mathfrak{S}}\tilde{S}_i^{n-1})$$

for $(n-1)$-dimensional volume on π_i; so by 9.7,

$$d((\mathbf{p}, \tilde{S}_i^{n-1})) = d((\mathbf{p}, Sd_{\mathfrak{S}}\tilde{S}_i^{n-1}))\ .$$

Adding these for $i = 0, 1, \cdots, n$, we obtain

$$d((\mathbf{p}, \partial\tilde{S})) = d((\mathbf{p}, Sd_{\mathfrak{S}}\partial(\tilde{S})))\ .$$

This completes the proof of our Main Theorem.

FACULTÉ DES SCIENCES, NANCY

REFERENCES

[1] P. ALEXANDROFF and H. HOPF, Topologie 1, Springer Verlag, Berlin, 1935.
[2] E. SCHMIDT, *Über die Darstellung der Lehre von Inhalt in der Integralrechung*, Math. **12**, 1922, 298-316.
[3] D. HILBERT, Grundlagen der Geometrie, 7, Aufl. Teubner, Berlin, 1930.

(Received November 18, 1968)

On the resolution of certain holomorphic mappings

By Arnold Kas

Let $f: X \to S$ be a flat morphism of reduced analytic spaces.

Definition. By a resolution of f, we mean a commutative diagram of reduced analytic spaces and morphisms

$$\begin{array}{ccc} X' & \xrightarrow{\psi} & X \\ \downarrow{\scriptstyle f'} & & \downarrow{\scriptstyle f} \\ T & \xrightarrow{\varphi} & S \end{array}$$

with the following properties

(i) $f': X' \to T$ is flat;

(ii) φ is surjective; ψ is proper and surjective;

(iii) each fibre X'_t of f' is non-singular, and

$$\psi \mid X'_t : X'_t \longrightarrow X_{\varphi(t)}$$

is a resolution of singularities in the sense of Hironaka [2].

THEOREM. (Brieskorn [1]). *Let $f: X \to S$ be a flat morphism as above. Assume*

(i) *X and S are both non-singular;*

(ii) $\dim X = 3$; $\dim S = 1$;

(iii) *each fibre X_s has only rational singularities. Then*

$$f: X \longrightarrow S$$

admits a resolution.

We recall that a 2-dimensional normal singularity (Y, y) is said to be rational if for some resolution

$$\pi: \tilde{Y} \longrightarrow Y ,$$

we have $R^1\pi * \mathcal{O}_{\tilde{Y}} = 0$.

It appears to be non-trivial to remove the assumptions (i), (ii) of Brieskorn's theorem. We are able to remove these assumptions only under the restrictions

$$\dim X_s = 2 \qquad \text{for all } s \in S ,$$

and each fibre X_s has no singularities other than binodes (the binodes are A_k, $k \geqq 1$ where A_k: $x^2 + y^2 + z^{k+1} = 0$). Our methods are mostly computational, and are strongly inspired by those of Brieskorn.

THEOREM. *Let $f: X \to S$ be a flat morphism of reduced analytic spaces. Assume that for some points $o \in S$, $p \in X$ with $f(p) = o$, that X_0 is 2-dimensional and reduced and that (X_0, p) is a binode. Then there exist neighborhoods U of p in X, and V of o in S such that $f(U) = V$ and such that $f|_U$: $U \to V$ admits a resolution.*

To prove this theorem, it is useful to introduce the concept of a maximal family. Let (Y_0, p) be an analytic space with an isolated singularity p. Let

$$f: (Y, p) \longrightarrow (S, o)$$

be a flat morphism $\big(f(p) = o\big)$ with $f^{-1}(o) = (Y_0, p)$.

Definition. $f: (Y, p) \to (S, o)$ is said to be a maximal family if it satisfies the universal property:

Let f': $(Y', p') \to (S', o')$ be a flat morphism such that

$$f'^{-1}(o') = (Y_0, p) .$$

Then there exist neighborhoods U of p' in Y', V of o' in S' with $f'(U) = V$, and a morphism

$$\varphi: (V, o') \longrightarrow (S, o)$$

such that the "family" f': $(U, p) \to (V, o')$ is isomorphic to the induced "family"

$$\varphi^*(Y, p) \longrightarrow (V, o') .$$

THEOREM (Schlessinger [3]). *Let (Y_0, p) be a complete intersection, and assume that Y_0 is defined in a neighborhood of the origin of $\mathbf{C}^n$ by equations*

$$g_1(z) = g_2(z) = \cdots = g_k(z) = 0 ,$$

where $n - k = \dim Y_0$, $p = 0 \in \mathbf{C}^n$. Let

$$\mathcal{O}_{Y_0,0} = \mathcal{O}_{\mathbf{C}^n,0}/(g_1, \cdots, g_k)$$

be the local ring of Y_0 at 0, and let M_0 be the submodule of $(\mathcal{O}_{Y_0,0})^k$

generated by

$$\left(\frac{\partial g_1}{\partial z_1}, \cdots, \frac{\partial g_k}{\partial z_1}\right), \left(\frac{\partial g_1}{\partial z_2}, \cdots, \frac{\partial g_k}{\partial z_2}\right), \cdots, \left(\frac{\partial g_1}{\partial z_n}, \cdots, \frac{\partial g_k}{\partial z_n}\right).$$

Let $\bar{P}_1, \bar{P}_2, \cdots, \bar{P}_r$ *be a basis of* $(\mathcal{O}_{Y_0,0})^k/M_0$ *(Clearly* $\dim_{\mathbf{C}} (\mathcal{O}_{Y_0,0})^k/M_0 < +\infty$*), and let* $P_1, P_2, \cdots, P_r \in \mathcal{O}^k_{\mathbf{C}^n,0}$ *be representatives of* $\bar{P}_1, \cdots, \bar{P}_r$*. Put*

$$P_i = (P_{i1}(z), \cdots, P_{ik}(z)) .$$

Let Y *be the analytic set defined in a neighborhood of the origin in* $\mathbf{C}^n(z) \times \mathbf{C}^r(s_1, \cdots, s_r)$ *by the equations*

$$\begin{aligned} g_1(z) + \textstyle\sum_{j=1}^r s_j P_{j1}(z) &= 0 \\ g_2(z) + \textstyle\sum_{j=1}^r s_j P_{j2}(z) &= 0 \\ &\vdots \\ g_k(z) + \textstyle\sum_{j=1}^r s_j P_{jk}(z) &= 0 . \end{aligned}$$

Let S *be a suitably small neighborhood of the origin in* $\mathbf{C}^r(s_1, \cdots, s_r)$*, and let*

$$f\colon (Y, 0) \longrightarrow (S, 0)$$

be the projection. Then $f\colon (Y, 0) \to (S, 0)$ *is a maximal family.*

We return now to the proof of our theorem. It is obvious that it suffices to prove the theorem for a maximal family

$$f\colon (X_0, p) \longrightarrow (S, 0) .$$

PROOF OF THE THEOREM. Since the binode A_k is defined by the equation

$$g(x, y, z) = x^2 + y^2 + z^{k+1} = 0 ,$$

it follows from Schlessinger's theorem that if X is defined in a neighborhood of the origin of

$$\mathbf{C}^3(x, y, z) \times \mathbf{C}^k(s_2, s_3, \cdots, s_{k+1})$$

by the equation

$$x^2 + y^2 + z^{k+1} + \textstyle\sum_{j=2}^{k+1} s_j z^{k+1-j} = 0 ,$$

and if $f\colon X \to S$ is the projection onto a neighborhood S of the origin of $\mathbf{C}^k(s_2, \cdots, s_{k+1})$, then

$$f\colon (X, 0) \longrightarrow (S, 0)$$

is a maximal family.

Let $T \subset \mathbf{C}^{k+1}(t_1, t_2, \cdots, t_{k+1})$ be the intersection of a neighborhood of the origin with the hyperplane

$$t_1 + t_2 + \cdots + t_{k+1} = 0 .$$

Define a holomorphic map $\varphi: T \to S$ by setting $s_j =$ the jth elementary symmetric function of $t_1, t_2, \cdots, t_{k+1}$. Then

$$z^{k+1} + \sum_{j=2}^{k+1} s_j(t) z^{k+1-j} = \prod_{j=1}^{k+1} (z + t_j) .$$

Let $V \subset \mathbf{P}^1_1 \times \mathbf{P}^1_2 \times \cdots \times \mathbf{P}^1_{k+1} \times C^3(x, y, z) \times T$ ($\mathbf{P}^1_j$ is the jth factor; $\mathbf{P}^1_j = \mathbf{P}^1$) be defined by the equations

$$\begin{aligned} &x^2 + y^2 + \prod_{\nu=1}^{k+1} (z + t_\nu) = 0 \\ &X_j(ix + y) = Y_j \prod_{\nu=j+1}^{k+1} (z + t_\nu) \\ &X_j \prod_{\nu=1}^{j} (z + t_\nu) = Y_j\, (ix - y) , \qquad j = 1, 2, \cdots, k+1 \end{aligned}$$

where (X_j, Y_j) are homogeneous coordinates in $\mathbf{P}^1_j$, and where $i = \sqrt{-1}$.

Let $X' =$ the closure of $V - \{t_\mu = t_\nu;\ \mu \neq \nu\}$. The projection

$$\mathbf{P}^1 \times \cdots \times \mathbf{P}^1 \times \mathbf{C}^3 \times T \longrightarrow T$$

induces a map $f': X' \to T$, and the composition

$$\mathbf{P}^1 \times \cdots \times \mathbf{P}^1 \times \mathbf{C}^3 \times T \longrightarrow \mathbf{C}^3 \times T \xrightarrow{1 \times \varphi} \mathbf{C}^3 \times S$$

induces a map $\psi: X' \to X$. Clearly the diagram

$$\begin{array}{ccc} X' & \xrightarrow{\psi} & X \\ \big\downarrow{\scriptstyle f'} & & \big\downarrow{\scriptstyle f} \\ T & \xrightarrow{\varphi} & S \end{array}$$

is commutative. We will show that this diagram is a resolution of

$$f: X \longrightarrow S .$$

Let $U_0, U_1, \cdots, U_{k+1}$ be the open subsets of $\mathbf{P}^1 \times \cdots \times \mathbf{P}^1 \times \mathbf{C}^3 \times T$ defined by

$$\begin{aligned} U_0&: X_1 \neq 0 , \\ U_\rho&: Y_\rho \neq 0,\ X_{\rho+1} \neq 0 , \qquad \text{for } \rho = 1, \cdots, k , \\ U_{k+1}&: Y_{k+1} \neq 0 . \end{aligned}$$

In U_ρ we let

$$\sigma_\rho = \frac{X_\rho}{Y_\rho}, \; \tau_{\rho+1} = \frac{Y_{\rho+1}}{X_{\rho+1}}; \qquad in\ X' \cap U_\rho$$

we can solve for the (X_j, Y_j), $j \neq \rho, \rho + 1$ by

$$(X_j, Y_j) = \left(\sigma_\rho \prod_{\nu=j+1}^{\rho} (z + t_\nu), 1\right), \qquad \text{for } j < \rho$$
$$(X_j, Y_j) = \left(1, \tau_{\rho+1} \prod_{\nu=\rho+2}^{j} (z + t_\nu)\right), \qquad \text{for } j > \rho + 2 .$$

By an elementary computation, it follows that $X' \cap U_\rho$ may be defined by the above equations together with the equations

$$ix + y = \tau_{\rho+1} \prod_{\nu=\rho+2}^{k+1} (z + t_\nu)$$
$$ix - y = \sigma_\rho \prod_{\nu=1}^{\rho} (z + t_\nu)$$
$$z + t_{\rho+1} = \sigma_\rho \tau_{\rho+1} .$$

It follows from these equations that X' is non-singular, and

$$f' \colon X' \longrightarrow T$$

has maximal rank, hence in particular f' is flat.

It follows easily from the definitions of φ, ψ that φ is surjective, and ψ is proper and surjective. By definition

$$X_s \colon x^2 + y^2 + z^{k+1} + \sum_{j=1}^{k+1} s_j z^{k+1-j} = 0 .$$

An easy computation shows that (x_0, y_0, z_0, s) is a singular point of X_s if and only if $x_0 = 0$, $y_0 = 0$, z_0 is a multiple root of the polynomial

$$z^{k+1} + \sum_{j=2}^{k+1} s_j z^{k+1-j} = 0 .$$

Hence if $t_1, \cdots, t_{k+1}$ are the roots of the above polynomial, then (x_0, y_0, z_0, s) is a *simple* point of X_s if and only if one of the following possibilities holds.

(1) $ix_0 + y_0$, $ix_0 - y_0$ are not both zero;

(2) For each j, $1 \leqq j \leqq k + 1$, the expressions $\prod_{\nu=1}^{j} (z + t_\nu)$ and $\prod_{\nu=j+1}^{k+1} (z + t_\nu)$ are not both zero.

Since $X'_t \subset \mathbf{P}^1 \times \cdots \times \mathbf{P}^1 \times C^3 \times \{t\}$ satisfies

$$X_j(ix + y) = Y_j \prod_{\nu=j+1}^{k+1} (z + t_\nu)$$
$$X_j \prod_{\nu=1}^{j} (z + t_\nu) = Y_j\,(ix - y) , \qquad j = 1, 2, \cdots, k + 1$$

in either of the two cases, (1) or (2), we can solve for (X_j, Y_j) for all j. It follows that

$$\psi \mid X'_t \colon X'_t \longrightarrow X_s$$

is a resolution of singularities. Hence

$$\begin{array}{ccc} X' & \xrightarrow{\psi} & X \\ \downarrow{\scriptstyle f'} & & \downarrow{\scriptstyle f} \\ T & \xrightarrow{\varphi} & S \end{array}$$

is a resolution.

UNIVERSITY OF CALIFORNIA, BERKELEY

REFERENCES

[1] E. BRIESKORN, *Über die Auflosung gewisser Singularitaten von holomorphen abbildungen*, Math. Ann. **166** (1966), 76-102.

[2] H. HIRONAKA, *Resolution of singularities of an algebraic variety over a field of characteristic zero*, Ann. of Math. **79** (1964), 109-203.

[3] M. SCHLESSINGER, Infinitesmal deformations of singularities, Ph.D. Thesis, Harvard University, Cambridge, Mass. (1964).

(Received October 10, 1968)

Harmonic integrals for differential complexes*

By J. J. KOHN

1. Introduction

The operators d (and ∂) on forms on compact riemannian (and hermitian) manifolds have been studied by means of the theory of harmonic integrals (see [3]). This theory is easily generalized to the case of "elliptic differential complexes". Consider vector bundles E, F, and G over a compact manifold M, then a differentiable complex is a sequence

$$\mathbf{E} \xrightarrow{A} \mathbf{F} \xrightarrow{B} \mathbf{G}\,, \tag{1.1}$$

Where $\mathbf{E}$, $\mathbf{F}$ and $\mathbf{G}$ denote the sheaves of germs of local C^∞ sections and A and B are first order differential operators for which $BA = 0$. For every $x \in M$ and every non-zero cotangent vectors η at x there is associated with (1.1) the *symbol sequence*

$$E_x \xrightarrow{\sigma_x(\eta,\, A)} F_x \xrightarrow{\sigma_x(\eta,\, B)} G_x\,, \tag{1.2}$$

where E_x, F_x and G_x are the fibers over x of E, F and G respectively and the $\sigma_x(\eta, A)$, $\sigma_x(\eta, B)$ are linear maps (defined in § 3) called the symbols of A and B. We say that the complex (1.1) is elliptic of the symbol sequence (1.2) is exact.

The "energy form"

$$Q(u,\ u) = \|\, A^*u \,\|^2 + \|\, Bu \,\|^2 + \|\, u \,\|^2 \tag{1.3}$$

plays an important role in the study of (1.1). Here u is an element of $\Gamma(F, M)$, the sections of F over M; A^* is the L_2-adjoint of A and $\|\quad\|$ is the L_2-norm. It is easy to prove that (1.1) is elliptic if and only if there exists a constant $C > 0$ such that

$$\|\, u \,\|_1^2 \leq CQ(u, u)\,, \tag{1.4}$$

for all $u \in \Gamma(F, M)$. We shall use $\|\quad\|_s$ to denote the Sobolev s-

* This work was partially supported by the N.S.F. through a research project at Brandeis University and by the Sloan Foundation.

norm (the sum of L_2-norms of derivatives of order s of all components, in case s is a non-negative integer). We set

$$\mathcal{H} = \{u \in \Gamma(F, M) \mid A^*u = 0 \quad \text{and} \quad Bu = 0\}. \tag{1.5}$$

If (1.1) is elliptic, then it follows by the standard L_2-theory of elliptic partial differential equations that:

(I) $\mathcal{H}$ is finite dimensional.

(II) $\Gamma(F, M)$ is decomposed into the sum of orthogonal subspaces by

$$\Gamma(F, M) = AA^*\Gamma(F, M) \oplus B^*B\Gamma(F, M) \oplus \mathcal{H}. \tag{1.6}$$

Furthermore for each $u \in \Gamma(F, M)$ there exists a unique $v \in \Gamma(F, M)$

$$u = AA^*v + B^*Bv + Hu, \tag{1.7}$$

where H: $\Gamma(F, M) \rightarrow \mathcal{H}$ is the orthogonal projection. Let N: $\Gamma(F, M) \rightarrow \Gamma(F, M)$ defined by $Nu = v$. Then we have:

(III) N is completely continuous in L_2 and in all the Sobolev norms.

We can then conclude that, given $f \in \Gamma(F, M)$ the equation

$$A\varphi = f \tag{1.8}$$

has a solution $\varphi \in \Gamma(F, M)$ if and only if $f \perp \mathcal{H}$ and $Bf = 0$, further the solution φ is unique if we also require $\varphi \perp \mathcal{H}$ and then it is given by:

$$\varphi = A^*Nf. \tag{1.9}$$

The conclusions (I), (II), and (III) hold under much weaker hypotheses. The following theorem is proven in [6].

THEOREM. 1.1. *If $Q(u, u)$ is completely continuous with respect to the L_2-norm, then* (I), (II), *and* (III) *hold.*

COROLLARY. *In particular if there exists constants $C > 0$ and $\varepsilon >$ such that*

$$\| u \|_\varepsilon^2 \leqq CQ(u, u) \tag{1.10}$$

for all $u \in \Gamma(F, M)$ then (I), (II), *and* (III) *hold.*

A large class of partial differential equations can be reduced to first order elliptic complexes by using the "Spencer resolution" (see [8]). Inequalities of the type (1.10) have been analyzed by Hörmander in [1]. Much attention has been given to establishing estimates of the type (1.10) on manifolds with boundary for u satisfying the Neumann boundary conditions; however, here we discuss only compact manifolds without boundary.

The main example that has been worked out of non-elliptic complexes which satisfy (1.10) is the case of a compact real submanifold M of a complex manifold; the differential complex is a "restriction" of the ∂-complex. In case the codimension of M is one, the best estimate that can be obtained is (1.10) with $\varepsilon = 1/2$ (see [4] and [5]). This example suggests that in case the operators A and B can be expressed locally in terms of a set of vector fields $X_1, \cdots, X_k$ then the crucial estimate to establish is

$$\sum_{i=1}^{k} \| X_i u \|^2 \leqq CQ(u, u) . \tag{1.11}$$

The corollary can then be applied if

$$\| u \|_\varepsilon \leqq \text{const.} \left(\sum \| X_i u \| + \| u \|\right) . \tag{1.12}$$

A sufficient condition for (1.12) has been found by Hörmander (see [2]). In fact the estimate proven by Hörmander is more delicate then (1.12). Here (in § 1) we give a simpler proof[1] of (1.12); however, our methods do not give the more delicate estimate.

2. The L^2-norm of vector fields

In this section we prove an inequality due to Hörmander (see [1]). Hörmander's proof uses special norms which are designed to measure differentiability along vector fields. His method obtains more precise results than ours. Here we use only elementary properties of pseudo-differential operators and we believe that our method will be useful in other problems.

Let $X_1, X_2, \cdots, X_k$ be vector fields defined in an open set $\Omega \subset \mathbf{R}^n$.

THEOREM. 2.1. *If the Lie algebra generated by* $X_1, \cdots, X_k$ *spans the tangent space at each point of* Ω, *then for each* $\Omega' \subset\subset \Omega$ *there exist* $\varepsilon > 0$ *and* $C > 0$ *such that*

$$\| u \|_\varepsilon \leqq C\left(\sum_{i=1}^{k} \| X_i u \| + \| u \|\right) . \tag{2.1}$$

for all $u \in C_0^\infty(\Omega')$.

Before proving this theorem we recall some elementary prop-

[1] This has been presented at the 1968 CIME Conference on pseudodifferential operators.

erties of pseudo-differential operators (for the proofs see [7]). First we define Λ^s by

(2.2) $$(\Lambda^s u)^\wedge(\xi) = (1 + |\xi|^2)^{s/2}\hat{u}(\xi)\,.$$

We will be concerned with the algebra generated by the operators Λ^s, multiplication by C^∞ functions, and the D_i. We will consider here only the operators in this algebra and their L_2-adjoints. These operators are all pseudo-differential operators. The following properties of pseudo-differential operators will be used here.

(A) A pseudo-differential operator T is of order s if for each real number r there exists a constant C_r such that

(2.3) $$\|Tu\|_r \leqq C_r \|u\|_{s+r}\,,$$

for all $u \in C_0^\infty$.

(B) If T, T' are pseudo-differential operators of orders s, s' respectively, then T^*, $T + T'$, TT', and $[T, T']$ are pseudodifferential operators of orders s, $\max(s, s')$, $s + s'$, and $s + s' - 1$, respectively.

(C) $\|u\|_s = \|\Lambda^s u\|$ and there exist positive constants C and C' such that

(2.4) $$C\sum_{i=1}^n \|D_i u\|_{s-1} \leqq \|u\|_s \leqq C'\sum_{i=1}^n \|D_i u\|_{s-1}$$

for all $u \in C_0^\infty$.

PROOF OF THEOREM 2.1. We wish to estimate $\|u\|_\varepsilon$, by (C) it suffices to estimate $\|D_i u\|_{\varepsilon-1}$. Let $\{F_\nu\}$ be a basis for the Lie algebra generated by the $\{X_i\}$ on Ω'. Since (by assumption) the D_i are linear combinations of the F_ν, it will suffice to estimate the $\|F_\nu u\|_{\varepsilon-1}$. Now we may choose for the $\{F_\nu\}$ elements of the form $[X_{i_1}[X_{i_2}, \cdots, [X_{i_{p-1}}, X_{i_p}]\cdots]$, this is easy to show by induction on p and the use of the Jacobi identity. Ket F^p be an element of the Lie algebra in the above form; we can then write $F^p = [X, F^{p-1}]$. We will show that

(2.5) $$\|F^p u\|_{\varepsilon-1} \leqq \text{const.}\,(\|F^{p-1}u\|_{2\varepsilon-1} + \|Xu\| + \|u\|)\,.$$

We will denote by T^s any PDO of order s, then we have

$$\begin{aligned}\|F^p u\|^2_{\varepsilon-1} &= (\Lambda^{\varepsilon-1}F^p u, \Lambda^{\varepsilon-1}F^p u) = (T^{2\varepsilon-1}F^p u, u)\\ &= (T^{2\varepsilon-1}XF^{p-1}u, u) - (T^{2\varepsilon-1}F^{p-1}Xu, u)\\ &= (T^{2\varepsilon-1}F^{p-1}u, X^*u) + ([T^{2\varepsilon-1}, X]\,F^{p-1}u, u)\\ &\quad - (T^{2\varepsilon-1}Xu, (F^{p-1})^*u) - ([T^{2\varepsilon-1}, F^{p-1}]\,Xu, u)\end{aligned}$$

Note that $[T^{2\varepsilon-1}, X]$ and $[T^{2\varepsilon-1}, F^{p-1}]$ are of order $2\varepsilon - 1$ and that $X^* = -X + T^\circ$ and $(F^{p-1})^* = -F^{p-1} + T^\circ$. Choosing ε so that $2\varepsilon - 1 \leqq 0$ we obtain the inequality (2.5). Now let p be the maximum of the number of brackets in the expressions for the F_ν; thus by repeated application of (2.5) we obtain

$$\| F_\nu u \|_{\varepsilon-1} \leqq \text{const.} \left(\sum \| X_i u \| + \| u \|\right) \tag{2.6}$$

for all $u \in C_0^\infty(\Omega')$ provided that we choose $2^p\varepsilon - 1 \leqq 0$. Hence (2.1) is true whenever $\varepsilon \leqq 2^{-p}$.

3. The symbol sequence

Definition 3.1. If A: $\mathbf{E} \to \mathbf{F}$ is the differential operator given by $(Au)^j = a_i^{jk} D_k u^i + a_i^j u^i$, then for each $x \in M$ and $\eta \in T_x^*$ we define the map $\sigma_x(\eta, A)$: $E_x \to F_x$ by

$$(\sigma_x(\eta, A)e)^j = a_i^{jk}(x)\eta_k e^i \,. \tag{3.1}$$

$\sigma_x(\eta, A)$ is called the *symbol of A*.

Observe that if we have a complex

$$\mathbf{E} \xrightarrow{A} \mathbf{F} \xrightarrow{B} \mathbf{G}\,, \tag{3.2}$$

where B is also a first order differential operator and

$$BA = 0\,, \tag{3.3}$$

then we have the *symbol sequence*

$$E_x \xrightarrow{\sigma_x(\eta, A)} F_x \xrightarrow{\sigma_x(\eta, B)} G_x \tag{3.4}$$

and

$$\sigma_x(\eta, B)\sigma_x(\eta, A) = 0 \tag{3.5}$$

for all $x \in M$ and all $\eta \in T_x^*$.

Definition 3.2. The complex (3.2) is called *elliptic* if the symbol sequence (3.4) is exact for all $\eta \in T_x^* - \{0\}$.

THEOREM 3.1. *The complex* (3.2) *is elliptic if and only if any of the following hold.*

(a) $\sigma_x(\eta, A)\big(\overline{\sigma_x(\eta, A)}\big)^t + \big(\overline{\sigma_x(\eta, A)}\big)^t\sigma_x(\eta, B)$ *is a positive definite form on* F_x *whenever* $\eta \neq 0$.

(b) *If* $\eta \in T_x^*$ *and* $\big(\overline{\sigma_x(\eta, A)}\big)^t f = \sigma_x(\eta, B)f = 0$ *for some* $f \in F_x$, $f \neq 0$, *then* $\eta = 0$.

(c) *There exists a positive constant C such that*

$$(3.6) \qquad \|u\|_1 \leqq C\{\|A^*u\| + \|Bu\| + \|u\|\}$$

for all $u \in \Gamma(F, M)$.

PROOF. It is clear that (a) and (b) are equivalent and that they are equivalent to the inequality

$$(3.7) \qquad |\eta||f| \leqq C\{|(\overline{\sigma_x(\eta, A)})^t f| + |\sigma_x(\eta, B)f|\},$$

where $C > 0$ can be chosen independently of x. Since

$$(3.8) \qquad (\overline{\sigma_x(\eta, A)})^t = \sigma_x(\eta, A^*),$$

it follows, from standard results of elliptic theory, that (3.7) is equivalent to (3.6) and hence that (a), (b), and (c) are equivalent. To complete the proof we show that ellipticity in the sense of definition 3.2 is equivalent to (b). To see this suppose that $f_0 \in F_x$ and that there exists $\eta \in T_x^*$, $\eta \neq 0$ such that

$$(3.9) \qquad \sigma_x(\eta, B)f_0 = 0,$$

we can decompose f_0 as follows

$$(3.10) \qquad f_0 = f_0' + f_0'',$$

where $f_0' \perp \sigma_x(\eta, A)E_x$ and $f_0'' \in \sigma_x(\eta, A)E_x$, then clearly we have

$$(3.11) \qquad (\overline{\sigma_x(\eta, A)})^t f_0' = 0$$

and by (3.5) $\sigma_x(\eta, B)f_0'' = 0$ so that from (3.9) and (3.10) we conclude that

$$(3.12) \qquad \sigma_x(\eta, B)f_0' = 0.$$

The exactness of the symbol sequence is equivalent to having $f_0 \in \sigma_x(\eta, A)E_x$ which is equivalent to $f_0' = 0$ which, by (3.11) and (3.12), is equivalent to (b).

The above theorem motivates the following definition.

Definition 3.3. Let $\mathring{T}_x^*(A, B)$ be the subset of $\mathring{T}_x^*$ defined by $\mathring{T}_x^*(A, B) = \{\eta \in T_x^* \mid \sigma_x(\eta, A)(\overline{\sigma_x(\eta, B)})^t + (\overline{\sigma_x(\eta, B)})^t\sigma_x(B, \eta)$ is not positive definite$\}$.

We call the elements of $\mathring{T}_x^*(A, B)$ the *non-elliptic co-tangent* vectors of the complex (3.2) at x.

Observe that the complex (3.2) is elliptic if and only if $\mathring{T}_x^* = \{0\}$ for all $x \in M$.

4. Elliptic tangent vectors

Definition 4.1. If A: $\mathbf{E} \rightarrow \mathbf{F}$ is a differential operator as in

Definition 3.1, we define $\mathcal{V}_x(A)$ the space of *characteristic cotangent vectors* of A by

$$\mathcal{V}_x(A) = \{\eta \in T_x^* \mid \sigma_x(\eta, A) = 0\} .$$

We say that A is *regular* if dim $\mathcal{V}_x(A)$ does not depend on x. Let $\mathcal{S}_x(A)$ be the annihilator of $\mathcal{V}_x(A)$, the elements of $\mathcal{S}_x(A)$ are called A-tangent vectors at x. In case A is regular, we can find local vector fields $X_1, \cdots, X_k$ which span $\mathcal{S}_x(A)$ at each x in their domain, we call these *A-vector fields.*

PROPOSITION 4.1. *If A is regular, then each point of M has a neighborhood U on which there are A-vector fields $X_1, \cdots, X_k$ and locally A can be expressed by*

$$(4.1) \qquad (Au)^i = \alpha_r^{ii} X_j + \alpha_r^i u^r ,$$

where α_m^{ij}, α_m^i are in C^∞.

PROOF. Let $0 \in U$. We will prove (4.1) in a neighborhood of 0. By the regularity assumption we can find a neighborhood V of 0 and vector fields $Y_1, \cdots, Y_{n-k}$ on V such that $X_1, \cdots, Y_k$, $Y_1, \cdots, {}_{n-k}$ form a basis of T_x at each $x \in V$. Then we can write

$$(4.2) \qquad (Au)^i = \alpha_r^{ij} X_j u^r + \gamma_r^{ij} Y_j u^r + \alpha_r^i u^r ,$$

where the coefficients are in $C^\infty(V)$. Let $P \in V$. We will show that $\gamma_r^{ij}(P) = 0$ which will prove (4.1). Let f be a function such that

$$(4.3) \qquad [X_s(f)]_P = 0 \quad \text{and} \quad [Y_s(f)]_P = \delta_{sj} .$$

Note that $(df)_P \in \mathcal{V}_P(A)$.

Hence $\sigma((df)_P, A) = 0$ but we also have $\sigma_P((df)_P, A) = \gamma_r^{ij}(P)$ thus completing the proof.

Definition 4.2. We say the complex (3.2) is *normal* if both A and B are regular and if $\mathcal{V}_x(A) = \mathcal{V}_x(B) = \mathring{T}_x^*(A, B)$. In that case $\mathcal{S}_x(A) = \mathcal{S}_x(B)$ is called the space of *elliptic tangent vectors* of the complex (3.2)

Definition. 4.3. We say that the complex (3.2) is *$\mathcal{S}$-elliptic* if it is normal and if for each $P \in M$ there exists a neighborhood U of P and a constant $C > 0$ such that

$$(4.4) \qquad \sum_{b=1}^k \| X_j u \| \leqq C(\| A^* u \| + \| Bu \| + \| u \|) .$$

for all $u \in C_0^\infty(U)$, where $X_1, \cdots, X_k$ is a basis of $\mathcal{S}_x(A, B)$ for each

$X \in U$.

Combining this with Theorem 2.1, we have

PROPOSITION 4.2. *If a complex* (3.2) *is* $\mathfrak{S}$*-elliptic and if each point has a neighborhood in which a basis for the vector fields that span* $\mathfrak{S}_x(A, B)$ *has the property that the Lie algebra generated by these vector fields is the Lie algebra of all vector fields, then there exists* $\varepsilon > 0$ *such that*

$$\| u \|_\varepsilon \leqq \text{const.} (\| A^* u \| + \| Bu \| + \| u \|) .$$

for all $u \in \Gamma(E)$.

5. The almost-complex case

In this section we assume that the differential complex is normal and further we assume that for each $x \in M$ there is a linear map $J_x \colon \mathfrak{S}_x \to \mathfrak{S}_x$ such that $J_x^2 = -I$, and that J_x depends differentiably (C^∞) on x. We denote by $\tilde{\mathfrak{S}}_x$ the complexification of $\mathfrak{S}_x$; i.e., $\tilde{\mathfrak{S}}_x = \mathfrak{S}_x \otimes \mathbf{C}$. Then there is a unique extension of J_x to $\tilde{\mathfrak{S}}$ and we have the splitting

$$\tilde{\mathfrak{S}}_x = \mathfrak{Z}_x \oplus \bar{\mathfrak{Z}}_x , \tag{5.1}$$

where

$$\mathfrak{Z}_x = \{ v \in \tilde{\mathfrak{S}}_x \mid J_x(v) = iv \}$$
$$\bar{\mathfrak{Z}}_x = \{ v \in \tilde{\mathfrak{S}}_x \mid J_x(v) = -iv \} .$$

Let $\tilde{\mathfrak{S}}$, $\mathfrak{Z}$ and $\bar{\mathfrak{Z}}$ denote the corresponding spaces of local vector fields over M. Then we have $\tilde{\mathfrak{S}} = \mathfrak{Z} \oplus \bar{\mathfrak{Z}}$ and conjugation map $\mathfrak{Z} \to \bar{\mathfrak{Z}}$ which is an isomorphism over the reals. We further assume that if $Z, Z' \in \mathfrak{Z}$ then $[Z, Z'] \in \mathfrak{Z}$ and hence if $\bar{Z}, \bar{Z}' \in \bar{\mathfrak{Z}}$, then $[\bar{Z}, \bar{Z}'] = \overline{[Z, Z']} \in \bar{\mathfrak{Z}}$.

Now observe that if $\mathfrak{S}_x$ is the whole tangent space then the J_x gives an integrable almost-complex structure on M. However, if $\mathfrak{S}_x$ is not the whole tangent space and if all local vector fields are contained in the Lie algebra generated by $\mathfrak{S}_x$ then there exist vector fields $Z_1, Z_2 \in \mathfrak{Z}$ such that $[Z_1, \bar{Z}_2] \notin \tilde{\mathfrak{S}}$.

Definition 5.1. Given $J \colon \mathfrak{S} \to \mathfrak{S}$ satisfying the above, where $\mathfrak{S}$ is associated with the complex (3.2), we say that (3.2) is a *J-complex* if the operators A and B can be expressed locally in terms of elements of $\mathfrak{Z}$.

Whenever M has a sub-bundle S of the tangent bundle with a J_x: $S_x \to S_x$ and $\mathfrak{S}$ the space of local sections of S satisfy the above conditions we associate certain J-complexes with $\mathfrak{S}$. These reduce to the ∂ and ∂-complexes $S - T$. We denote by $R_x^{p,q}$ the set of maps

$$\varphi\colon \underbrace{\mathfrak{Z}_x \times \cdots \times \mathfrak{Z}_x}_{p\text{-times}} \times \underbrace{\bar{\mathfrak{Z}}_x \times \cdots \times \bar{\mathfrak{Z}}_x}_{q\text{-times}} \longrightarrow \mathbf{C}\,,$$

where φ is multilinear over the reals and for $c \in \mathbf{C}$ we have

$$\varphi(cV_1, V_2, \cdots, V_p, W_1, \cdots, W_q) = c\varphi(V_1, \cdots, V_p, W_1, \cdots, W_q)$$

$$\varphi(V_1, \cdots, V_p, cW_1, W_2, \cdots, W_q) = \bar{c}\varphi(V_1, \cdots, V_p, W_1, \cdots, W_q)$$

and if π is a permutation of $\{1, \cdots, p\}$ and σ is a permutation of $\{1, \cdots, q\}$ then

$$\begin{aligned}\varphi(V_{\pi(1)}, \cdots, V_{\pi(p)}, W_{\sigma(1)}, \cdots, W_{\sigma(q)}&\\ = \operatorname{sgn}\pi \operatorname{sgn}\sigma\varphi(V_1, \cdots, V_p, W_1, \cdots, W_q)&\,.\end{aligned}$$

We set $R^{p,q} = U_x R_x^{p,q}$ so that in a natural way $R^{p,q}$ is a bundle over M. Now we define

$$D\colon \Gamma(R^{p,q}, U) \longrightarrow \Gamma(R^{p+1,q}, U)$$

by

$$\begin{aligned}&D\varphi(V_1, \cdots, V_{p+1}, W_1, \cdots, W_q)\\ &\quad= \textstyle\sum_{i=1}^{p}(-1)^{i+1}V_i(\varphi(V_1, \cdots, \hat{V}_i, \cdots, V_p, W_1, \cdots, W_q)\\ &\qquad+ \textstyle\sum_{i,j=1}^{p}(-1)^{i+j+1}\varphi\\ &\qquad\qquad [V_i, V_j], V_1, \cdots, \hat{V}_i, \cdots, \hat{V}_j, \cdots, V_p, W_1, \cdots, W_q)\,.\end{aligned} \tag{5.2}$$

We thus have the complex

$$\mathbf{R}^{p-1,q} \xrightarrow{D} \mathbf{R}^{p,q} \xrightarrow{D} \mathbf{R}^{p,q+1}\,. \tag{5.3}$$

Analogously, we define the operator $\bar{D}$: $\mathbf{R}^{p,q} \xrightarrow{\bar{D}} \mathbf{R}^{p,q+1}$ and we have the complex

$$\mathbf{R}^{p,q-1} \xrightarrow{\bar{D}} \mathbf{R}^{p,q} \xrightarrow{\bar{D}} \mathbf{R}^{p,q+1}\,. \tag{5.4}$$

From (5.2) it easily follows that $D^2 = 0$. To compute the symbol of D we define the map ρ_x: $T_x^* \to R_x^{1,0}$ as follows, if $\eta \in T_x^*$ we define $\rho_x(\eta) \in R_x^{1,0}$ by

(5.5) $$\rho_x(\eta)(V) = \eta(V)$$

for all $V \in \mathcal{Z}_x$.

Note that the null space of ρ_x equals the annihilator of S_x. For each $\eta \in T_x^*$, $\eta \neq 0$ the symbol $\sigma_x(D, \eta)$: $R_x^{p,q} \to R_x^{p+1,q}$ is given by

(5.6) $$\sigma_x(D, \eta)\theta = \rho_x(\eta) \wedge \theta ,$$

so that the symbol sequence is exact if and only if $\rho_x(\eta) \neq 0$.

Definition 5.2. If ν is a real one-form in a neighborhood U such that $\rho_x(\nu) = 0$ for all $x \in U$ then we associate with ν a hermitian form on $\mathcal{Z}_x$ denoted by $\mathcal{L}_x^\nu$ and defined by:

$$\mathcal{L}_x^\nu(Z_1, Z_2) = \sqrt{-1}(d\nu)_x(Z_1, \bar{Z}_2) .$$

These forms will be called the *Levi forms*.

Observe that if Z_1 and Z_2 are vector fields over U and are elements of $\mathcal{Z}$, then for each $x \in U$ we have

(5.7) $$\mathcal{L}_x^\nu(Z_1, Z_2) = \sqrt{-1}([Z_1, \bar{Z}_2]) .$$

It is easy to construct a hermitian metric on the complexified tangent bundle $\mathbf{C}T$ such that for each $x \in M$ the map J_x: $\mathcal{S}_x \to \mathcal{S}_x$ is an isometry, and $\mathcal{Z}_x$ is orthogonal to $\bar{\mathcal{Z}}_x$. Each point in M has a neighborhood U on which there exist orthonormal vector fields $Z_1, \cdots, Z_k, \bar{Z}_1, \cdots, \bar{Z}_k, N_1, \cdots, N_m$ such that $2k + m = n$ and for each $x \in U$ $Z_1(x), \cdots, Z_k(x)$ is a basis of $\mathcal{Z}_x$ and $\bar{Z}_1(x), \cdots, \bar{Z}_k(x)$ is a basis of $\bar{\mathcal{Z}}_x$. Let $\zeta^1, \cdots, \zeta^k, \bar{\zeta}^1, \cdots, \bar{\zeta}^k, \nu^1, \cdots, \nu^m$ be the dual basis of one-forms. Thus we have $\rho_x(\nu^j) = 0$ for $x \in U$ and $j = 1, \cdots, m$. Furthermore on U we have

(5.8) $$[Z_i, \bar{Z}_j] = \sum_1^k a_{ij}^h Z_h + \sum_1^k b_{ij}^h \bar{Z}_h + \sum_1^m c_{ij}^h N_h ,$$

if we choose the N_h to be purely imaginary vector fields then the (c_{ij}^h) will be hermitian matrices that correspond to the Levi forms associated with the $\sqrt{-1}\,\nu^h$.

If $\varphi \in \Gamma(R^{p,q}, U)$, we can represent φ by

(5.9) $$\varphi = \sum \varphi_{I\bar{J}} \zeta^{I\bar{J}} ,$$

where the I run through the ordered p-tuples $1 \leqq i < \cdots < i_p \leqq k$ and the J through the ordered q-tuples $1 \leqq j_1 < \cdots < j_q \leqq k$ and

$$\zeta^{I\bar{J}} = \zeta^{i_1} \wedge \cdots \wedge \zeta^{i_p} \wedge \bar{\zeta}^{j_1} \wedge \cdots \wedge \bar{\zeta}^{j_q} .$$

Then we have

$$\bar{D}\varphi = \sum \varepsilon^{ij}_{H}\bar{Z}^{j}(\varphi^{I\bar{J}})\zeta^{I\bar{H}} + 0(\|\varphi\|) \tag{5.10}$$

and

$$\bar{D}^{*}\varphi = -\sum \varepsilon^{J}_{jK}Z_{i}(\varphi_{I\bar{J}})\zeta^{I\bar{K}} + 0(\|\varphi\|)\,, \tag{5.11}$$

where

$$\varepsilon^{jJ}_{H} = \begin{cases} 0 & \text{if } \{jJ\} \neq H \\ \text{sign of permutation } \binom{jJ}{H} & \text{if } \{jJ\} = H\,. \end{cases}$$

If $u \in C^{\infty}(U)$ and u has compact support contained in U then, integrating by parts twice, we obtain

$$\begin{aligned}(Z_{i}u, Z_{j}u) = (\bar{Z}_{j}u, \bar{Z}_{i}u) &+ \textstyle\sum_{h=1}^{m}(c^{h}_{ij}N_{h}u, u) \\ &+ O(\textstyle\sum \|\bar{Z}_{j}u\|\,\|u\| + \|u\|^{2})\,.\end{aligned} \tag{5.12}$$

And combining (5.10), (5.11) and (5.12) we obtain

$$\begin{aligned}\|\bar{D}\varphi\|^{2} + \|\bar{D}^{*}\varphi\|^{2} + \|\varphi\|^{2} = \textstyle\sum \|\bar{Z}_{j}\varphi_{I\bar{J}}\|^{2} \\ + \textstyle\sum_{i,j\in J}\operatorname{Re}(c^{h}_{ij}N_{h}\varphi_{I\bar{J}}, \varphi_{I\bar{J}}) \\ + O(\textstyle\sum \|\bar{Z}_{j}\varphi_{I\bar{J}}\|\,\|\varphi\| + \|\varphi\|^{2})\,,\end{aligned} \tag{5.13}$$

Where the support of φ lies in U.

Note that in case the null space of ρ_{x} is one-dimensional then the space of Levi-forms is one-dimensional and hence the number of eigenvalues at each point of M of the same sign is an invariant of the structure. We then have the following result.

PROPOSITION 5.1. *If the null space of ρ_{x} is one-dimensional and if for $x_{0} \in M$ the Levi-form at x_{0} has at least* $\max(n-q, q+1)$ *non-zero eigenvalues of the same sign, then there exists a neighborhood U of x_{0} and a constant C such that*

$$\|\bar{D}\varphi\|^{2} + \|\bar{D}^{*}\varphi\|^{2} + \|\varphi\|^{2} \geqq C(\|\varphi\|^{2}_{z} + \|\varphi\|^{2}_{\bar{z}}) \tag{5.14}$$

for all $\varphi \in \Gamma(R^{p,q}, U)$ with support of φ containe in U, where $\|\varphi\|_{z}$ and $\|\varphi\|_{\bar{z}}$ are defined by

$$\|\varphi\|^{2}_{z} = \textstyle\sum_{i,I,J}\|Z_{i}\varphi_{I\bar{J}}\| \quad \textit{and} \quad \|\varphi\|^{2}_{\bar{z}} = \textstyle\sum_{i,I,J}\|\bar{Z}_{i}\varphi_{I\bar{J}}\|^{2}\,.$$

PROOF. First we choose the $Z_{1}, \cdots, Z_{n-1}, \bar{Z}_{1}, \cdots, \bar{Z}_{n-1}$ so that the Levi-form is diagohal; i.e., $c^{1}_{ij} = \delta_{ij}\lambda_{ij}$. For any $f \in C^{\infty}$ and $u \in C^{\infty}$ with support in U we have

$$\begin{aligned}|\operatorname{Re}(fNu, u)| \leqq C(\max\nolimits_{U}|f|\,(\|u\|^{2}_{z} + \|u\|^{2}_{\bar{z}}) \\ + \max\nolimits_{U}\{|Z_{i}f|, |\bar{Z}_{i}f|\}\,\|u\|^{2}\end{aligned} \tag{5.15}$$

so that, in particular, we have

$$\text{(5.16)}\qquad \begin{aligned}\operatorname{Re}(\lambda_i Nu, u) &= \lambda_i(x_0)\operatorname{Re}(Nu, u)\\ &\quad + O(\text{small const. } (\|u\|_z^2 + \|u\|_{\bar z}^2)\,,\end{aligned}$$

where the "small constant" may be made as small as we wish by choosing U sufficiently small, here we set $N = N_1$. From (5.12) we then obtain

$$\text{(5.17)}\qquad \begin{aligned}(Z_i u, Z_j u) &= (\bar Z_j u, \bar Z_i u) + \delta_{ij}\lambda_i(x_0)\operatorname{Re}(Nu, u)\\ &\quad + O(\text{s.c. } (\|u\|_z^2 + \|u\|_{\bar z}^2)\,,\end{aligned}$$

where s.c. stands for a "small constant" as above. Then (5.13) combined with (5.16) gives

$$\text{(5.18)}\qquad \begin{aligned}&\|\bar D\varphi\|^2 + \|\bar D^*\varphi\|^2 + \|\varphi\|^2\\ &\quad = \|\varphi\|_z^2 + \textstyle\sum_{I,J}\sum_{j\in J}\lambda_j(x_0)\operatorname{Re}(N\varphi_{I\bar J}, \varphi_{I\bar J})\\ &\qquad + O(\text{s.c. } (\|\varphi\|_z^2 + \|\varphi\|_{\bar z}^2) + \|\varphi\|^2)\,.\end{aligned}$$

Let $0 < \alpha < 1$ and let $\mathcal{V}$ be any subset of triples $\{(I, J, h)\}$ where $1 \leqq h \leqq n - 1$. Using (5.17) we obtain

$$\text{(5.19)}\qquad \begin{aligned}\|\varphi\|_{\bar z}^2 &\geqq \alpha\|\varphi\|_{\bar z}^2 + (1-\alpha)\textstyle\sum_{(I,J,h)\in\mathcal{V}}\|\bar Z_h\varphi_{IJ}\|^2\\ &\geqq \alpha\|\varphi\|_{\bar z}^2 - (1-\alpha)\textstyle\sum_{(I,J,h)\in\mathcal{V}}\lambda_h(x_0)\operatorname{Re}(N\varphi_{IJ}, \varphi_{IJ})\\ &\quad + O(\text{s.c. } (\|\varphi\|_z^2 + \|\varphi\|_{\bar z}^2) + \|\varphi\|^2)\,.\end{aligned}$$

For and two disjoints subsets $\mathcal{P}$ and $\mathcal{N}$ of the set $\{(I, J)\}$ we let $R^{p,q}[\mathcal{P}, \mathcal{N}]$ be the subset of $\Gamma(R^{p,q}, U)$ defined by

$$\text{(5.20)}\qquad \begin{aligned}R^{p,q}[\mathcal{P}, \mathcal{N}] = \{\varphi \in \Gamma(R^{p,q}U) \mid\\ \operatorname{Re}(N\varphi_{I\bar J}, \varphi_{I\bar J}) > 0 &\qquad \text{when } (I, J) \in \mathcal{P}\\ \operatorname{Re}(N\varphi_{I\bar J}, \varphi_{I\bar J}) > 0 &\qquad \text{when } (I, J) \in \mathcal{N}\}\,.\end{aligned}$$

Since there are finitely many of these subsets and since $\Gamma(R^{p,q}, U)$ is their union, it suffices to prove (5.14) for all $\varphi \in R^{p,q}[\mathcal{P}, \mathcal{N}]$ whose support lies in a suitably small neighborhood of x_0. By renumbering the $Z_1, \cdots, Z_{n-1}$ and by replacing N by $-N$ if necessary, we can assume that $\lambda_h(x_0) > 0$ if $1 \leqq h \leqq r$, $\lambda_h(x_0) < 0$ if $r < h \leqq m$, and $\lambda_h(x_0) = 0$ if $m < h \leqq n - 1$. Then by assumption we have $r \geqq \max(n - q, q + 1)$. For fixed $\mathcal{P}$ and $\mathcal{N}$ we define $\mathcal{V}$ by

$$\begin{aligned}\mathcal{V} = \{(I, J, h) \mid 1 \leqq h \leqq r &\qquad \text{when } (I, J) \in \mathcal{N}\,,\\ r < h \leqq m &\qquad \text{when } (I, J) \in \mathcal{P}\}\,.\end{aligned}$$

So that when $\operatorname{Re}(N\varphi_{I\bar J}, \varphi_{I\bar J}) > 0$ we choose h so that $\lambda_h(x_0) < 0$;

and when Re $(N\varphi_{I\bar{J}}, \varphi_{I\bar{J}}) < 0$, we choose h so that $\lambda_h(x_0) > 0$. Thus, combining (5.18) and (5.19) we obtain

$$\begin{aligned} \| \bar{D}\varphi \|^2 + \| \bar{D}^*\varphi \|^2 + \| \varphi \|^2 \geqq \alpha \| \varphi \|_{\bar{z}}^2 \\ + \sum c_{IJ} \operatorname{Re} (N\varphi_{I\bar{J}}, \varphi_{I\bar{J}}) \\ + 0O(\text{s.c.} (\| \varphi \|_z^2 + \| \varphi \|_{\bar{z}}^2) + \| \varphi \|^2) , \end{aligned} \tag{5.21}$$

Where

$$C_{IJ} = \begin{cases} \sum_{j \in J, j \leqq r} \lambda_j(x_0) - \alpha \sum_{h \in J, h > r} \lambda_h(x_0) & \text{when } (I, J) \in \mathcal{P} \\ -(1 - \alpha) \sum_{h \in J, h \leqq r} \lambda_h(x_0) + \alpha \sum_{h \in J, h \leqq r} \lambda_h(x_0) & \\ & \text{when } (I, J) \in \mathfrak{N} \end{cases}$$

Thus, choosing α sufficiently small we conclude that $c_{IJ} \neq 0$ and c_{IJ} Re $(N\varphi_{I\bar{J}}, \varphi_{I\bar{J}}) \geqq 0$. Setting $i = j$ in (5.17) and summing we obtain

$$\| u \|_z^2 \leqq C(\| u \|_{\bar{z}}^2 + | (Nu, u) | + \| u \|^2) . \tag{5.22}$$

From (5.21) we have

$$\begin{aligned} \| \varphi \|_z^2 + \sum | (N\varphi_{I\bar{J}}, \varphi_{I\bar{J}}) | \leqq \text{const.} \{\| \bar{D}\varphi \|^2 + \| \bar{D}^*\varphi \|^2 + \| \varphi \|^2\} \\ + \text{s.c.} \{\| \varphi \|_z^2 + \| \varphi \|_{\bar{z}}^2\} . \end{aligned} \tag{5.23}$$

The desired estimate (5.14) then follows from (5.22) and (5.23) when the s.c. is small enough. Hence the proposition is proved, since the s.c. can be made arbitrarily smally by choosing a sufficiently small U.

Since, under the hypothesis of Proposition 5.1 the vector fields Re Z_j, Im Z_j, and [Re Z_i, Im Z_j] span the tangent space at each point, we can conclude that Theorem 2.1 holds with $\varepsilon = 1/2$, that is

$$\| u \|_{1/2} \leqq C(\| u \|_z + \| u \|_{\bar{z}} + \| u \|) . \tag{5.24}$$

Thus we obtain, under these hypotheses, the existence regularity and compactness properties mentioned in the Introduction.

A natural example of the J-complexes discussed here is obtained by "restricting" the $\bar{\partial}$-complex. That is, let M be a real submanifold of a complex manifold M' then for $x \in M$ we let $\mathcal{Z}_x$ be the space of holomorphic tangent vectors in M' which are tangent *to* M. We can then define the complexes (5.3) as above. It is not known whether every complex (5.3) is locally equivalent to the restriction described here.

PRINCETON UNIVERSITY

REFERENCES

[1] L. HORMANDER, *Hypoelliptic second order differential equations*, Acta Math. **119** (1967), 147-171.

[2] ———, "*Pseudo-differential operators and hypoelliptic equations*", in Amer. Math. Soc. Symp. Pure Math., vol. **10** (1967), 138-183.

[3] K. KODAIRA, *Harmonic fields in riemannian manifolds*, Ann. of Math. **50** (1949), 587-665.

[4] J. J. KOHN, "Boundaries of complex manifolds", in Proceedings of the Conference on Complex Analysis, Minneapolis, 1964 81-94, Springer Verlag, Berlin, 1965.

[5] ——— and H. ROSSI, *On the extension of holomorphic functions from the boundary of a complex manifold*, Ann. of Math. **81** (1965), 451-472.

[6] J. J. KOHN and L. NIRENBERG, *Non-coercive boundary value problems*, Comm. Pure Appl. Math. **18** (1965), 443-492.

[7] ———, *An algebra of pseudodifferential operators*, Comm. Pure Appl. Math. **18** (1965), 269-305.

[8] D. C. SPENCER, *De Rham theorems and Neumann decompositions associated with linear partial differential equations*, Ann. Inst. Fourier, **14** (1964), 1-20.

(Received March 29, 1969)

A note on families of complex structures

BY MASATAKE KURANISHI

This paper consists of two remarks on complex analytic families of complex structures. The first is about the definition of families on complex structures, and the second is about automorphisms of the locally complete families. Both of these results were apparently known to Douady. The first appears without proof in [1] in a somewhat different form. In any case, we could not find any proofs in the literature.

When the writer developed a theory of complex families in [3], we put in (for technical reasons) side conditions in the definition of complex analytic families. However, it turns out that we can easily eliminate the conditions. This will be done in § 1.

Let M be a complex analytic structure on a compact C^∞ manifold M. Denote $\mathfrak{M}_T = \{M_t; t \in T\}$ the locally complete family of deformations of M with parameter space T which was constructed in [3]. Let M_S be a complex analytic family of deformations of M with parameter space S. Then M_S is (locally) induced by a holomorphic map $\tau: S \to T$. Wavrik, elsewhere in this volume, develops a general theory, which in particular implies that τ is uniquely determined (as a germ of maps at the reference point) provided the dimension of the automorphism group of complex structure M_t is independent of t on a neighborhood of the reference point t_0. This suggested the following theorem (to be proved in § 2) which is a little stronger than the above.

THEOREM 1. *Under the above assumption, we can find a neighborhood* $\mathfrak{U}$ *of the identity map in the group of diffeomorphisms of* M *(with Soboler* $\|_k$*-norm topology for sufficiently large* k*) and a neighborhood* T_1 *of* t_0 *in* T *such that, for any* $f \in \mathfrak{U}$ *and* t, t' *in* T_1, f *is not a holomorphic map of* M_t *to* $M_{t'}$.

1. Definition of complex analytic families

Let S be an analytic space. Take a holomorphic map ρ of an analytic space $\mathfrak{V}$ to S. We say that ρ is locally trivial if and only if the following conditions are satisfied. For any $s \in S$ and $x \in \rho^{-1}(s)$, we can find an open neighborhood $\mathfrak{U}$ of x in $\mathfrak{V}$, an open neighborhood W of s in S, a domain U of $\mathbf{C}^n$, and an analytic homeomorphism $\zeta: \mathfrak{U} \to U \times W$ such that $\rho(y) = \rho_w(\zeta(y))$ for all y in $\mathfrak{U}$, where ρ_w is the projection of W of $U \times W$. Such ζ will be called a chart of the locally trivial map ρ. Since ρ is holomorphic, $\rho^{-1}(s)$ is an analytic space. If ρ is locally trivial, $\rho^{-1}(s)$ is a complex analytic manifold, because U in the above is a domain in $\mathbf{C}^n$. Moreover, $\zeta_s: \mathfrak{U} \cap \rho^{-1}(s) \to U$ defined by $\zeta(y) = \zeta_s(y) \times s$ is a chart of the complex analytic manifold $\rho^{-1}(s)$.

Definition 1. *By a complex analytic family of compact complex structures over S (or with parameter space S) we mean a triple $(\mathfrak{V}, S, \rho)$ where ρ is a proper and locally trivial holomorphic map $\mathfrak{V} \to S$.*

Let M be a complex analytic structure on a compact C^∞ manifold $\mathbf{M}$. Let s_0 be a point S.

Definition 2. *By a complex analytic family of deformations of M over S with reference point s_0, we mean quadruple (V, S, ρ, γ_0) where (V, S, ρ) is a complex analytic family of compact complex structures and where γ_0 is a holomorphic homeomorphism of M to $\rho^{-1}(s_0)$.*

Let $\rho: \mathfrak{V} \to S$ be a proper and locally trivial holomorphic map. Assume that $\rho^{-1}(s_0)$ is compact. Then we can find a finite number of charts of ρ, say ζ_j with domain $\mathfrak{U}_j$, such that $\bigcup \mathfrak{U}_j \supseteqq \rho^{-1}(s_0)$. We claim that $\bigcup \mathfrak{U}_j \supseteqq \rho^{-1}(s)$ for s sufficiently near s_0. To see this, take $\mathfrak{V}_j \subset \mathfrak{U}_j$ such that $\bar{\mathfrak{V}}_j \subset \mathfrak{U}_j$ and such that $\bigcup V_j \supseteqq \rho^{-1}(s_0)$. If our contention is not true, the assumption that ρ is proper implies that we can find a sequence y_r in $\mathfrak{V}$ outside of $\bigcup \mathfrak{U}_j$ which converges to a point $y \in \rho^{-1}(s_0)$. Then y must belong to $\mathfrak{V}_j$ for an index j. Hence y_r for large r must belong to $\mathfrak{U}_j \supset \bar{\mathfrak{V}}_j$. This contradicts our choice of the sequence y_r.

PROPOSITION 1. *Let $(\mathfrak{V}, S, \rho, \gamma_0)$ be a complex analytic family of deformations of M over S with reference point s_0. Then there is a bijective map $F: \mathbf{M} \times S \to \mathfrak{V}$ (provided we replace S by a neighborhood of s_0) satisfying the following*

(1) $F(y, s_0) = \gamma_0(s_0)$,

(2) *for each s, $F \mid \mathbf{M} \times s$ is a diffeomorphism γ_s of $\mathbf{M}$ to the C^∞ manifold $\rho^{-1}(s)$,*

(3) *when we denote by* M_s *the complex analytic structure on* M *which makes* γ_s *a holomorphic map,* $\{M_s: s \in S\}$ *is a complex analytic family of deformations of* M *according to the definition in* [3].

PROOF. Take finite number of charts of ρ, say ζ_j with domain $\mathfrak{U}_j$, such that $\bigcup \mathfrak{U}_j = \mathfrak{V}$ (replacing S be a small neighborhood of s_0 if necessary). Set $V_j(s) = \mathfrak{U}_j \cap \rho^{-1}(s)$ and $U_j(0) = U_j(s_0)$. Write $\zeta_j(y) = (z_j(y), \rho(y))$ where $z_j: \mathfrak{U}_j \to U_j$. z_j restricted to $V_j(s) \times s$ gives a chart $z_j(s): U_j(s) \to U_j$ of $\rho^{-1}(s)$. Then ζ_j induces a diffeomorphism $k_j(s): U_j(s) \to U_j(0)$ defined by the formula $(z_j(s_0)) \circ (k_j(s)) = z_j(s)$. The collection $\{k_j(s)\}$ of course may not match together; i.e., $k_j(s)$ and $k_i(s)$ may not be equal on $U_i(s) \cap U_j(s)$. However, we can construct a diffeomorphism by means of them. Note that $k_i(s_0)$ is the identity map.

Take an imbedding h_0 of $\rho^{-1}(s_0)$ in a euclidean space $\mathbf{R}^m$ for sufficiently large m. Let $\mathbf{M}'$ be the image submanifold. Take a small tubular neighborhood A of $\mathbf{M}'$ and a C^∞ map $a: A \to \mathbf{M}'$ such that $a \mid \mathbf{M}'$ is the identity map. Let $\{f_j\}$ be a partition of unity with respect to the covering $\{\mathfrak{U}_j\}$ of V. Set

$$h(y) = \textstyle\sum_j f_j(y) h_0((k_j(s))(y)) \qquad (y \in \rho^{-1}(s))$$

where the sum is taken in $\mathbf{R}^m$. $h(y)$ is in the convex hull determined by $h_0(k_j(s)(y))$ for all j with $y \in \mathfrak{U}_j$. The distances between them tend to zero when s tend to s_0, because $k_j(s_0)$ is the identity map. Hence, $h(y)$ will be in A if s is sufficiently close to s_0. Therefore, we have a C^∞ map $a \circ h: \mathfrak{V} \to M'$. $a \circ h$ restricted to $\rho^{-1}(s_0)$ is equal to h_0. Hence, $k = k_0^{-1} \circ a \circ h: \mathfrak{V} \to \rho^{-1}(s_0)$ is of class C^∞ and the restriction to $\rho^{-1}(s_0)$ is the identity map. By observing the jacobian, we see easily that k restricted to $\rho^{-1}(s)$ gives a diffeomorphism $\rho^{-1}(s) \to \rho^{-1}(s_0)$. Denote by γ'_s the inverse of the map and set $\gamma_s = \gamma'_s \circ \gamma_0$. The collection $\{\gamma_s\}$ induces $\mathbf{F}: M \times S \to \mathfrak{V}$. It is clear that F satisfies our conditions (1) and (2). Since γ_s is constructed from charts ζ_j it is easy to see that our condition (3) is also satisfied.

2. Automorphisms of families

We denote by A^p the vector space of C^∞ differential forms of type $(0, p)$ with values in the vector bundle of tangent vectors of type $(1, 0)$ with respect to the complex structure M. If ω is in A^1, we define $\bar\partial_\omega: A^p \to A^{p+1}$ by

$$\bar\partial_\omega(\theta) = \bar\partial\theta + 2[\omega, \theta] .$$

w also represents an almost complex structure M_ω. We denote by A^p_ω the vector space as above with respect to M_ω. If, moreover, ω is sufficiently close to o, we have an isomorphism $\mu_\omega\colon A^p \to A^p_\omega$ such that μ_ω corresponds to $\bar\partial$ with respect to M_ω ([4, 1, Ch. II]). Moreover, μ_ω depends infinitely differentiably on ω and μ_0 is the identity map. For $\zeta \in A^1$ and sufficiently small, we set

$$\mathbf{e}_\omega(\zeta) = \exp\big(\mu_\omega(\zeta) + \overline{\mu_\omega(\zeta)}\big)$$

where exp is the exponential map in differential geometry (not that of Riemann geometry). $\mathbf{e}_\omega(\zeta)$ is a diffeomorphism of $\mathbf{M}$.

$\bar\partial_\omega$ is not exact unless ω satisfies the integrability condition, but we can still consider its formal adjoint $\mathfrak{S}_\omega$ and Laplace operator $\Box_\omega = \bar\partial_\omega\delta_\omega + \delta_\omega\bar\partial_\omega$. Let λ be the smallest non-zero eigenvalue of $\Box_0 = \Box$ on A^0. Denote by $H^\sharp_\omega$ the sum of the eigenspaces of $\Box_\omega$ on A^0 with eigenvalues $< \lambda$. $H^\sharp_\omega$ is finite dimensional. Then, on a neighborhood of 0, $H^\sharp_\omega$ depends differentiably on ω when ω is restricted to a finite dimensional submanifold of A^1 ([2, Th. 3, p. 48]). If ω satisfies the integrability condition, and if dimension of the holomorphic automorphism group of M_ω is equal to that of M, then $H^\sharp_\omega = H^0_\omega$, the space of harmonic elements of A^0 with respect to $\Box_\omega$. Hence, $\mu_\omega(H^\sharp_\omega)$ is the space of holomorphic vector fields on M_ω and $\mathbf{e}_\omega(\zeta)$ cover a neighborhood of the identity in the group of holomorphic automorphism of M_ω.

We recall that we defined $\mathbf{e}(\xi)$ for $\xi \in A^0$, and $\omega \circ f$ (f being a diffeomorphism of $\mathbf{M}$), in [3]. e is an isomorphism of a neighborhood of origin in A^0 with a neighborhood of the group of diffeomorphisms of $\mathbf{M}$, both with Sobolov $||_k$-norm topology for large k. In the following, we always consider A^0 with the above topology. $e(\xi)$ is the exponential of $\xi + \bar\xi$ in Riemann geometry. Denote by H^p the space of harmonic elements with respect to $\Box$ on A^p. For sufficiently small $t \in H^1$, we denote by $\varphi(t)$ the element in A^1 constructed in [3]. We define F_t: a neighborhood of zero in $H^*_{\varphi(t)} \oplus {}^\perp A^0 \to A^0$, where ${}^\perp A^0$ is the orthogonal complement of H^0 in A^0, as follows:

$$e\big(F_t(\zeta, \xi)\big) = \mathbf{e}_{\varphi(t)}(\zeta) \circ e(\xi) .$$

By means of standard arguments involving Sobolev $||_k$-norm, we see easily that F_t is a differentiable map, depending differentiably in t. $H^\sharp_\omega \oplus {}^\perp A^0$ can be naturally identified with A^0. Under this identification, it is clear that dF_0 at the origin is the identity map. Therefore, by the implicit function theory, we have the following.

PROPOSITION 2. *Let* $\mathbf{W}$ *be a neighborhood of origin in* ${}^{\perp}A^0$. *Then we can find a neighborhood* $\mathbf{V}$ *of the identity in the group of diffeomorphism of* $\mathbf{M}$ *and a neighborhood* Y *of origin in* H^1 *such that, for any* f *in* $\mathbf{V}$ *and* t *in* Y, *we can write*

$$f = \mathbf{e}_{\varphi(t)}(\zeta) \circ e(\xi)$$

for suitable ξ *in* $\mathbf{W}$ *and* $\zeta \in H^{\sharp}_{\varphi(t)}$

Theorem 1 (announced at the end of introduction) is a corollary of Proposition 2. Namely, take $\mathbf{W}$ in Proposition 2 such that Proposition in [3] holds. Assume that $\varphi(t_2) = \varphi(t_1) \circ f$ with $f \in \mathbf{V}$. Write $f = h \circ e(\xi)$, where $h = \mathbf{e}_{\varphi(t_1)}(\zeta)$ is holomorphic with respect to $M_{\varphi(t_1)}$ and ξ in $\mathbf{W}$. Then $\varphi(t_2) = \varphi(t_1) \circ f = \varphi(t_1) \circ e(\xi)$. Hence by Proposition in [3], $\varphi(t_2) = \varphi(t_1)$.

COLUMBIA UNIVERSITY

REFERENCES

[1] A. DOUADY, Le problem des modules pour les variétés analytiques complexes, Seminaire Bourbaki, 1964/65, §277.

[2] K. KODAIRA and D. C. SPENCER, *On deformations of complex analytic structures* III, Ann. of Math., **71** (1960) 43-76.

[3] M. KURANISHI, "New proof for existence of locally complete families of complex structures", in Proceeding of the Conference on Complex Analysis, Minneapolis, 1964, 142-154, Springer, Verlag, Berlin, 1965.

[4] ———, *On a type of family of complex structures*, Ann. of Math. **74** (1961), 262-328.

[5] J. WAVRIK, Obstructions to the existence of a space of Moduli, (in this volume).

(Received September 27, 1968)

A survey of some results on complex Kähler manifolds

By JAMES A. MORROW

One of the first questions that one asks about a given complex manifold is "What are the other complex structures on the underlying differentiable manifold?" The whole theory of classification of (complex) surfaces is motivated by this question, and the Kodaira-Spencer theory of deformations can be thought of as a study of the complex structures close to a given one. Sometimes one has uniqueness (e.g. S^2), and it is conjectured that $\mathbf{P}^n$ has only one complex structure. Emery Thomas [9] has computed the total Chern class for all possible almost complex structures on $\mathbf{P}^n$, $1 \leqq n \leqq 4$. Kodaira and Hirzebruch [7] have proved that the only Kähler complex structure on $\mathbf{P}^n$, n odd, is the standard one. More recently the author discovered a similar result for quadrics which, in fact, had already been published by Brieskorn [3] some years ago. It is conjectured that a similar result holds for the (complex) grassmannians. We intend to give here a statement and outline of proof for the results of Kodaira-Hirzebruch and Brieskorn, and an indication of what troubles arise for the grassmannians.

We give the statement of the theorem of Kodaira-Hirzebruch.

THEOREM 1. *Let X be an n-dimensional compact Kähler manifold, such that X is homeomorphic to $\mathbf{P}^n$. If n $(= \dim X)$ is odd, then X is complex analytically equivalent to $\mathbf{P}^n$. Let $g \in H^2(X, \mathbf{Z})$ generate $H^2(X)$ and be such that g belongs to the cohomology class of a Kähler metric on X. Then if n is even and $c_1(X) \neq -(n+1)g$, $X = \mathbf{P}^n$.*

We next give the statement of the theorem of Brieskorn and Van de Ven (unpublished proof by the latter author).

THEOREM 2. *Let X be a compact complex Kähler manifold which is homeomorphic to a (n-complex dimensional) quadric $\mathbf{Q}_n$* of $\mathbf{P}^{n+1}$. *We suppose $n > 2$. Then*

(i) *if n is odd $X = \mathbf{Q}_n$ complex analytically;*

(ii) *if n is even $(n \neq 2)$,*

let $g \in H^2(X, \mathbf{Z})$ $(\cong \mathbf{Z})$ *belong to the class of a Kähler metric on* X *and generate* $H^2(X, \mathbf{Z})$. *Then* $c_1(X) = \pm ng$; *and if* $c_1(X) = ng$, *then* $X = \mathbf{Q}_n$ *complex analytically.*

We should remark that the Hirzebruch surfaces give counter-examples for the case $n = 2$ (Hirzebruch [6]). He has constructed infinitely many different complex structures on $S^2 \times S^2$ each of which is projective algebraic and hence Kähler. We also notice that $H^2(S^2 \times S^2, \mathbf{Z}) \cong \mathbf{Z} \oplus \mathbf{Z}$ and hence the condition makes no sense as stated. In fact using the techniques of Borel and Hirzebruch [2] one can show that all of the Hirzebruch structures have the same Chern class; so we can not distinguish between them with c_1.

Before we sketch the result of Kodaira and Hirzebruch, we remark that they assumed X was diffeomorphic to $\mathbf{P}^n$, an assumption which is no longer necessary, by using the result of Novikov [8] established in 1965, which states that the rational Pontrjagin classes are invariant under homeomorphisms and the fact that $\mathbf{P}^n$ has no torsion. Similarly $\mathbf{Q}_n$ has no torsion so the Brieskorn result can be stated under the assumption of homeomorphism. As a first step in the proof of Theorem 1, we prove

THEOREM 3. *Let* X *be a compact Kähler manifold homeomorphic to* $\mathbf{P}^n$. *If* $g \in H^2(X, \mathbf{Z})$ *is the positive generator, then there is a holomorphic divisor* D *on* X *such that* $c(E) = g$ *where* $E = [D]$ *is the bundle of the divisor* D *and* $c(E)$ *is the (first) Chern class of* E. *Then*

(1) *if* n *is odd,* $c_1(X) = (n+1)g$,

(2) *if* n *is even,* $c_1(X) = \pm(n+1)g$,

(3) *if* $c_1(X) = (n+1)g$, and $E^s = E \otimes \cdots \otimes E$ s-times, *then we have*

$$\dim H^\circ(X, E^s) = \binom{n+s}{n}$$

PROOF. We remark that X is automatically algebraic since it is Kähler and $H^2(X, \mathbf{Z}) \cong \mathbf{Z}$ implies that X has positive line bundles. For the proof we recall the Hirzebruch Riemann-Roch formula,

$$(1) \qquad \chi(X, D) = \Bigl(\sum\nolimits_{r+2s=n} \frac{\left(d + \frac{1}{2} c_1\right)^r}{r!} \hat{A}s(p_1, \cdots, p_s)\Bigr)[X] \,,$$

where D is a divisor, $d = c(D)$, $c_1 = c_1(X)$, the p_i are the Pontrjagin classes of X, $\chi(X, D)$ is the analytic characteristic

$$\chi(X, D) = \sum_{i=0}^{n}(-1)^i \dim H^i(X, D) ,$$

and the $\hat{A}_s$ are a multiplicative sequence of polynomials corresponding to the function $1/2\sqrt{z}/\sinh(\sqrt{z}/2)$ (Hirzebruch [5]). To prove (1) and (2), we show that the Todd genus $T(X)$ of X is 1. But from (1) we get

$$(2) \qquad T(X) = \sum_{r+2s=n} \frac{\left(\frac{1}{2}c_1\right)^r}{r!} \hat{A}_s(p_1, \cdots, p_s)[X] .$$

Since the p_i are invariants, we try to solve this equation for c_1, obtaining (1) and (2).

First let us show that $T(X) = 1$. Let $h^{p,q}$ be the dimension of the space of harmonic forms of type (p, q) and b_r be the rth Betti number. Then

$$(3) \qquad \dim H^k(X, \theta) = g_k = h^{k,0} = h^{0,k} ,$$

where θ is the sheaf of germs of holomorphic functions, and

$$(4) \qquad b_r = \sum_{p+q=r} h^{p,q} .$$

Then it is easy to see that $b_{2p} = 1$, $b_{2p+1} = 0$, and $h^{p,p} \geqq 1$ implies that $g_k = 0$ for $k \neq 0$ and clearly $g_0 = 1$. Hence $T(X) = 1$.

Next we let h be the positive generator of $H^2(\mathbf{P}^n, \mathbf{Z})$. Then it is easy to see (using Novikov's result) that the polynomials $\hat{A}_s(p_1(X), \cdots, p_s(X))$, which are certain polynomials in g^2, are the same as the polynomials $\hat{A}_s(p_1(\mathbf{P}^n), \cdots, p_s(\mathbf{P}^n))$ which are certain polynomials in h^2. We also find, using the fact that the second Stiefel-Whitney class w_2 is a topological invariant, that

$$(5) \qquad c_1(X) = \lambda g \qquad \text{with } \lambda \equiv n + 1 \pmod 2 .$$

Using (1) we get

$$(6) \qquad 1 = e^{\lambda/2\lambda g} \cdot \hat{A}(g)[X] ,$$

where

$$\hat{A}(g) = \sum_{k=0}^{\infty} \hat{A}_k(p_1(X), \cdots, p_k(X)) .$$

Let $\lambda = 2s + n + 1$. To evaluate the right side of (6) we need to compute the coefficient of g^n in $e^{(s+(n+1)/2)g} \cdot \hat{A}(g)$. This can be done by using the Riemann-Roch theorem for $\mathbf{P}^n$. Then (6) becomes

$$(7) \qquad 1 = \binom{n+s}{n} ,$$

where

$$\binom{n+s}{n} = \frac{(n+s)\cdots(1+s)}{n(n-1)\cdots 2\cdot 1}.$$

We conclude that $s = 0$ if n is odd and $s = 0$ or $-(n+1)$ if n is even, thus yielding parts (1) and (2) of Theorem 3. To find a holomorphic divisor D such that $c(D) = g$ we notice that every element of $H^2(X, \mathbf{Z})$ is of type (1, 1), hence occurs as the Chern class of a bundle; and X algebraic implies that every bundle is the bundle of a divisor. Then part (3) easily follows by applying (1) and the Kodaira vanishing theorem, q.e.d.

Now we want to use D to construct a candidate for an analytic homeomorphism onto $\mathbf{P}^n$. As before, let $E = [D]$. Then if $\Gamma(E^\nu)$ is the space of holomorphic sections of E^ν we have $\binom{n+\nu}{n} = \dim \Gamma(E^\nu)$ so $\dim \Gamma(E) = n + 1$. Let $\{\varphi_0, \cdots, \varphi_n\}$ be a basis for $\Gamma(E)$. Then the map $\varphi: X \rightarrow \mathbf{P}^n$ given by $\varphi(z) = (\varphi_0(z), \cdots, \varphi_n(z))$ is meromorphic, but may not be everywhere holomorphic.

LEMMA 1. *$\varphi(X)$ is a (closed) subvariety of dimension n in $\mathbf{P}^n$ and hence $\varphi(X) = \mathbf{P}^n$, where by $\varphi(X)$ we mean the closure of the image of the points where φ is well-defined.*

PROOF. We give a brief sketch and refer to Kodaira-Hirzebruch [7] if the reader desires more details. Let $\lambda \in \mathbf{P}_1^n$ (another copy of $\mathbf{P}^n$). Then the zeroes of the section $\lambda_0\varphi_0 + \cdots + \lambda_n\varphi_n$, $\lambda = (\lambda_0, \cdots, \lambda_n)$ define a divisor D_λ on X with $c(D_\lambda) = g$. We can describe D_λ as $\varphi^{-1}(H_\lambda)$ where H_λ is the hyperplane $H_\lambda = \{z \in \mathbf{P}^n \mid z_0\lambda_0 + \cdots + z_n\lambda_n = 0\}$ of $\mathbf{P}^n$. Let $B = \bigcap \{z \in D_\lambda \mid \lambda \in \mathbf{P}_1^n\}$. Then $f = \varphi$ restricted to $X - B$ is a holomorphic map and $\varphi(X) = \overline{f(X - B)}$ where the bar denotes topological closure. Let $r = \dim \varphi(X)$. Then we want to show that a general linear space of dimension $d = n - r$ cuts $\varphi(X)$ in one point. For this we need a preliminary lemma.

LEMMA 2. *Let $D_1, \cdots, D_s$ be effective divisors on X with $c(D_i) = g$. Then the D_i are irreducible and their intersection $D_1 \cap \cdots \cap D_s$ is irreducible and hence connected.*

PROOF. We first notice that if $W = \sum m_i\theta_i$ with the θ_i irreducible varieties and $m_i > 0$, with $c(W) = g^r$ where $c(W)$ is the Poincaré dual of W and $\dim W = n - r$, then W must be an irreducible variety of X. This follows by noticing $c(\Theta_i) = k_i g^r$ for some $k_i > 0$ and hence $g^r = \sum m_i k_i g^r$. This is only possible when $W = \Theta_1$ (say). This already proves that each D_i is irreducible. The second statement is proved by induction on s. We have done $s = 1$. Assume the result for $l = s$, and let $W^{(l)} = D_1 \cap \cdots \cap D_l$ with $r' = n - \dim W^{(l)}$. Then if $W^{(l)} \supseteq D_{l+1}$, we are done. If not $W^{(l)} \not\subset D_{l+1}$, and then $D_{l+1} \not\subset W^{(l)}$; since otherwise $W^{(l)} = D_{l+1}$ because both are irreducible and $\dim D_{l+1} = n - 1$ and this is a contradiction. But then $W^l \cap D_{l+1} = W^l \cdot D_{l+1}$ (intersection of cycles) and

$$c(W^l \cap D_{l+1}) = c(W^l \cdot D_{l+1}) = g^{r'} \cdot g = g^{r'+1} .$$

Thus $W^{(l+1)} = W^l \cdot D_{l+1}$ is irreducible by the first part of the proof. q.e.d.

We return to the proof of Lemma 1. First of all notice that $\varphi_0, \cdots, \varphi_n$ linearly independent implies that $\varphi(X)$ is not contained in any hyperplane. Let $x \in X - B$ be a point such that $f(x) \in \varphi(X)$ is a manifold point. Let $r = \dim \varphi(X)$. Let T be the tangent space to $\varphi(X)$ through $f(x)$ considered as a linear subspace of $\mathbf{P}^n$. We will show $f(X - B) \subseteqq T$ and hence $\varphi(X) \subseteqq \mathrm{T}$. This together with the remark just made tells us that $\dim \varphi(X) = n$. Assume there is a point $y \in X - B$ such that $f(y) \notin T$. Then there is a linear subspace L of $\mathbf{P}^n$ which has dimension $n - r$ passing through $f(x)$ and $f(y)$ which is transversal to T. So $f(x)$ is an isolated point of $f(X - B) \cap L$. Let L be defined by the intersection of hyperplanes $H_1, \cdots, H_r$. Let $D_i = \varphi^{-1}(H_i)$ and $W = D_1 \cap \cdots \cap D_r$. We have assumed x and y are not in B and clearly $x \in W$, $y \in W$. Thus $B \subset W$, $B \neq W$. Since W is irreducible $\dim B < \dim W$, and $W - B$ is connected. This implies $f(W - B) \subseteqq L \cap f(X - B)$ is connected. Thus there is an arc in $L \cap f(X - B)$ connecting $f(x)$ and $f(y)$. This contradicts $f(x)$ being isolated in $f(X - B) \cap L$. q.e.d.

We can now complete the proof of Theorem 1. Let $\mathscr{H}_\nu$ be the (vector) space of homogeneous polynomials $p(z_0, \cdots, z_n)$ of degree ν in $n + 1$ variables. Let $\{p_0, \cdots, p_N\}$ be a basis for $\mathscr{H}_\nu$ where

$$(8) \qquad N + 1 = \dim H_\nu = \binom{n + \nu}{n} .$$

Then the map $P_\nu : \mathbf{P}^n \to \mathbf{P}^N$ given by $P_\nu(z) = (p_0(z), \cdots, p_N(z))$ is an embedding. To each p_α we associate a section ψ_α of E^ν given by

$$(9) \qquad \psi_\alpha(x) = p_\alpha(\varphi_0(x), \cdots, \varphi_n(x)) \qquad x \in X .$$

We claim $\{\psi_0, \cdots, \psi_N\}$ is a basis for $\Gamma(E^\nu)$. This follows from the fact that $\overline{f(X - B)}$ is equal to $\mathbf{P}^n$. Then since $c(E^\nu) = \nu g > 0$, we know that for large enough ν the map $\psi_\nu : X \to \mathbf{P}^N$ given by $\psi_\nu(x) = (\psi_0(x), \cdots, \psi_N(x))$ is an embedding (by the Kodaira embedding theorem). This already proves $B = \emptyset$ and φ is everywhere defined. We have the following commuting diagram

$$\begin{array}{ccc} X & \xrightarrow{\varphi} & \mathbf{P}^n \\ & \searrow^{\psi_\nu} & \downarrow P_\nu \\ & & \mathbf{P}^N \end{array}$$

The maps P_ν and ψ_ν are embeddings. Hence φ is an analytic homeomorphism (we already know $\varphi(X) = \mathbf{P}^n$), q.e.d.

We proceed with the proof of Theorem 2. The reader will see that its proof is quite similar to the proof of Theorem 1. We shall be quite brief, stressing only those parts of the proof of Theorem 1 which need essential modification, and referring the reader to Brieskorn's paper for more details. The cohomology ring of a quadric $\mathbf{Q}_n$ is only slightly more complicated than that of $\mathbf{P}^n$. As a group $H^*(\mathbf{Q}_n)$ can be computed easily by using the Lefschetz theorem on hyperplane sections and by knowing the Euler characteristic of $\mathbf{Q}_n$. If one wants to find the structure of $H^*(\mathbf{Q}_n)$ as a ring, one should notice that as a differentiable manifold,

$$\mathbf{Q}_n = SO(n + 2, \mathbf{R})/SO(n, \mathbf{R}) \times SO(2, \mathbf{R}) , \tag{10}$$

where $SO(m, \mathbf{R})$ is the special orthogonal group in m real variables. Using this, one could use the methods of A. Borel [1] to compute $H^*(\mathbf{Q}_n, K_p)$ where K_p is a field of characteristic p and $p \neq 0, 2$. However it seems that to find $H^*(\mathbf{Q}_n, \mathbf{Z})$ one needs to refer to the results of Ehresmann [4]. To compute Pontrjagin classes we can use formula (10) and the techniques of Borel-Hirzebruch [2]; in fact it is done on page 525 of their paper. From the Ehresmann paper we need the following facts. First we have the cohomology of $\mathbf{Q}_n$, $n = 2k + 1$. As a group it is the same as $H^*(\mathbf{P}^n, \mathbf{Z})$ with generators $e_i \in H^{2i}(\mathbf{Q}_n, \mathbf{Z})$ whicn are the Poincaré duals of algebraic subvarieties of $\mathbf{Q}_n$ and $g = e_1$ is a positive class (in the sense of Kodaira). Let juxtaposition denote cup product. Then we have

$$\begin{aligned} ge_j &= e_{j+1} && \text{if } j > k, \text{ or } j < k - 1 \\ g^j &= 2e_j && \text{if } j > k . \end{aligned} \tag{11}$$

In particular,

$$g^k e_{k+1} = g^{k-1} e_{k+2} = \cdots = e_{2k+1} = e_n \, . \tag{12}$$

For $\mathbf{Q}_n$, $n = 2k$ we have $H^{2i}(\mathbf{Q}_n, \mathbf{Z})$ generated by e_i if $i \neq k$; and $H^{2k}(\mathbf{Q}_n, \mathbf{Z})$ which is isomorphic to $\mathbf{Z} \oplus \mathbf{Z}$ has generators e_k, f_k. Again all generators are duals of algebraic subvarieties of $\mathbf{Q}_n$, and $g = e_1$ is a positive class. For the cup product we have the relations

$$\begin{cases} ge_j = e_{j+1} & \text{if } j > k, \text{ or } j < k-1 \\ g^j = 2e_j & \text{if } j > k \\ g^k = e_k + f_k \, . \end{cases} \tag{13}$$

In strict analogy to the proof of Theorem 1 we first prove the following theorem.

THEOREM 4. *Let X be a compact Kähler manifold homeomorphic to $\mathbf{Q}_n$ where $n > 2$. If $g \in H^2(X, \mathbf{Z})$ is the positive generator, then there is a holomorphic divisor D on X such that $c(E) = g$ where $E = [D]$. Then*

(1) *if n is odd $c_1(X) = ng$,*

(2) *if n is even $c_1(X) = \pm\, ng$,*

(3) *if $c_1(X) = ng$ then*

$$\dim H^\circ(X, E^s) = \binom{n+s+1}{n+1} - \binom{n+s-1}{n+1}$$

PROOF. The proof is similar to that of Theorem 3. We give only the briefest of sketches. The existence of the holomorphic divisor D is clear. In computing the Pontrjagin classes one finds, again, that they depend on g^2. The formula

$$c_1(X) = (2s + n)g \qquad \text{for some } s \in \mathbf{Z}\, , \tag{14}$$

is again verified. Using the Riemann-Roch formula we find

$$1 = \binom{n+s+1}{n+1} - \binom{n+s-1}{n+1} \, . \tag{15}$$

This equation depends on knowing that $T(X)$ (the Todd genus of X) is 1, a fact which is easily verified because of the simplicity of $H^*(X, \mathbf{Z})$. As before, using (15) we get parts (1) and (2). Part (3) follows from the Riemann-Roch formula and the Kodaira vanishing theorem, q.e.d.

We find a candidate for embedding X into $\mathbf{P}^{n+1}$ by using the sections of E. Let $\{\varphi_0, \cdots, \varphi_{n+1}\}$ be a basis for $\Gamma(E)$. We get a meromorphic map $\varphi\colon X \to \mathbf{P}^{n+1}$.

LEMMA 3. *$\varphi(X)$ is a (closed) subvariety of dimension n of order 2 in $\mathbf{P}^{n+1}$ (where by $\varphi(X)$ we mean the closure of image of the set of points where φ is well defined).*

PROOF. We use the notation of Lemma 1. We first remark that if A is an analytic cycle of dimension l of X and $c(A)$ is its Poincaré dual, then $g^l c(A)[X] > 0$ since

$$0 < \int_A g^l = \int_X g^l c(A) = g^l c(A)[X] .$$

Then it follows (as in the proof of Lemma 2) that if A is an effective l-dimensional cycle (and assuming $l \neq k$ when $n = 2k$) with $c(A) = ae_{n-l}$, that A has at most a components. Then for the proof of Lemma 3 we need another result.

LEMMA 4. *Let $D_1, \cdots, D_l$ be effective divisors on X with $c(D_i) = g$. Then the D_i are irreducible and $D_1 \cap \cdots \cap D_l$ has at most 2 connected components.*

PROOF. We notice that the D_i are irreducible by the remark. We have thus begun a proof by induction on l. We might as well assume that each D_i contributes to $D_1 \cap \cdots \cap D_l$; e.g., $D_1 \cap \cdots \cap D_r \cap D_{r+1} \neq D_1 \cap \cdots \cap D_r$. First assume $D_1 \cap \cdots \cap D_r$ has exactly one connected component with multiplicity 1, $r \leqq k$, and $c(D_1 \cap \cdots \cap D_r) = g^r$. If $D_1 \cap \cdots \cap D_{r+1} = \varnothing$ we are done. If not, $c(D_1 \cap \cdots \cap D_{r+1}) = g^{r+1}$. If it is irreducible we have the induction. If not, the only way it can have several components is for $n = 2k + 1$, $r = k$, and, $g^{r+1} = 2e_{r+1}$; or $n = 2k$, $r = k - 1$, and $g^{r+1} = e_k + f_k$. Consider the first case first. Then $D_1 \cap \cdots \cap D_{r+1}$ has exactly 2 components E and F with $c(E) = c(F) = e_{r+1}$. Then we can see that $D_1 \cap \cdots \cap D_{r+2}$ has exactly two components and so on up to $D_1 \cap \cdots \cap D_l$. For

$$D_1 \cap \cdots \cap D_{r+2} = E \cap D_{r+2} + F \cap D_{r+2} ,$$

with $c(E \cap D_{r+2}) = c(F \cap D_{r+2}) = e_{r+2}$; so each term is irreducible. Since $c(E \cap D_{r+2} \cap D_{r+3}) = e_{r+3}$, etc., it follows that $D_1 \cap \cdots \cap D_l$ has exactly two components. In the second case suppose $D_1 \cap \cdots \cap D_{r+1} = \sum k_i G_i = W$ with the G_i irreducible and $k_i > 0$. Then $c(W) = g^k$ so

$$g^k c(W) = g^{2k} = \sum k_i g^k c(G_i) . \tag{16}$$

Then

$$(17) \qquad 2 = \sum k_i g^k c(G_i)[X]\ ,$$

and by a previous remark $g^k c(G_i)[X] > 0$. Thus (17) is only possible when $D_1 \cap \cdots \cap D_{r+1} = G_1 + G_2$ (say), and $g^k c(G_1) = g^k c(G_2) = e_{2k} = e_n$. It is easy to check that this implies $gc(G_1) = gc(G_2) = e_{k+1}$. Thus $D_{r+2} \cap G_1$ and $D_{r+2} \cap G_2$ are irreducible with $c(D_{r+2} \cup G_1) = e_{k+1}$, etc. Thus continuing, and repeating previous arguments, we see again that $D_1 \cap \cdots \cap D_l$ has at most 2 irreducible components, q.e.d.

We return to the proof of Lemma 3. We define B as in the proof of Lemma 1. Let $d = \dim \varphi(X)$. We know that an irreducible variety of order 1 in $\mathbf{P}^{n+1}$ of dimension d is a $\mathbf{P}^d$. Since $\varphi_0, \cdots, \varphi_{n+1}$ are independent and $d \leqq n$, we know $\varphi(X)$ is not of order 1. Suppose the order of $\varphi(X)$ is $\geqq 3$. Then we can find a linear subspace L of dimension $n + 1 - d$ which cuts $\varphi(X - B)$ in at least three isolated points p_1, p_2, p_3. Then $L = H_1 \cap \cdots \cap H_d$ with the H_1 hyperplanes in $\mathbf{P}^{n+1}$. So $D_i = \varphi^{-1}(H_i)$ is a divisor on X with $c(D_i) = g$. Then $D_1 \cap \cdots \cap D_d$ has at most two connected components. Let $p'_1, p'_2, p'_3 \in D_1 \cap \cdots \cap D_d - B$ be such that $\varphi(p'_i) = p_i$. Since $D_1 \cap \cdots \cap D_d - B \neq \phi$, $D_1 \cap \cdots \cap D_d - B$ has at most two components. Thus two of the points, p'_1, p'_2, say, belong to a connected component of $D_1 \cap \cdots \cap D_d - B$ and thus p_1, p_2 are connected in $L \cap \varphi(X - B)$. This contradicts p_1, p_2, p_3 being isolated, proving our lemma, q.e.d.

At last we return to the proof of Theorem 2. We know that any irreducible subvariety of $\mathbf{P}^{n+1}$ of order 2 and dimension d is contained in some linear space H of dimension $d + 1$. Since the φ_i are independent, $\varphi(X)$ is an irreducible (possibly singular) quadric *of dimension* n in $\mathbf{P}^{n+1}$. From now on our proof proceeds as in Theorem 1. Again we use polynomials to construct embeddings $P_\nu\colon \mathbf{P}^{n+1} \to \mathbf{P}^N$, where here $\mathscr{H}_\nu$ is the space of homogeneous polynomials of order ν in $n + 2$ variables, and $N + 1 = \dim \mathscr{H}_\nu$. Then if $p \in \mathscr{H}_\nu$, $p(\varphi_0, \cdots, \varphi_{n+1}) \in \Gamma(E^\nu)$; and $p(\varphi_0, \cdots, \varphi_{n+1}) \equiv 0$ if and only if p vanishes on $\varphi(X)$. Thus

$$\begin{aligned} \dim \{p(\varphi_0, \cdots, \varphi_{n+1}) \mid p \in \mathscr{H}_\nu\} &= \binom{n + \nu + 1}{n + 1} - \binom{n + \nu - 1}{n + 1} \\ &= m + 1\ . \end{aligned}$$

Then choose a basis $\{p_0, \cdots, p_m, \cdots, p_N\}$ for $\mathscr{H}_\nu$ such that $\{p_1(\varphi(x)), \cdots, p_m(\varphi(x))\}$ gives a basis for $\Gamma(E^\nu)$, and $p_{m+1}, \cdots, p_N$ are identically zero on $\varphi(X)$. Then we define $\Phi_\nu\colon X \to \mathbf{P}^m \subseteq \mathbf{P}^N$ by

$$\Phi_\nu(x) = (p_1(\varphi(x)), \cdots, p_m(\varphi(x)), p_{m+1}(\varphi(x)), \cdots, p_N(\varphi(x))) .$$

For large enough ν this is an embedding. Thus $B = \varnothing$. We have the commuting diagram (remember $p_j(\varphi(x)) = 0$ if $j > m$)

$$\begin{array}{ccc} X & \xrightarrow{\varphi} & \mathbf{P}^{n+1} \\ & \searrow^{\Phi_\nu} & \downarrow P_\nu \\ & & \mathbf{P}^N \end{array}$$

The proof proceeds as before, showing φ is an embedding onto a non-singular quadric in $\mathbf{P}^{n+1}$, q.e.d.

We remark that the proof of Lemma 4 depends crucially on the "simplicity" of the ring $H^*(\mathbf{Q}_n, \mathbf{Z})$. One would like to generalize this result to Kähler manifolds homeomorphic to a grassmannian $G_{m,n}$. Several obstacles confront a straightforward generalization of this proof. First one must obtain $T(X) = 1$; and this does not seem to follow so easily, since some of the Betti numbers of $G_{m,n}$ will be large. Even assuming $T(X) = 1$, one must prove a "Lemma 4" or something like it. Again this is not obvious for a grassmannian since the cohomology ring is now complicated.

UNIVERSITY OF CALIFORNIA, BERKELEY

REFERENCES

[1] A. BOREL, *Sur la cohomologie des espaces fibres principaux et des espaces homogenes de groupes de Lie compacts,* Ann. of Math. **57** (1953), 115-207.

[2] ______ and F. Hirzebruch, *Characteristic classes and homogeneous spaces:* I, Amer, J. Math. **80** (1958), 458-538.

[3] E. BRIESKORN, *Ein Satz über die komplexen Quadriken,* Math. Ann. **155** (1964), 184-193.

[4] C. EHRESMANN, *Sur la topologie des certains espaces homogènes,* Ann. of Math. **35** (1934), 396-443.

[5] F. HIRZEBRUCH, Neue topologische Methoden in der algebraischen Geometrie, Springer, 1962.

[6] ______, *Uber eine Klasse von einfach-zusammenhangenden komplexen Mannigfaltigkeiten,* Math. Ann. **124** (1951), 77-86.

[7] ______ and K. Kodaira, *On the complex projective spaces,* J. Math. Pures Appl. **36** (1957), 201-216.

[8] S. P. NOVIKOV, *On manifolds with free abelian fundamental group and their application,* Izv. Akad. Nauk SSSR, Ser. Mat. **30** (1966), 207-246.

[9] E. THOMAS, *Complex structures on real vector bundles,* Amer. J. Math. **89** (1967), 887-908.

(Received January 28, 1969)

Enriques' classification of surfaces in char p: I

By DAVID MUMFORD

The principal assertion in Enriques' classification of surfaces is:

THEOREM. *Let F be a non-singular projective surface, without exceptional curves of the 1st kind. Let K_F be the canonical divisor class on F. Then*

(i) *if* $|12K_F| = \emptyset$, *then F is ruled,*

(ii) *if* $|12K_F| \neq \emptyset$, *then either* $12K_F \equiv 0$, *or else* $|nK_F|$ *is a linear system without base points for some n.*

As a Corollary, if we introduce the notations:

(a) $\mathcal{K} = \operatorname{tr} d_k \sum_{n=0}^{\infty} H^0(F, o_F(nK)) - 1$

(b) F *elliptic* if $\exists$ a morphism $f: F \to C$, C a curve, with almost all fibres non-singular elliptic curves,

(c) F *quasi-elliptic* if $\exists$ a morphism $f: F \to C$, C a curve, with almost all fibres singular rational curves E with $p_a(E) = 1$,[1]

(d) F of *general type* if $\exists f: F \to F_0$, f birational F_0 normal with K_{F_0} ample. Then we find that there are 4 types of surfaces F,

$\mathcal{K} = -1$	F ruled
$\mathcal{K} = 0$	$12K_F \equiv 0$
$\mathcal{K} = 1$	F elliptic, or quasi-elliptic, with K_F of positive degree
$\mathcal{K} = 2$	F of general type .

All this, plus a detailed analysis of the case $\mathcal{K} = 0$, has been proven in char 0; it is due essentially to Enriques [1], and has been worked out in detail by Kodaira [4], [5], in Safarevic's seminar [7], and in Zariski's seminar at Harvard. The purpose of this note and its sequel is to supply some new ideas which make the proof work in char p. Some of the steps supply new proofs of parts of the theorem in char 0, which have, I believe, some interest.

---

[1] In view of the results of Tate [11], such surfaces can only occur if char $(k) = 2$ or 3; moreover, almost all fibres E have a single ordinary cusp.

List of notation

F = non-singular projective surface
K = canonical divisor class on F
c_2 = 2nd Chern class of F (a number)
q = dimension of Albanese of F
B_2 = 2nd Betti number of F, dim $H^2(F, \mathbf{R})$
ρ = Base number of F
Ω_F^i = sheaf of i-forms on F,
$h^{p,q} = \dim H^q(\Omega_F^p)$
$p_g = h^{2,0} = h^{0,2}$
$p_a = h^{0,2} - h^{0,1} = \chi(o_F) - 1$
$p_a(D) = 1 + \dfrac{(D\cdot D + K)}{2}$, if D is a curve on F

1. $(K\cdot D) < 0$ for some effective divisor D

The situation $(K\cdot D) < 0$, some D, is known classically as the case "*adjunction terminates.*" In this case, we shall prove that F is ruled.

Step (I). *There is an ample H such that* $(K\cdot H) < 0$.

PROOF. In fact, first replacing D by a suitable component of itself, we find an *irreducible* D such that $(K\cdot D) < 0$. If $(D^2) < 0$, then since

$$(K\cdot D) + (D^2) = 2p_a(D) - 2 \geqq -2 ,$$

it follows that $(K\cdot D) = (D^2) = -1$, i.e., D is exceptional of 1st kind. This has been excluded, so $(D^2) \geqq 0$. Let H_1 be any ample divisor on F. Then for all $n \geqq 0$, $nD + H_1$ is ample by the Nakai-Moisezon criterion of ampleness [2]. But if $n \gg 0$, $(K\cdot nD + H_1) < 0$. q.e.d.

COROLLARY OF *Step* I. $|nK| = \emptyset$, *all* $n \geqq 1$.

Step (II). *If* $(K^2) > 0$, *then F is rational, hence ruled.*

PROOF. Use the general formulas

$$12(\chi(o_F)) = (K^2) + c_2$$
$$c_2 = 2 - 4q + B_2 .$$

In our case, $p_g = \dim |K| + 1 = 0$, hence it follows that the Picard scheme of F is reduced [6, lecture 26]. Therefore

$$\begin{aligned}\dim H^1(o_F) &= \dim(\text{tangent space to Picard scheme})\\ &= \dim(\text{Picard scheme})\\ &= q\ ,\end{aligned}$$

and so

$$\begin{aligned}\chi(o_F) &= \dim H^0(o_F) - \dim H^1(o_F) + \dim H^2(o_F)\\ &= 1 - q + p_g\\ &= 1 - q\ .\end{aligned}$$

Therefore

$$12 - 12q = (K^2) + 2 - 4q + B_2\ ,$$

or

$$(*)\qquad 10 = 8q + (K^2) + B_2\ .$$

But if $\rho =$ base number of F, then $B_2 \geqq \rho > 0$ (Igusa [12]), so if $(K^2) > 0$, it follows that $q = 0$ or 1. If $q = 1$, then F admits a morphism onto an elliptic curve, hence $B_2 \geqq 2$ and $(*)$ still cannot be satisfied. But if $q = 0$, then since $|2K| = \varnothing$, the hypotheses of Castelnuovo's criterion are met, and by Zariski [9], it follows that F is rational. q.e.d.

Step III. (Kodaira). *If* $(K^2) \leqq 0$, *then for all n there are effective divisors D on F such that*

(a) $|D + K| = \varnothing$

(b) $\dim |D| \geqq n$.

PROOF. Let H be an ample divisor such that $(K\cdot H) < 0$. For all n, note that $(nH + mK. H) < 0$ and hence $nH + mK$ cannot be linearly equivalent to an effective divisor, if $m \gg 0$. Let m_n be a non-negative integer such that

$$\begin{aligned}|nH + m_nK| &\neq \varnothing\\ |nH + (m_n + 1)K| &= \varnothing\ .\end{aligned}$$

Let $D_n \in |nH + m_nK|$. Write $D_n = D'_n + D''_n$, where D'_n and D''_n are positive and the components E of D'_n satisfy $(E.K) < 0$, while those of D''_n satisfy $(E\cdot K) \geqq 0$. Note that $(E\cdot K) < 0 \Rightarrow (E^2) \geqq 0$ (cf. *Step* I) so $(D'^2_n) \geqq 0$. Next, note that $|K - D'_n| = \varnothing$. In fact if not, K itself would be effective, since it would be the sum of D'_n and an effective divisor in $K - D'_n$. So by Serre duality, $H^2(o_F(D'_n)) = (0)$. Now use Riemann-Roch.

$$\begin{aligned}\dim|D'_n| &= \dim H^0(o_F(D'_n)) - 1\\ &\geqq \chi(o_F(D'_n)) - 1\\ &= \frac{(D'_n\cdot D'_n - K)}{2} + \chi(o_F) - 1\\ &\geqq -\frac{(D'_n\cdot K)}{2} + \chi(o_F) - 1\\ &\geqq -\frac{(D_n\cdot K)}{2} + \chi(o_F) - 1\\ &= -n\frac{(H\cdot K)}{2} - m_n\frac{(K^2)}{2} + \chi(o_F) - 1\\ &\geqq \frac{n}{2} + \chi(o_F) - 1\,.\end{aligned}$$

Since $|K + D'_n| = \varnothing$, D'_n has all the required properties. q.e.d.

Now comes the new idea to take care of char p.

Key Step IV. *If D is an effective divisor such that $|K+D| = \varnothing$, then the natural map*

$$\mathrm{Pic}^0_F \longrightarrow \mathrm{Pic}^0_D$$

is surjective.

PROOF. $H^0(o_F(K + D)) = (0)$ implies $H^2(o_F(-D)) = (0)$ by Serre duality. Using the exact sequence

$$0 \longrightarrow o_F(-D) \longrightarrow o_F \longrightarrow o_D \longrightarrow 0\,,$$

it follows that the natural map $H^1(o_F) \to H^1(o_D)$ is surjective. But these vector spaces are the tangent spaces at 0 to the connected and reduced schemes group Pic^0_F and Pic^0_D. (Pic^0_F is reduced since $p_g = 0$!). Hence Pic^0_F maps onto Pic^0_D. q.e.d.

Step (V). *If D is an effective divisor such that $|K + D| = \varnothing$, and if $D = \sum n_iE_i$, then*

(i) *all E_i are non-singular*

(ii) *if $n_i \geqq 2$, then*

(a) *E_i is rational or*

(b) *E_i is elliptic, $(E_i^2) = 0$ with non-trivial normal bundle or*

(c) $(E_2^i) < 0$

(iii) *the E_i are connected together without loops.*

PROOF. In fact, by Step (IV), Pic^0_D is an abelian variety. Since the natural map $\mathrm{Pic}^0_D \to \mathrm{Pic}^0_{E_i}$ is surjective, $\mathrm{Pic}^0_{E_i}$ is abelian too, so

E_i is non-singular. If (iii) were false, Pic^0_D would have subgroups of type $\mathbf{G}_m$ coming from the loops. To check (ii), use the fact that if $k_i \geqq 2$, $\mathrm{Pic}^0_D \to \mathrm{Pic}^0_{2E_i}$ is surjective, hence $\mathrm{Pic}^0_{2E_i}$ is abelian, hence $\mathrm{Pic}^0_{E_i} = \mathrm{Pic}^0_{2E_i}$. Looking at tangent spaces at 0, this implies that $H^1(o_{2E_i}) \cong H^1(o_{E_i})$. But *via* the exact sequence

$$0 \longrightarrow o_{E_i}(-E_i^2) \longrightarrow o_{2E_i} \longrightarrow o_{E_i} \longrightarrow 0 ,$$

this implies

$$H^1\big(o_{E_i}(-E_i^2)\big) = (0) .$$

But if $(E_i^2) \geqq 0$, $o_{E_i}\big(-(E_i^2)\big)$ has degree $\leqq 0$, hence $H^1 \neq (0)$ except in the cases E_i rational, $o_{E_i}\big(-(E_i^2)\big)$ of degree -1, 0; or E_i elliptic, $(E_i^2) = 0$ and $o_{E_i}\big(-(E_i^2)\big) \not\cong o_{E_i}$. q.e.d.

Step (VI). *There is a non-singular rational curve on F passing through every point of F.*

PROOF. Suppose to the contrary. Then there are only a finite number of non-singular rational curves on F of each degree. Moreover, on any surface there are only a finite number of non-singular curves E of each degree such that either

(a) $(E^2) < 0$, or

(b) E elliptic, $(E^2) = 0$, with *non-trivial* normal bundle.

Taking all these curves together, we get a countable set of curves, which cannot exhaust F. Let $P_1, \cdots, P_q$ be distinct points of F on none of these curves. Let D be a divisor satisfying the requirements of *Step* (III) with $\dim |D| \geqq 3q$. It follows that there is a divisor $D' \in |D|$ with double points at each P_i, $1 \leqq i \leqq q$. Let $D' = \sum n_i E_i$. If $n_i \geqq 2$, then by *Step* (V), E_i would be in the countable set of curves just described, so E_i does not contain any P_j. In other words, each P_i lies only on *simple* components of D', and since each E_i is non-singular (*Step* (V)), each P_i lies on 2 components. Since the E_i's are connected together as a tree, this shows that there are in all at least $q + 1$ E_i's containing some P_j. Note that they are all curves of genus at least 1. Therefore

$$q = \dim(\mathrm{Pic}^0_F) \geqq \dim(\mathrm{Pic}^0_{D'}) \geqq \textstyle\sum_i \dim(\mathrm{Pic}^0_{E_i}) \geqq q + 1 .$$

So our assumption was false. q.e.d.

Step (VII). *F is ruled.*

PROOF. This follows by any number of methods. If $q = 0$, there must be a *linear* system of positive dimension of non-singular rational curves. Apply Tsen's theorem[2] to a pencil of such curves. If $q > 0$, look at the Albanese map $\pi: F \to A$. Since π maps all rational curves on F to points, it follows from *Step* (VI) that π factors

$$F \xrightarrow{\pi'} C \hookrightarrow A$$

where C is a curve, and the fibres of π' are rational. Therefore F is ruled by Tsen's theorem. q.e.d.

2. $(K \cdot D) \geqq 0$, all effective divisors D; $(K^2) = 0$

In this section we will show that

$$\left\{\begin{array}{l}(K^2) = 0 \\ (K \cdot D) \geqq 0, \text{ eff. } D\end{array}\right\} \Longrightarrow \left\{\begin{array}{l}\text{either } 2K \equiv 0\,, \\ \text{or } \exists \text{ a pencil of curves on } F \text{ with } p_a = 1\end{array}\right\}$$

(Note: in char 2 or 3, the pencil may consist entirely of rational curves with cusps; Bertini's theorem tells us only that almost all the curves are irreducible, cf. [8].)

First make the definition.

DEFINITION. A curve $D = \sum n_i E_i$ on F is *of canonical type* if $(K \cdot E_i) = (D \cdot E_i) = 0$, all i. D is *indecomposable of canonical type*, if D is connected, and g.c.d. $(n_i) = 1$.

Step (I). *Either* $2K \equiv 0$ *or* F *contains at least one indecomposable curve of canonical type.*

PROOF. (Enriques; cf. also Šafarevič [7, Lemmas 9, 10, pp. 71–73)]. If $D \in |2K|$, then D is either 0 or of canonical type. In fact, let $D = \sum n_i E_i$. Then

$$0 = (K \cdot D) = \sum n_i (K \cdot E_i)\,.$$

Since $(K \cdot E_i) \geqq 0$, all i, it follows that $(K \cdot E_i) = 0$, all i; hence $(D \cdot E_i) = 2(K \cdot E_i) = 0$, too. Decomposing D, we find an indecomposable curve of canonical type. On the other hand, suppose $|2K| = \emptyset$. In this case, $q = h^{0,1}$ and $q \neq 0$ or else F would be rational by Castelnuovo; and $q \geqq 2$ would contradict the formula

$$\chi(o_F) = 1 - \dim H^1(o_F) + \dim H^2(o_F) \leqq 1 - q$$
$$12\chi(o_F) = (K^2) + (c_2) = 2 - 4q + B_2 \geqq 3 - 4q\,.$$

There remains the one really subtle case of $q = 1$. Consider the

[2] If $\pi: F \to C$ is a morphism of a surface to a curve, with fibres of $p_a = 0$, then $k(F) \cong k(C)(X)$.

Albanese morphism $\pi\colon F \to \varepsilon$, ε an elliptic curve. Let $f = \pi^{-1}(P)$, some $P \in \varepsilon$, be an irreducible fibre of π. If $p_a(f) = 1$, f is of canonical type and we are through. If $p_a(f) > 1$, then $(K \cdot f) = 2p_a(f) - 2 \geqq 2$. We then use the following argument of Enriques: for all $Q \in \varepsilon$, $Q \neq P$ consider the exact sequence

$$0 \longrightarrow o_F(2K + \pi^{-1}(Q) - f) \longrightarrow o_F(2K + \pi^{-1}(Q)) \longrightarrow o_f(2K \cdot f) \longrightarrow 0 .$$

Note that $H^2(o_F(2K + \pi^{-1}(Q) - f)) = (0)$, since if $|f - \pi^{-1}(Q) - K|$ contained an effective divisor, then $(K \cdot f) \leqq 0$. Check by Riemann-Roch that $\chi(o_F(2K + \pi^{-1}(Q) - f)) = 0$. Therefore, for all $Q \neq P$, either

(i) $|2K + \pi^{-1}(Q) - f| \neq \emptyset$, or

(ii) $H^i(o_F(2K + \pi^{-1}(Q) - f)) = (0)$, all i, hence

$$H^0(o_F(2K + \pi^{-1}(Q))) \xrightarrow{\sim} H^0(o_f(2K \cdot f)) .$$

Suppose $|2K + \pi^{-1}(Q) - f| = \emptyset$ for all Q. Then fix a non-zero $s \in H^0(o_f(2K \cdot f))$ and let A be the Cartier divisor $s = 0$ on f. For all Q, s lifts to a unique $s'_Q \in H^0(o_F(2K + \pi^{-1}(Q)))$, and let D_Q be the divisor $s'_Q = 0$. This is a 1-dimensional algebraic family of divisors, such that $D_Q \cdot f = A$ for all $Q \neq P$. Moreover, all the D_Q's are distinct since they are not even linearly equivalent. Therefore,

$$F = \text{closure of } \bigcup_{Q \neq P} D_Q .$$

In particular, as $Q \to P$, D_Q must specialize to a divisor D_P containing the whole fibre f. Then

$$D_P = f + D_P^*$$
$$D_P \in |2K + f| .$$

So $|2K| \neq \emptyset$. So finally in all cases, $|2K + \pi^{-1}(Q) - f| \neq \emptyset$ for some $Q \in \varepsilon$ (possibly $Q = P$). But a divisor $D \in |2K + \pi^{-1}(Q) - f|$ is of canonical type just as before. q.e.d.

We would like to assert next that on any surface F and for any indecomposable D of canonical type, $\dim H^0(o_F(nD)) > 1$ and hence $|nD|$ is composite with a pencil, for some n. Unfortunately, this is false, as one sees by considering an elliptic ruled surface obtained by completing a line bundle of infinite order over an elliptic curve, the 0-section of the bundle being taken as the curve of canonical type. The rest of the proof is an exercise in avoiding this case. The following will be useful.

LEMMA. *Let* $D=\sum n_i E_i$ *be an indecomposable curve of canonical type, and let* L *be an invertible sheaf on* D, *(regarded as a 1-dimensional scheme). If* $\deg(L\otimes o_{E_i}) = 0$, *all* i, *then*

$$H^0(D, L) \neq (0) \Longleftrightarrow L \cong o_F$$

and moreover $H^0(D, o_D) = k$.

PROOF. Let $s\in H^0(D, L)$. It suffices to show that if $s\neq 0$, s generates L hence defines an isomorphism of o_D with L; in particular, if $L = o_D$, this shows that all non-zero elements of the algebra $H^0(D, o_D)$ are units, i.e., $H^0(D, o_D)$ is a field, hence is k. But consider the induced section of $L\otimes o_{E_i}$. Since this sheaf has degree 0, a section either generates $L\otimes o_{E_i}$ or is identically 0. If s vanishes on one E_i, it vanishes on one point of each E_j meeting E_i; since D is connected, s either generates L everywhere, or vanishes on all E_i's. Assume that s vanishes on all E_i's; let k_i = order of vanishing of s on E_i. Then $1\leqq k_i\leqq n_i$. Whenever $k_i < n_i$, s defines a non-zero section of

$$L\otimes \left[o_F(-k_iE_i)/o_F\left(-(k_i+1)E_i\right)\right].$$

This section vanishes to order at least $I\left(E_i, \sum_{j\neq i} k_jE_j; P\right)$ at every $P\in E_i$ (I = intersection multiplicity). It follows that if $k_i < n_i$, then

$$\begin{aligned}\left(E_i.\ \textstyle\sum_{j\neq i} k_jE_j\right) &\leqq \deg\left\{L\otimes o_F(-k_iE_i)/o_F\left((-k_i+1)E_i\right)\right\}\\ &= \deg\left[o_F(-E_i)/o_F(-E_i^2)\right]^{k_i} \qquad (1)\\ &= -k_i(E_i^2)\,.\end{aligned}$$

Note that if $k_i = n_i$, then since $(E_i\cdot D) = 0$,

$$\left(E_i\cdot\textstyle\sum_j k_jE_j\right) = -\left(E_i\cdot\sum_j (n_j-k_j)E_j\right) \leqq 0\,. \qquad (2)$$

So if $D_1 = \sum_j k_jE_j$, then *for all* i, using (1) or (2), according as $k_i < n_i$ or $k_i = n_i$, $(E_i\cdot D_1)\leqq 0$. On the other hand,

$$\sum n_i(E_i\cdot D_1) = (D\cdot D_1) = \sum k_i(D\cdot E_i) = 0\,,$$

so it follows that $(E_i\cdot D_1) = 0$, all i. This shows that $D - D_1$ and D_1 are of canonical type too. I claim that D and D_1 are both multiples of a 3rd divisor of canonical type, which will show that D is decomposable, unless $D = D_1$, i.e., $s\equiv 0$. To prove this, we must show that k_i/n_i is independent of i. Let $a/b = \max(k_i/n_i)$ and let $Z = aD - bD_1$. Then Z is an effective cycle and E_i occurs in Z if and only if $k_i/n_i < a/b$. Then if $k_i/n_i = a/b$, E_i is not a component of Z, and since $(Z\cdot E_i) = a(D\cdot E_i) - b(D_1\cdot E_i) = 0$, E_i does not even meet Z; so Z does not contain any E_j's meeting E_i. Since D is connected, this shows that $Z = 0$, hence k_i/n_i independent of i as needed. q.e.d.

COROLLARY 1. *If D is indecomposable of canonical type, then*

$$o_F(K + D) \otimes o_F \cong o_D .$$

PROOF. Let $\omega = o_F(K + D) \otimes o_D$: then ω is the dualizing sheaf on the Cohen-Macauley scheme D, so by Serre duality $\dim H^1(D, \omega) = \dim H^0(D, o_D) = 1$. *Via* the exact sequence

$$0 \longrightarrow o_F(K) \longrightarrow o_F(K + D) \longrightarrow \omega \longrightarrow 0 ,$$

we see that

$$\begin{aligned} \chi(\omega) &= \chi\big(o_F(K + D)\big) - \chi\big(o_F(K)\big) \\ &= \frac{(K + D\cdot D)}{2} \qquad \text{(Riemann-Roch on } F) \\ &= 0 , \end{aligned}$$

hence $\dim H^0(D, \omega) = 1$ too. Hence $\omega \cong o_D$ by the lemma. q.e.d.

COROLLARY 2. *If $D = \sum n_i E_i$ is an indecomposable divisor of canonical type, and D' is any effective divisor on F such that $(D'\cdot E_i) = 0$ all i, then*

$$D' = nD + D''$$

where $n \geqq 0$ and D'' is an effective divisor disjoint from D.

Incidentally, a complete list of all curves of canonical type can be found in Kodaira [3], [4], and the lemma could be checked case by case.

Step (II). *If $p_g = 0$ (and* $(K^2) = 0$, $(K\cdot D) \geqq 0$ all eff. *Das always) and D is an indecomposable curve of canonical type, then $|nD|$ is composite with a pencil of curves of canonical type, for some n.*

PROOF. First, look at the sequences

$$0 \longrightarrow o_F\big(nK + (n - 1)D\big) \longrightarrow o_F(nK + nD) \longrightarrow o_D \longrightarrow 0$$

obtained by applying Corollary 1. If $n \geqq 2$, then

$$H^2\big(o_F(nK + (n - 1)D)\big) = H^2\big(o_F(nK + nD)\big) = (0) .$$

Since $H^1(o_D) \neq (0)$, we find $H^1\big(o_F(nK + nD)\big) \neq (0)$. But

$$\chi\big(o_F(nK + nD)\big) = \chi(o_F) = 0 \quad \text{or} \quad 1 \qquad \text{(cf. } \textit{Step I}) .$$

Therefore $H^0(o_F(nK + nD)) \neq (0)$. This shows that there is a divisor $D_n \in |nK + nD|$.

D_n is of canonical type. Note that if $D = \sum n_i E_i$, then

$$(D_n \cdot E_i) = n(K \cdot E_i) + n(D \cdot E_i) = 0 .$$

So by Corollary 2, $D_n = aD + \sum k_i F_i$, where the F_i are disjoint from D. Now $(K \cdot F_i) \geqq 0$ for all i, while

$$\begin{aligned}\sum k_i(K \cdot F_i) &= \left(K \cdot \sum k_i F_i + aD\right)\\ &= (K \cdot nK + nD)\\ &= 0\end{aligned}$$

so $(K \cdot F_i) = 0$, all i. Finally

$$(D_n \cdot F_i) = n(K \cdot F_i) + n(D \cdot F_i) ,$$

and this is 0 since D does not meet F_i.

It would seem to follow that we have produced at least one indecomposable curve of canonical type disjoint from D. But in one case this would not be so; namely, if each D_n was a multiple of D. In that case, though, K itself would be a multiple of D, hence $p_g \neq 0$ contrary to hypothesis.

Now assume that D and D' are disjoint indecomposable curves of canonical type. Then look at the sequence

$$0 \longrightarrow o_F(2K + D + D') \longrightarrow o_F(2K + 2D + 2D') \longrightarrow o_D \oplus o_{D'} \longrightarrow 0$$

(again using Corollary 1). As before, all H^2's vanish, so now

$$\dim H^1(o_F(2K + 2D + 2D')) \geqq 2 ,$$

hence calculating Euler characteristics, we find

$$\dim H^0(o_F(2K + 2D + 2D')) \geqq 2$$

so $|2K + 2D + 2D'|$ is composite with pencil of curves of canonical type. Since the intersection multiplicity $(D \cdot 2K + 2D + 2D') = 0$, it follows that one of the fibres of this pencil must be a multiple of D. q.e.d.

The final step is a new approach, not in Enriques.

Step (III). If $p_g > 0$, and D is an indecomposable curve of canonical type, then $|nD|$ is composite with a pencil of curves of canonical type, for some n.

PROOF. Let $\mathcal{F}_n$ be the quotient sheaf $o_F(nD)/o_F$. In view of the exact sequences

$$H^0(o_F(nD)) \longrightarrow H^0(\mathcal{F}_n) \longrightarrow H^1(o_F) ,$$

it will suffice to show that $\dim H^0(\mathcal{F}_n) \to \infty$ as $n \to \infty$ in order to establish *Step* (III). Let L be invertible sheaf $\mathcal{F}_1$ on the scheme D. For all n, $\mathcal{F}_{n-1}$ is a subsheaf of $\mathcal{F}_n$ with quotient L^n:

$$(*) \qquad 0 \longrightarrow \mathcal{F}_{n-1} \longrightarrow \mathcal{F}_n \longrightarrow L^n \longrightarrow 0 .$$

This proves

(A) $\dim H^0(\mathcal{F}_n)$ *is a non-decreasing function of* n.

Next, by the Riemann-Roch theorem on F, you see that $\chi(o_F(nD)) = \chi(o_F)$. Therefore $\chi(\mathcal{F}_n) = 0$. Now use the exact sequence

$$H^1(\mathcal{F}_n) \longrightarrow H^2(o_F) \longrightarrow H^2(o_F(nD)) .$$

Since D is effective, $|K - nD|$ is empty for large n, hence by Serre duality, $H^2(o_F(nD)) = (0)$. But since $p_g > 0$, $H^2(o_F) \neq (0)$, hence $H^1(\mathcal{F}_n) \neq (0)$, for large n. This proves

(B) $\dim H^0(\mathcal{F}_n) > 0$ *for* $n \gg 0$, *and* $\chi(\mathcal{F}_n) = 0$, *all* n.

Now assume that $\dim H^0(\mathcal{F}_n)$ is bounded above, and let n be the largest integer for which $\dim H^0(\mathcal{F}_{n-1}) < \dim H^0(\mathcal{F}_n)$ (there is at least one such n by (B) and the fact that $\mathcal{F}_0$ is the 0 sheaf). Using exact sequence $(*)$, it follows that L^n has a non-0 section. But since D is of canonical type, L^n has degree 0 on each component of D. So by Corollary 1, $L^n \cong o_D$. Therefore $\mathcal{F}_n$ has a section s which generates L^n everywhere as an o_D-module, hence s generates $\mathcal{F}_n$ everywhere as o_F-module. In other words, s defines an isomorphism

$$o_F/o_F(-nD) \cong \mathcal{F}_n = o_F(nD)/o_F .$$

Taking powers of s, we obtain isomorphisms

$$o_F/o_F(-nD) \cong o_F(nmD)/o_F(n(m-1)D) = \mathcal{F}_{nm}/\mathcal{F}_{n(m-1)} .$$

Now consider the diagram

$$\begin{array}{ccccccccc}
 & & & & 0 & & 0 & & \\
 & & & & \downarrow & & \downarrow & & \\
0 & \longrightarrow & o_F & \longrightarrow & o_F(n(m-1)D) & \longrightarrow & \mathcal{F}_{n(m-1)} & \longrightarrow & 0 \\
 & & \| & & \downarrow & & \downarrow & & \\
0 & \longrightarrow & o_F & \longrightarrow & o_F(nmD) & \longrightarrow & \mathcal{F}_{nm} & \longrightarrow & 0 \\
 & & & & \downarrow & & \downarrow & & \\
 & & & & \mathcal{F}_n & = & \mathcal{F}_n & & \\
 & & & & \downarrow & & \downarrow & & \\
 & & & & 0 & & 0 & &
\end{array}$$

set up by means of the above isomorphisms. Taking cohomology, we get

$$\begin{array}{ccccccc} H^1(\mathcal{F}_{n(m-1)}) & \longrightarrow & H^1(\mathcal{F}_{nm}) & \longrightarrow & H^1(\mathcal{F}_n) & \longrightarrow & 0 \\ \downarrow & & \downarrow & & & & \\ H^2(o_F) & = & H^2(o_F) & & & & \\ \downarrow & & \downarrow & & & & \\ 0 & & 0 & & & & \end{array}$$

Since $H^2(o_F) \neq (0)$, it follows that the map from $H^1(\mathcal{F}_{n(m-1)})$ to $H^1(\mathcal{F}_{nm})$ is not zero. Therefore $\dim H^1(\mathcal{F}_{nm}) > \dim H^1(\mathcal{F}_n)$. Since $\chi(\mathcal{F}_{nm}) = \chi(\mathcal{F}_n) = 0$, this shows $\dim H^0(\mathcal{F}_{nm}) > \dim H^0(\mathcal{F}_n)$, contradicting our hypothesis on n. Hence $\dim H^0(\mathcal{F}_n)$ is unbounded. q.e.d.

This establishes the existence of an elliptic or quasi-elliptic pencil on F in all cases except when $2K \equiv 0$. A propos the general question raised by this last step—namely, given a curve C such that $(C^2) = 0$, when does nC lie in a pencil, for some n?—there is one curious result that shows the situation in char p is "better" than that in char 0.

PROPOSITION. *Let C be an irreducible curve on a non-singular projective surface F. Assume* $(C^2) = 0$ *and* char $(k) \neq 0$.

$$\left\{\begin{array}{l} nC \textit{ lies in a pencil} \\ \quad \textit{for some } n \end{array}\right\} \Longleftrightarrow \left\{\begin{array}{l} o_F(nC)/o_F \cong o_F/o_F(-nC) \\ \quad \textit{for some } n \end{array}\right\}.$$

The proof is left as a curiosity to the reader. To get a counter example in char 0, let F = completion of the non-trivial G_a-bundle over an elliptic curve, let C = line at ∞.

3. All but one of the remaining steps

Note first that if $(K \cdot D) \geqq 0$, for all effective divisors D, then $(K^2) \geqq 0$ (cf. [6]). Therefore, Enriques' theorem will follow if we show

(a) $[(K \cdot D) \geqq 0$, all eff D, $(K^2) > 0] \Rightarrow |2K| \neq \emptyset$ and $|nK|$ has no base points for some n.

(b) [F elliptic or quasi-elliptic, for the morphism $f\colon F \to C$] $\Rightarrow$ $nK \equiv f^*(A)$ for some n and for some 0-cycle A on C.

(c) [F elliptic and $(K \cdot D) \geqq 0$, all eff D] $\Rightarrow |12K| \neq \emptyset$

(d) [F quasi-elliptic and $(K \cdot D) \geqq 0$, all eff D] $\Rightarrow$ either F elliptic too or $|2K| \neq \emptyset$.

PROOF OF (a). Suppose that $|2K| = \emptyset$. Then $p_g = 0$, hence $q = h^{0,1}$, and by the Riemann-Roch theorem and Serre duality,

$$(*) \qquad \begin{aligned} (K^2) + \chi(o_F) &= \chi(o_F(2K)) \\ &= \dim H^0(o_F(2K)) + \dim H^0(o_F(-K)) \\ &\quad - \dim H^1(o_F(2K)) \\ &\leqq 0 . \end{aligned}$$

But $\chi(o_F) = 1 - q$, and by the Riemann-Roch theorem,

$$\begin{aligned} 12(1 - q) &= 12\chi(o_F) \\ &= (K^2) + c_2 \\ &= (K^2) + 2 - 4q + B_2 , \end{aligned}$$

hence,

$$(**) \qquad 10 = (K^2) + 8q + B_2 .$$

Equation $(**)$ shows that $q \leqq 1$, hence $\chi(o_F) \geqq 0$, hence by $(*)$, $(K^2) \leqq 0$ which contradicts our hypotheses. Finally, the fact that in case (a), $|nK|$ has no base points for some n was shown in my appendix to [10].

PROOF OF (b). Each of the fibres of f is a connected curve of canonical type, and almost all are irreducible with multiplicity 1. A finite number of them, say $f^{-1}(x_1), \cdots, f^{-1}(x_n)$, are of the form

$$f^{-1}(x_i) = k_i D_i , \qquad k_i > 1$$

where D_i is an indecomposable curve of canonical type. Let $k =$ l.c.m. $(k_1, \cdots, k_n)$.

Now K intersects the generic fibre of f in a divisor linearly equivalent to 0. Therefore K can be represented by a divisor E_i whose support is disjoint from the generic fibre, i.e.,

$$\operatorname{supp}(E_i) \subset f^{-1}(y_1, \cdots, y_m) .$$

Therefore for a suitable 0-cycle A on C,

$$K + f^{-1}(A) \equiv E_2 ,$$

where E_2 is effective and is a sum of components of various fibres of f. Applying Corollary 2 § 2 above, it follows that

$$E_2 = a_1 D_1 + \cdots + a_n D_n + f^{-1}(A')$$

for some integers a_i, and a 0-cycle A' on C. Therefore

$$kK \equiv f^{-1}\Big(-KA + KA' + \frac{ka_1}{k_1}x_1 + \cdots + \frac{ka_n}{k_n}x_n\Big).$$

PROOF OF (c). This is the step which we defer to the sequel of this paper!

PROOF OF (d). Assume $|2K| = \emptyset$. Then if $q = 0$, F is rational by Castelnuovo's criterion and this contradicts the hypotheses. And $q \leqq 1$ just as in the proof of (a). Therefore $q = 1$. Let

$$f: F \longrightarrow \varepsilon$$

be the Albanese map. Since the fibres of the quasi-elliptic fibration are rational, they are mapped to points under f, hence f must be the quasi-elliptic fibration. Since almost all fibres of f have a unique singular point, the set of points $x \in F$ where $df = 0$ contains a curve $\varepsilon' \subset F$ mapped generically one-to-one to ε. Now res (f): $\varepsilon' \rightarrow \varepsilon$ is flat and inseparable. Let its degree be p^n ($p = 2$ or 3 of course). Therefore

$$p^n = (\varepsilon' \cdot f^{-1}(x)), \qquad \text{all } x \in \varepsilon.$$

Now a non-singular branch can meet a cusp only with multiplicity 2 or 3, so choosing a non-singular point of ε', it follows that $(\varepsilon' \cdot f^{-1}(x)) = 2$ or 3. Therefore $n = 1$. Moreover, if ε' had a singular point y, then y would also be singular on $f^{-1}(f(y))$, so ε' would meet $f^{-1}(f(y))$ with multiplicity at least 4. Therefore ε' is non-singular and elliptic: especially it is of canonical type! So now apply the results of *Steps* II and III, § 2. We conclude that $|n\varepsilon'|$ varies in a pencil for some n, hence F is an elliptic surface.

HARVARD UNIVERSITY

REFERENCES

[1] F. ENRIQUES, Le Superficie Algebriche, Bologna, 1949.
[2] S. KLEIMAN, *A note on the Nakai-Moišezon test for ampleness of a divisor*, Amer. J. Math. **87** (1965), 221.
[3] K. KODAIRA, "On compact analytic surfaces," in Analytic Functions, Princeton Univ. Press. 1960.
[4] ———, *On Compact complex analytic surfaces*, I, II, III, Ann. of Math., **71**, **77**, **78** (1960, 1963), 11/563/1.
[5] ———, *On the structure of compact complex analytic surfaces*, I, II, III, IV, Amer. J. Math., 86 and 88 (1964 and 1966).
[6] D. MUMFORD, Lectures on curves on an algebraic surface, Annals of Math. Studies 59, Princeton, 1966.
[7] I. ŠAFAREVIČ and others, *Algebraic surfaces*, Proc. of Steklov Inst. of Math., Moscow, 1965 or Am. Math. Soc., 1967.

[8] O. ZARISKI, *Proof of a theorem of Bertini*, Trans. Amer. Math. Soc., 50 (1941), 48.
[9] ———, *On Castelnuovo's criterion of rationality in the theory of algebraic surfaces*, III. J. Math. **2** (1958).
[10] ———, *The theorem of Riemann-Roch for high multiples of an effective divisor*, Ann. of Math. **76** (1962), 560.
[11] J. J. TATE, *Genus change in inseparable extensions of function fields*, Proc. Amer. Math. Soc. **3** (1952), 400.
[12] J. I. IGUSA, *Betti and Picard numbers of abstract algebraic surfaces*, Proc. Nat. Acad. Sci. U.S.A. **46** (1960), 724.

(Received October 21, 1967)

On modular imbeddings of a symmetric domain of type (IV)*

BY ICHIRO SATAKE

The purpose of this note is to give some supplementary results concerning the holomorphic imbeddings of a symmetric domain of type (IV) into a Siegel space constructed in [8]. Namely, in § 1, we will give a simple geometric interpretation of such an imbedding along with an "analytic expression" of it (in the sense as explained in the Appendix); and, in § 2, some results concerning the relationship between the fields of automorphic functions on these domains with respect to suitable arithmetic groups. Similar results for other kinds of holomorphic imbeddings have been obtained by Klingen, Hammond, Freitag, and K. Iyanaga (see References). An Appendix will provide some basic notions pertinent to our theory.

1. The modular imbedding $\tilde{\rho}_a$

1.1. Let V be a vector-space over $\mathbf{R}$ and S a non-degenerate symmetric bilinear form (or quadratic form) on V of signature $(p, 2)$ $(p > 2)$. A symmetric domain $\mathfrak{D}$ of type (IV) and the corresponding compact hermitian space M attached to the pair (V, S) are constructed as follows. Call M the quadratic hypersurface defined by $S = 0$ in the projective space associated to $V_{\mathbf{C}}$, i.e., M is the totality of the (complex) 1-dimensional, totally isotropic subspaces W of $V_{\mathbf{C}}$ (with respect to the natural extension of S to $V_{\mathbf{C}} = V \otimes_{\mathbf{R}} \mathbf{C}$). Put, further, $F(x, y) = S(\bar{x}, y)$ for $x, y \in V_{\mathbf{C}}$. It is known that the open subset $\tilde{\mathfrak{D}}$ of M formed of all W such that $F|W < 0$ decomposes into two (mutually conjugate) connected components, of which either one can be taken as a realization of $\mathfrak{D}$. Thus, in the projective space $P(V_{\mathbf{C}})$, $\tilde{\mathfrak{D}} = \mathfrak{D} \cup \bar{\mathfrak{D}}$ is defined by the conditions

$$S|W = 0, \qquad F|W < 0. \tag{1}$$

* Partially supported by NSF grant GP-6654.

Let L be a lattice in V such that $S(x)(=S(x, x)) \in \mathbf{Z}$ for all $x \in L$ (i.e., the symmetric bilinear form S is "half-integral"). Furthermore, we consider the following condition:

(A) There exist $f, f^* \in L$ such that $S(f) = S(f^*) = 0$, $S(f, f^*) = 1/2$.

When (A) is satisfied, denote by V_1 the orthogonal complement of $\{f, f^*\}_{\mathbf{R}}$ in V and put $L_1 = L \cap V_1$. Then it is immediate that one has the following direct decompositions

$$V = \{f, f^*\}_{\mathbf{R}} + V_1 , \qquad L = \{f, f^*\}_{\mathbf{Z}} + L_1 .$$

The condition (1) for W is equivalent to saying that W is generated (over C) by a vector of the form $z + \zeta f + f^*$ with $z \in V_{1\mathbf{C}}$ such that

$$S(\mathrm{Im}(z)) < 0 \tag{1'}$$

and $\zeta = -S(z)$.

1.2. A similar realization of the Siegel space is given as follows. Let V' be a vector-space of dimension $2n'$ over $\mathbf{R}$ with a non-degenerate alternating bilinear form A' on it. Putting $F'(x, y) = \sqrt{-1}A'(\bar{x}, y)$ for $x, y \in V'_{\mathbf{C}}$, one obtains a hermitian form F' on $V'_{\mathbf{C}}$ of signature (n', n'). One defines M' as the totality of the (complex) n'-dimensional, totally isotropic subspaces W' of $V'_{\mathbf{C}}$ and $\mathfrak{D}'$ as a subset of M' formed of all W' such that $F' | W' < 0$. Then M' is again a (non-singular) algebraic submanifold of a grassmanian manifold formed of all (complex) n'-dimensional subspaces of $V_{\mathbf{C}}$ and $\mathfrak{D}'$ is an open connected submanifold of M'. Thus, in that grassmannian, $\mathfrak{D}'$ is defined by the conditions

$$A' | W' = 0 , \qquad F' | W' < 0 . \tag{2}$$

Let L' be a lattice in V' such that $A'(x, y) \in \mathbf{Q}$ for all $x, y \in L'$. Then the following property is well-known (Frobenius).

(B) There exist two (real) n'-dimensional totally isotropic subspaces U and U^* of V' such that one has the direct decompositions:

$$V' = U + U^* , \qquad L' = L' \cap U + L' \cap U^* .$$

Moreover, such a pair (U, U^*) is unique up to the equivalence defined by the action of $\mathrm{Sp}(L', A')$.

In the following, U^* will be identified with the dual of U by means of the inner product defined by

$$\langle x, y\rangle = -A'(x, y) \qquad \text{for } x \in U, y \in U^* .$$

Then one can find a dual basis $\mathfrak{B}$ and $\mathfrak{B}^*$ of U and U^*, respectively, such that $L' = \{\mathfrak{B}\delta, \mathfrak{B}^*\}_{\mathbf{Z}}$, where $\delta = \mathrm{diag}(d_1, \cdots, d_{n'})$, d_i's being positive rational numbers such that d_i is a divisor of d_{i+1} for $1 \leqq i \leqq n' - 1$. By the uniqueness in (B), δ is uniquely determined, independently of the choice of $(\mathfrak{B}, \mathfrak{B}^*)$, and is called the "elementary divisor" of (L', A'). We also call the basis $(\mathfrak{B}, \mathfrak{B}^*)$ of V' a "canonical basis" with respect to (L', A'). In terms of these notations, the condition (2) for W' is equivalent to saying that W' is of the form $(Z + 1)U_{\mathbf{C}}^*$ with a C-linear map Z of $U_{\mathbf{C}}^*$ into $U_{\mathbf{C}}$ such that

$$(2') \qquad {}^tZ = Z , \qquad \mathrm{Im}(Z) > 0$$

with respect to the inner product defined above. (Taking the matrix expression of Z in terms of a canonical basis, one obtains the usual representation of Z by a point of a Siegel's upper half space.)

1.3. For a given (V, S, L) (as in 1.1), we consider (V', A'_a, L') defined as follows:

$$(3) \qquad \begin{cases} V' = C^+(V, S) & \text{(the even part of the Clifford algebra)} , \\ A'_a(x, y) = \mathrm{tr}(ax^\iota y) & \text{for } x, y \in V' , \\ L' = C^+(L, S) , \end{cases}$$

where ι is the canonical involution in the Clifford algebra, tr the trace of the regular representation of C^+, and a is an invertible, skew-symmetric element in $C_{\mathbf{Q}}^+ = L' \otimes_{\mathbf{Z}} \mathbf{Q}$. It is clear that A'_a thus defined is an alternating bilinear form on V' satisfying all the assumptions in 1.2 and, if one puts $G = \mathrm{Spin}(V, S)$ $(\subset C^+(V, S))$ and $G' = \mathrm{Sp}(V', A'_a)$, the left multiplication in the Clifford algebra defines a representation ρ of G into G'. Moreover, if one puts $\Gamma_0 = G \cap C^+(L, S)$, Γ_0 is an arithmetic subgroup of G and one has $\rho(\Gamma_0) \subset \Gamma' = \mathrm{Sp}(L', A'_a)$. Whenever a is to be specified, we will write G'_a, $\mathfrak{D}'_a$, M'_a, etc., instead of G', $\mathfrak{D}'$, M', etc. It was shown in [8] that ρ satisfies the analyticity condition (H_2) (with respect to the given complex structures on $\mathfrak{D}$ and $\mathfrak{D}'_a$) if and only if

(4) the bilinear form tr $(ax^\iota e_- y)$ $(x, y \in V')$ is symmetric and positive-definite,

where e_- is an element in $C^+ = C^+(V, S)$ such that $(1/2)e_-$ is an "H-element" in the Lie algebra of G (see [8]). Note that the set of all skew-symmetric elements a in C^+ such that $\mathrm{tr}(ax^\iota a^{-1}y)$ is positive-definite is a union of two (mutually opposite) open convex cones in the vector-space of all skew-symmetric elements, and condition (4) is equivalent to saying that a is in the cone containing $-e_-$. Thus, for an $a \in C^+_{\mathbf{Q}}$ satisfying condition (4), one obtains a "modular imbedding" $\tilde{\rho}_a = (\rho, V', A'_a, I_0, L')$ of (G, K, Γ_0) (in the sense as defined in the Appendix), where K is the maximal compact subgroup of G corresponding to e_- and $I_0 = \rho(e_-)$. Two modular imbeddings $\tilde{\rho}_{a_1}$ and $\tilde{\rho}_{a_2}$ are equivalent if and only if one has $\lambda a_1 = v a_2 v^\iota$ with a positive rational number λ and a unit v of the order $C^+(L, S)$.

Now, to obtain a geometric interpretation of ρ, we need the following

LEMMA 1. *Let f, f^* be elements of V such that $S(f) = S(f^*) = 0$, $S(f, f^*) = 1/2$. Then, the right ideals fC^- and f^*C^- are totally isotropic subspaces of $V' = C^+$ (with respect to A'_a) of dimension $n' = 2^p$, and one has a direct decomposition $V' = fC^- + f^*C^-$, where C^- is the odd part of the Clifford algebra.* (Replacing V by $V_{\mathbf{C}}$ or L, one obtains a complex or integral version of the Lemma.)

In fact, by the assumption, one has $f^2 = f^{*2} = 0$, $ff^* + f^*f = 1$, so that ff^* and f^*f are mutually orthogonal idempotent elements in C^+. Therefore, one has $fC^- = ff^*C^+$, $f^*C^- = f^*fC^+$, and $V' = fC^- + f^*C^-$ (direct sum). On the other hand, fC^- and f^*C^- are clearly totally isotropic with respect to A'_a, so that one must have $\dim fC^- = \dim f^*C^- = n'$.

PROPOSITION. *The holomorphic imbedding of $\mathfrak{D}$ into $\mathfrak{D}'_a$ induced by the representation ρ defined above is given by the correspondence*

$$W \longmapsto W' = W \cdot C^-_{\mathbf{C}} . \tag{5}$$

In fact, by Lemma 1 applied to $C^+(V_{\mathbf{C}}, S)$ one sees that $W \in M$ implies $W' = W \cdot C^-_{\mathbf{C}} \in M'_a$, The correspondence $W \mapsto W'$ is clearly a holomorphic (rational) mapping of M into M'_a, which is compatible with the action of $G_{\mathbf{C}}$ (resp. $G'_{a\mathbf{C}}$) on M (resp. M'_a). In order to show that $\mathfrak{D}$ is mapped into $\mathfrak{D}'_a$, it is therefore enough to show that an origin o of $\mathfrak{D}$ is mapped to an origin o' of $\mathfrak{D}'_a$. Let $o = W^0$ be the origin of $\mathfrak{D}$ corresponding to e_-. Then, in the notation of [8], one has $e_- = e_{p+1}e_{p+2}$, $W^0 = \{e_{p+1} + \sqrt{-1}e_{p+2}\}_{\mathbf{C}}$, so that $W'^0 = W^0 C^-_{\mathbf{C}}$ is the $(-\sqrt{-1})$-eigenspace in $C^+_{\mathbf{C}}$ of the complex structure $I_0 = \rho(e_-)$, i.e., $o' = W'^0$ is the origin of $\mathfrak{D}'$ corresponding to I_0. This completes the proof.

When the lattice L satisfies the condition (A) in **1.1**, put

$$U = fC^- , \qquad U^* = f^*C^- .$$

Then, by Lemma 1 applied to $C^+(V, S)$ and to $C^+(L, S)$, it is easy to see that the pair (U, U^*) satisfies the condition stated in (B). In the expressions of $\mathfrak{D}$ and $\mathfrak{D}'_a$ thus obtained, let $z \leftrightarrow W$, $Z = Z(z) \leftrightarrow W' = W \cdot C^-_{\mathrm{C}}$. Then by the definitions one has

$$(z + \zeta f + f^*) \cdot C^-_{\mathrm{C}} = (Z + 1) f^* C^-_{\mathrm{C}} .$$

Comparing the components in U_{C} and in U^*_{C}, one has

$$(ff^*z + \zeta f) \cdot x = Z(f^*y) ,$$
$$(f^*fz + f^*) \cdot x = f^*y \qquad (x, y \in C^-_{\mathrm{C}}) ,$$

whence one has

$$(6) \qquad Z(f^*y) = (zf)(f^*y) \qquad \text{for all } y \in C^-_{\mathrm{C}} .$$

Thus, *the linear map $Z = Z(z)$ of U^*_{C} into U_{C} is simply given by the left multiplication of zf.*

1.4. From the above proposition, it is also easy to determine the correspondence under ρ of the boundary components of $\mathfrak{D}$ and $\mathfrak{D}'_a$. Recall first that the boundary components of $\mathfrak{D}$ are one of the following two types.

(i) for a 2-dimensional, totally isotropic subspace U of V, one puts

$$\tilde{F}(U) = \{W \in M \mid W + \bar{W} = U_{\mathrm{C}}\} .$$

Then $\tilde{F}(U)$ has two connected components, one of which, denoted by $F(U)$, is a "first" boundary component of $\mathfrak{D}$. ($F(U)$ is analytically equivalent to the unit disc.)

(ii) For a 1-dimensional, totally isotropic subspace W_{R} of V, one obtains a "second" boundary component $F(W_{\mathrm{R}})$ consisting of just one point $W = W_{\mathrm{R}} \otimes \mathrm{C}$.

On the other hand, for any totally isotropic subspace U' of V' of dimension r ($0 \leqq r \leqq n'$), one obtains an rth boundary component of $\mathfrak{D}'_a$ defined by

$$F'(U') = \{W' \in M'_a \mid F' \mid W' \leqq 0,\ W' \cap \bar{W}' = U'_{\mathrm{C}}\} .$$

Then, it can be shown that, under the holomorphic imbedding ρ, $F(U)$ is mapped into a $2^{p-1\,\text{th}}$ boundary component $F'(U')$ with $U' = (x_1x_2)C^+$, where $U = \{x_1, x_2\}_{\mathrm{R}}$, and the point boundary component $F(W_{\mathrm{R}})$ of $\mathfrak{D}$ is mapped to the point boundary component $F'(W_{\mathrm{R}} \cdot C^-)$ of $\mathfrak{D}'_a$ (cf. [9, Prop. 4]).

1.5. Under a more restrictive assumpiton, we can further reduce the analytic expression of $\tilde{\rho}_a$. Namely, in addition to (A), we first assume that there exist $f_1, f_2 \in L_1$ such that $S(f_1) = S(f_2) = 0$, $S(f_1, f_2) = -1/2$. (For instance, if $(L, 2S)$ is "unimodular", this assumption is satisfied.) Then, denoting by V_0 the orthogonal complement of $\{f_1, f_2\}_{\mathbf{R}}$ in V_1 and putting $L_0 = L \cap V_0$, one has the direct decompositions

$$V_1 = \{f_1, f_2\}_{\mathbf{R}} + V_0\,, \qquad L_1 = \{f_1, f_2\}_{\mathbf{Z}} + L_0\,.$$

The condition (1′) for $z = z_1 f_1 + z_2 f_2 + z_0$ $(z_1, z_2 \in \mathbf{C}, z_0 \in V_{0\mathbf{C}})$ reads as

$$\text{(1″)} \qquad \operatorname{Im}(z_1)\cdot\operatorname{Im}(z_2) - S_0(\operatorname{Im}(z_0)) > 0\,,$$

where S_0 is the restriction of S on V_0, which is positive-definite. In the following, we shall take $\mathfrak{D}$ to be the connected component defined by $\operatorname{Im}(z_1) > 0$, $\operatorname{Im}(z_2) > 0$.

Put

$$\varepsilon_1 = -f_1 f_2 f^* f\,, \qquad \varepsilon_2 = -f_1 f_2 f f^*\,,$$
$$\varepsilon_3 = -f_2 f_1 f^* f\,, \qquad \varepsilon_4 = -f_2 f_1 f f^*\,.$$

Then ε_i's are mutually orthogonal idempotent elements in $L' = C^+(L)$ such that $\sum \varepsilon_i = 1$. Hence one has decompositions of V' and L' into the direct sum of left ideals.

$$(7) \qquad V' = \sum_{i=1}^{4} V'\varepsilon_i\,, \qquad L' = \sum_{i=1}^{4} L'\varepsilon_i\,.$$

Our second assumption is that a is given by $a_0 = (f_1 + f_2)(f^* - f)$, which surely satisfies condition (4). (Actually, $-(1/2)a_0$ is an H-element for G.) The $V'\varepsilon_i$'s are then mutually orthogonal with respect to A'_{a_0}, so that the modular imbedding $\tilde{\rho}_{a_0}$ decomposes into the direct sum of those defined by $V'\varepsilon_i$, which we call $\tilde{\rho}_{a_0}^{(i)}$ $(1 \leqq i \leqq 4)$. Since they can be treated quite similarly, let us consider only $\tilde{\rho}_{a_0}^{(1)}$.

For $L'\varepsilon_1$, one has a direct decomposition *à la Frobenius* as follows:

$$L'\varepsilon_1 = U \cap L'\varepsilon_1 + U^* \cap L'\varepsilon_1\,,$$

where

$$U \cap L'\varepsilon_1 = fC^-(L)\varepsilon_1 = C^+(L_0) f_2 f + C^-(L_0) f_1 f_2 f\,,$$
$$U^* \cap L'\varepsilon_1 = f^* C^-(L)\varepsilon_1 = C^+(L_0) f_1 f_2 f^* f + C^-(L_0) f_2 f^* f\,.$$

Let $\mathfrak{B}_0^{\pm}$ be a basis of $C^{\pm}(L_0)$; then a basis of $U \cap L'\varepsilon_1$ and $U^* \cap L'\varepsilon_1$ are given respectively by

$$\mathfrak{B}^{(1)} = (\mathfrak{B}_0^+ f_2 f, \mathfrak{B}_0^- f_1 f_2 f) , \qquad \mathfrak{B}^{(1)*} = (\mathfrak{B}_0^+ f_1 f_2 f^* f, \mathfrak{B}_0^- f_2 f^* f) .$$

If one puts

$$\tilde{S}_0^{\pm} = \left(\mathrm{tr}_0(x_0^\iota y_0)\right)_{x_0, y_0 \in \mathfrak{B}_0^{\pm}} ,$$

where tr_0 denotes the (regular) trace in $C^+(V_0)$, a simple calculation shows that one has

$$\left(A'_{a_0}(x, y)\right)_{x \in \mathfrak{B}^{(1)}, y \in \mathfrak{B}^{(1)*}} = \begin{pmatrix} 4\tilde{S}_0^+ & 0 \\ 0 & 4\tilde{S}_0^- \end{pmatrix} .$$

It follows that the elementary divisor of $(L'\varepsilon_1, A'_{a_0})$ is equal to $\mathrm{diag}(\delta^+, \delta^-)$, where $\delta^{\pm}$ are the elementary divisors of $4\tilde{S}_0^{\pm}$, and that a canonical basis of $V'\varepsilon_1$ is given by

$$\left(-\frac{1}{4}(\mathfrak{B}_0^+ f_2 f)\tilde{S}_0^{+-1}, \frac{1}{4}(\mathfrak{B}_0^- f_1 f_2 f)\tilde{S}_0^{--1}, \mathfrak{B}_0^+ f_1 f_2 f^* f, -\mathfrak{B}_0^- f_2 f^* f\right) .$$

In terms of this basis, the left multiplication of zf is expressed by the matrix

$$(8) \qquad Z^{(1)}(z) = \begin{pmatrix} z_2\tilde{S}_0^+ & \tilde{S}_0^+ E_{z_0}^- \\ \tilde{S}_0^- E_{z_0}^+ & z_1\tilde{S}_0^- \end{pmatrix} ,$$

where $E_{z_0}^{\pm}$ are matrices defined by

$$z_0(\mathfrak{B}_0^{\pm}) = (\mathfrak{B}_0^{\mp})E_{z_0}^{\pm} .$$

It is easy to see that an analytic expression of $\tilde{\rho}_{a_0}^{(3)}$ may be obtained simply by interchanging z_1 and z_2 in (8), while the modular imbeddings $\tilde{\rho}_{a_0}^{(2)}$ and $\tilde{\rho}_{a_0}^{(4)}$ are equivalent to $\tilde{\rho}_{a_0}^{(3)}$ and $\tilde{\rho}_{a_0}^{(1)}$, respectively. All the modular imbeddings $\tilde{\rho}_{a_0}^{(i)}$ have the same elementary divisor (δ^+, δ^-). (In principle, $\delta^{\pm}$ are computable. The only primes possibly appearing in $\delta^{\pm}$ are 2 and the prime divisors of $\det(2S_0)$.)

2. Relation between the fields of automorphic functions

2.1. The notation being as in § 1, we denote by $\mathfrak{V}^*$ the standard compactification of the quotient space $\mathfrak{V} = \Gamma_0\backslash\mathfrak{D}$, which carries a natural structure of (projective) algebraic variety. The function field $\mathbf{C}(\mathfrak{V}^*)$ of $\mathfrak{V}^*$ may be identified with the field of automorphic functions on $\mathfrak{D}$ with respect to Γ_0. Similarly, let $\mathfrak{V}'^*$ denote the standard compactification of $\mathfrak{V}' = \mathfrak{V}'_a = \Gamma'_a\backslash\mathfrak{D}'_a$. It is known that the holomorphic mapping ρ of $\mathfrak{V}$ into $\mathfrak{V}'$ can be extended to a morphism (of algebraic variety), denoted again by ρ, of $\mathfrak{V}^*$ into $\mathfrak{V}'^*$ [9]. Hence $\rho(\mathfrak{V}^*)$ is an algebraic subvariety of $\mathfrak{V}'^*$, and for a function $f' \in C(\mathcal{V}'^*)$, which is defined (generically) at $\rho(\mathcal{V}^*)$, one has $f = f' \circ \rho \in C(\mathcal{V}^*)$. The correspondence $f' \mid \rho(\mathcal{V}^*) \mapsto f' \circ \rho$ gives clearly an isomorphism of $C(\rho(\mathcal{V}^*))$ onto a subfield of $C(\mathcal{V}^*)$, which may also be interpreted as the field of all automorphic functions on $\mathfrak{D}$ with respect to Γ_0 which are extendible to an automorphic function on $\mathfrak{D}'$ with respect to Γ'. In the following, we will always identify $C(\rho(\mathcal{V}^*))$ with this subfield of $C(\mathcal{V}^*)$. Our purpose here is to investigate the extension $C(\mathcal{V}^*)/C(\rho(\mathcal{V}^*))$.

For that purpose, we introduce the following notation:

$$N(\rho(\mathfrak{D})) = \{g' \in G' \mid g'\rho(\mathfrak{D}) = \rho(\mathfrak{D})\} ,$$
$$Z(\rho(\mathfrak{D})) = \{g' \in N(\rho(\mathfrak{D})) \mid g' \mid \rho(\mathfrak{D}) = \text{id.}\} ,$$

and put

$$(9) \qquad \Gamma'_{\rho(\mathfrak{D})} = \Gamma' \cap N(\rho(\mathfrak{D}))/\Gamma' \cap Z(\rho(\mathfrak{D})) .$$

$\Gamma'_{\rho(\mathfrak{D})}$ is a group acting discontinuously on $\rho(\mathfrak{D})$ and contains

$$\overline{\rho(\Gamma_0)} = \rho(\Gamma_0)/\rho(\Gamma_0) \cap Z(\rho(\mathfrak{D}))$$

as a subgroup of finite index. Using the results in [9], one can show easily (and actually in a more general setting) that *there exist a standard compactification $\mathcal{V}_1^*$ of $\Gamma'_{\rho(\mathfrak{D})}\backslash\rho(\mathfrak{D})$ and a morphism ρ_1 of $\mathcal{V}_1^*$ onto $\rho(\mathcal{V}^*)$, which is a birational correspondence, such that one can factor the morphism ρ into a covering map $\mathcal{V}^* \longrightarrow \mathcal{V}_1^*$ and ρ_1* (cf. also [2]). Thus an automorphic function on $\mathfrak{D}$ with respect to Γ_0 is extensible to an automorphic function on $\mathfrak{D}'$ with respect to Γ' if and only if it is invariant under $\Gamma'_{\rho(\mathfrak{D})}$. In particular, one has

$$(10) \qquad [C(\mathcal{V}^*) : C(\rho(\mathcal{V}^*))] = [\Gamma'_{\rho(\mathfrak{D})} : \overline{\rho(\Gamma_0)}] .$$

2.2. In order to calculate the right hand side of (10), we first determine $\Gamma' \cap Z(\rho(\mathfrak{D}))$. Denoting by K'_Z the stabilizer of $Z \in \mathfrak{D}'$ in G', one has

$$\begin{aligned} Z(\rho(\mathfrak{D})) &= \textstyle\bigcap_{z \in \mathfrak{D}} K'_{\rho(z)} \\ &= \textstyle\bigcap_{g \in G} \rho(g) K' \rho(g)^{-1} \\ &= Z_{G'}(\{\rho(g) I_0 \rho(g)^{-1} \mid g \in G\}) , \end{aligned}$$

where K' is the maximal compact subgroup of G' corresponding to I_0 and $Z_{G'}(\ )$ denotes the centralizer in G'. Hence, by Kuga's lemma (see Appendix), one has

$$\Gamma' \cap Z(\rho(\mathfrak{D})) = \Gamma' \cap Z_{G'}(\rho(G)) .$$

On the other hand, since $\rho(G)$ is linearly dense in C^+, one has

$$Z_{G'}(\rho(G)) = \{R_v \mid v \in C^+, \; vav^\iota = a\} ,$$

R_v denoting the right multiplication of $v \in C^+$. Therefore, if one puts

$$\hat{K}_a = \{v \in C^+ \mid vav^\iota = a\} , \qquad \Delta_a^0 = C^+(L)^\times \cap \hat{K}_a ,$$

$C^+(L)^\times$ denoting the unit group of the order $C^+(L)$, Δ_a^0 is a finite group and the natural correspondence $v \mapsto R_v$ gives an isomorphism

$$\Gamma' \cap Z(\rho(\mathfrak{D})) \cong \Delta_a^0 . \tag{11}$$

Remark. Using the argument in the proof of [8, Prop. 4], one can actually prove that $Z(\rho(\mathfrak{D})) = Z_{G'}(\rho(G))$. Note also that $\hat{K}_a$ is the "orthogonal group" of a positive involution $v \mapsto av^\iota a^{-1}$ and thus a maximal compact subgroup of the multiplicative group of C^+.

2.3. Next, let us denote by $H(\mathfrak{D})$ the group of all holomorphic automorphisms of $\mathfrak{D}$. The connected component of the identity of $H(\mathfrak{D})$ is identified with $\bar{G} = G/Z$, Z denoting the center of G. More precisely, one has

$$[H(\mathfrak{D}) : \bar{G}] = \begin{cases} 1 & \text{if } p \equiv 1 \pmod 2 , \\ 2 & \text{if } p \equiv 0 \pmod 2 . \end{cases}$$

By a natural correspondence, $N(\rho(\mathfrak{D}))/Z(\rho(\mathfrak{D}))$ is isomorphic to a subgroup of $H(\mathfrak{D})$, and, under this isomorphism, one has

$$\rho(G) \cdot Z(\rho(\mathfrak{D}))/Z(\rho(\mathfrak{D})) \cong \bar{G} .$$

Therefore, if p is odd, one has $N(\rho(\mathfrak{D})) = \rho(G) \cdot Z(\rho(\mathfrak{D}))$. In case p is even, it can easily be seen that there actually exists an element in $N(\rho(\mathfrak{D}))$ inducing on $\mathfrak{D} \approx \rho(\mathfrak{D})$ an automorphism not belonging to $\bar{G}$ (e.g., the automorphism $(z_1, z_2, z_0) \mapsto (z_2, z_1, z_0)$ in the expression (1″)). Thus the natural homomorphism $N(\rho(\mathfrak{D})) \to H(\mathfrak{D})$ is always surjective and one has

$$N(\rho(\mathfrak{D}))/Z(\rho(\mathfrak{D})) \cong H(\mathfrak{D}) . \tag{12}$$

Now we pass to the determination of $\Gamma' \cap N(\rho(\mathfrak{D}))$ and first assume that p is odd. Then, by the *Remark* in 2.2, every element $g' \in N(\rho(\mathfrak{D})) = \rho(G) \cdot Z(\rho(\mathfrak{D}))$ can be written in the form

$$g' = \rho(g) \cdot R_v \qquad \text{with } g \in G, \; v \in \hat{K}_a ,$$

and the pair (g, v) is uniquely determined up to (ζ, ζ^{-1}) with $\zeta \in Z$. Moreover, for such a g', one has

$$g' \in \Gamma' \Longleftrightarrow g'(C^+(L)) = g \cdot C^+(L) \cdot v = C^+(L)$$
$$\Longleftrightarrow \begin{cases} g \cdot C^+(L) \cdot g^{-1} = C^+(L), \text{ and} \\ gv \in C^+(L)^\times . \end{cases}$$

The first condition for g is equivalent to saying that $gLg^{-1} = L$ by virtue of the following Lemma.

LEMMA 2. *Let $L_i (i = 1, 2)$ be two lattices in V such that $S(L_i) \subset \mathbf{Z}$. If $C^+(L_1) \subset C^+(L_2)$ in $C^+(V)$, then one has $L_1 \subset L_2$.*

PROOF. Let $(f_j^{(i)})_{1 \leqq j \leqq n}$ be a basis of L_i and put $(f_j^{(1)}) = (f_j^{(2)})P$ with $P \in \mathrm{GL}(n, \mathbf{R})$. Then, $(f_{j_1}^{(i)} \cdots f_{j_r}^{(i)})$ $(1 \leqq j_1 < \cdots < j_r \leqq n$, r: even) forms a basis of $C^+(L_i)$. Put

$$(f_{j_1}^{(1)} \cdots f_{j_r}^{(1)}) = (f_{j_1}^{(2)} \cdots f_{j_r}^{(2)})\tilde{P}$$

with $\tilde{P} \in \mathrm{GL}(2^{n-1}, \mathbf{R})$. Then, by an easy induction one can show that $\tilde{P}$ is of the form

$$\tilde{P} = \begin{bmatrix} 1 & * & * & \cdots \\ 0 & \bigwedge_2(P) & * & \cdots \\ 0 & 0 & \bigwedge_4(P) & \cdots \\ \vdots & \vdots & \vdots & \ddots \end{bmatrix}.$$

Since, by the assumption, $\tilde{P}$ is integral, $\bigwedge_2(P)$ is integral, and hence so is also P. Therefore, one has $L_1 \subset L_2$, q.e.d.

Now, let Γ be the subgroup of G formed of all $g \in G$ such that $gLg^{-1} = L$, and put

$$\Delta_a = (\Gamma \cdot C^+(L)^\times) \cap \hat{K}_a .$$

Then we have seen that the correspondence $g' \mapsto v$ defines a homomorphism of $\Gamma' \cap N(\rho(\mathfrak{D}))$ onto Δ_a/Z. The kernel of this homomorphism is clearly equal to $\{g' = \rho(\gamma) \mid \gamma \in \Gamma_0\}$. Thus one has

$$\Gamma' \cap N(\rho(\mathfrak{D}))/\rho(\Gamma_0) \cong \Delta_a/Z . \tag{13}$$

Under this isomorphism, one has by (11)

$$\rho(\Gamma_0) \cdot (\Gamma' \cap Z(\rho(\mathfrak{D})))/\rho(\Gamma_0) \cong \Delta_a^0 \cdot Z/Z .$$

On the other hand, the isomorphism (12) induces an isomorphism

$$\rho(\Gamma_0) \cdot (\Gamma' \cap Z(\rho(\mathfrak{D})))/\Gamma' \cap Z(\rho(\mathfrak{D})) \cong \bar{\Gamma}_0 (= \Gamma_0/\Gamma_0 \cap Z) .$$

Hence, $\Gamma'_{\rho(\mathfrak{D})}$ may be identified with a discrete subgroup of $\bar{G}$, containing $\bar{\Gamma}_0$ as a normal subgroup, and one has

(14) $$\Gamma'_{\rho(\mathfrak{D})}/\bar{\Gamma}_0 \cong \Delta_a/\Delta_a^0\cdot Z .$$

In case p is even, we put

$$G^{\pm} = \{g \in C^{\pm} \mid g^{\iota}g = 1, gVg^{-1} = V\} \qquad (G^{+} = G) ,$$
$$\hat{K}_a^{\pm} = \{v \in C^{\pm} \mid vav^{\iota} = a\} \qquad (\hat{K}_a^{+} = \hat{K}_a) .$$

Then it is easy to see that the linear transformation g' of V' defined by

$$g'(x) = gxv \qquad \text{with } g \in G^{\pm} \text{ and } v \in \hat{K}_a^{\pm}$$

(where the $\pm$'s in $G^{\pm}$ and $\hat{K}_a^{\pm}$ are concordant) belongs to $N(\rho(\mathfrak{D}))$ and (g, v) is uniquely determined by g' up to (ζ, ζ^{-1}) with $\zeta \in Z$. If we call N_1 the subgroup of $N(\rho(\mathfrak{D}))$ formed of all such g', the correspondence $g' \mapsto g$ gives rise to an isomorphism

$$N_1/Z\big(\rho(\mathfrak{D})\big) \cong (G^{+} \cup G^{-})/Z(\supsetneqq \bar{G}) .$$

Therefore, one must have $N(\rho(\mathfrak{D})) = N_1$. (Incidentally this again proves (12).) We put further $\Gamma^{\pm} = \{g \in G^{\pm} \mid gLg^{-1} = L\}$ and

$$\Delta_a^{\pm} = \big(\Gamma^{\pm}\cdot C^{+}(L)^{\times}\big) \cap \hat{K}_a^{\pm} .$$

Then, by the same argument as above, one has

(13′) $$\Gamma' \cap N\big(\rho(\mathfrak{D})\big)/\rho(\Gamma_0) \cong (\Delta_a^{+} \cup \Delta_a^{-})/Z ,$$

and hence

(14′) $$\Gamma'_{\rho(\mathfrak{D})}/\bar{\Gamma} \cong \big(\Delta_a^{+} \cup \Delta_a^{-}\big)/\Delta_a^0\cdot Z .$$

Remark. Γ^{-} and so Δ_a^{-} may be empty. If there exists an element $e \in L$ such that $S(e) = \pm 1$, then $e \in \Gamma^{-}$ and so $\Gamma^{-} \neq \varnothing$. The above argument shows that there exists an element in $\Gamma' \cap N(\rho(\mathfrak{D}))$ inducing on $\mathfrak{D} \approx \rho(\mathfrak{D})$ a holomorphic automorphism not belonging to $\bar{G}$, if and only if $\Delta_a^{-} \neq \varnothing$.

2.4. Summing up, we obtain the following results.

THEOREM. *The notation being as above, $\rho(\mathfrak{V}^*)$ is birationally equivalent to the quotient variety of $\mathfrak{V}^*$ by a finite group of automorphisms which is isomorphic to $\Delta_a/\Delta_a^0\cdot Z$ if p is odd, and to $(\Delta_a^{+} \cup \Delta_a^{-})/\Delta_a^0\cdot Z$ if p is even.*

COROLLARY. *Suppose a is taken in a general position. Then, if $p \neq 0 \pmod 4$, $\mathfrak{V}^*$ is birationally equivalent to $\rho(\mathfrak{V}^*)$. In case $p \equiv 0 \pmod 4$, $\mathbf{C}(\mathfrak{V}^*)/\mathbf{C}(\rho(\mathfrak{V}^*))$ is a cyclic extension.*

Put $\tilde{Z} = \{\zeta \in \text{center of } C^{+} \mid \zeta\zeta^{\iota} = 1\}$. Then one has

$$\left.\begin{array}{l}\bigcap_a \hat{K}_a \\ \bigcap_a (\hat{K}_a^+ \cup \hat{K}_a^-)\end{array}\right\} = \tilde{Z} \qquad \begin{array}{l}\text{if } p \text{ is odd ,} \\ \text{if } p \text{ is even}\end{array}$$

and

$$\tilde{Z} = \begin{cases} \{\pm 1\} & \text{if } p \equiv 1\,(2)\,, \\ \{(\pm 1, \pm 1)\} & \text{if } p \equiv 2\,(4)\,, \\ \{(\zeta, \zeta^{-1}) \mid \zeta \in \mathbf{C}, \zeta\bar{\zeta} = 1\} & \text{if } p \equiv 0\,(4)\,. \end{cases}$$

Hence, in case p is odd, for a in a general position, one has

$$\Delta_a = \big(\Gamma \cdot C^+(L)^\times\big) \cap \tilde{Z} = Z\,,$$

and the group $\Delta_a/\Delta_a^0 \cdot Z$ reduces to the identity element. The same is also true for the case $p \equiv 2\,(4)$. (For p even, one has $\Delta_a^- = \varnothing$.) In case $p \equiv 0\,(4)$, one has

$$\Delta_a^+ \cup \Delta_a^- = (\Gamma \cdot C^+(L)^\times) \cap \tilde{Z}\,,$$
$$\Delta_a^0 \cdot Z = \big(C^+(L)^\times \cap \tilde{Z}\big) \cdot Z\,,$$

so that $(\Delta_a^+ \cup \Delta_a^-)/\Delta_a^0 \cdot Z$ is a certain finite cyclic group depending only on the choice of L. This completes the proof.

Appendix

Let G be a connected semi-simple linear Lie group of hermitian type, K a maximal compact subgroup of G, Γ a discrete subgroup of G such that the volume of $\Gamma\backslash G$ is finite. By a *modular imbedding* of the triple (G, K, Γ), we mean a 5-tuple $\tilde{\rho} = (\rho, V', A', I_0, L')$ satisfying the following conditions:

- V': a $2n'$-dimensional vector-space over $\mathbf{R}$,
- A': a non-degenerate alternating bilinear form on V',
- I_0 : a complex structure on V' such that $A'I_0$ is symmetric and positive-definite,
- L' : a lattice in V' such that $A'(L', L') \subset \mathbf{Q}$,
- ρ : a symplectic representation $G \longrightarrow G' = \mathrm{Sp}(V', A')$ satisfying the condition (H_1) for the given H-element H_0 corresponding to K and for $H_0' = (1/2)I_0$ and such that $\rho(\Gamma) \subset \mathrm{Sp}(L', A')$.

We say that two modular imbeddings $\tilde{\rho}_i = (\rho_i, V_i', A_i', I_{0i}, L_i')$ ($i = 1, 2$) are *equivalent* if there is a linear isomorphism φ of V_1' onto V_2' and $\lambda \in \mathbf{R}$ such that one has

$$(\mathrm{E})\qquad \begin{cases} \varphi\rho_1(g)\varphi^{-1} = \rho_2(g) & \text{for all } g \in G\,, \\ \lambda\,{}^t\varphi^{-1}A_1'\varphi^{-1} = A_2'\,, \\ \varphi I_{01}\varphi^{-1} = I_{02}\,, \\ \varphi L_1' = L_2'\,. \end{cases}$$

Note that if one denotes by δ_i the elementary divisor of (L_i', A_i'), one has $\lambda\delta_1 = \delta_2$; in particular, λ is a positive rational number. We also consider the following (apparently) weaker condition of equivalence

$$(\mathrm{E}')\qquad \begin{cases} \varphi\rho_1(g)I_{01}\rho_1(g)^{-1}\varphi^{-1} = \rho_2(g)I_{02}\rho_2(g)^{-1} & \text{for all } g \in G\,, \\ \lambda\,{}^t\varphi^{-1}A_1'\varphi^{-1} = A_2'\,, \\ \varphi L_1' = L_2'\,. \end{cases}$$

LEMMA (Kuga). *When G has no compact simple component, the conditions* (E) *and* (E′) *are equivalent.*

Obviously it is sufficient to show that (E′) implies the first condition in (E). Suppose (E′) is satisfied by φ and λ, and call Φ the set of all φ's satisfying (E′) for the same λ. Then Φ is a non-empty finite set, which is invariant under the following action of $\gamma \in \Gamma$:

$$\Phi \ni \varphi \longmapsto \rho_2(\gamma)^{-1}\varphi\rho_1(\gamma) \in \Phi\,.$$

Hence there is a subgroup Γ_0 of Γ of finite index such that one has

$$\rho_2(\gamma)^{-1}\varphi\rho_1(\gamma) = \varphi \qquad \text{for all } \varphi \in \Phi,\ \gamma \in \Gamma_0\,.$$

Since the volume of $\Gamma_0\backslash G$ is finite, it follows, by a density theorem of Borel, that one has

$$\rho_2(g)^{-1}\varphi\rho_1(g) = \varphi \qquad \text{for all } \varphi \in \Phi,\ g \in G\,,$$

which proves our assertion.

Remark. Given a modular imbedding $\tilde{\rho} = (\rho, V', A', I_0, L')$ such that $A'(L', L') \subset \mathbf{Z}$, one obtains a (uniformized) analytic family of polarized abelian varieties over $\mathfrak{D} = G/K$:

$$\{P_z = (V'/L',\ \rho(g)I_0\rho(g)^{-1},\ A')\ (z = gK \in \mathfrak{D})\}$$

(see [7, Appendix]). The above condition (E′) is equivalent to saying that the two families over $\mathfrak{D}$ obtained from $\tilde{\rho}_1$ and $\tilde{\rho}_2$ in this manner are equivalent (over the identity map of $\mathfrak{D}$) in an obvious sense.

Taking a canonical basis $(\mathfrak{B}, \mathfrak{B}^*)$ of V' with respect to (L', A') and representing $\rho(z)$ by a point in the Siegel's upper half space of degree n' (as explained in **1.2**), one obtains an *analytic expression* $Z = Z(z)$ $(z \in \mathfrak{D})$ of the modular imbedding $\tilde{\rho}$. It follows immediately from Kuga's lemma that, if $Z_i = Z_i(z)$ $(z \in \mathfrak{D})$ are analytic expressions of two modular imbeddings $\tilde{\rho}_i$ $(i = 1, 2)$ (of the same degree n') and if δ_i are the respective elementary divisors, then $\tilde{\rho}_1$ and $\tilde{\rho}_2$ are equivalent if and only if one has

$$(\mathrm{E}'')\qquad \begin{cases} \lambda\delta_1 = \delta_2 , \\ \lambda Z_1 = (AZ_2 + B)(CZ_2 + D)^{-1} , \end{cases}$$

where A, B, C, D are n' by n' matrices such that

$$\begin{pmatrix} A & B \\ C & D \end{pmatrix} \in \mathrm{Sp}(2n', \mathbf{R})$$

and

$$\begin{pmatrix} \delta_2 & 0 \\ 0 & 1 \end{pmatrix}^{-1} \begin{pmatrix} A & B \\ C & D \end{pmatrix} \begin{pmatrix} \delta_2 & 0 \\ 0 & 1 \end{pmatrix} \in \mathrm{GL}(2n', \mathbf{Z}) .$$

UNIVERSITY OF CALIFORNIA, BERKELEY

REFERENCES

[1] M. EICHLER, Quadratische Formen und orthogonale Gruppen, Springer-Verlag, Berlin-Göttingen-Heidelberg, 1952.

[2] E. FREITAG, *Fortsetzung von automorphen Funktionen*, Math. Ann. **177** (1968), 231-247.

[3] W. F. HAMMOND, *The modular groups of Hilbert and Siegel*, Amer. J. Math. **88** (1966), 497-516.

[4] H. KLINGEN, *Über einen Zusammenhang zwischen Siegelschen und Hermitschen Modulfunktionen*, Abh. Math. Sem. Univ. Hamburg **27** (1964), 1-12.

[5] K. IYANAGA, *Arithmetic of special unitary groups and their symplectic representations*, J. Fac. Sci., Univ. of Tokyo, Sec. I, **15** (1968), 35-69.

[6] I. SATAKE, *Holomorphic imbeddings of symmetric domains into a Siegel space*, Amer. J. Math. **87** (1965), 425-461.

[7] ———, *Symplectic representations of algebraic groups satisfying a certain analyticity condition*, Acta Math. **117** (1967), 215-279.

[8] ———, *Clifford algebras and families of abelian varieties*, Nagoya Math. J. **27** (1966), 435-446; *Correction*, *ibid.* **31** (1968), 295-296.

[9] ———, *A note on holomorphic imbeddings and compactification of symmetric domains*, Amer. J. Math. **90** (1968), 231-247.

(Received August 6, 1968)

The curvature of 4-dimensional Einstein spaces*

I. M. SINGER AND J. A. THORPE

It is well known that the diagonalization or normal form of a self adjoint linear transformation T on a real inner product space $\{V, \langle , \rangle\}$ is equivalent to the analysis of the critical point behavior of the function $v \to \langle Tv, v \rangle$ on the unit sphere (actually projective space) of V. It is also classical that the curvature tensor at a point of a riemannian manifold is completely determined by the sectional curvature function σ on the Grassman manifold of 2-planes at the point.

It seems natural, then, to attack the problem of a "normal form" for the curvature tensor by analyzing the critical point behavior of the sectional curvature function σ. We take a small step in this direction when the curvature tensor comes from a 4-dimensional oriented Einstein manifold where a canonical form (Petrov) is known. In this case, the function σ on each 2-plane is equal to its value on the orthogonal complement. Using this characterization, we show that the curvature function σ is completely determined by its critical point behavior and we show what the locus of critical points looks like. This leads to the Petrov canonical form for the curvature tensor and we relate the parameters in the canonical form to the critical values of σ.

In higher dimensions, the analogue would be to take the Weyl curvature tensor part of the curvature tensor and ask whether or not the critical point behavior of its sectional curvature function determines the tensor. We state this problem then: if two such functions have the same critical points and same values at these points, are the corresponding Weyl curvature tensors equal?

1. The space of curvature tensors

Let V be an n-dimensional vector space over the reals with inner product $\langle , \rangle$. Let Λ^2 denote the space of 2-vectors of V, with

* Research partially supported by the National Science Foundation.

inner product given by

$$\langle u_1 \wedge u_2, v_1 \wedge v_2 \rangle = \det[\langle u_i, v_j \rangle], \qquad (u_i, v_i \in V)$$

A *curvature tensor* on V is a symmetric linear transformation $R: \Lambda^2 \to \Lambda^2$. The curvature tensors on V form a vector space $\mathcal{R}$ with inner product given by $\langle R, S \rangle = \operatorname{trace} RS$. Using the usual isomorphisms defined by the inner product, a curvature tensor on V may be regarded as a 2-form on V with values in the vector space of skew symmetric endomorphisms of V. The *sectional curvature* σ_R of $R \in \mathcal{R}$ is the real valued function defined on Grassmann manifold G of oriented 2-dimensional subspaces of V by $\sigma_R(P) = \langle RP, P \rangle$ $(P \in G \subset \Lambda^2)$ where G is identified with the set of decomposable 2-vectors of length 1. The orthogonal group $O(V)$ of V acts isometrically on Λ^2 by $g(u \wedge v) = (gu) \wedge (gv)$, $g \in O(V)$, $u, v \in V$. $O(V)$ also acts isometrically on $\mathcal{R}$ by $g(R) = g^{-1}Rg: \Lambda^2 \to \Lambda^2$.

Consider now the following linear maps: the *Bianchi map* $b: \mathcal{R} \to \mathcal{R}$ defined by

$$[b(R)](u_1, u_2)(u_3) = \sum_\alpha R(u_{\alpha(1)}, u_{\alpha(2)})(u_{\alpha(3)})$$

where α runs through all cyclic permutations of (1, 2, 3); the *Ricci contraction* $r: \mathcal{R} \to \mathcal{T}$, where $\mathcal{T}$ is the space of symmetric linear transformations of V, defined by

$$\langle r(R)(v), w \rangle = \operatorname{tr}\{u \to R(v, u)(w)\};$$

and the *trace functional* $\operatorname{tr}: \mathcal{R} \to \mathbf{R}$. Each of these maps is equivariant with respect to the $O(V)$-action on the spaces involved ($O(V)$ acts trivially on $\mathbf{R}$) and hence the kernel of each of these maps is an $O(V)$-invariant subspace of $\mathcal{R}$. Since $O(V)$ acts isometrically on $\mathcal{R}$, orthogonal complements of invariant subspaces are invariant. Set

$$\mathcal{R}_1 = (\operatorname{Ker} b)^\perp, \qquad \mathcal{R}_2 = (\operatorname{Ker} \operatorname{tr})^\perp,$$

$$\mathcal{R}_3 = (\operatorname{Ker} r) \cap (\operatorname{Ker} b), \qquad \mathcal{R}_4 = (\operatorname{Ker} r)^\perp \cap (\operatorname{Ker} \operatorname{tr}).$$

THEOREM 1.1. $\mathcal{R} = \sum_{i=1}^4 \oplus \mathcal{R}_i$. *Moreover,*

(i) $R \in \mathcal{R}_1 \Leftrightarrow$ *the sectional curvature of R is identically zero.*

(ii) $R \in \mathcal{R}_1 \oplus \mathcal{R}_2 \Leftrightarrow$ *the sectional curvature of R is constant.*

(iii) $R \in \mathcal{R}_1 \oplus \mathcal{R}_3 \Leftrightarrow$ *the Ricci tensor of R is zero.*

(iv) $R \in \mathcal{R}_1 \oplus \mathcal{R}_2 \oplus \mathcal{R}_3 \Leftrightarrow$ *the Ricci tensor of R is a scalar multiple of the identity.*

(v) $R \in \mathcal{R}_1 \oplus \mathcal{R}_3 \oplus \mathcal{R}_4 \Leftrightarrow$ *the scalar curvature of R is zero.*

(vi) $R \in \mathcal{R}_2 \oplus \mathcal{R}_3 \oplus \mathcal{R}_4 \Leftrightarrow R$ *satisfies the first Bianchi identity.*

PROOF. (i) $R \in \mathcal{R}_1$ implies $\operatorname{tr} RS = \langle R, S \rangle = 0$ for all $S \in \mathcal{R}$ with $b(S) = 0$. For $P \in G$, the tensor $S_P \in \mathcal{R}$ defined by $S_P(P) = P$ and $S_P(Q) = 0$ whenever $\langle Q, P \rangle = 0$ has $b(S_P) = 0$ and so $0 = \operatorname{tr} RS_P = \sigma_R(P)$. Conversely, if $R \in \mathcal{R}$ is such that $\sigma_R = 0$ then $R = R_1 + R'$ where $R_1 \in \mathcal{R}_1$ and $b(R') = 0$. Then $\sigma_{R'} = \sigma_R - \sigma_{R_1} = 0$, so $R' = 0$ by the usual proof that zero sectional curvature implies zero curvature tensor in the presence of the Bianchi identity.

(ii) First note that $\mathcal{R}_1 \subset \operatorname{Ker} r \subset \operatorname{Ker} \operatorname{tr}$: the first inclusion is because

$$\langle R(v, u)w, u \rangle = \frac{1}{2}[\langle R(v + w, u)(v + w), u \rangle - \langle R(v, u)v, u \rangle - \langle R(w, u)w, u \rangle]$$

and each term on the right is zero for $R \in \mathcal{R}_1$; the second is because $\operatorname{tr} r(R) = 2 \operatorname{tr} R$. Using these inclusions, it is an elementary verification that $\mathcal{R}$ is the orthogonal direct sum $\sum \oplus \mathcal{R}_i$. Furthermore, since $\operatorname{Ker} \operatorname{tr}$ has codimension 1, $\dim \mathcal{R}_2 = 1$ and hence $\mathcal{R}_2$ consists of all scalar multiples of the identity map $I: \Lambda^2 \to \Lambda^2$. But for $b(R) = 0$ it is well known that $R = aI$ if and only if $\sigma_R \equiv a$.

(iii) and (vi) are clear since $\mathcal{R}_1 \subset \operatorname{Ker} r \subset \operatorname{Ker} \operatorname{tr}$.

(iv) clear, since r maps $\mathcal{R}_2$ isomorphically onto the multiples of the identity in $\mathcal{T}$ and $\operatorname{Ker} r = \mathcal{R}_1 \oplus \mathcal{R}_3$.

Remark. The subspaces $\mathcal{R}_i$ are the minimal invariant subspaces of $\mathcal{R}$ under the action of $O(V)$. (See [6, p. 157].) The $\mathcal{R}_3$-component of a curvature tensor is its Weyl conformal curvature tensor.

Remark. Since a curvature tensor R coming from a riemannian manifold satisfies the Bianchi identity, we have $R \in \mathcal{R}_1^\perp$. From Theorem 1.1 (i), R_1 contributes nothing to the sectional curvature function. We can assume, then, that all our curvature tensors lie in $\mathcal{R}_1^\perp = \mathcal{R}_2 \oplus \mathcal{R}_3 \oplus \mathcal{R}_4$.

Remark. Let $\mathcal{E} = \mathcal{R}_2 \oplus \mathcal{R}_3$. Statement (iv) of the above theorem says that a riemannian manifold M has curvature tensor in $\mathcal{E}$ at each point if and only if M is an *Einstein space.*

THEOREM 1.2. *For $n > 2$, the Ricci contraction r maps $\mathcal{R}_2 \oplus \mathcal{R}_4$ isomorphically onto $\mathcal{T}$. The inverse map $s: \mathcal{T} \to \mathcal{R}_2 \oplus \mathcal{R}_4$ is given by*

$$s(T)(u_1, u_2)(u_3) = \frac{1}{n-2}[\langle u_1, u_3\rangle Tu_2 - \langle u_2, u_3\rangle Tu_1 + \langle Tu_1, u_3\rangle u_2 - \langle Tu_2, u_3\rangle u_1 - \frac{1}{n-1}(\operatorname{tr} T)(\langle u_1, u_3\rangle u_2 - \langle u_2, u_3\rangle u_1)]\,. \qquad (T \in \mathcal{T},\ u_i \in V)$$

PROOF. Since $\mathcal{R}_2 \oplus \mathcal{R}_4$ is the orthogonal complement of Ker r, it is clear that r maps $\mathcal{R}_2 \oplus \mathcal{R}_4$ isomorphically onto the image of r. An elementary computation verifies that $r \circ s$ is the identity and hence the image of r equals $\mathcal{T}$. To complete the proof, it suffices check that the image of s is contained in $\mathcal{R}_2 \oplus \mathcal{R}_4$; i.e., that, for $T \in \mathcal{T}$, $s(T) \perp \mathcal{R}_1 \oplus \mathcal{R}_3$. To check this, let $\{e_1, \cdots, e_n\}$ be an orthonormal basis for V consisting of eigenvectors of T. Let $\lambda_1, \cdots, \lambda_n$ denote the corresponding eigenvalues. Then, for $1 \leqq i \leqq j \leqq n$,

$$(*) \qquad s(T)(e_i \wedge e_j) = \frac{1}{n-2}\left[\lambda_i + \lambda_j - \frac{1}{n-1}(\operatorname{tr} T)\right] e_i \wedge e_j$$

so $\{e_i \wedge e_j \mid i < j\}$ is an orthonormal basis for Λ^2 consisting of eigenvectors of $s(T)$. Using the fact that $r(R) = 0$ for $R \in \mathcal{R}_1 \oplus \mathcal{R}_3$, a computation with respect to this basis shows that $\langle s(T), R\rangle = \operatorname{tr} s(T)R = 0$ for all $R \in \mathcal{R}_1 \oplus \mathcal{R}_3$.

Remark. Under the isomorphism r, $\mathcal{R}_2$ is mapped onto the scalar multiples of the identity and R_4 is mapped onto $\{T \in \mathcal{T} \mid \operatorname{tr} T = 0\}$.

Now suppose $n = 4$ and let V be given a definite orientation. The *star operator* $*: \Lambda^2 \rightarrow \Lambda^2$ is defined by $\langle *\alpha, \beta\rangle\omega = \alpha \wedge \beta$ where ω is the generator for Λ^4 determined by the inner product and the orientation of V. For $P \in G$, $*P$ is the oriented orthogonal complement $P^\perp$ of P in V. $*$ is symmetric, $*^2 = I$, and $\Lambda^2 = \Lambda^+ \oplus \Lambda^-$ (orthogonal direct sum) where $\Lambda^\pm$ are the (3-dimensional) eigenspaces of $*$ corresponding to the eigenvalues ± 1.

THEOREM 1.3. *Let V be oriented and of dimension 4. Then $\mathcal{R}_1$ consists of all scalar multiples of $*$; $\mathcal{R}_2$ consists of all scalar multiples of I; $R \in \mathcal{R}_3 \Leftrightarrow *R = R*$, $\operatorname{tr} R = 0$, and $\operatorname{tr} *R = 0$; and $R \in \mathcal{R}_4 \Leftrightarrow *R = -R*$. In particular, $R \in \mathcal{E} \Leftrightarrow R* = *R$ and $b(R) = 0$.*

PROOF. Restriction to Λ^+ gives an isomorphism of

$$\mathcal{A} = \{R \in \mathcal{R} \mid *R = -R*\}$$

with Hom (Λ^+, Λ^-) and hence

$$\dim \mathcal{A} = 9 = \dim \mathcal{T} - 1 = \dim \mathcal{R}_4\,.$$

Also, by formula $(*)$ in the proof of Theorem 1.2, it is clear that $s(T)* = -*s(T)$ for all $T \in \mathcal{T}$ with $\mathrm{tr}\, T = 0$ so $\mathcal{R}_4 \subset \mathcal{A}$. Thus $\mathcal{A} = \mathcal{R}_4$; i.e., $R \in \mathcal{R}_4 \Leftrightarrow *R = -R*$. It then follows that $\mathcal{R}_4^{\perp} = \{R \in \mathcal{R} \mid *R = R*\}$. The proofs of the remaining statements are elementary, using the fact $\mathrm{tr}\, *R = 2b(R)$.

Remark. Under the action of the special orthogonal group $SO(V)$, the 10-dimensional $O(V)$-invariant subspace $\mathcal{R}_3$ splits further into the direct sum of the 5-dimensional $SO(V)$-invariant subspaces $\mathcal{R}_3^{\pm} = \{R \in \mathcal{R} \mid R* = \pm R\}$.

THEOREM 1.4. *Let M be a 4-dimensional riemannian manifold. Then M is an Einstein space if and only if for each tangent 2-plane P, $\sigma(P^{\perp}) = \sigma(P)$.*

PROOF. By Theorem 1.3, M is an Einstein space if and only if its curvature tensor R satisfies $R* = *R$. (Note that, although $*$ is defined in terms of an orientation at each point, this condition is independent of the orientation used.) Since

$$\sigma_R(P^{\perp}) = \sigma_R(*P) = \langle R*P, *P\rangle = \langle *R*P, P\rangle ,$$

it follows that $\sigma_R(P^{\perp}) = \sigma_R(P)$ if and if $\langle(*R* - R)P, P\rangle = 0$. If $R* = *R$, this is clearly the case for all P. Conversely, if this condition is satisfied for all P then the curvature tensor $*R* - R$ has sectional curvature identically zero and has

$$b(*R* - R) = \frac{1}{2}\,\mathrm{tr}\,[*(*R* - R)] = \frac{1}{2}\,\mathrm{tr}\,(R* - *R) = 0$$

so $*R* - R = 0$; that is, $R* = *R$.

Remark. Similarly, a riemannian 4-manifold has curvature tensor in $\mathcal{R}_4$ at each point if and only if $\sigma(P^{\perp}) = -\sigma(P)$ for each 2-plane P.

Remark. For a compact oriented riemannian 4-manifold M, the Euler-Poincaré characteristic $\mathcal{X}(M)$ is given by the generalized Gauss-Bonnet formula [3] which in our setting reads

$$\mathcal{X}(M) = \frac{3}{4\pi^2}\int_M \mathrm{tr}\,(*R)^2 dV .$$

Since for M Einstein $R* = *R$, we have

$$\mathrm{tr}\,(*R)^2 = \mathrm{tr}\, R^2 = \langle R, R\rangle \geqq 0$$

which implies Berger's result [1] that the Euler-Poincaré characteristic of a compact 4-dimensional Einstein space is non-negative and is in fact strictly positive unless M is flat.

2. Critical planes and normal forms

Henceforth V will denote an oriented 4-dimensional real inner product space, R will denote a curvature tensor on V, and σ will denote its sectional curvature function. A 2-plane $P \in G$ is a *critical plane* of R if P is a critical point of $\sigma: G \to \mathbf{R}$. In order to study the critical point theory of σ we need the following description of $G \subset \Lambda^2$.

LEMMA. *G is a product of two 2-spheres, namely the spheres of radius $1/\sqrt{2}$ in the ± 1 eigenspaces $\Lambda^{\pm}$ of the star operator.*

PROOF. First note that a 2-vector $\xi \in \Lambda^2$ is decomposable if and only if $\xi \wedge \xi = 0$. For if $\xi \wedge \xi = 0$ then the linear map $T_\xi: \Lambda^2 \to \Lambda^3$ given by $T_\xi(v) = \xi \wedge v$ has rank 2 (this is easily checked using bases) and $\xi = \pm u \wedge v$ where $\{u, v\}$ is any orthonormal basis for Ker T. The converse is clear. Thus

$$\begin{aligned} G &= \{\xi \in \Lambda^2 \mid \xi \wedge \xi = 0, \langle \xi, \xi \rangle = 1\} \\ &= \{\xi \in \Lambda^2 \mid \langle \xi, *\xi \rangle = 0, \langle \xi, \xi \rangle = 1\} \\ &= \{\alpha + \beta \mid \alpha \in \Lambda^+, \beta \in \Lambda^-, \|\alpha\| = \|\beta\| = 1/\sqrt{2}\}\,. \end{aligned}$$

PROPOSITION. *$P \in G$ is a critical plane of R if and only if*

$$RP = \lambda P + \mu P^\perp$$

for some real numbers λ and μ. The number λ is the (critical) value of σ at P.

PROOF. Consider the three real valued functions defined on Λ^2 by $f(\xi) = \langle R\xi, \xi \rangle$, $g(\xi) = \langle \xi, \xi \rangle$, and $h(\xi) = \langle \xi, *\xi \rangle$. Then σ is the restriction of f to $G = g^{-1}(1) \cap h^{-1}(0)$. Hence $P \in G$ is a critical point of σ is and only if, at P,

$$\operatorname{grad} f = \lambda \operatorname{grad} g + \mu \operatorname{grad} h$$

where λ and μ are Lagrange multipliers. But the gradient at P of a quadratic form $\langle A\xi, \xi \rangle$ is just $2AP$. Applying this to the quadratic forms f, g and h completes the proof.

Now for $\{e_1, \cdots, e_4\}$ an orthonormal basis for V, the 2-planes $\{e_1 \wedge e_2, e_1 \wedge e_3, e_1 \wedge e_4, e_3 \wedge e_4, e_4 \wedge e_2, e_2 \wedge e_3\}$ form a basis for Λ^2 and the entries in the matrix $[R]$ for $R: \Lambda^2 \to \Lambda^2$ relative to this basis are just the components R_{ijkl} of the curvature tensor R relative to the basis $\{e_1, \cdots, e_4\}$ for V. If this basis is chosen so that $e_1 \wedge e_2$ is a critical plane, then the above proposition says that the first row and first column of $[R]$ have the form $(\lambda, 0, 0, \mu, 0, 0)$; that is, $R_{12kl} = 0$ for $(k, l) \neq (1, 2)$ or $(3, 4)$. This fact has been used previously in the work of Chern [3] and of Bishop and Goldberg [2]. The next theorem shows that, if $R \in \mathfrak{E}$, then one can choose a basis for Λ^2 consisting entirely of critical planes and obtain a corresponding simplification in the matrix $[R]$ of curvature components.

THEOREM 2.1. *Let $R \in \mathfrak{E}$. Then there exists an orthonormal basis for V such that each pair of vectors in this basis spans a critical plane of R. Relative to this basis, the matrix of curvature components has the form*

$$[R] = \begin{bmatrix} A & B \\ B & A \end{bmatrix}, \text{ where } A = \begin{bmatrix} \lambda_1 & & \\ & \lambda_2 & \\ & & \lambda_3 \end{bmatrix} \text{ and } B = \begin{bmatrix} \mu_1 & & \\ & \mu_2 & \\ & & \mu_3 \end{bmatrix}.$$

PROOF. Since $R* = *R$, the self-adjoint operators $*$ and R can be simultaneously diagonalized. Let then $\xi_i (1 \leqq i \leqq 6)$ be an orthonormal basis for Λ^2 such that $*\xi_i = \xi_i$ for $i \leqq 3$, $*\xi_i = -\xi_i$ for $i > 3$, and $R\xi_i = a_i\xi_i (1 \leqq i \leqq 6)$. Then $\xi_i \in \Lambda^+$ for $i \leqq 3$, and $\xi_i \in \Lambda^-$ for $i > 3$, and so $P_i = (\xi_i + \xi_{3+i})/\sqrt{2} \in G(1 \leqq i \leqq 3)$ and $P_i^\perp = (\xi_i - \xi_{3+i})/\sqrt{2}$. Moreover, $\{P_1, P_2, P_3, P_1^\perp, P_2^\perp, P_3^\perp\}$ is an orthonormal basis for Λ^2 consisting of critical planes. Indeed, $RP_i = \lambda_i P_i + \mu_i P_i^\perp$ and $RP_i^\perp = \lambda_i P_i^\perp + \mu_i P_i (i \leqq 3)$ where $\lambda_i = (a_i + a_{3+i})/2$ and $\mu_i = (a_i - a_{3+i})/2$. The matrix $[R]$ relative to this basis is then of the required form.

There still remains the problem of constructing the appropriate basis for V. Since $P_1 \wedge P_2 = \langle P_1, *P_2 \rangle \omega = 0$ it follows that $P_1 \cap P_2 \neq (0)$. Let $e_1 \in P_1 \cap P_2$ be a unit vector. Let $e_2, e_3 \in V$ be such that $\{e_1, e_2\}$ and $\{e_1, e_3\}$ are oriented orthonormal bases for P_1 and P_2 respectively. Then $P_1 = e_1 \wedge e_2$, $P_2 = e_1 \wedge e_3$, and $\langle e_2, e_3 \rangle = 0$ since $\langle P_1, P_2 \rangle = 0$. Choosing e_4 to complete an orthonormal basis consistent with the orientation of V, it follows from the orthogonality of the $\{P_i, P_i^\perp\}$ that P_3 is a linear combination of $e_1 \wedge e_4$ and $e_2 \wedge e_3$ and from the condition $\langle P_3, *P_3 \rangle = 0$ that in fact P_3 is a multiple of one of these. This implies that $e_1 \wedge e_4$ is $\pm P_3$ or $\pm P_3^\perp$. But $-P_3$ and $\pm P_3^\perp$ are also critical planes of R with the same Lagrange multipliers λ_3 and μ_3 as P_3. Thus the basis $\{e_1, \cdots, e_4\}$ has the required properties.

Remarks. (1) The normal form for $[R]$ given in Theorem 2.1 is the riemannian version of the Petrov canonical forms. Petrov

[4, 5] studied these forms within the framework of Lorentz geometry and they have been widely used in general relativity theory.

(2) Curvature tensors which lie in the invariant subspaces $\mathcal{R}_i$ of $\mathcal{E}$ can be recognized by their normal forms: for $R \in \mathcal{R}_2$, $A = \lambda I$ and $B = 0$; for $R \in \mathcal{R}_3$, $\sum \lambda_i = \sum \mu_i = 0$; and for $R \in \mathcal{R}_3^{\pm}$, $\mu_i = \pm\lambda_i$ and $\sum \lambda_i = 0$.

(3) For $R \in \mathcal{R}_4$ the proof of Theorem 1.2 shows that, in contrast to the $R \in \mathcal{E}$ case, the matrix of curvature components can be diagonalized by an appropriate choice of basis for V; namely, a basis consisting of eigenvectors of the Ricci tensor $r(R)$. Of course, this works in all dimensions.

(4) The normal form for a given curvature tensor in $\mathcal{E}$ is not unique. In fact, if the eigenvalues of R are distinct, then it is clear from the proof of Theorem 2.1 that, neglecting orientations, there are exactly nine distinct critical plane pairs $(P_{ij}, P_{ij}^{\perp})$, one corresponding to each pair (ξ_i, ξ_{3+j}) $(i, j \leqq 3)$ of eigenvectors of R. Moreover, P_{ij} and P_{kl} are orthogonal if and only if $i \neq k$ and $j \neq l$ It follows that, if the eigenvalues of R are distinct, then there are exactly six distinct normal forms for $[R]$ (here we do not regard two matrices which differ merely by permutations of rows and columns as distinct).

As points in the cartesian plane, the nine Lagrange multiplier pairs

$$(\lambda_{ij}, \mu_{ij}) = \big((a_i + a_{3+j})/2, (a_i - a_{3+j})/2\big)$$

are the points of intersection of six lines, three with slope $+1$ and three with slope -1. (See Figure 1.) Moreover, a selection of three (λ, μ)-pairs corresponds to an orthonormal basis of critical planes if and only if no two of these pairs lie on a common line of Figure 1.

In the case when the eigenvalues of R are not distinct, then the nine critical values of σ may be not distinct, pairs of the above lines may coincide, and we may get less than six distinct normal forms. However, there are then entire submanifolds of critical planes. The geometric configuration in $G = S^2 \times S^2 \subset \Lambda^+ \oplus \Lambda^-$ of the critical plane set is evident from the fact that this set consists of all ordered pairs (ξ, η) where $\xi \in \Lambda^+$ and $\eta \in \Lambda^-$ are eigenvectors of R of length $1/\sqrt{2}$.

Observe (when the eigenvalues of R are distinct) that if P and $Q \neq P^{\perp}$ are critical planes, then Q lies on a geodesic from P halfway

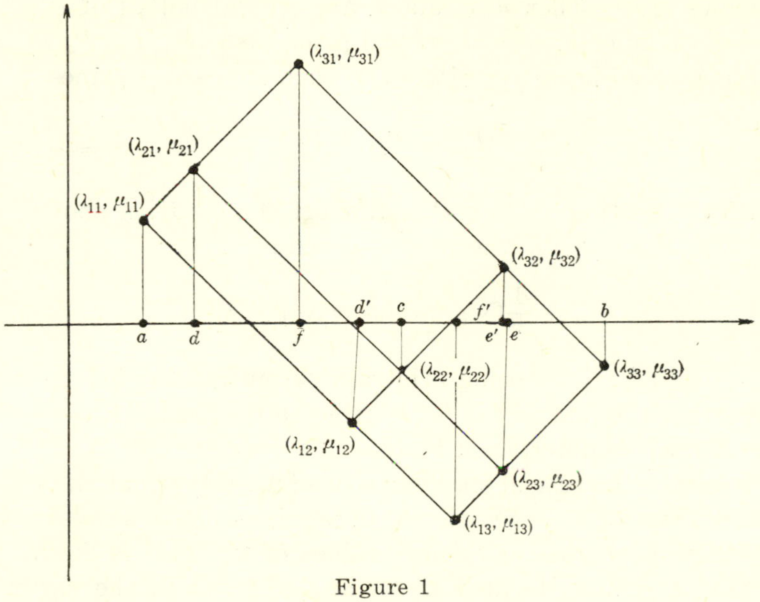

Figure 1

to the first conjugate point of P. This observation may be useful in higher dimensions.

The orbit space of $\mathfrak{S}$ under the action of $O(V)$ is homeomorphic with the orbit space of $\mathbf{R}^5$ under an action of the finite group $(S_3 \times S_3) \dot{\times} S_2$, where S_k denotes the permutation group on k letters and $\dot{\times}$ denotes semi-direct product. Indeed, one may takes as parameters describing the orbit of $R \in \mathfrak{S}$ the eigenvalues $(a_1, \cdots, a_6)$ of R subject to the condition $B = \sum_{i=1}^3 a_i - \sum_{i=1}^3 a_{3+i} = 0$ (the Bianchi identity). Since $SO(V)$ acts on Λ^2 as isometries of Λ^+ and Λ^-, $SO(V)$ acts on the parameter space $\mathbf{R}^3 \times \mathbf{R}^3/(B) = \mathbf{R}^5$ as a pair of permutations, one on each factor $\mathbf{R}^3$. Finally, since orientation reversal in V interchanges the spaces $\Lambda^{\pm}$, it acts on the parameter space by interchanging the two factors.

THEOREM 2.2. *Suppose $R \in \mathfrak{S}$. Then R is completely determined by the critical planes and the values of the sectional curvature on these planes. Moreover, given two such curvature tensors with the same set of critical values of sectional curvature, there exists an isometry of V mapping one onto the other.*

Proof. Let $a = \lambda_{11}$ and $b = \lambda_{33}$ denote the minimum and maximum values of σ and let $c = \lambda_{22}$ denote the real number such that $a + b + c = 1/4$ (scalar curvature of R) $= 1/3$ (sum of all the critical values of σ). Then a, c, and b are critical values of σ corresponding to an orthonormal critical plane basis $\{P_{11}, P_{22}, P_{33}, P_{11}^{\perp}, P_{22}^{\perp}, P_{33}^{\perp}\}$ for Λ^2 (see Figure 1). Moreover, each of the nine planes

$$P_{ij} = \frac{1}{2}(P_{ii} + P_{ii}^{\perp} + P_{jj} - P_{jj}^{\perp}) \qquad (1 \leqq i, j \leqq 3)$$

is critical. Setting $\lambda_{ij} = \sigma(P_{ij})$, a short computation shows that $RP_{ij} = \lambda_{ij}P_{ij} + \mu_{ij}P_{ij}^{\perp}$ where μ_{ij} is given by

$$\mu_{ij} = \frac{1}{3}(\lambda_{ij'} + \lambda_{ij''} - \lambda_{i'j} - \lambda_{i''j}) ,$$

(i, i', i'') and (j, j', j'') being arbitrary permutations of (1, 2, 3). It follows that the matrix for R (relative to the basis $\{P_{11}, \cdots, P_{33}^{\perp}\}$) hence R itself is determined by the λ_{ij}'s.

To prove the second statement, we first show that the (λ, μ)-diagram (Figure 1) is completely determined, up to a reflection in the x-axis, by the set of critical values of σ. This is due to the fact that a rectangle tilted at an angle of $\pi/4$ in the cartesian plane is determined, modulo a motion in the group H generated by vertical translations and reflection in the x-axis, by the x-coordinates of its vertices. To apply this fact, let a, b, and c be as above. Let d denote the critical value adjacent to a (see Figure 1) and let d' be such that $d - a = c - d'$. Then the x-coordinates a, d, d', c determine (mod H) the small rectangle at the left of the (λ, μ)-diagram. Similarly, using b instead of a determines (mod H) the small rectangle at the right of the diagram, with corresponding x-coordinates c, e', e, b. a and b, together with the two remaining critical values f and f', determine (mod H) the outside rectangle. It is now easy to check that these rectangles fit together in only one way (mod H) into the configuration of Figure 1. Since $0 = b(R) = 1/3 \sum \mu_{ij}$, the center of mass of the vertex set must lie on the x-axis. Hence the diagram is complety determined up to a reflection in the x-axis, as claimed.

Now suppose $R, R' \in \mathfrak{S}$ are such that σ_R and $\sigma_{R'}$ have the same set of critical values. Then, up to a possible reflection in the x-axis, R and R' have the same (λ, μ) diagram. Since a change of orientation on V causes such a reflection of the diagram, there exists $h \in O(V)$ such that $R'' = h^{-1}R'h$ and R have the same diagram. Let (λ_i, μ_i), $(1 \leqq i \leqq 3)$, correspond in this diagram to an orthonormal critical plane basis for Λ^2. Then there exist bases $\{e_1, \cdots, e_4\}$

and $\{e'_1, \cdots, e'_4\}$ for V such that both $[R]_{e_i}$, the matrix for R relative to $\{e_i\}$, and $[R'']_{e'_i}$, that for R'' relative to $\{e'_i\}$, are equal to

$$\begin{bmatrix} A & B \\ B & A \end{bmatrix} \text{ where } A = \begin{bmatrix} \lambda_1 & & \\ & \lambda_2 & \\ & & \lambda_3 \end{bmatrix} \text{ and } B = \begin{bmatrix} \mu_1 & & \\ & \mu_2 & \\ & & \mu_3 \end{bmatrix}.$$

Let $g \in O(V)$ be such that $ge_i = e'_i$. Then

$$[R]_{e_i} = [R'']_{e'_i} = [R'']_{ge_i} = [g^{-1}R''g]_{e_i}$$

and hence $R = g^{-1}R''g = (hg)^{-1}R'(hg)$.

MASSACHUSETTS INSTITUTE OF TECHNOLOGY
HAVERFORD COLLEGE AND INSTITUTE FOR ADVANCED STUDY

REFERENCES

[1] M. BERGER, "Sur les variétés d'Einstein compactes", in C. R. IIIe Réunion du Groupment des Mathematicien d'Expression Latine, Louvain-Belgique (1966), 35–55.

[2] R. L. BISHOP and S. GOLDBERG, *Some implications of the generalized Gauss-Bonnet theorem*, Trans. Amer. Math. Soc. **112** (1964), 508–535.

[3] S. S. CHERN, *On the curvature and characteristic classes of a riemannian manifold*, Abh. Math. Sem. Univ. Hamburg **20** (1956), 117–126.

[4] A. Z. PETROV, *Classification of spaces defining gravitational fields*, Sci. Notes Kazan State University **114** (1954), 55–69 (in Russian).

[5] ———, *Einstein-Raume*, Akademie-Verlag, Berlin, 1964.

[6] H. WEYL, The Classical Groups, Princeton University Press, Princeton. 1939.

(Received July 22, 1968)

On deformation of pseudogroup structures

BY D. C. SPENCER

1. Introduction

In his papers ([10(a), (b)]) the author developed a general mechanism for the local deformation of structures on manifolds defined by transitive continuous pseudogroups. Meanwhile additional papers on this subject have appeared, notably those by Griffiths [3] and a paper by Guillemin and Sternberg [5] in which an approach is given based on the use of the fundamental form of the structure, which incorporates in an elegant way many of the formal aspects of the author's approach. Recently B. Malgrange found a proof for the existence of local coordinates compatible with an integrable structure defined by an elliptic transitive, continuous pseudogroup (generalization of the Newlander-Nirenberg theorem for complex analytic structure). Still more recently, Ngô Van Quê ([8]) has approached deformation theory from the point of view of Lie groupoids and has generalized the author's method.

Since the author's original treatment is unnecessarily complicated, a new and simplified version of the author's method is obviously desirable. The simplifying idea, on which this paper is based, is to express the differential operator D of the second complex, applied to an element u, in the form of a bracket $[d, u]$, where d is an exterior differential operator, and then to define, for a transformation h, the operator $\mathfrak{D}h$ by the formula $Adh^{-1} \cdot d = d - \mathfrak{D}h$, i.e., $\mathfrak{D}h$ measures the amount of twist introduced by h and it vanishes if and only if $h = j_m(f)$ where f is a transformation of the pseudogroup. A draft of this manuscript was sent to B. Malgrange, who pointed out to the author a more effective method of carrying out this program; his method is expressed in terms of the Grothendieck prolongation spaces along the diagonal of $M \times M$ and, in the author's opinion, supersedes the older approach involving frame bundles. A new version of the theory of deformations based on these prolongation spaces has meanwhile been developed in collaboration with Malgrange and will appear in a joint paper in the *Bull. Soc. Math. France.* Thus there is no longer any need to try to carry out in detail the original program of twisting "d" in the context of frame bundles, and we shall therefore give here only an outline of the method expressed in the language of frame bundles (in terms of which this paper was originally written) and refer the reader to the forthcoming joint paper with Malgrange for a more definitive treatment. It is hoped that the present paper may supply a bridge connecting the old and new approaches.

Sections 2–4 of the paper summarize the apparatus required. For economy of notation, we use the same letters for vector bundles and their sheaves of sections. The relevant interpretation is obvious from the context—if differentiation or a parameter is involved it is a sheaf—otherwise it is a bundle in most cases. For simplicity, we introduce a single parameter t of deformation; the generalization to several parameters is trivial.

2. Derivations of differential forms

We assemble here the facts we shall require about the derivations of differential forms. These derivations were first determined by A. Frölicher and A. Nijenhuis [1], but we shall reformulate their results in a way better suited to our purpose and we shall follow the presentation given in §1 of the paper [7].

Let M be a differential manifold of dimension n, and denote by $T = T(M)$, $T^* = T^*(M)$ the tangent and cotangent bundles of M respectively. For simplicity of notation we shall, in this section, denote by $A = \bigoplus A^i$ the sheaf of exterior algebras of germs of differential forms, where A^i is the sheaf of sections of $\wedge^i T^*$ (i-tuple exterior product of T^*), and we shall denote by $B = \bigoplus B^i$ the sheaf of germs of vector-valued differential forms where B^i is the sheaf of sections of $\wedge^i T^* \otimes T$.

Definition 2.1. A derivation of degree r of A is a linear transformation $u\colon A \to A$ (over the real numbers $\mathbf{R}$) satisfying $u\colon A^s \to A^{s+r}$ and the condition

$$u(\sigma \wedge \tau) = u\sigma \wedge \tau + (-1)^{rs}\sigma \wedge u\tau\ , \qquad \text{for } \sigma \in A^s, \tau \in A\ . \tag{2.1}$$

Let $x = (x^1, x^2, \cdots, x^n)$ be a local coordinate of M, and let u be a derivation of degree r. Since each x^i is a local differentiable function, $u(x^i)$ and $u(dx^i)$ are defined and we set

$$\text{(2.2)} \qquad \begin{cases} \varphi = \sum_{i=1}^{n} \varphi^i \dfrac{\partial}{\partial x^i}\,, & \text{where } \varphi^i = u(x^i)\,, \\ \xi = \sum_{i=1}^{n} \xi^i \dfrac{\partial}{\partial x^i}\,, & \text{where } \xi^i = (-1)^r u(dx^i)\,. \end{cases}$$

Clearly φ and ξ are local vector-valued (differential) forms of respective degrees r and $r + 1$. The following proposition is easily proved.

PROPOSITION 2.1. *For any $\sigma \in A^s$, the explicit expression for $u\sigma$ is given by either of the following expressions*:

$$\text{(2.3)} \qquad \begin{cases} u\sigma = \sum \varphi^i \wedge \dfrac{\partial \sigma}{\partial x^i} + (-1)^r \sigma \barwedge \xi\,, \\ u\sigma = d_\varphi \sigma - (-1)^r \sigma \barwedge (d\varphi - \xi)\,. \end{cases}$$

Here the operations $\barwedge \varphi$ and d_φ (contraction with a vector-valued form and Lie derivation along the vector-valued form φ, respectively) are defined as follows. Let

$$\sigma = \sum \sigma_{j_1 \cdots j_s}(x) dx^{j_1} \wedge \cdots \wedge dx^{j_s}\,,$$

where the summation is over $j_1, \cdots, j_s$ satisfying $j_1 < \cdots < j_s$. Then

$$\text{(2.4)} \qquad \sigma \barwedge \varphi = \sum \sigma_{j j_1 \cdots j_{s-1}}(x) \varphi^j \wedge dx^{j_1} \wedge \cdots \wedge dx^{j_{s-1}}$$

where the summation is over j and $j_1, \cdots, j_{s-1}$ satisfying $j_1 < \cdots < j_{s-1}$. The Lie derivative of σ along φ is defined by the formula

$$\text{(2.5)} \qquad d_\varphi \sigma = d\sigma \barwedge \varphi + (-1)^r d(\sigma \barwedge \varphi)\,.$$

Finally the term $\partial\sigma/\partial x^i$ occurring in (2.3) is

$$\partial\sigma/\partial x^i = \sum (\partial \sigma_{j_1 \cdots j_s}(x)/\partial x^i) dx^{j_1} \wedge \cdots \wedge dx^{j_s}\,.$$

We infer from (2.3) that

$$\text{(2.5)}' \qquad d_\varphi \sigma = \sum \varphi^i \wedge \partial\sigma/\partial x^i + (-1)^r \sigma \barwedge d\varphi\,.$$

Finally let $y = f(x)$ be a change of local coordinate. Then

$$\text{(2.6)} \qquad \begin{cases} \varphi^j(y) = \sum \dfrac{\partial y^j}{dx^i} \varphi^i(x) \\ \xi^j(y) = \sum \left\{ \dfrac{\partial y^j}{\partial x^i} \xi^i(x) + d\left(\dfrac{\partial y^j}{dx^i}\right) \wedge \varphi^i(x) \right\}\,. \end{cases}$$

In fact, using (2.3), we obtain

$$\varphi^j(y) = u(y^j) = \sum \varphi^i \frac{\partial y^j}{\partial x^i}$$

$$\begin{aligned}\xi^j(y) &= (-1)^r u(dy^j) = (-1)^r u\Big(\sum \frac{\partial y^j}{\partial x^l} dx^l\Big)\\ &= (-1)^r \sum\nolimits_{l,m} \varphi^m(x) \frac{\partial^2 y^j}{\partial x^l \partial x^m} dx^l + \sum\nolimits_l \frac{\partial y^j}{\partial x^l} \xi^l(x)\\ &= \sum \Big\{ \frac{\partial y^j}{\partial x^i} \xi^i(x) + d\Big(\frac{\partial y^j}{\partial x^i}\Big) \wedge \varphi^i(x) \Big\} .\end{aligned}$$

The formulas (2.6) show that φ and $d\varphi - \xi$ are vector-valued forms of degrees r and $r + 1$, respectively.

We observe that locally the derivation u of degree r is represented by the pair (φ, ξ) where φ, ξ have respective degrees r and $r + 1$. It follows that the only non-vanishing derivations have degrees r in the range $-1 \leqq r \leqq n$.

Now let J^r denote the sheaf over M of germs of derivations of degree r. We have the map

$$J^r \otimes J^s \longrightarrow J^{r+s} ,$$

linear over the constants, where $u \otimes v$ goes into the bracket defined by

$$[u, v] = uv - (-1)^{rs} vu \tag{2.7}$$

These maps impart to $\bigoplus J^r$ a structure of graded Lie algebra.

The exterior derivative d is clearly a derivation of degree 1 on the algebra A. It can be identified with the pair $(1, 0) \in J^1$ where $1\colon T \to T$, the identity map of the tangent bundle T of M, is an element of B^1. In terms of a local coordinate $x = (x^1, x^2, \cdots, x^n)$ of M, the element 1 has the form

$$\sum\nolimits_{i=1}^n dx^i \otimes \frac{\partial}{\partial x^i} . \tag{2.8}$$

We see immediately from (2.5)′ that

$$d = d_1 . \tag{2.9}$$

Definition 2.2. For $u \in J^r$ we define

$$Du = [d, u] . \tag{2.10}$$

We infer from the Jacobi identity, and the obvious identity

$[d, d] = 0$, that $D^2 = 0$. Moreover, as an immediate consequence of the Jacobi identity, we obtain the important formula

$$D[u, v] = [Du, v] + (-1)^p[u, Dv], \qquad u \in J^p, v \in J^q . \tag{2.11}$$

The following formulas are easily verified. Let $u = (\varphi, \xi) \in J^r$; then

$$Du = (d\varphi - \xi, -d\xi) . \tag{2.12}$$

Let $u = (\varphi, \xi) \in J^r$, $v = (\psi, \eta) \in J^s$; then

$$[u, v] = (\omega, \zeta)$$

where

$$\begin{cases} \omega = d_\varphi\psi - (-1)^{rs}d_\psi\varphi \\ \qquad\qquad - (-1)^r\psi \barwedge (d\varphi - \xi) + (-1)^{rs+s}\varphi \barwedge (d\psi - \eta) , \\ \zeta = (-1)^r d_\varphi\eta - (-1)^{rs+s}d_\psi\xi \\ \qquad\qquad - \eta \barwedge (d\varphi - \xi) + (-1)^{rs}\xi \barwedge (d\psi - \eta) . \end{cases} \tag{2.13}$$

We define

$$\begin{cases} i\colon B^r \longrightarrow J^r \\ k\colon B^{r+1} \longrightarrow J^r \end{cases}$$

where

$$\begin{cases} i\varphi = (\varphi, d\varphi) , \\ k\varphi = (0, \varphi) . \end{cases} \tag{2.14}$$

Then we have the decomposition

$$J^r = iB^r \oplus kB^{r+1} . \tag{2.15}$$

Let

$$\pi\colon J^r \longrightarrow B^r \tag{2.16}$$

where $\pi u = \pi(\varphi, \xi) = \varphi$. Clearly $\pi i\varphi = \varphi$, and we have by (2.12)

$$i\varphi = -Dk\varphi . \tag{2.17}$$

Frölicher and Nijenhuis have defined on $B = \oplus B^r$ a structure of graded Lie algebra with the bracket

$$[\varphi, \psi] = \pi[i\varphi, i\psi] \qquad \varphi, \psi \in B . \tag{2.18}$$

By (2.11) we have

$$[i\varphi, i\psi] = i[\varphi, \psi] , \tag{2.19}$$

i.e., the injection i is a (graded) Lie algebra homomorphism. From

(2.13) we obtain the explicit expression for the bracket $[\varphi, \psi]$ of vector-valued forms $\varphi \in B^r$, $\psi \in B^s$:

$$[\varphi, \psi] = d_\varphi \psi - (-1)^{rs} d_\psi \varphi . \tag{2.20}$$

If φ, ψ are vector fields (i.e., both of degree zero), this bracket coincides with the usual Lie bracket.

Let

$$J = \bigoplus_{r \geq 0} J^r \tag{2.21}$$

be the sheaf of germs of derivations of non-negative degrees. We suppress the derivations of degree -1 since they are simply contractions of differential forms with vector fields. We infer from (2.17) that the sequence

$$0 \longrightarrow B^0 \xrightarrow{i} J^0 \xrightarrow{D} J^1 \xrightarrow{D} \cdots \xrightarrow{D} J^n \longrightarrow 0 \tag{2.22}$$

is exact.

Finally, let V be a finite-dimensional vector space, and consider the differential forms with values in the trivial (product) bundle $M \times V$, i.e., the sections of $\wedge^* T^* \otimes (M \times V)$, which are simply the V-valued forms. The derivations of V-valued forms are obviously precisely the same as those we have considered.

3. Principal bundles and differential forms

Let P denote an arbitrary (differential) principal bundle over M with fibre and group a Lie group G, and let Q be a vector bundle over M with fibre V which is associated with P. We let

$$q: P \times V \longrightarrow Q = P \times_G V$$

be the projection used in defining Q in terms of a representation of G on V, where

$$q(pg, g^{-1}v) = q(p, v) , \qquad p \in P, v \in V, g \in G . \tag{3.1}$$

Let $\alpha: P \to M$ be the projection, and let $\alpha_*: T(P) \to T(M)$ be the projection induced by α. A (differentiable) section σ of Q over an open set U of M determines uniquely a (differentiable) V-valued function $\sigma_{(e)}$ on $\alpha^{-1}(U)$ which is equivariant under the right operations of G, namely we have (by (3.1))

$$\sigma_{(e)}(pg) = g^{-1}\sigma_{(e)}(p) .$$

We denote by r_g the right operation of g on P.

Now let σ be a differential form of degree j on an open set U of M with values in Q, i.e., σ is a section of $\wedge^j T^* \otimes Q$ over U. Then σ defines a unique V-valued differential form $\sigma_{(e)}$ of degree j on $\alpha^{-1}(U)$ which satisfies the following two conditions:

$$r_g^* \sigma_{(e)} = g^{-1} \sigma_{(e)} \; ; \tag{3.2}$$

if $\eta \in T(P)$ (tangent bundle of P) and $\alpha_* \eta = 0$ (i.e., η is tangent to a fibre of P), then

$$\sigma_{(e)} \barwedge \eta = 0 \, . \tag{3.3}$$

Condition (3.2) is equivariance, and we shall describe (3.3) by saying (with some abuse of language) that $\sigma_{(e)}$ is "horizontal" (no connection implied). Let $E(U) = \bigoplus E^j(U)$ denote the graded vector space of V-valued differential forms on $\alpha^{-1}(U)$ which satisfy (3.2) and (3.3). Then $U \to E(U)$ defines over M a sheaf $E = \bigoplus E^j$, which we call the sheaf (over M) of equivariant V-valued forms on P. We have the isomorphism of sheaves

$$\varepsilon \colon \wedge^* T^* \otimes Q \longrightarrow E \tag{3.4}$$

where, according to our convention made at the end of §1, $\wedge^* T^* \otimes Q$ denotes the sheaf of sections of the bundle denoted by the same letters.

Next, let $F(P)$ be the sub-bundle of the tangent bundle $T(P)$ of P composed of the vectors tangent to the fibers of P (bundle along the fibers of P). We remark that $F(P)$ is a trivial bundle, namely

$$F(P) = P \times \mathfrak{g}$$

where $\mathfrak{g}$ is the Lie algebra of left invariant vector fields on G. We have the exact sequence of vector bundles

$$0 \longrightarrow F(P) \longrightarrow T(P) \xrightarrow{\alpha_*} T(M) \longrightarrow 0 \, .$$

The right operations of G on P induce right operations of G on $T(P)$ and $F(P)$ and we denote by $T(P)/G$, $F(P)/G$ the vector bundles over M obtained from $T(P)$, $F(P)$, respectively, by identifying points which are transformed into one another by the right operations of G. We have over M the exact sequence of vector bundles:

$$0 \longrightarrow F(P)/G \longrightarrow T(P)/G \xrightarrow{\pi} T(M) \longrightarrow 0 \, . \tag{3.5}$$

A section σ of $T(P)/G$ over an open set $U \subset M$ determines uniquely a vector field $\sigma_{(i)}$ on $\alpha^{-1}(U)$ (section over $\alpha^{-1}(U)$ of $T(P)$) which is invariant under the right action of G, i.e.,

$$(r_g)_* \sigma_{(i)}(P) = \sigma_{(i)}(pg) .$$

Let σ be a differential form of degree j on U with values in $T(P)/G$, i.e., a section of $\wedge^j T^* \otimes T(P)/G$ over U. Then σ defines uniquely a vector-valued differential form $\sigma_{(i)}$ of degree j on $\alpha^{-1}(U)$ (section of $\wedge^j T^*(P) \otimes T(P)$ over $\alpha^{-1}(U)$) which is "horizontal," i.e., satisfies (3.3), and in addition the condition

$$r_g^* \sigma_{(i)} = \sigma_{(i)} . \tag{3.6}$$

Let $I(U) = \bigoplus I^j(U)$ be the graded space of vector-valued differential forms on $\alpha^{-1}(U)$ satisfying (3.3) and (3.6). Then $U \to I(U)$ defines over M a sheaf $I = \bigoplus I^j$, which we call the sheaf over M of invariant vector-valued forms on P. We have the isomorphism of sheaves

$$\iota\colon \wedge^* T^* \otimes T(P)/G \longrightarrow I . \tag{3.7}$$

We define (see [6], [9] and [10(a)]) an operation of the sheaf $\wedge^* T^* \otimes T(P)/G$ on the sheaf $\wedge^* T^* \otimes Q$ by means of maps

$$(\wedge^r T^* \otimes T(P)/G) \otimes (\wedge^s T^* \otimes Q) \longrightarrow \wedge^{r+s} T^* \otimes Q ,$$

which send $\sigma \otimes \tau$ into $\sigma \cdot \tau$ where

$$\sigma \cdot \tau = \varepsilon^{-1}(i(\iota(\sigma)) \cdot \varepsilon(\tau)) \tag{3.8}$$

and $i(\iota(\sigma))$ operates on $\varepsilon(\tau)$ as a derivation in the sense of the first formula (2.14), i.e., as Lie derivation of $\varepsilon(\tau)$ along $\iota(\sigma)$. We then obtain on the sheaf $\wedge^* T^* \otimes T(P)/G$ a structure of graded Lie algebra with bracket defined, for elements σ_1, σ_2 of degrees r and s by the formula (compare Nickerson [9])

$$[\sigma_1, \sigma_2]\tau = \sigma_1 \cdot (\sigma_2 \cdot \tau) - (-1)^{rs} \sigma_2 \cdot (\sigma_1 \cdot \tau) , \quad \tau \in \wedge^* T^* \otimes Q . \tag{3.9}$$

In this algebra, $\wedge^* T^* \otimes F(P)/G$ is a sub-sheaf of ideals of $\wedge^* T^* \otimes T(P)/G$, and the sequence of sheaves of graded Lie algebras

$$\begin{aligned} 0 \longrightarrow \wedge^* T^* \otimes F(P)/G \longrightarrow \wedge^* T^* \otimes T(P)/G \\ \longrightarrow \wedge^* T^* \otimes T(M) \longrightarrow 0 \end{aligned} \tag{3.10}$$

is exact, where $\wedge^* T^* \otimes T(M)$ has the graded Lie algebra induced from the Lie bracket of vector fields.

Suppose now that P is the principal bundle of the tangent

bundle $T = T(M)$ of M. Then, taking $Q = T$ and writing (as in § 2) $B^r = \wedge^r T^* \otimes T$, the sheaf $\wedge^* T^* \otimes T(P)/G$ operates on $B = \bigoplus B^r$. There is a distinguished element $1 \in B^1$, namely the identity map of T.

Definition 3.1. The torsion of an element σ of the sheaf $\wedge^r T^* \otimes T(P)/G$ is the element $\mathcal{T}\sigma \in B^{r+1}$ defined by

$$\mathcal{T}\sigma = \sigma \cdot 1 . \tag{3.11}$$

Let $x = (x^1, x^2, \cdots, x^n)$ be a local coordinate of M. The element 1, expressed in terms of this coordinate, has the form (see (2.8))

$$\textstyle\sum_{i=l}^{n} dx^i \otimes \frac{\partial}{\partial x^i} .$$

Let σ be represented locally by the pair $(\pi\sigma, \lambda)$ where

$$\pi\sigma = (\pi\sigma^1, \cdots, \pi\sigma^n) \in B^r$$

and

$$\lambda = (\lambda_l^j)_{j,l=1,2,\cdots,n}$$

is locally a differential form of degree r with values in the Lie algebra of the group G of P. Then

$$(\mathcal{T}\sigma)^j = (-1)^r (d(\pi\sigma)^j - \textstyle\sum_{l=1}^{n} dx^l \wedge \lambda_l^j) , \quad j = 1, 2, \cdots, n . \tag{3.12}$$

The sheaf $\wedge^* T^* \otimes T(P)/G$ also operates on the exterior algebra $A = \wedge^* T^*$ of differential forms on M (provided that appropriate identifications are made—see [9]). Namely, we have the maps

$$(\wedge^r T^* \otimes T(P)/G) \otimes A^s \longrightarrow A^{r+s}$$

where $\sigma \otimes \eta$ goes into $\sigma * \eta$ and

$$\sigma * \eta = d_{\pi\sigma}\eta - \eta \,\overline{\wedge}\, \mathcal{T}\sigma . \tag{3.13}$$

We thus obtain another structure of graded Lie algebra on $\wedge^* T^* \otimes T(P)/G$ with bracket defined, for elements σ_1, σ_2 of degrees r and s, by the formula

$$[\sigma_1, \sigma_2]\eta = \sigma_1 * (\sigma_2 * \eta) - (-1)^{rs} \sigma_2 * (\sigma_1 * \eta) , \qquad \eta \in A . \tag{3.14}$$

In this algebra $\Lambda = \bigoplus \Lambda^r$ is an ideal, where

$$\Lambda^r = \{\lambda \in \wedge^* T^* \otimes F(P)/G \mid \mathcal{T}\lambda = 0\} , \tag{3.15}$$

and the following proposition is easily seen to be valid.

PROPOSITION 3.1. *Assume that $G = GL(n, \mathbf{R})$. Then the sequence*

$$0 \longrightarrow \Lambda \longrightarrow \wedge^* T^* \otimes T(P)/G \xrightarrow{\sharp} J \longrightarrow 0 \tag{3.16}$$

is an exact sequence of sheaves of graded Lie algebras, where $J = \bigoplus_{r \geqq 0} J^r$ is the sheaf of derivations (of non-negative degrees) of the differential forms on M and, if $\sigma \in \wedge^r T^* \otimes T(P)/G$,

$$\sharp\sigma = (\pi\sigma, d(\pi\sigma)) - (-1)^r (0, \mathcal{T}\sigma) \in J^r .$$

If the group G of the principal bundle P of the tangent bundle $T = T(M)$ is a proper subgroup of $GL(n, \mathbf{R})$ (case where M has a G-structure), (3.16) is to be replaced by the sequence

$$0 \longrightarrow \Lambda \longrightarrow \wedge^* T^* \otimes T(P)/G \longrightarrow J_0 \longrightarrow 0 \tag{3.17}$$

where J_0 a sheaf of sub-algebras of derivations of the sheaf J.

We recall that a connection is a splitting of (3.5), i.e., it is an element χ of $T^* \otimes T(P)/G$ satisfying $\pi\chi = 1 \in B^1$. The following proposition is obvious.

PROPOSITION 3.2. *The derivation d belongs to J_0 in* (3.17) *if and only if there is a torsionless connection $\chi \in T^* \otimes T(P)/G$.*

In fact, if χ is a connection, we have by (2.9) and (3.13),

$$\chi * \eta = d\eta - \eta \barwedge \mathcal{T}\chi .$$

Hence $\chi * \eta = d\eta$ if and only if $\mathcal{T}\chi = 0$.

Another formulation of Proposition 3.2 is the following. Let χ be a connection, and let

$$D_0 \colon \wedge^r T^* \otimes T(P)/G \longrightarrow \wedge^{r+1} T^* \otimes T(P)/G$$

be the sheaf map defined by setting

$$D_0\sigma = [\chi, \sigma] . \tag{3.18}$$

From the Jacobi identity we obtain the following two formulas.

$$D_0[\sigma, \tau] = [D_0\sigma, \tau] + (-1)^r[\sigma, D_0\tau] , \quad \sigma \in \wedge^r T^* \otimes T(P)/G , \tag{3.19}$$

$$D_0^2\sigma = \frac{1}{2}\,[[\chi, \chi], \sigma] . \tag{3.20}$$

Then $[\chi, \chi] \in \Lambda^2$ if and only if χ is torsionless and, if it is, the operator D_0 defined by (3.18) passes to the quotient J_0 and defines there the operator (2.10). This assertion is easily verified.

4. Sheaves of germs of 1-parameter families

In order to avoid interrupting the development at a later stage, we place here some definitions and notation which will be required in § 9.

We let $B \to M$ be a (differentiable) fibre bundle which will be either a vector bundle or a principal bundle. We set $\mathcal{B} = B \times \mathbf{R}$, $\mathcal{V} = M \times \mathbf{R}$, and let $\pi\colon \mathcal{B} \to \mathbf{R}$, $\varpi\colon \mathcal{V} \to \mathbf{R}$ be the projections. We identify the bundle $B \to M$ with the fibre $\pi^{-1}(0)$ of $\mathcal{B}$, and we identify the base manifold M with $\varpi^{-1}(0)$. We denote by t a point of $\mathbf{R}$ and call it the parameter (of deformation). Finally we let p be a non-negative integer.

Definition 4.1. Suppose that B is a principal bundle with structure group G. Then we denote by $H = H(B)$ the sheaf of groups over $M = \varpi^{-1}(0)$ of germs of bidifferentiable maps of $\mathcal{B}$ into $\mathcal{B}$ which induce the identity on $\mathbf{R}$ (but not necessarily on M) commute with the right operations of G on $\mathcal{B}$ and reduce to the identity on $\pi^{-1}(0)$.

If B is a principal bundle, a germ of $H = H(B)$ over a point of M is represented by a 1-parameter family of bidifferentiable maps $h = h_t\colon B \mid U \to B \mid U_t$, where t is small, U is a neighborhood of the point in question, U_t a slightly deformed neighborhood of the same point, h commutes with the right operations of G and reduces to the identity map of $B \mid U$ for $t = 0$. Thus $h = h_t$ induces a bidifferentiable map $f = f_t\colon U \to U_t$ such that the following diagram commutes:

$$\begin{array}{ccc} B \mid U & \xrightarrow{h_t} & B \mid U_t \\ \downarrow & & \downarrow \\ U & \xrightarrow{f_t} & U_t \end{array}$$

Definition 4.2. If B is a vector bundle, we denote by $B_{(p)}$ the sheaf over $M = \varpi^{-1}(0)$ of germs of (differentiable) sections of B over U which vanish on $M = \varpi^{-1}(0)$ with the order p (where order 0 means that the sections do not necessarily vanish on $M = \varpi^{-1}(0)$).

5. Transitive continuous pseudogroups (infinitesimal transformations)

Let Γ denote a transitive continuous pseudogroup of order

k, M a Γ-manifold of dimension n (for definitions see [5], [7] and [10(a), (b)]). Let P_m be the principal bundle over M of jets of order m of maps of Γ from $\mathbf{R}^n$ (real n-spaces) into M with source at the origin of $\mathbf{R}^n$. For $m = 0$ we set $P_0 = M$. The structure group G_m of P_m is the group of all m-jets of transformations of Γ with source and target at the origin of $\mathbf{R}^n$.

For simplicity we write $R_m = T(P_m)/G_m$ for $m \geqq 1$, $R_0 = T(M)$. Let $J_m(T)$ be the vector bundle over M of jets of order m (m-jets) of (differentiable) sections of $T = T(M)$, where $J_0(T) = T = T(M)$. Then R_m can be identified with a vector sub-bundle of $J_m(T)$ and we write $F_m = J_m(T)/R_m$. We have the usual map j_m taking sections of T into sections of $J_m(T)$ by differentiation up to and including the order m. Following our convention (see § 1) of denoting bundles and their sheaves of germs of sections by the same letters, we let

$$\nabla \colon T \longrightarrow F_k$$

be the differential operator of order k (the order of Γ) which makes the triangle

$$\begin{array}{ccc} T & \xrightarrow{\;j_k\;} & J_k(T) \\ & \searrow^{\nabla} & \big\downarrow p \\ & & F_k \end{array}$$

commutative, where $p\colon J_k(T) \rightarrow F_k$ is the natural projection. Let Θ be the sheaf of Lie algebras over M of the Γ-vector fields (infintesimal transformations of Γ). Then the sequence

$$0 \longrightarrow \Theta \longrightarrow T \xrightarrow{\;\nabla\;} F_k$$

is exact, i.e. ∇ is the differential operator whose homogeneous solutions are the Γ-vector fields.

Example. Let G be a linear Lie group, $G \subset GL(n, \mathbf{R})$, and denote by Γ_G the pseudogroup of local differentiable transformations of $\mathbf{R}^n$ whose jacobian matrices belong to G. The manifold M is a (flat) Γ_G-manifold if it can be covered by local coordinates whose transformations belong to Γ_G. The pseudogroup Γ_G has order $k = 1$ and the vector bundle F_1 associated with Γ_G has $\mathfrak{gl}(n, \mathbf{R})/\mathfrak{g}$ as fibre, where $\mathfrak{gl}(n, \mathbf{R})$ and $\mathfrak{g}$ are respectively the Lie algebras of $GL(n, \mathbf{R})$ and G. Expressed in terms of the coordinate

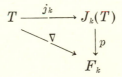

$$x = (x^1, x^2, \cdots, x^n)$$

of $\mathbf{R}^n$ on which Γ_G operates, a vector field $\theta = (\theta^1, \theta^2, \cdots, \theta^n)$ satisfies $\nabla\theta = 0$ if and only if the (nxn)-matrix $(\partial\theta^i/\partial x^j)$ belongs to the Lie algebra $\mathfrak{g}$ of G.

For each non-negative integer l, let $p_l = p_l(\nabla)\colon J_{k+l}(T) \to J_l(F_k)$ (l^{th} prolongation of $p = p_0$) be the homomorphism of vector bundles such that the following (sheaf) diagram commutes:

$$\begin{array}{ccc} J_{k+l}(T) & \xrightarrow{p_l} & J_l(F_k) \\ \uparrow{\scriptstyle j_{k+l}} & & \uparrow{\scriptstyle j_l} \\ T & \xrightarrow{\nabla} & F_k \end{array} .$$

Then $R_{k+l} = T(P_{k+l})/G_{k+l}$ is the kernel of p_l, i.e.,

$$0 \longrightarrow R_{k+l} \longrightarrow J_{k+l}(T) \xrightarrow{p_l} J_l(F_k) ,$$

and R_{k+l} can be identified with the bundle of jets of Γ-vector fields of order $k + l$.

Next, let $\pi_{k+l-1}\colon J_{k+l}(T) \to J_{k+l-1}(T)$ be the projection, and denote by $S^{k+l}T^*$ the $(k + l)$-tuple symmetric product of the cotangent bundle $T^* = T^*(M)$ of M. For $l \geqq 1$ we define

$$\sigma_l = \sigma_l(\nabla)\colon S^{k+l}T^* \otimes T \longrightarrow S^lT^* \otimes F_k$$

to be the morphism of vector bundles such that the following diagram is exact and commutative:

$$\begin{array}{ccc} 0 & & 0 \\ \downarrow & & \downarrow \\ S^{k+l}T^* \otimes T & \xrightarrow{\sigma_l} & S^lT^* \otimes F_k \\ \downarrow & & \downarrow \\ J_{k+l}(T) & \xrightarrow{p_l} & J_l(F_k) \\ \downarrow{\scriptstyle \pi_{k+l-1}} & & \downarrow{\scriptstyle \pi_{l-1}} \\ J_{k+l-1}(T) & \xrightarrow{p_{l-1}} & J_{l-1}(F_k) \\ \downarrow & & \downarrow \\ 0 & & 0 \end{array}$$

If $l = 0$, we define $\sigma = \sigma_0 = \sigma_0(\nabla)$ to be the restriction of

$$p = p_0 = p_0(\nabla)$$

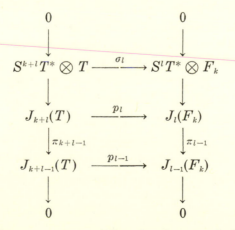

to the sub-bundle $S^kT^* \otimes T$ of $J_k(T)$. The resulting map

$$\sigma = \sigma(\nabla): S^kT^* \otimes T \longrightarrow F_k$$

is the *symbol* of the differential operator ∇ and $\sigma_l = \sigma_l(\nabla)$ is its lth prolongation. We denote by g_{k+l} the kernel of σ_l, i.e.,

$$0 \longrightarrow g_{k+l} \longrightarrow S^{k+l}T^* \otimes T \xrightarrow{\sigma_l} S^lT^* \otimes F_k \,.$$

Then g_{k+l} is also the kernel of the projection

$$\pi_{k+l-1}: R_{k+l} \longrightarrow R_{k+l-1} \,.$$

For convenience we set $g_{k-l} = S^{k-l}T^* \otimes T$ for $1 \leqq l \leqq k$.

Definition 5.1. A covector $\xi \in T^*$ is characteristic (for ∇) if and only if there is a non-zero tangent vector $u \in T$ such that

$$\xi^k \otimes u \in g^k \,.$$

It is easily seen that $\xi \in T^*$ is characteristic if and only if there is a tangent vector $u \in T$, $u \neq 0$, such that $\xi^{k+l} \otimes u \in g_{k+l}$ for some or all $l \geqq 0$.

Definition 5.2. The operator ∇ is elliptic if and only if there are no *real* characteristic covectors different from zero. We say that the pseudogroup Γ is elliptic if the operator ∇ defining its Γ-vector fields is elliptic.

Example. The pseudogroup Γ_G is elliptic if and only if the Lie algebra $\mathfrak{g}$ of G contains no non-trivial *real* subalgebra generated by elements of rank 1.

In the definition of ellipticity the word "real" is important since the condition that ∇ has no non-zero complex characteristics is equivalent to the pseudogroup being of finite type, i.e., $g_m = 0$ for all sufficiently large m (see [4]). If G is a compact group, Γ_G is always of finite type.

Definition 5.3. We say that the pseudogroup is of analytic type if, for some choice of local coordinates in the neighborhood of each point of M, the operator ∇ is (real) analytic.

6. The fundamental form

The fibre V_m of $R_m = T(P_m)/G_m$ is the space of m-jets of Γ-vector fields at the origin of $\mathbf{R}^n$ and it is isomorphic to $\mathbf{R}^n \oplus \mathfrak{g}_m$ where $\mathfrak{g}_m$ is the Lie algebra of the group G_m of P_m. We regard the bundle R_m as associated with P_{m+1}.

Let $E_m = \bigoplus E_m^j$ be the sheaf over M of equivariant V_{m-1}-valued differential forms on P_m, $I_m = \bigoplus I_m^j$ the sheaf of invariant vector-valued differential forms on P_m, in the sense of § 3 (where both types of forms are annihilated by vectors tangent to the fibers of P_m). We have the isomorphisms of sheaves over M (see (3.4) and (3.7))

$$\begin{cases} \varepsilon_m\colon \wedge^* T^* \otimes R_{m-1} \longrightarrow E_m \,, \\ \iota_m\colon \wedge^* T^* \otimes R_m \longrightarrow I_m \,, \end{cases} \tag{6.1}$$

and, in particular, for degree 0, the isomorphisms

$$\begin{cases} \varepsilon_m^0\colon R_{m-1} \longrightarrow E_m^0 \,, \\ \iota_m^0\colon R_m \longrightarrow I_m^0 \,. \end{cases} \tag{6.1°}$$

From (6.1) we have the isomorphism

$$E_{m+1} \cong I_m \tag{6.2}$$

Moreover, the projection $\pi_m\colon \wedge^* T^* \otimes R_{m+1} \to \wedge^* T^* \otimes R_m$ induces projections (which we denote by the same letter)

$$\begin{cases} \pi_m\colon E_{m+1} \longrightarrow E_m \,, \\ \pi_m\colon I_{m+1} \longrightarrow I_m \,. \end{cases} \tag{6.3}$$

Definition 6.1. The mth fundamental form ω_m for the Γ-structure of M is the morphism of vector bundles

$$\omega_m\colon T(P_m) \longrightarrow P_m \times V_{m-1} \tag{6.4}$$

which makes the square

$$\begin{array}{ccc} R_m & \xrightarrow{\ \iota_m^0\ } & I_m^0 \\ \downarrow{\scriptstyle \pi_{m-1}} & & \downarrow{\scriptstyle \omega_m} \\ R_{m-1} & \xrightarrow{\ \varepsilon_m^0\ } & E_m^0 \end{array}$$

commutative, where $\omega_m\colon I_m^0 \to E_m^0$ is the sheaf map induced by (6.4).

The 1-form ω_m on P_m is uniquely determined by its values on I_m^0. We see easily that ω_m extends to a map $\omega_m\colon I_m \to E_m$, sending the element $\sigma_{(i)}$ of I_m into $\omega_m \barwedge \sigma_i$ (see § 2), such that the square

$$\begin{array}{ccc} \wedge^* T^* \otimes R_m & \xrightarrow{\ \iota_m\ } & I_m \\ \downarrow{\scriptstyle \pi_{m-1}} & & \downarrow{\scriptstyle \omega_m} \\ \wedge^* T^* \otimes R_{m-1} & \xrightarrow{\ \varepsilon_m\ } & E_m \end{array} \tag{6.5}$$

is commutative.

We note that the following diagrams are exact and commutative:

$$\begin{array}{ccccccccc} 0 \longrightarrow & \wedge^* T^* \otimes g_{m+1} & \longrightarrow & \wedge^* T^* \otimes R_{m+1} & \xrightarrow{\pi_m} & \wedge^* T^* \otimes R_m & \longrightarrow 0 \\ & \downarrow & & \downarrow \iota_{m+1} & & \downarrow \varepsilon_{m+1} & \\ 0 \longrightarrow & \iota_{m+1}(\wedge^* T^* \otimes g_{m+1}) & \longrightarrow & I_{m+1} & \xrightarrow{\omega_{m+1}} & E_{m+1} & \longrightarrow 0 \end{array} \tag{6.6}$$

$$\begin{array}{ccc} I_{m+1} & \xrightarrow{\omega_{m+1}} & E_{m+1} \\ \downarrow \pi_m & & \downarrow \pi_m \\ I_m & \xrightarrow{\omega_m} & E_m \end{array} \tag{6.7}$$

We have a natural bracket operation on V_m defined by maps

$$V_m \otimes V_m \longrightarrow V_{m-1} \tag{6.8}$$

where $x_1 \otimes v_1$ goes into the bracket $[\![v_1, v_2]\!]$, which is defined as follows. Let θ_1, θ_2 be Γ-vector fields defined in a neighborhood of the origin 0 of $\mathbf{R}^n$ such that $v_1 = j_m\theta_1(0)$, $v_2 = j_m\theta_2(0)$. Then

$$[\![v_1, v_2]\!] = -j_{m-1}[\theta_1, \theta_2](0) \tag{6.9}$$

where $[\theta_1, \theta_2]$ is the usual Lie bracket of vector fields. By combining this bracket with the exterior multiplication of differential forms, we obtain a bracket operation for the V_m-valued differential forms. Namely, if φ_1, φ_2 are (scalar-valued) differential forms we define (in the usual way)

$$[\![\varphi_1 \otimes v_1, \varphi_2 \otimes v_2]\!] = (\varphi_1 \wedge \varphi_2) \otimes [\![v_1, v_2]\!] .$$

In particular, the bracket of two elements of E_m is a V_{m-2}-valued differential form on P_m and we therefore have the map

$$(\wedge^* T^* \otimes R_{m-1}) \otimes (\wedge^* T^* \otimes R_{m-1}) \longrightarrow \wedge^* T^* \otimes R_{m-2} .$$

Let

$$\begin{cases} \mu_m \colon F(P_m) = P_m \times \mathfrak{g}_m \longrightarrow T(P_m) , \\ \nu_m \colon P_m \times \mathfrak{g}_m \longrightarrow P_m \times V_m , \end{cases}$$

be the inclusions, and let

$$\begin{cases} \varpi_{m-1} \colon V_m \longrightarrow V_{m-1} \\ \varphi_{m-1} \colon P_m \times \mathfrak{g}_m \longrightarrow P_m \times \mathfrak{g}_{m-1} \end{cases}$$

be the projections.

The fundamental form has the following well-known properties (see, e.g., Guillemin and Sternberg [5]).

PROPOSITION 6.1. (i) *The fundamental form is equivariant*, i.e.,

$$r_g^*\omega_m = g^{-1}\omega_m\,, \qquad g \in G_m\,. \tag{6.10}$$

(ii) *The square.*

$$\begin{array}{ccc} P_m \otimes \mathfrak{g}_m & \xrightarrow{\mu_m} & T(P_m) \\ \big\downarrow \varphi_{m-1} & & \big\downarrow \omega_m \\ P_m \times \mathfrak{g}_{m-1} & \xrightarrow{\nu_{m-1}} & P_m \times V_{m-1} \end{array} \tag{6.11}$$

is commutative.

(iii) The integrability condition

$$d(\varpi_{m-2}\omega_m) - \frac{1}{2}[\![\omega_m, \omega_m]\!] = 0 \tag{6.12}$$

is satisfied.

An additional property of the fundamental form is the following (see [5]).

PROPOSITION 6.2 (first fundamental theorem). *Suppose that* $m \geqq k$ (order of Γ). *Then a local bidifferentiable bundle isomorphism h of P_m* (i.e., *a local bidifferentiable map of P_m which commutes with the right operations of G_m on P_m*) *satisfies $h = j_m(f)$, $f \in \Gamma$, if and only if $h^*\omega_m = \omega_m$.*

7. First complex attached to the pseudogroup

Let

$$\alpha_m : P_{m+1} \longrightarrow P_m$$

be the projection; then the triangle

$$\begin{array}{ccc} E_{m+1} & \xrightarrow{\varpi_{m-1}} & \alpha_m^*(E_m) \\ & \searrow^{\pi_m} & \big\uparrow \alpha_m^* \\ & & E_m \end{array}$$

is commutative where (see § 6) $\varpi_{m-1}: E_{m+1} \longrightarrow \alpha_m^*(E_m)$ and π_m is the first projection (6.3). Here $\alpha_m^*: E_m \longrightarrow \alpha_m^*(E_m)$ is an isomorphism and we denote its inverse by $\mathcal{H}_m$; then

(7.1) $$\pi_m = \mathcal{H}_m \circ \varpi_{m-1}\colon E_{m+1} \to E_m \, .$$

We shall frequently denote elements of I_m and E_m by attaching subscripts i, e, respectively. For example, if $\sigma \in \wedge^* T^* \otimes R_m$, we may write $\sigma_{(i)} = \iota_m(\sigma) \in I_m$, or

$$\sigma_{(e)} = \omega_m \barwedge \sigma_{(i)} \in E_m \, .$$

We have the map

$$L_m\colon I^r_{m+1} \longrightarrow E^{r+1}_m$$

sending $\sigma_{(i)}$ into $L_m \sigma_{(i)}$ where

(7.2) $$L_m \sigma_{(i)} = (-1)^r \pi_m (d_{\sigma_{(i)}} \omega_{m+1})$$

and $d_{\sigma_{(i)}} \omega_{m+1}$ is the Lie derivative of ω_{m+1} along $\sigma_{(i)}$. In fact, writing $\sigma_{(e)} = \omega_{m+1} \barwedge \sigma_{(i)} \in E^r_{m+1}$, we have by (2.5) and (7.1)

$$\begin{aligned} L_m \sigma_{(i)} &= \pi_m (d(\omega_{m+1} \barwedge \sigma_{(i)}) + (-1)^r d\omega_{m+1} \barwedge \sigma_{(i)}) \\ &= \mathcal{H}_m (d(\varpi_{m-1} \sigma_{(e)}) + (-1)^r d(\varpi_{m-1} \omega_{m+1}) \barwedge \sigma_{(i)}) \\ &= \mathcal{H}_m \Big(d(\varpi_{m-1} \sigma_{(e)}) + \frac{(-1)^r}{2} [\![\omega_{m+1}, \omega_{m+1}]\!] \barwedge \sigma_{(i)} \Big) \end{aligned}$$

by (6.12), and hence

$$L_m \sigma_{(i)} = \mathcal{H}_m \big(d(\varpi_{m-1} \sigma_{(e)}) - [\![\omega_{m+1}, \sigma_{(e)}]\!] \big)$$

where the right member belongs to E^{r+1}_m and depends only on

$$\sigma_{(e)} = \omega_{m+1} \barwedge \sigma_{(i)} \in E^r_{m+1} \, .$$

Therefore we obtain, for each r, a differential operator

$$D_{(e)}\colon E^r_{m+1} \longrightarrow E^{r+1}_m$$

where, for $\sigma_{(e)} \in E^r_{m+1}$ (compate [5]),

(7.3) $$D_{(e)} \sigma_{(e)} = \mathcal{H}_m \big(d(\varpi_{m-1} \sigma_{(e)}) - [\![\omega_{m+1}, \sigma_{(e)}]\!] \big) \, ,$$

or, equivalently,

(7.4) $$D_e \sigma_{(e)} = (-1)^r \pi_m (d_{\sigma_{(i)}} \omega_{m+1})$$

where $\sigma_{(i)}$ is any element of I^r_{m+1} whose projection $\pi_m(\sigma_{(i)})$ corresponds to $\sigma_{(e)}$ in the isomorphism (6.2) (compare (6.7)).

Finally we define the operator

(7.5) $$D\colon \wedge^r T^* \otimes R_m \longrightarrow \wedge^{r+1} T^* \otimes R_{m-1}$$

such that the rectangle

$$\begin{array}{ccc} \wedge^r T^* \otimes R_m & \xrightarrow{D} & \wedge^{r+1} T^* \otimes R_{m-1} \\ \downarrow \varepsilon_{m+1} & & \downarrow \varepsilon_m \\ E_{m+1} & \xrightarrow{D_{(e)}} & E_m \end{array}$$

is commutative. We verify easily that $D^2 = 0$. Thus we have the following complex (first complex attached to the differential operator ∇ in the sense of [10(a)]; see also [5], [10(c)])

$$(7.6)\quad 0 \longrightarrow \Theta \xrightarrow{j_m} R_m \xrightarrow{D} T^* \otimes R_{m-1} \xrightarrow{D} \wedge^2 T^* \otimes R_{m-2} \xrightarrow{D} \cdots \xrightarrow{D} \wedge^n T^* \otimes R_{m-n} \longrightarrow 0$$

where, if $m \geqq k$ (order of Γ), this sequence is exact at R_m.

The properties of the first sequence are well known, and we recall how the operator D defines another operator δ which is linear over the functions (i.e., it is a bundle morphism). We have the exact commutative diagram

$$\begin{array}{ccccccccc} 0 & \longrightarrow & \wedge^r T^* \otimes g_{m+1} & \longrightarrow & \wedge^r T^* \otimes R_{m+1} & \xrightarrow{\pi_m} & \wedge^r T^* \otimes R_m & \longrightarrow & 0 \\ & & \downarrow -\delta & & \downarrow D & & \downarrow D & & \\ 0 & \longrightarrow & \wedge^{r+1} T^* \otimes g_m & \longrightarrow & \wedge^{r+1} T^* \otimes R_m & \xrightarrow{\pi_{m-1}} & \wedge^{r+1} T^* \otimes R_{m-1} & \longrightarrow & 0 \end{array}$$

where

$$\delta \colon \wedge^r T^* \otimes g_{m+1} \longrightarrow \wedge^{r+1} T^* \otimes g_m$$

is a morphism of vector bundles. We thus obtain the complex

$$(7.7)\quad \begin{array}{l} 0 \longrightarrow g_m \xrightarrow{\delta} T^* \otimes g_{m-1} \xrightarrow{\delta} \wedge^2 T^* \otimes g_{m-2} \xrightarrow{\delta} \\ \cdots \xrightarrow{\delta} \wedge^{m-k} T^* \otimes g_k \xrightarrow{\delta} \wedge^{m-k+1} T^* \otimes S^{k-1} T^* \otimes T \end{array}$$

where $m \geqq k$, and we denote by $H^{m-j,i} = H^{m-j,i}(g_k)$ the cohomology of this sequence at $\wedge^j T^* \otimes g_{m-j}$. The following theorem is well known.

THEOREM 7.1. *There is an integer $\mu = \mu(n, k)$, depending only on n (the dimension of M) and k (the order of Γ) such that $H^{m,i} = 0$ for all $m \geqq \mu$ and $i \geqq 0$.*

Finally, suppose that $m \geqq \mu$ and choose a splitting of the sequence

$$0 \longrightarrow \wedge^* T^* \otimes g_{m+1} \longrightarrow \wedge^* T^* \otimes R_{m+1} \xrightarrow{\pi_m} \wedge^* T^* \otimes R_m \longrightarrow 0\,,$$

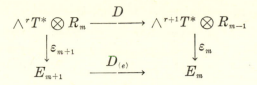

namely a map of degree zero (inducing the identity on $\wedge^* T^*$)

(7.8) $$P\colon \wedge^* T^* \otimes R_m \longrightarrow \wedge^* T^* \otimes R_{m+1}$$

satisfying $\pi_m \circ P = 1$ (identity map). We then obtain an operator

$$D_0 = D_{0,m}\colon \wedge^r T^* \otimes R_m \longrightarrow \wedge^{r+1} T^* \otimes R_m$$

where

(7.9) $$D_0 = D \circ P$$

and, if $\sigma \in \wedge^r T^* \otimes R_m$,

(7.10) $$\pi_{m-1} D_0 \sigma = D\sigma, \quad DD_0 = 0, \quad D_0^2 \sigma = R\sigma \in \delta(\wedge^{r+1} T^* \otimes g_{m+1})$$

where R is the curvature.

Since the transition functions are obtained from the jacobians of the Γ-coordinate transformations by differentiation, the vector bundle R_{m-1} can be regarded as having either the group G_m of P_m or the first prolongation $G_{m-1}^{(1)}$ of the group G_{m-1} of P_{m-1}. We take the first point of view and regard R_{m-1} as having P_m for its associated principal bundle. Then we have the operation of $\wedge^* T^* \otimes R_m$ on $\wedge^* T^* \otimes R_{m-1}$ as sheaves. Namely, if

$$\sigma \in \wedge^r T^* \otimes R_m, \tau \in \wedge^* T^* \otimes R_{m-1},$$

we have by (3.8),

$$\sigma \cdot \tau = \varepsilon_m^{-1}(i_m(\iota_m(\sigma)) \cdot \varepsilon_m(\tau))$$

where i_m has the meaning of the first formula (2.14) and the operation is that of Lie derivation along the vector-valued form $\iota_m(\sigma) \in I_m^r$. The commutators of these operations define on $\wedge^* T^* \otimes R_m$ a structure of graded Lie algebra with bracket $[\cdots]$ (see 3.9).

Finally, let $\sigma \in \wedge^r T^* \otimes R_m$; then

$$\varepsilon_{m+2}(P\sigma) \in E_{m+2}, \sigma_{(e)} = \varepsilon_{m+1}(\sigma) \in E_{m+1},$$

and we set

(7.11) $$D_{0,(e)} \sigma_{(e)} = d\sigma_{(e)} - \mathcal{H}_{m+1} [\![\omega_{m+2}, \varepsilon_{m+2}(P\sigma)]\!] .$$

Then $D_0 \sigma = \varepsilon_{m+1}^{-1}(D_{0,(e)} \sigma_{(e)})$.

8. Second complex attached to the pseudogroup

Let $\mathbf{J}_m = \bigoplus_{r \geq 0} \mathbf{J}_m^r$ where

$$\mathbf{J}_m^r = i_m(I_m^r) \oplus k_m(I_m^{r+1})$$

and i_m, k_m are defined by (2.14) with $B = \bigoplus B^r$ replaced by $I_m = \bigoplus I^r_m$. The projection $\pi_m: I_{m+1} \to I_m$ induces a projection (which we denote by the same symbol)

$$\pi_m: \mathbf{J}_{m+1} \longrightarrow \mathbf{J}_m .$$

Let

$$\mathbf{L}_m: \mathbf{J}^r_{m+1} \longrightarrow E^{r+1}_m$$

be the map sending $u = (\sigma_{(i)}, \xi_{(i)})$ into $\mathbf{L}_m u$ where

$$\mathbf{L}_m u = (-1)^r \pi_m(u \cdot \omega_{m+1}) \tag{8.1}$$

and u operates on ω_{m+1} as a derivation in the sense of § 2. By (2.3), (7.2), (7.4) and (6.7), we have

$$\begin{aligned} \mathbf{L}_m u &= L_m \sigma_{(i)} - \pi_m(\omega_{m+1} \barwedge (d\sigma_{(i)} - \xi_{(i)})) \\ &= D_{(e)}\sigma_{(e)} - \omega_m \barwedge (\pi_m(d\sigma_{(i)} - \xi_{(i)})) \end{aligned}$$

where $\sigma_{(e)} = \omega_{m+1} \barwedge \sigma_{(i)}$. Thus, $\mathbf{L}_m$ defines a map

$$\mathbf{D}: \mathbf{J}^r_m \longrightarrow E^{r+1}_m \tag{8.2}$$

where, for $u = (\sigma_{(i)}, \xi_{(i)}) \in \mathbf{J}^r_m$,

$$\mathbf{D}u = \varepsilon_m D(\iota^{-1}_m \sigma_{(i)}) - \omega_m(\pi D u) \tag{8.3}$$

and $\pi: \mathbf{J}^r_m \to I^r_m$ is the projection sending $u = (\sigma_{(i)}, \xi_{(i)})$ to $\sigma_{(i)}$,

$$Du = (d\sigma_{(i)} - \xi_{(i)}, - d\xi_{(i)})$$

(see § 2). Or, equivalently,

$$\mathbf{D}u = (-1)^r \pi_m(u_{m+1} \cdot \omega_{m+1}) \tag{8.4}$$

where u_{m+1} is any element of J^r_{m+1} such that

$$\pi_m(u_{m+1}) = u_m = u \in J^r_m .$$

Now let $\tilde{J}_{m-1} = \bigoplus_{r \geq 0} \tilde{J}^r_{m-1}$ where

$$\tilde{J}^r_{m-1} = \{\pi_{m-1}(u) \mid u \in \mathbf{J}^r_m, \mathbf{D}u = 0\} . \tag{8.5}$$

Then, by (8.4), $\tilde{J}_{m-1}$ is an algebra of derivations of the equivariant V_{m-1}-valued forms E_m on P_m. Moreover, we infer from (8.3) that, if $u = (\sigma_{(i)}, \xi_{(i)}) \in \mathbf{J}^r_m$ and $\mathbf{D}u = 0$, then $\pi_{m-1}(u) \in \mathbf{J}^r_{m-1}$ is completely determined by $\sigma_{(i)} \in I^r_m$ and, in fact,

$$\begin{aligned} \pi_{m-1}(u) &= i_{m-1}(\pi_{m-1}(\sigma_{(i)})) - k_{m-1}\iota_{m-1}D(\iota^{-1}_m \sigma_{(i)}) \\ &= (\pi_{m-1}(\sigma_{(i)}), d\pi_{m-1}(\sigma_{(i)}) - (0, \iota_{m-1}D(\iota^{-1}_m \sigma_{(i)})) . \end{aligned} \tag{8.6}$$

Thus we have the surjective map

(8.7) $$\sharp\colon \wedge^r T^* \otimes R_m \longrightarrow \tilde{J}^r_{m-1}$$

where

(8.8) $$\sharp\sigma = (\iota_{m-1}(\pi_{m-1}\sigma), d\iota_{m-1}(\pi_{m-1}\sigma)) - (0, \iota_{m-1}(D\sigma)) .$$

The kernel of (8.7) is $\delta(\wedge^{r-1}T^* \otimes g_{m+1})$, i.e., we have the exact sequence

(8.9) $$0 \longrightarrow \delta(\wedge^{r-1}T^* \otimes g_{m+1}) \longrightarrow \wedge^r T^* \otimes R_m \xrightarrow{\ \sharp\ } \tilde{J}^r_{m-1} \longrightarrow 0 .$$

Let m be a fixed integer, $m \geqq \mu$ (integer of Theorem 7.1), and define $C = C_{m-1} = \otimes C^r$ where

(8.10) $$\begin{cases} C^0 = C^0_{m-1} = R_m , \\ C^{r+1} = C^{r+1}_{m-1} = (\wedge^{r+1}T^* \otimes R_m)/\delta(\wedge^r T^* \otimes g_{m+1}) . \end{cases}$$

An element u of C^r can be represented by a pair $u = (\sigma, \xi)$ (in the sense of § 2) where $\sigma \in \wedge^r T^* \otimes R_{m-1}$ and $\xi = d\sigma - D\sigma_m$ where σ_m is an element of $\wedge^r T^* \otimes R_m$ with $\pi_{m-1}(\sigma_m) = \sigma_{m-1} = \sigma$. We have the isomorphism of vector bundles

(8.11) $$C^r_{m-1} \cong \tilde{J}^r_{m-1} .$$

In fact, the map $C^r_{m-1} \to \tilde{J}^r_{m-1}$ taking (σ, ξ) into

$$(\iota_{m-1}\sigma, d(\iota_{m-1}\sigma)) - (0, \iota_{m-1}(d\sigma - \xi))$$

is linear over the functions. Moreover, we have the exact sequence of vector bundles

(8.12) $$0 \longrightarrow \delta(\wedge^{r-1}T^* \otimes g_{m+1}) \longrightarrow \wedge^r T^* \otimes R_m \xrightarrow{\ \sharp\ } C^r_{m-1} \longrightarrow 0$$

where

(8.13) $$\begin{aligned} \sharp\sigma &= (\pi_{m-1}\sigma, d(\pi_{m-1}\sigma)) - (0, D\sigma) \\ &= -(Dk_{m-1}\pi_{m-1}\sigma + k_{m-1}D\sigma) \end{aligned}$$

and $Dk_{m-1}\pi_{m-1}\sigma = D(0, \pi_{m-1}\sigma) = -(\pi_{m-1}\sigma, d(\pi_{m-1}\sigma))$.

We remark finally that (8.11) induces on $C = C_{m-1}$ a structure of graded Lie algebra.

We have the operations

$$(\wedge^r T^* \otimes R_m) \otimes E^s_{m-1} \longrightarrow E^{r+s}_{m-1}$$

where $\sigma \otimes \eta_{(e)}$ goes into $\sigma * \eta_{(e)} = (\sharp\sigma)\cdot\eta_{(e)}$ and $\sharp\sigma$ operates on $\eta_{(e)}$ as a derivation of $\tilde{J}^r_{m-1}$. For $\sigma_1 \in \wedge^r T^* \otimes R_m$, $\sigma_2 \in \wedge^s T^* \otimes R_m$, we define the bracket $[\sigma_1, \sigma_2]$ to be the element of $\wedge^{r+s}T^* \otimes R_m$ such that

$$(8.14) \qquad [\sigma_1, \sigma_2] * \eta_{(e)} = \sigma_1 * (\sigma_2 * \eta_{(e)}) - (-1)^{rs} \sigma_2 * (\sigma_1 * \eta_{(e)})$$

for $\eta_{(e)} \in E_{m-1}$. This bracket imparts a structure of algebra to $\wedge^* T^* \otimes R_m$ and $\delta(\wedge^* T^* \otimes g_{m+1})$ is an ideal of it. Thus we have the exact sequence of sheaves of algebras

$$0 \longrightarrow \delta(\wedge^* T^* \otimes g_{m+1}) \longrightarrow \wedge^* T^* \otimes R_m \xrightarrow{\sharp} C_{m-1} \longrightarrow 0 .$$

In particular, if $\sigma_1, \sigma_2 \in \wedge^* T^* \otimes R_m$,

$$(8.15) \qquad \sharp[\sigma_1, \sigma_2] = [\sharp\sigma_1, \sharp\sigma_2] .$$

Finally, choose a splitting (7.8) and let D_0 be defined by (7.9). It follows from (7.10) and (8.13) that

$$\sharp(D_0\sigma) = (D\sigma, dD\sigma) , \qquad \sigma \in \wedge^r T^* \otimes R_m ,$$

and hence $\sharp(D_0\sigma) \in C^{r+1}_{m-1}$ is independent of the choice of the splitting (7.8). On the other hand, if $\sharp\sigma = u$ is regarded as belonging to $\tilde{J}^r_{m-1}$ then, by (2.12) and (8.8),

$$D(\sharp\sigma) = Du = (\iota_{m-1}(D\sigma), d\iota_{m-1}(D\sigma)) ,$$

and we see that $D(\sharp\sigma)$ corresponds (in the isomorphism (8.11)) to the element $\sharp(D_0\sigma)$, i.e.,

$$\sharp D_0 = D\sharp .$$

Thus $\tilde{J}_{m-1}$ is stable under the operator D and $C = C_{m-1}$ has the corresponding operator

$$(8.16) \qquad D\colon C^r_{m-1} \longrightarrow C^{r+1}_{m-1}$$

where, if $u = (\sigma, \xi) \in C^r_{m-1}$

$$(8.17) \qquad Du = D(\sigma, \xi) = (d\sigma - \xi, -d\xi) .$$

There is a connection Ω, not belonging to $\wedge^* T^* \otimes R_m$, such that $[\Omega, P\sigma] = D_0\sigma$ and $\sharp\Omega = d$, where $[d, \sharp\sigma] = D(\sharp\sigma)$. This connection is a reflection of (7.3) in which Ω corresponds to $d - [\omega, \cdots]$ where d is total (exterior) derivation, $[\omega, \cdots]$ is vertical derivation, and the difference is horizontal derivation. For a more precise formulation, see the forthcoming paper by Malgrange and the author.

Passing to the quotient we have

$$(8.18) \qquad Du = [d, u] , \qquad u \in C^r_{m-1} ,$$

where $u = \sharp\sigma$ and d is the image of ω, $d = d_{m-1} = \sharp\omega$. Clearly $[d, d] = 0$ and hence (see § 2) $D^2 = 0$. Moreover, it follows from

(8.18) and the Jacobi identity (compare (2.11)) that we have the formula

(8.19) $$D[u, v] = [Du, v] + (-1)^r[u, Dv]\ , \qquad u \in C^r_{m-1},\ v \in C_{m-1}\ .$$

Finally, $d = \operatorname{Ad} h^{-1}\cdot d$ if and only if $h = j_m f,\ f \in \Gamma$.

We thus obtain the second complex attached to the operator ∇ (see [10(a), (c)], [2])

$$0 \longrightarrow \Theta \xrightarrow{j_m} C^0 \xrightarrow{D} C^1 \xrightarrow{D} C^2 \xrightarrow{D} \cdots \xrightarrow{D} C^n \longrightarrow 0$$

where $C^r = C^r_{m-1}$, $m \geqq \mu$ (integer of Theorem 7.1). If $m \geqq \mu + n$, the cohomologies of the first and second sequences are the same (see [10(a), (c)], [11]).

9. Operations of the sheaf of groups

For simplicity we suppose from the outset that m is a fixed integer, $m \geqq \mu + n$ (where μ is the integer of Theorem 7.1).

Now let $H = H_m = H(P_m)$ be the sheaf of groups over M of 1-parameter families of bundle isomorphisms of P_m in the sense of Definition 4.1. For each integer l, $0 \leqq l < m$, we have the projection $H_m \rightarrow H_l$ and its kernel $H_{m,l}$ is a sheaf of normal subgroups of H_m, i.e., we have the exact sequence of sheaves of groups over M

(9.1) $$1 \longrightarrow H_{m,l} \longrightarrow H_m \longrightarrow H_l \longrightarrow 1\ .$$

In particular, H_0 is the sheaf $S(GP)$ of groups of 1-parameter families of transformations of the general pseudogroup GP of all local bidifferentiable maps. We denote by $S(\Gamma)$ the sheaf of subgroups of $S(GP)$ composed of the families of transformations belonging to the given pseudogroup Γ. We have the map

(9.2) $$j_m\colon S(\Gamma) \longrightarrow H = H_m$$

which maps the germ f_t of $S(\Gamma)$ into its m-jet $j_m(f)$ (where $j_m(f)$ involves differentiation with respect to the variables of M but not with respect to the parameter t). The monomorphism (9.2) defines $S(\Gamma)$ as a sheaf of left operators on $H = H_m$.

Let $C_{(p)} = C_{m-1,(p)}$ be the sheaf of 1-parameter families of sections of the corresponding bundle which vanish with order p (at least) in the parameter t, as described in Definition 4.2. We shall denote by Ad h the operations on

$$C_{(p)} = C_{m-1,(p)}\ , \quad \text{or} \quad (\wedge^* T^* \otimes R_m)_{(p)}\ ,$$

defined by an element h of H.

The operations of H_m on $C_{m-1,(p)}$ commute with the bracket, i.e., if $h \in H_m$,

$$\text{Ad}\, h \cdot [u, v] = [\text{Ad}\, h \cdot u, \text{Ad}\, h \cdot v]\,, \qquad u, v \in C_{m-1,(p)}\,. \tag{9.3}$$

Hence, in particular (see (8.18)),

$$\text{Ad}\, h \cdot [d, u] = [\text{Ad}\, h \cdot d, \text{Ad}\, h \cdot u]\,, \qquad u \in C_{m-1,(p)}\,. \tag{9.4}$$

Now let $D\colon C^r_{(p)} \longrightarrow C^{r+1}_{(p)}$ be the differential operator defined in the preceding section. Then we have (see (8.18)) $Du = [d, u]$, $u \in C^r_{(p)}$. Let h be an element of $H = H_m$, and define

$$\mathfrak{D}\colon H = H_m \longrightarrow C_{(1)} = C^1_{m-1,(1)} \tag{9.5}$$

to be the map where

$$\mathfrak{D}h = d - \text{Ad}\, h^{-1} d\,. \tag{9.6}$$

We assert that $\mathfrak{D}h$ is invariant under the left operations of $S(\Gamma)$ defined by (9.2), i.e.,

$$\mathfrak{D}(j_m f \cdot h) = \mathfrak{D}h, f \in S(\Gamma)\,,$$

and it is an easy consequence of (9.6) that

$$\mathfrak{D}(h_1 h_2) = \text{Ad}\, h_2^{-1} \cdot \mathfrak{D}h_1 + \mathfrak{D}h_2\,, \qquad h_1, h_2 \in H\,. \tag{9.7}$$

Moreover, $\mathfrak{D}h$ belongs to $C^1_{(1)}$. The following proposition is immediate.

Proposition 9.1. *We have the formula*

$$D\, \text{Ad}\, hu - \text{Ad}\, hDu = -\text{Ad}\, h \cdot [\mathfrak{D}h, u]\,, \qquad u \in C_{m-1,(p)}\,. \tag{9.8}$$

In fact, if $a \in C_{m-1,(p)}$, we have by (8.18) and (9.6)

$$\begin{aligned}(D\, \text{Ad}\, h - \text{Ad}\, hD)u &= [d, \text{Ad}\, h \cdot u] - \text{Ad}\, h \cdot [d, u] \\ &= \text{Ad}\, h \cdot [\text{Ad}\, h^{-1} \cdot d - d, u] \\ &= -\text{Ad}\, h \cdot [\mathfrak{D}h, u]\,.\end{aligned}$$

Now let

$$\mathfrak{D}'\colon C^1_{(1)} \longrightarrow C^2_{(1)} \tag{9.9}$$

where, for $v = v \in C^1_{(1)}$,

$$\mathfrak{D}'v = Dv - \frac{1}{2}[v, v]\,, \tag{9.10}$$

and let

(9.11) $$D_t\colon C^r_{(0)} \longrightarrow C^{r+1}_{(0)}$$

be the operator where

(9.12) $$D_t u = Du - [v_t, u]\ .$$

We have $D_t^2 = 0$ if and only if $\mathfrak{D}'v = \mathfrak{D}'v_t = 0$. Since $D_0 = D$, D_t is a perturbation of D.

The main features of the mechanism of deformation, as originally developed in the papers [10(a), (b)], are incorporated in the following theorem.

THEOREM 9.1. (i) *The sequence*

(9.13) $$1 \longrightarrow S(\Gamma) \xrightarrow{j_m} H \xrightarrow{\mathfrak{D}} \mathfrak{D}(H) \longrightarrow 0$$

is exact, and the right operation of H on $\mathfrak{D}(H)$ (isomorphic to $H/S(\Gamma)$) is given by the formula

(9.14) $$(\mathfrak{D}h_1)^{h_2} = \mathfrak{D}(h_1h_2) = \mathrm{Ad}\, h_2^{-1}\cdot\mathfrak{D}h_1 + \mathfrak{D}h_2\ , \qquad h_1, h_2 \in H\ .$$

In particular, if $h \in H$, $f \in S(\Gamma)$, we have

(9.15) $$\mathfrak{D}h^{-1} = -\mathrm{Ad}\, h\cdot\mathfrak{D}h\ .$$

(ii) *The Frechet derivative of $\mathfrak{D}$ is the negative of the operator $D\colon C^0 \to C^1$. In fact, let*

(9.16) $$(d/dt)_0\colon H \longrightarrow C^0$$

be differentiation at $t = 0$, i.e., $(d/dt)_0(h) = (dh/dt)_{t=0} \in R_m = C^0$. *Then*

(9.17) $$(d/dt)_0(\mathfrak{D}h) = D(d/dt)_0(h)\ .$$

(iii) *The sequence*

(9.18) $$1 \longrightarrow S(\Gamma) \xrightarrow{j_m} H \xrightarrow{\mathfrak{D}} C^1_{(1)} \xrightarrow{\mathfrak{D}'} C^2_{(1)}$$

is a complex, in particular

(9.19) $$\mathfrak{D}' \circ \mathfrak{D} = 0\ ,$$

and it is formally exact and exact except possibly at $C^1_{(1)}$. The exactness at $C^1_{(1)}$ is equivalent to the exactness of the sequence

(9.20) $$C^0_{(0)} \xrightarrow{D_t} C^1_{(0)} \xrightarrow{D_t} C^2_{(0)}$$

under the assumption that $D_t^2 = 0$. The right operation of H on $C^1_{(1)}$ in (9.18) *is defined, for $h \in H$ and $u \in C^1_{(1)}$, by the formula*

$$u^h = \mathrm{Ad}\, h^{-1} \cdot u + \mathfrak{D}h \tag{9.21}$$

where

$$(u^{h_1})^{h_2} = u^{h_1 \cdot h_2}\,, \qquad h_1, h_2 \in H\,, \tag{9.22}$$

and

$$\mathfrak{D}' u^h = \mathrm{Ad}\, h^{-1} \cdot \mathfrak{D}' u\,. \tag{9.23}$$

Hence ker $\mathfrak{D}'$ *is stable under the right operations of* H *on* $C^1_{(1)}$, *and the cohomology of* (9.18) *at* $C^1_{(1)}$ *is the set of orbits of* ker $\mathfrak{D}'$ *under these operations of* H.

(iv) *Suppose that* M *is compact* (*without boundary*). *Then* $H^1(M, S(\Gamma))$ *is the set of germs of deformations of the* Γ*-structure of* M (*with distinguished element* 1 *corresponding to the given* Γ*-structure of* M), *and* $H^1(M, H) = 1$ (*the deformations of the principal bundle* P_m *are trivial*). *The cohomology sequence of* (9.13), *namely*

$$1 \longrightarrow H^0(M, S(\Gamma)) \xrightarrow{j_m} H^0(M, H) \xrightarrow{\mathfrak{D}} H^0(M, \mathfrak{D}(H)) \xrightarrow{\partial^*} H^1(M, S(\Gamma)) \longrightarrow 1 \tag{9.24}$$

is exact in the usual sense, i.e., *two elements of* $H^0(M, \mathfrak{D}(H))$ *have the same image in* $H^1(M, S(\Gamma))$ *if and only if they are congruent modulo the right operations of* $H^0(M, H)$ *on* $H^0(M, \mathfrak{D}(H))$.

Formulas entirely similar to those of Theorem 9.1 hold for the first complex (7.6) with bracket $[\![\cdots]\!]$ defined by (6.9), where both the operator D (defined in § 7) and the bracket depress the order by a unit (see [5]).

The equivalence of the exactness of (9.18) at $C^1_{(1)}$ with that of (9.20) is proved in [10(b)] by using the formulas of Theorem 9.1, which are the formulas listed in Proposition 10.1 of [10(b)]. The author has little to add to that proof.

We begin with part (i) of the theorem. The exactness of (9.13) is a consequence of the invariance of $\mathfrak{D}(H)$ under left operations of $S(\Gamma)$, and hence $\mathfrak{D}(H)$ is isomorphic to $H/S(\Gamma)$. Formula (9.14) is a restatement of (9.7) and (9.15) is a trivial consequence of it.

As for part (ii), formula (9.17) is easily checked. We observe that differentiation of (9.18) with respect to t at $t = 0$ yields the sequence

$$0 \longrightarrow \Theta \xrightarrow{j_m} C^0 \xrightarrow{D} C^1 \xrightarrow{D} C^2\,.$$

In fact, setting $(d/dt)_0 v = u$ we have, by (9.10), $(d/dt)_0 \mathfrak{D}' \mathrm{v} = Du$ since the bracket vanishes to second order in t.

Consider then part (iii). We have, by (9.3) and (9.6),

$$\begin{aligned} 0 &= \operatorname{Ad} h^{-1} \cdot [d, d] = [\operatorname{Ad} h^{-1} \cdot d, \operatorname{Ad} h^{-1} \cdot d] = [d - \mathfrak{D}h, d - \mathfrak{D}h] \\ &= [d, d] - 2[d, \mathfrak{D}h] + [\mathfrak{D}h, \mathfrak{D}h] = -2D(\mathfrak{D}h) + [\mathfrak{D}h, \mathfrak{D}h] \end{aligned}$$

and this is (9.19). Next, formula (9.22) is an immediate consequence of (9.21). As for (9.23), let $u \in C^1_{(1)}$, $h \in H$, and write $v = u^h$, Then

$$u = \operatorname{Ad} h \cdot (v - \mathfrak{D}h) ,$$

and we have by (9.8)

$$\begin{aligned} Du - \frac{1}{2}[u, u] &= D \operatorname{Ad} h \cdot (v - \mathfrak{D}h) - \frac{1}{2} \operatorname{Ad} h [v - \mathfrak{D}h, v - \mathfrak{D}h] \\ &= \operatorname{Ad} h \Big(D(v - \mathfrak{D}h) - [\mathfrak{D}h, v - \mathfrak{D}h] \\ &\qquad - \frac{1}{2}[v - \mathfrak{D}h, v - \mathfrak{D}h] \Big) \\ &= \operatorname{Ad} h \Big(Dv - \frac{1}{2}[v, v] - (D\mathfrak{D}h) - \frac{1}{2}[\mathfrak{D}h, \mathfrak{D}h] \Big) \\ &= \operatorname{Ad} h \Big(Dv - \frac{1}{2}[v, v] \Big) \end{aligned}$$

by (9.19), and this is (9.23). Finally the sequence (9.18) can be shown to be exact at $C^1_{(1)}$ if $\Gamma = GP$ (general pseudogroup of all local bidifferentiable transformations), more generally if Γ is a multifoliate pseudogroup as defined in [7], or in the real-analytic case. We shall omit here a proof of the formal exactness (in the sense of jets).

Part (iv) of Theorem 9 is a consequence of well-known theorems.

PRINCETON UNIVERSITY

REFERENCES

[1] A. FRÖLICHER and A. NIJENHUIS, *Theory of vector-valued differential forms*, Part I. *Derivations in the graded ring of differential forms*, Proc, Kon. Ned. Akad. Wet. Amsterdam **59** (1956), 338-359.

[2] H. GOLDSCHMIDT, *Existence theorems for analytic linear partial differential equations*, Ann. of Math. **86** (1967), 246-270.

[3] P. A. GRIFFITHS: (a) *Deformations of G-structures* I, II, Math. Annalen **155** (1964), 292-315; (b) *On theory of variation of structures defined by transitive, continuous pseudogroups*, Osaka Journ, Math. **1** (1964), 175-199.

[4] V. GUILLEMIN, D. QUILLEN, and S. STERNBERG, *The classification of the complex primitive infinite pseudogroups*, Proc. Nat. Acad. Sci. USA **55** (1966), 687-690.

[5] V. GUILLEMIN and S. STERNBERG, *Deformation theory of pseudogroup structures*, Memoirs of the Amer. Math. Soc., No. 64 (1966), 80 pages.

[6] C. J. Henrich, *Derivations on an arbitrary vector bundle*, Trans. Amer. Math. Soc. **109** (1963), 411-419.

[7] K. KODAIRA and D. C. SPENCER, *Multifoliate structures*, Ann. of Math. **74** (1961), 52-100.

[8] NGÔ VAN QUÊ, *Non-abelian Spencer cohomology and deformation theory* Journ. of Differential Geometry (to appear).

[9] H. K. NICKERSON, *Differential operators and connections*, Trans. Amer. Math. Soc. **99** (1961), 509-539.

[10] D. C. SPENCER: (a) *Deformation of structures on manifolds defined by transitive, continuous pseudogroups*; I-II, Ann. of Math. **76** (1962), 306-445; (b) *Deformation of structures on manifolds defined by transitive, continuous pseudogroups*: III, Ann. of Math. **81** (1965), 389-450. (c) *Overdetermined systems of linear partial differential equations*, Bull. Amer. Math. Soc. **75** (1969), 179-239.

[11] D. G. QUILLEN, *Formal properties of over-determined systems of linear partial differential equations*, Thesis, Harvard Uuiversity, 1964 (unpublished).

Received December 6, 1968

Sur les variétés d'ordre fini

Par René Thom

Les problèmes classiques de Géométrie Finie (au sens de Juel, O. Haupt, de Marchaud, etc) relèvent probablement de la théorie des singularités affines des plongements différentiables. Le présent article vise à préciser ce rapport.

Définition 1. Un ensemble A de $\mathbf{R}^n$ sera *dit de k-degré fini m*, si tout k-plan de R^n rencontre A en au plus m points.

Exemple. Un ensemble algébrique A de $\mathbf{R}^n$, de codimension k, est de k-degré fini (majoré par le degré algébrique de A pris en tant que variété affine), à moins qu'il ne contienne une variété linéaire.

On se propose d'établir le théorème:

Théorème 1. *Soit V une variété différentiable compacte de dimension n, $C^\infty(V, R^m)$ l'espace fonctionnel des applications différentiables de V dans $\mathbf{R}^m$ muni de la C^r topologie $(m > n)$. Il existe dans $C^\infty(V, \mathbf{R}^m)$ un ouvert U partout dense, tel que, pour toute application $g \in U$, l'ensemble image $g(V)$ est de $(m-n)$ degré fini.*

Remarque. Ce degré n'est pas borné sur U, car on peut se donner une application f: $V \longrightarrow \mathbf{R}^m$ dont l'image a avec un k-plan un nombre arbitrairement grand de points d'intersection transversaux, donc stables.

Définition 2. *Degré local.* Soit V^n une variété différentiablement plongée dans $\mathbf{R}^{n+k}$, O un point de V. On appelle *degré local* de V en O la limite supérieure du nombre des points d'intersection de V par un k-plan H, lorsque la distance de H à O tend vers zéro.

Lemme 1. *Etant donné un jet régulier $z \in J^r(n, n+k)$, soit C la contre-image de z dans $J^{r+s}(n, n+k)$ par l'application canonique, l'entier positif s ne dépendant que de n et k. Il existe alors un vrai sous-ensemble algébrique S de C, tel que tout jet s du complémentaire $C - S$ a la propriété suivante: si $j^{r+s}(f)(O) = s$, alors l'ensemble image $f(R^n)$ est de k-degré local fini dans $\mathbf{R}^{n+k}$.*

Pour des raisons de commodité d'écriture, on démontrera le

Lemme 1 dans le cas d'une hypersurface ($k = 1$). Comme le jet z est régulier, on définira une hypersurface issue de O par son équation locale, écrite sous forme $\langle F(x_1, x_2, \cdots, x_n) - u = 0\rangle$. Supposons que le degré local en O soit exactement $q+1$; ceci veut dire qu'une droite de représentation paramétrique $x_i = x_i^0 + pX_i$, $u = u^0 + pU$ (1) coupe $F = 0$ en $q+1$ points au plus, et qu'il en existe qui coupent $F = 0$ en $q+1$ points exactement pour x_i^0, u^0 arbitrairement petits. Formons le développement de Taylor de F où l'on a substitué la représentation (1):

$$\begin{aligned} 0 &= F(x_i^0 + pX_i^0) - (u_0 + pU_0) \\ &= F(x_i^0)^{-u_0} + p \sum X_i F_i(x_i^0) + p^2 \sum X_i X_j F x_i x_j + \cdots . \end{aligned}$$

Le degré local étant $q+1$, ceci implique que tous les coefficients de p jusqu'à celui de p^q inclus s'annulent lorsqu'on fait $x_i^0 = 0$. Mais alors on peut faire choix des dérivées d'ordre $q+1$ de manière à ce que le coefficient de p^{q+1} soit non nul quels que soient les X_i, par exemple, par un choix convenable de F^{q+1}. Dans ces conditions, $F = 0$ est de degré local q strict, aucune droite locale autour de 0 ne peut rencontrer un représentant F du $(q+1)$-jet en plus de $(q+1)$ points. En particulier une hypersurface de $\mathbf{R}^n$ est génériquement de degré local $(n+1)$, puisqu'il faut n équations pour déterminer les n paramètres X_i, U.

COROLLAIRE. I *Il existe dans l'espace des jets réguliers $J^\infty(n, n+k)$ un pro-ensemble S de codimension infinie, tel que, si le jet de f est en dehors de S, la variété image par f est de degré local fini autour de O.*

La démonstration du Théorème 1 est alors évidente; on déforme l'application f de manière que le plongement du graphe de f ne présente en aucun point un jet situé dans l'ensemble S. Le graphe de f est alors de degré local fini dans le produit: Source $\times$ But, et son image, par compacité, est de degré fini dans le but.

Comme on l'a vu plus haut, le degré total ne peut être borné sur l'espace fonctionnel $C^\infty(V, \mathbf{R})$. On peut cependant conjecturer que, si l'on se donne une borne supérieure au volume de l'*application de Gauss*, alors il existe une borne supérieure au degré fonction de ce seul volume. En effet, on peut penser que si une variété linéaire coupe V en un grand nombre de points, la mesure totale de l'application de Gauss restant petite, on a localement une situation du type d'une "oscillation plate" pour une courbe, comme la transformation

$y = x^2 - x \to y = x^3 + x$ pour une courbe plane; on pourrait donc diminuer le degré local par une C^m déformation. Toutefois la théorie des singularités affines, et de leurs bifurcations, est à peine ébauchée, et, sauf dans le cas des courbes, on sera encore loin de pouvoir attaquer ce type de problèmes.

Il est clair que le degré local générique d'une n-variété plongée va en croîssant avec la codimension. Aussi, pour une variété V^n compacte, il existe un entier k tel que le degré global d'un plongement de degré minimum dans $\mathbf{R}^{n+k}$ n'excède pas le degré générique local. Pour une surface compacte, cette situation est déjà atteinte pour $k = 1$; il existe pour toute surface orientée compacte un plongement de degré total 4 dans R^3 (comme me l'a fait remarquer E. Calabi); par exemple le voisinage tubulaire d'un graphe tel que celui de la Figure 1. Par contre, une surface non orientée (tore de

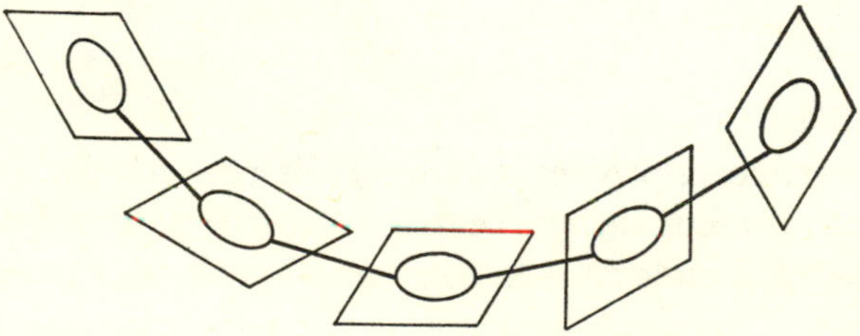

Fig. 1

Klein ou plan projectif immergé) semble requérir le degré six au moins. Pour un plongement local de $\mathbf{R}^2$ dans $\mathbf{R}^3$, le degré local est deux pour un plongement convexe; trois pour un plongement à courbures opposées générique (branches de l'asymptotique convexes). Il est trois aussi pour le point parabolique normal (modèle $z = x^2 + y^3$). Il est quatre pour des points hyperboliques exceptionnels où l'asymptotique a un point d'inflexion et aussi pour les points paraboliques exceptionnels de modèle $z = x^2 + y^4$.

Un problème intéressant est de caractériser, en toute dimension, les variétés dont le degré total est inférieur au degré local générique (comme les surfaces de degré trois dans $\mathbf{R}^3$). Ceci implique évidemment une grande rigidité pour ces variétés, et il faut voir là l'origine des théorèmes connus en Géométrie Finie.

Plongements analytiques

Le problème ici considéré est le suivant: soit V^{2n} une variété de dimension $2n$, compacte, plongée dans l'espace projectif complexe $PC(n+k)$. On suppose qu'il existe dans la grossmannienne G_k des k-plans complexes de $PC(n+k)$ un ouvert U *partout dense* tel que tout k-plan $u \in U$ rencontre V en un nombre fixe m de points. Cette propriété implique-t-elle que V est une variété algébrique complexe? On va proposer ici une démonstration, incomplète toutefois en ce qu'elle fait appel à des hypothèses de généricité difficiles à justifier.

Utilisant le fait bien connu qu'une fonction complexe C^∞, $f(z_1, z_2, \cdots, z_n)_1$ analytique en chacune des variables z_i est analytique en l'ensemble des variables, il suffit, par section plane, de démontrer la proposition pour $n = 1$, et, par projection, pour $n = 1$ et $k = 1$. Soit donc V une surface dans un plan projectif complexe $PC(2)$, rencontrée par presque toute droite projective en m points isolés. Enonçons-le:

LEMME 2. *Pour presque toute projection affine complexe* q: $\mathbf{C}^2 \to \mathbf{C}$, *l'ensemble critique de la restriction* $q \mid \bar{V}$ *ne comporte que des points isolés.*

Si, en effet, cette projection admettait une courbe critique, ce qui est la situation générique, cette courbe ne pourrait donner lieu à une courbe pli-simple dans le but, car deux droites projectives de la forme $q^{-1}(a)$, $q^{-1}(b)$, a, b de part et d'autre de la courbe de contour apparent couperaient V en un nombre de points différant localement par deux unités. La courbe c de contour apparent, si elle existe dans le but $\mathbf{C}$, est donc nécessairement une courbe double, projection par q de deux (ou d'un nombre pair O de courbes plis-simples γ, γ', cf.

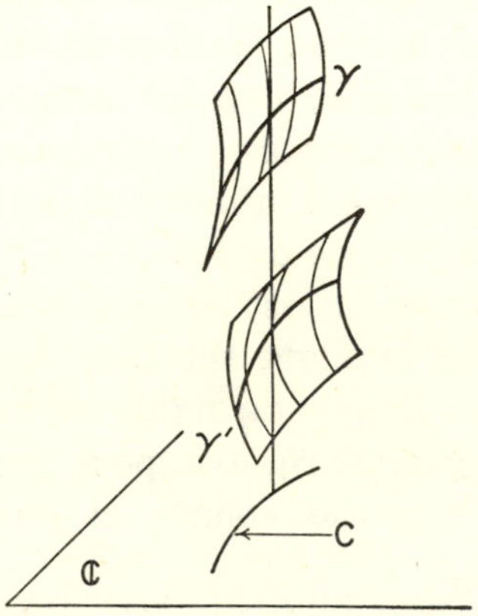

Fig. 2

Figure 2). Je dis que cette situation exceptionnelle ne peut se réaliser pour tout un ouvert de directions complexes q. S'il en était ainsi, on pourrait trouver pour la surface V une famille à trois paramètres de droites bitangentes à V; ceci impliquerait que toutes les droites tangentes à V au voisinage d'un point w de V sont bitangentes; en fixant l'autre point de contact, on voit de suite que le plan tangent ne varie pas le long d'un réseau à deux paramètres de courbes sur la surface V, ce qui exige que V soit un plan (éventualité triviale que nous excluons). Dans ces conditions pour presque toute direction complexe q, les points critiques de $q \mid V$ sont isolés, et le degré local de l'application $q\colon V \to \mathrm{C}$ (le signe du jacobien) ne change pas. Par suite V est une surface orientée, dont certains points ont la propriété que le plan tangent y est complexe. L'ensemble W de ces points est non vide, fermé, car les conditions de Cauchy-Riemann définissent un fermé. Je dis que ce fermé est dense dans V. Soit en effet un point w de la frontière de W, q la direction complexe du plan tangent à V en w. Pour toute direction q' assez voisine de q, il y a des points critiques isolés voisins de w; ceci se voit par des considérations de degré local de q pour un voisinage U' de w dans C. Ce degré local est positif et non nul, et même si d'autres points critiques viennent se confondre avec w au cours de la variation $q \to q'$, leurs degrés locaux, étant de même signe, ne peuvent que s'ajouter l'application de Gauss dans la grassmannienne complexe est donc localement ouverte: il existe des points critiques sur un ouvert partout dense de V pour les directions complexes; l'image de l'application de Gauss est donc toute entière contenue dans la grassmannienne complexe, et la surface V est analytique, donc algébrique, par le théorème de Chow.

Il resterait donc à se libérer des hypothèses de généricité faites sur l'ensemble critique. On doit s'attendre d'ailleurs à ce que le théorème soit valable même en affaiblissant les hypothèses de différentiabilité.

I.H.E.S., BURES-SUR-YVETTE

(Received October 8, 1968)

Obstructions to the existence of a space of moduli*

BY JOHN J. WAVRIK

Introduction

Kuranishi has shown [8], [9], [1] that, if X_0 is a compact complex analytic manifold, a complete family of deformations of X_0 always exists. Let Θ_0 denote the sheaf of germs of holomorphic vector fields on X_0. If $H^0(X_0, \Theta_0) = 0$, the Kuranishi space is a (local) space of moduli for X_0 [1]. If $H^0(X_0, \Theta_0) \neq 0$, X_0 may still have a space of moduli (e.g. X_0 = a complex torus). Examples are known, however, of manifolds not having a space of moduli (e.g. X_0 = a Hirzebruch surface). In this paper we will show that the existence of a space of moduli is equivalent to the solvability of an extension problem [3] for vertical vector fields on the Kuranishi family. Thus there is a sequence of subspaces L_i of $H^1(X_0, \Theta_0)$ which vanish if and only if X_0 has a space of moduli. L_1 is the image of the bracket operation of $H^0(X_0, \Theta_0)$ on $H^1(X_0, \Theta_0)$. If the Kuranishi space (T, o) is reduced and if $\dim H^0(X_t, \Theta_t)$ is independent of t near o, all obstructions vanish and X_0 has a space of moduli. Kuranishi subsequently obtained the latter result by a different method [10]; it was also known to Douady who indicated to me a proof based on the method of [1]. However, no mention of this theorem has been found in the literature.

We rely on the terminology of [5]. By a family of compact complex analytic manifolds we mean a triple (X, π, S) of analytic spaces X, S and a proper, simple morphism $\pi: X \to S$. A family of deformations of a fixed compact complex manifold X_0 is a family (X, π, S) with distinguished point $s \in S$, together with an isomorphism $i: X_0 \to X_s$. A morphism of families of deformations is required to be compatible with the given isomorphisms. If (X, π, S, s, i) is as above and $f: (S', s') \to (S, s)$ is a morphism of pointed spaces, $X_f = X \times_S S'$ becomes, in a natural way, a family of deformations of X_0 over S'. This family is called the

* This work has been partially supported by the National Science Foundation under grant NSF-GP-8988.

family induced by f. In this paper we will adopt an abuse of language and refer to the family as X or X/S (the projection map, distinguished point, and isomorphism being understood). The distinguished point of any space S will be denoted o. All families are understood to be families of deformations of X_0.

We are really only interested in behavior near $o \in S$ (thus in "germs of families"). Consequently, a family X/S will be identified with its restriction to any open neighborhood of o and two families X/S, X'/S will be considered isomorphic if their restrictions to some open neighborhood of o are isomorphic. A (germ of) family X/S is called complete if for any family X'/S' there is a morphism f of some open neighborhood of o in S' into S such that X' and X_f are isomorphic (as germs of families of deformations). Thus, f is a morphism of germs of analytic spaces $f: (S', o) \to (S, o)$. If X/S is complete and the germ f is unique (i.e., if X_f and X_g are isomorphic, then $f = g$) X/S is called a modular family and (S, o) a space of moduli for X_0.

1. Preliminaries

THEOREM 1.1. *If a space of moduli exists for X_0, it must be the Kuranishi space.*

PROOF. Let (S, o) denote the space of moduli and (T, o) the Kuranishi space. Both of these spaces have a Zariski tangent space of dimension $\dim H^1(X_0, \Theta_0)$ (see [5, 16–07] and [1]). Furthermore, we have morphisms $\alpha: (S, o) \to (T, o)$ and $\beta: (T, o) \to (S, o)$ since both spaces are complete, while $\beta \circ \alpha = 1$ since (S, o) is a space of moduli. Under these circumstances β is a isomorphism. This is a consequence of the following lemma.

LEMMA. *Let (S, o), (T, o) be germs of analytic spaces having the same tangential dimension. Let $\alpha: (S, o) \to (T, o)$ and $\beta: (T, o) \to (S, o)$ be morphisms such that $\beta \circ \alpha = 1$. Then β is an isomorphism.*

PROOF. Choose "neat imbeddings" [6, p. 153] g, h of (S, o), (T, o) respectively into $(\mathbf{C}^a, o)$. Extend the morphisms α, β to morphisms $\tilde{\alpha}, \tilde{\beta}: (\mathbf{C}^a, o) \to (\mathbf{C}^a, o)$. Let α^*, β^*, etc. denote the corresponding homomorphisms of local rings at o. It suffices to show that β^* is an isomorphism [5, 13–03]. The *implicit function theorem* shows that $\tilde{\beta}^*$ is an isomorphism (in particular surjective); g^*, h^* are surjective, and $\beta^* g^* = h^* \tilde{\beta}^*$, so β^* is surjective. On the other hand, $\alpha^* \beta^* = 1$ so β^* is injective.

Let X/T be a family of deformations of X_0. Since our considerations are local with respect to T, we may assume that T is chosen so small that X is covered by open sets $\mathfrak{U}_i$ each of which is isomorphic over T to an open set of the form $U_i \times T$ with U_i a polydisk in $\mathbf{C}^m$. X is obtained by patching together the $U_i \times T$ by means of isomorphisms $F_{ij}: N_{ij} \to N_{ji}$ over T (where N_{ij} is open in $U_j \times T$). Let $p_2: U_j \times T \to T$ denote the projection. We must have $F_{ij} = (f_{ij}, p_2)$ where $f_{ij}: N_{ij} \to \mathbf{C}^m$. If $f: S \to T$ is a morphism, $X_f = X \times_T S$ is obtained by patching together the $U_i \times S$ by means of $G_{ij} = (g_{ij}, p_2)$ where $g_{ij} = f_{ij} \circ (1 \times f)$.

Let S be an analytic space, $o \in S$. We will denote by $\mathfrak{M}$ (or $\mathfrak{M}_S$) the sheaf of germs of holomorphic functions on S which vanish at o.

Definition 1.1 [5]. The nth infinitesimal neighborhood of o in S is the analytic subspace $S^{(n)}$ of S defined by the ideal $\mathfrak{M}^{n+1}$. The underlying topological space of $S^{(n)}$ is the point o. The structure sheaf is the finite dimensional C-algebra $\mathcal{O}/\mathfrak{M}^{n+1}$.

If X/S is a family of deformations of X_0, $X^{(n)}/S^{(n)}$ will denote the restriction of this family to $S^{(n)}$. $X^{(n)}$ is X_0 equipped with a different structure sheaf (viz. $\mathcal{O}_X/\mathfrak{M}^{n+1}\mathcal{O}_X$), $X^{(0)} = X_0$. Families of the type $X^{(n)}/S^{(n)}$ will be of prime importance in our discussion; thus we are ultimately dealing with problems about sheaves of local C-algebras on X_0.

A family of deformations of X_0 over $S^{(n)}$ (as for example $X^{(n)}$) can be described sheaf theoretically as follows. Let $\mathcal{F}^{(n)}$ (or $\mathcal{F}_S^{(n)}$) denote the sheaf on X_0 whose sections over an open set U are the automorphisms of $U \times S^{(n)}$ over $S^{(n)}$ which are the identity on $U \times S^{(0)} = U$. $\mathcal{F}^{(n)}$ is a sheaf of (non-abelian) groups and an isomorphism class of deformations of X_0 over $S^{(n)}$ corresponds exactly to an element of $H^1(X_0, \mathcal{F}^{(n)})$ (a cohomology set, see [4]). The restriction to $U \times S^{(n-1)}$ of an automorphism of $U \times S^{(n)}$ over $S^{(n)}$ is an automorphism of $U \times S^{(n-1)}$ over $S^{(n-1)}$. We have, therefore, an exact sequence of sheaves

$$0 \longrightarrow \mathcal{K}^{(n)} \longrightarrow \mathcal{F}^{(n)} \longrightarrow \mathcal{F}^{(n-1)} \longrightarrow 1 \, .$$

A method sketched below can be used to show that $\mathcal{K}^{(n)}$ is a sheaf of abelian groups in the center of $\mathcal{F}^{(n)}$ which is isomorphic to $\Theta_0 \otimes_{\mathbf{C}} \mathfrak{M}^n/\mathfrak{M}^{n+1}$. Thus a family of deformations over $S^{(n-1)}$ can be extended to a family of deformations over $S^{(n)}$ if and only if a certain element of $H^2(X_0, \Theta_0) \otimes \mathfrak{M}^n/\mathfrak{M}^{n+1}$ vanishes [4]. Two extensions differ by an element of $H^1(X_0, \Theta_0) \otimes \mathfrak{M}^n/\mathfrak{M}^{n+1}$ (cf. [5, 14–24]).

We will primarily be concerned with a problem of the following type. We are given some sort of object on $X^{(n-1)}/S^{(n-1)}$ (usually an automorphism or vector field) and the problem will be to extend this to an object on $X^{(n)}/S^{(n)}$. The objects on $X^{(n)}/S^{(n)}$ will be sections of sheaves of groups $\mathcal{A}^{(n)}$ on X_0, where we have an exact sequence

$$0 \longrightarrow \mathcal{K}^{(n)} \longrightarrow \mathcal{A}^{(n)} \longrightarrow \mathcal{A}^{(n-1)} \longrightarrow 1\,.$$

Thus a section of $\mathcal{A}^{(n-1)}$ is extendible if and only if its image in $H^1(X_0, \mathcal{K}^{(n)})$ vanishes. This image, therefore, represents the "obstruction to extending the section". Many problems in the theory of deformations of structure are of this type. This explains the occurrence of an abundance of conditions of a cohomological nature in deformation problems. The point of view expressed here is more or less explicit in [3], [5], and [11]; but it is also implicit in the earlier work of Kodaira and Spencer in their use of power series methods. The results obtained by reasoning of this type are formal and it is usually a matter of some delicacy to show that the formal construction "converges" to an actual one (which problem happily does not arise in the present paper).

We will now indicate a method, based upon a generalization of Taylor series expansion, which can be used in discussions of the above type. This method describes the sheaf morphisms and kernels non-intrinsically, but has the advantage of providing fairly explicit information reducing most proofs to simple calculations.

A basis for $\mathcal{O}_S/\mathfrak{M}_S^{n+1}$ may be obtained in the following way. Fix a neat imbedding of S in $\mathbf{C}^a$ (with $o \in S$ going to $0 \in \mathbf{C}^a$). Denote by $s_1, \cdots, s_a$ the coordinate functions on $\mathbf{C}^a$ (or their restrictions to an appropriate subspace, depending on the context). $\mathfrak{M}^k$ is generated by the s^I with $|I| = k$ (we employ multi-index notation). Choose, for each k, a set $\mathcal{I}_k$ of multi-indices with $|I| = k$ so that the images, of the $s^I(\mathrm{I} \in \mathcal{I}_k)$ provide a basis for $\mathfrak{M}^k/\mathfrak{M}^{k+1}$. $\{s^I \mid I \in \mathcal{I}_k; k = 0, \cdots, n\}$ is a basis for $\mathcal{O}/\mathfrak{M}^{n+1}$. If U is open in $\mathbf{C}^m$,

$$H^0(U \times S^{(n)}, \mathcal{O}) \cong H^0(U, \mathcal{O}) \otimes \mathcal{O}_S/\mathfrak{M}_S^{n+1}$$

[5, 10–10]. Thus every holomorphic function f on $U \times S^{(n)}$ may be uniquely written in the form $f = \sum f_I s^I$ where f_I is a holomorphic function on U and where the sum extends over $I \in \mathcal{I}_k$ for $k = 0, \cdots, n$. The restriction of f to $U \times S^{(n-1)}$ is obtained by omitting the terms with $I \in \mathcal{I}_n$. This representation of a holomorphic function is the desired generalization of the Taylor polynomial. The reader is warned that, if S is singular at o, the "Taylor polynomials" do not multiply like polynomials; the s_i are not independent on S.

For the sheaves $\mathcal{A}^{(n)}$ of interest to us, a section over an open set U will be described in local coordinates by morphisms

$$U \cap U_i \times S^{(n)} \longrightarrow \mathbf{C}^m$$

which are required to satisfy compatibility conditions on

$$U \cap U_i \cap U_j \times S^{(n)}$$

(which may depend on the given family X/S). A section of $\mathcal{A}^{(n-1)}$ over $U \cap U_i$ can be extended to a section of $\mathcal{A}^{(n)}$ by using the "Taylor polynomial" representation and adding a linear combination of the $s^I(I \in \mathcal{I}_n)$ with coefficients holomorphic on $U \cap U_i$. The sheaf morphism $\mathcal{A}^{(n)} \rightarrow \mathcal{A}^{(n-1)}$ will therefore be surjective. If a section of $\mathcal{A}^{(n)}$ restricts to the identity on $\mathcal{A}^{(n-1)}$ the coefficients of the $s^I(I \in \mathcal{I}_n)$ in the local expressions will be required (by the compatibility conditions on $\mathcal{A}^{(n)}$) to describe a section of a sheaf $\mathcal{K}^{(n)}$. $\mathcal{K}^{(n)}$ is therefore the kernel of the morphism $\mathcal{A}^{(n)} \rightarrow \mathcal{A}^{(n-1)}$.

2. Automorphisms and the moduli problem

We remind the reader that "automorphism of X/S" will mean automorphism *of the family of deformations* (and hence will be the identity on X_0).

Let X/T be the Kuranishi family and $f, g\colon (S, o) \rightarrow (T, o)$ germs of morphisms such that X_f and X_g are isomorphic (as germs of families of deformations of X_0). We will show that if any automorphism of $X_g^{(n-1)}$ can be extended to $X_g^{(n)}$ we must have $f = g$. If $f^{(n)}$, $g^{(n)}$ denote the restrictions of f, g to $S^{(n)}$, then $f = g$ if and only if $f^{(n)} = g^{(n)}$ for all n. Our proof will show $f^{(n)} = g^{(n)}$ by induction on n ($f^{(0)} = g^{(0)}$ is obvious). The following theorem will be useful.

THEOREM 2.1. *Let $g^{(n)}\colon S^{(n)} \rightarrow T^{(n)}$. There is a $1-1$ correspondence between the set of morphisms $f^{(n)}\colon S^{(n)} \rightarrow T^{(n)}$ such that $f^{(n-1)} = g^{(n-1)}$ and*

$$(\mathfrak{M}_T/\mathfrak{M}_T^2)^* \otimes_{\mathbf{C}} (\mathfrak{M}_S^n/\mathfrak{M}_S^{n+1}) \,.$$

$g^{(n)}$ corresponds to 0 *and serves to "fix the origin".*

PROOF. We will suppress the superscript $^{(n)}$. The underlying topological maps are obvious, so f, g may be considered as homomorphisms of local C-algebras $f, g\colon \mathcal{O}_T/\mathfrak{M}_T^{n+1} \to \mathcal{O}_S/\mathfrak{M}_S^{n+1}$. These are completely determined by the homomorphisms which they induce $\mathfrak{M}_T/\mathfrak{M}_T^{n+1} \to \mathfrak{M}_S/\mathfrak{M}_S^{n+1}$. By the hypothesis, the compositions of the latter with the natural homomorphism $\mathfrak{M}_S/\mathfrak{M}_S^{n+1} \to \mathfrak{M}_S/\mathfrak{M}_S^{n}$ coincide. Thus $f = g + jh$ for some homomorphism $h\colon \mathfrak{M}_T/\mathfrak{M}_T^{n+1} \to \mathfrak{M}_S^n/\mathfrak{M}_S^{n+1}$ (j is the inclusion $\mathfrak{M}_S^n/\mathfrak{M}_S^{n+1} \to \mathfrak{M}_S/\mathfrak{M}_S^{n+1}$). We have the diagram

$$\begin{array}{ccccccc} 0 \longrightarrow \mathfrak{M}_T^2/\mathfrak{M}_T^{n+1} & \xrightarrow{\alpha} & \mathfrak{M}_T/\mathfrak{M}_T^{n+1} & \xrightarrow{\beta} & \mathfrak{M}_T/\mathfrak{M}_T^2 \longrightarrow 0 \\ & & \big\downarrow h & & \\ & & \mathfrak{M}_S^n/\mathfrak{M}_S^{n+1} \,. & & \end{array}$$

Since f, g are algebra homomorphisms, it is easily seen that $h^\alpha = 0$. Thus we have a vector space homomorphism $\sigma\colon \mathfrak{M}_T/\mathfrak{M}_T^2 \to \mathfrak{M}_S^n/\mathfrak{M}_S^{n+1}$ such that $h = \sigma\beta$. Conversely if σ is given, $f = g + j\sigma\beta$ has the desired properties.

Let X/S be a family of deformations of X_0. An automorphism of $X^{(n)}/S^{(n)}$ is a section of a sheaf $\mathcal{G}^{(n)}$ over X_0. A section of $\mathcal{G}^{(n)}$ over an open set U is described by automorphisms G_i of $U \cap U_i \times S^{(n)}$, which satisfy

(2.1) $G_i \circ F_{ij} = F_{ij} \circ G_j$ as automorphisms of $U \cap U_i \cap U_j \times S^{(n)}$,

(2.2) $G_{i|0} =$ identity,

where $G_{i|0}$ is the restriction of G_i to $U \cap U_i \times S^{(0)} = U \cap U_i$. $G_i = (g_i, p_2)$ where $g_i\colon U \cap U_i \times S^{(n)} \to \mathbf{C}^m$. A local extension of g_i, hence of G_i, is always possible. We have, therefore, an exact sequence

$$0 \longrightarrow \mathcal{K}^{(n)} \longrightarrow \mathcal{G}^{(n)} \longrightarrow \mathcal{G}^{(n-1)} \longrightarrow 1$$

and the method described in § 1 shows that $\mathcal{K}^{(n)} = \Theta_0 \otimes \mathfrak{M}_S^n/\mathfrak{M}_S^{n+1}$; (cf. [5, 14–22]).

THEOREM 2.2. *Let X/T be the Kuranishi family, $f, g\colon (S, o) \to (T, o)$ germs of morphisms inducing isomorphic families. If, for all n, any automorphism of $X_g^{(n-1)}$ can be extended to $X_g^{(n)}$, then $f = g$.*

PROOF. The isomorphism $\Phi\colon X_f/S \to X_g/S$ induces isomorphisms $\Phi^{(n)}\colon X_f^{(n)} \to X_g^{(n)}$. If we assume $f^{(n-1)} = g^{(n-1)}$, the changes of coordinates for $X_f^{(n-1)}$ and $X_g^{(n-1)}$ are the same. $\Phi^{(n-1)}$ may therefore be considered an automorphism of $X_g^{(n-1)}$. Let $\omega \in H^1(X_0,\Theta_0) \otimes \mathfrak{M}_S^n/\mathfrak{M}_S^{n+1}$ denote the obstruction to extending $\Phi^{(n-1)}$ to $X_g^{(n)}$. Using $X_g^{(n)}$ to "fix the origin" we obtain an element $\gamma \in H^1(X_0, \Theta_0) \otimes \mathfrak{M}_S^n/\mathfrak{M}_S^{n+1}$ corresponding to $X_f^{(n)}$. Let σ correspond to $f^{(n)}$ as in Theorem 2.1, and let $\rho: (\mathfrak{M}_T/\mathfrak{M}_T^2)^* \to H^1(X_0, \Theta_0)$ be the Kodaira-Spencer map (which is an isomorphism in this case). It can be shown that $\omega = \gamma = (\rho \otimes 1)(\sigma)$. By hypothesis $\omega = 0$ thus $\sigma = 0$. Theorem 2.1 shows $f^{(n)} = g^{(n)}$, which completes the induction.

Remark. Suppose, on the other hand, $X_g^{(n-1)}$ has a non-extendible automorphism $\Phi^{(n-1)}$. Let ω be the obstruction to extending $\Phi^{(n-1)}$ and $\sigma = (\rho \otimes 1)^{-1}(\omega)$. σ determines a morphism $f^{(n)}: S^{(n)} \to T^{(n)}$, $f^{(n)} \neq g^{(n)}$, which induces a family isomorphic to $X_g^{(n)}$. The hypothesis of Theorem 2.2 is therefore also a necessary condition for X_0 to have a space of moduli.

Conclusion. X_0 has a space of moduli if and only if every family X/S of deformations of X_0 has the property that automorphisms of $X^{(n-1)}/S^{(n-1)}$ can always be extended to automorphisms of $X^{(n)}/S^{(n)}$.

3. Automorphisms and vector fields

In this section we will establish a correspondence between automorphisms of $X^{(n)}/S^{(n)}$ and vertical vector fields on $X^{(n)}/S^{(n)}$ which vanish on X_0. This permits us to replace the extension problem for automorphisms by a more manageable extension problem for vertical vector fields. $\mathcal{G}^{(n)}$ will denote the sheaf of germs of automorphisms of $X^{(n)}/S^{(n)}$ defined in § 2.

Let $\Theta^{(n)}$ denote the sheaf of germs of vertical vector fields on $X^{(n)}/S^{(n)}$. $\Theta^{(n)}$ is a sheaf of abelian groups on X_0 whose sections over an open set U are described by morphisms $\varphi_i: U \cap U_i \times S^{(n)} \to \mathbf{C}^m$ satisfying

$$\varphi_i \circ F_{ij} = (\partial f_{ij}/\partial z_j)\cdot \varphi_j \,, \tag{3.1}$$

where $F_{ij} = (f_{ij}, p_2)$ and matrix notation is employed. We will be interested in the subsheaf $\mathcal{V}^{(n)}$ of germs of vertical vector fields vanishing on X_0, i.e., vertical vector fields such that

$$\varphi_{i|0} = 0 \,. \tag{3.2}$$

Let $\varphi_i: U \cap U_i \times S^{(n)} \to \mathbf{C}^m$ be a morphism satifying (3.2). We define a 1-parameter family of morphisms $g_{t,i}: U \cap U_i \times S^{(n)} \to \mathbf{C}^m$ by

$$\begin{aligned} g_{t,i} &= p_1 + \int_0^t \varphi_i \circ (g_{\tau,i}, p_2)d\tau \\ g_{t,i|0} &= \text{identity} \,, \end{aligned} \tag{3.3}$$

where $p_1: U \cap U_i \times S^{(n)} \to \mathbf{C}^m$ is the projection. A word must be said about the meaning of (3.3). If $h_i: U \cap U_i \times S^{(k)} \to U \cap U_i$ is given $(0 \leqq k \leqq n-1)$, $\varphi_i \circ (h_i, p_2)$ is *a priori* only a morphism $U \cap U_i \times S^{(k)} \to \mathbf{C}^m$. h_i, however, may be extended (in many ways) to a morphism

$$\tilde{h}_i: U \cap U_i \times S^{(k+1)} \longrightarrow U \cap U_i \,.$$

$\varphi_i \circ (\tilde{h}_i, p_2)$ is a morphism $U \cap U_i \times S^{(k+1)} \to \mathbf{C}^m$ which, because of (3.2), is independent of the choice of $\tilde{h}_i$. Thus $\varphi_i \circ (h_i, p_2)$ may itself be considered a morphism $U \cap U_i \times S^{(k+1)} \to \mathbf{C}^m$. In view of these remarks, (3.3) is an inductive definition of $g_{t,i}$.

THEOREM 3.1. *If $\varphi \in H^0(U, \mathcal{V}^{(n)})$ is represented by φ_i, then the $G_{t,i} = (g_{t,i}, p_2)$ represent an element $G_t \in H^0(U, \mathcal{G}^{(n)})$ for all t.*

PROOF. The proof is by induction on n, the assertion being obvious for $n = 0$. For $t = 0$, $G_t =$ identity so the assertion is true for $t = 0$. We substitute $G_{t,i}$ into (2.1) and differentiate with respect to t. On the left we obtain $\varphi_i \circ G_{t,i} \circ F_{ij}$, and on the right $(\partial f_{ij}/\partial z_j \cdot \varphi_j) \circ G_{t,j}$. To complete the proof we need only show that these coincide. By induction it may be assumed that $G_{t,i} \circ F_{ij}$ and $F_{ij} \circ G_{t,j}$ coincide on $U \cap U_i \cap U_j \times S^{(n-1)}$. As observed above, this implies that $\varphi_i \circ G_{t,i} \circ F_{ij} = \varphi_i \circ F_{ij} \circ G_{t,j}$ on $U \cap U_i \cap U_j \times S^{(n)}$. The result now follows from (3.1) by composing both sides with $G_{t,j}$.

In particular, $\varphi \mapsto G_1$ defines a morphism of sheaves of sets $\alpha: \mathcal{V}^{(n)} \to \mathcal{G}^{(n)}$ (which commutes with the restrictions to $X^{(n-1)}$).

THEOREM 3.2. *α is an isomorphism.*

PROOF. We show that $\alpha_U: H^0(U, \mathcal{V}^{(n)}) \to H^0(U, \mathcal{G}^{(n)})$ is bijective for $U \subset U_i$. Using the "Taylor polynomial" representations for φ_i, g_i, (3.3) defines the coefficient of s^I in g_i recursively as a function of the coefficients for φ_i and "lower order" coefficients for g_i. Thus, finally, we can define the coefficients for g_i as functions of the coefficients for φ_i. These relations are invertible, however, and allow us to compute the coefficients for φ_i as functions of the coefficients for g_i. Hence, for $U \subset U_i$, α_U is bijective.

Conclusion. $H^0(X_0, \mathcal{G}^{(n)}) \to H^0(X_0, \mathcal{G}^{(n-1)})$ is surjective if and only if $H^0(X_0, \mathcal{V}^{(n)}) \to H^0(X_0, \mathcal{V}^{(n-1)})$ is surjective. Thus X_0 has a space of moduli if and only if the latter homomorphism is surjective for every family X/S of deformations of X_0.

4. Obstructions and the moduli problem

Definition 4.1. We say that the 1st extension problem is solvable for X/S if $H^0(X_0, \mathcal{V}^{(n)}) \to H^0(X_0, \mathcal{V}^{(n-1)})$ is surjective for all n.

Definition 4.2. We say that the 2nd extension problem is solvable for X/S if $H^0(X_0, \Theta^{(n)}) \to H^0(X_0, \Theta_0)$ is surjective for all n.

We have shown in the preceding sections that X_0 has a space of moduli if and only if the 1st extension problem is solvable for every family of deformations of X_0. In this section we will show that this is equivalent to the requirement that the 2nd extension problem be solvable for the Kuranishi family. We then examine conditions for the solvability of the 2nd extension problem.

PROPOSITION 4.1. *If the* 2nd *extension problem is solvable for* X/S, *then*

$$H^0(X_0, \Theta^{(n)}) \longrightarrow H^0(X_0, \Theta^{(n-1)})$$

is surjective for all n.

PROOF. If $n = 1$, the assertion follows from the hypothesis. Assume that the proposition has been proved for $n - 1$. Since $H^0(X_0, \Theta^{(n)}) \to H^0(X_0, \Theta_0)$ is surjective, assume that $H^0(X_0, \Theta^{(n)}) \to H^0(X_0, \Theta^{(k-1)})$ is surjective. Denote by $\mathcal{V}_k^{(n)}$ the kernel of the morphism $\Theta^{(n)} \to \Theta^{(k-1)}$. We have the following commutative diagram with exact rows.

$$\begin{array}{ccccccccc} 0 & \longrightarrow & H^0(X_0, \mathcal{V}_k^{(n)}) & \longrightarrow & H^0(X_0, \Theta^{(n)}) & \longrightarrow & H^0(X_0, \Theta^{(k-1)}) & \longrightarrow & 0 \\ & & \downarrow & & \downarrow & & \| & & \\ 0 & \longrightarrow & H^0(X_0, \mathcal{K}^{(k)}) & \longrightarrow & H^0(X_0, \Theta^{(k)}) & \longrightarrow & H^0(X, \Theta^{(k-1)}) & \longrightarrow & 0 . \end{array}$$

$H^0(X_0, \mathcal{K}^{(k)}) \cong H^0(X_0, \Theta_0) \otimes \mathfrak{M}^k/\mathfrak{M}^{k+1}$ while $H^0(X_0, \mathcal{V}_k^{(n)})$ is the group of vertical vector fields on $X^{(n)}/S^{(n)}$ which vanish on $X^{(k-1)}$. If we use "Taylor polynomials", the first vertical homomorphism can be seen, under the hypothesis, to be surjective. It follows that the second vertical homomorphism is surjective. The proof is now complete by induction.

COROLLARY. *If the* 2nd *extension problem is solvable for* X/S, *then the* 1st *extension problem is also solvable for* X/S.

PROPOSITION 4.2. *If the* 2nd *extension problem is solvable for* X/T *and* $f\colon (S, o) \to (T, o)$, *then the* 2nd *extension problem is solvable for* X_f/S.

Remark. The corresponding assertion for the 1st extension problem is false.

PROOF. f induces a natural morphism $f^\#\colon \Theta_X^{(n)} \to \Theta_{X_f}^{(n)}$. The proof follows from the commutative diagram

$$\begin{array}{ccccc} H^0(X_0, \Theta_X^{(n)}) & \longrightarrow & H^0(X_0, \Theta_0) & \longrightarrow & 0 \\ \downarrow & & \| & & \\ H^0(X_0, \Theta_{X_f}^{(n)}) & \longrightarrow & H^0(X_0, \Theta_0) \,. & & \end{array}$$

COROLLARY. *The* 2nd *extension problem is solvable for every family if and only if it is solvable for the Kuranishi family.*

PROPOSITION 4.3. *The* 1st *extension problem is solvable for every family if and only if the* 2nd *extension problem is solvable for every family.*

PROOF. The sufficiency is a consequence of the Corollary to Proposition 4.1. Suppose the 2nd extension problem is not solvable for a family X/S. Let $T = S \times \mathbf{C}$, f the projection to the 1st factor, t the projection to the 2nd factor. If φ is a non-extendible vector field on X_0 for family X/S, $tf^{\#}(\varphi)$ is a non-extendible vector field for the family X_f/T which vanishes on X_0. The 1st extension problem is, therefore, not solvable for X_f/T.

As a consequence of the preceding results, we obtain the following theorem.

THEOREM 4.1. *X_0 has a space of moduli if and only if the* 2nd *extension problem is solvable for the Kuranishi family X/T.*

COROLLARY. *If $H^0(X_0, \Theta_0) = 0$, then X_0 has a space of moduli.*

The 2nd extension problem is an extension problem of the type studied by Griffiths in [3]. If $\varphi^{(n)}$ is a vertical vector field on $X^{(n)}/T^{(n)}$, the obstruction to extending $\varphi^{(n)}$ to a vertical vector field on $X^{(n+1)}/T^{(n+1)}$ is an element $\omega \in H^1(X_0, \Theta_0) \otimes \mathfrak{M}_T^n/\mathfrak{M}_T^{n+1}$. Thus $\omega = \sum \omega_I t^I (I \in \mathcal{I}_n)$ with $\omega_I \in H^1(X_0, \Theta_0)$. Let L_{n+1} be the subspace of $H^1(X_0, \Theta_0)$ generated by the ω_I for various $\varphi^{(n)}$. L_1 is the image of the bracket operation

$$[\ ,\ ]: H^0(X_0, \Theta_0) \otimes H^1(X_0, \Theta_0) \longrightarrow H^1(X_0, \Theta_0)\,, \qquad \text{(see [3])}.$$

$L_1 = L_2 = \cdots = 0$ is necessary and sufficient for X_0 to have a space of moduli. If (T, o) is non-singular we have $L_1 \subset L_2 \subset \cdots$.

THEOREM 4.2. *If the Kuranishi space (T, o) is reduced, then* $\dim H^0(X_t, \Theta_t)$ *independent of t near o is sufficient for the existence of a space of moduli for X_0.*

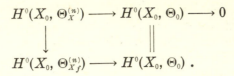

PROOF. By [2, p. 63] any vector field on X_0 can be extended to a vertical vector field on $X|_U$ for some neighborhood U of o in T. In particular an extension to $X^{(n)}/T^{(n)}$ is possible for all n.

COROLLARY. *If* (T, o) *is reduced and* $\dim H^1(X_t, \Theta_t)$ *is independent of* t *near* o, X_0 *has a space of moduli.*

Remark. 1. The condition given in the corollary is essentially that given by Griffiths for the solvability of the 2nd extension problem. When (T, o) is non-singular, it is equivalent to the requirement that X/T be "effectively parametrized" used by Kodaira and Spencer in their definition of space of moduli [7, p. 382]. It is stronger than the condition given in the theorem.

2. Douady has informed me that it is possible to prove the following generalization of Grauert's theorem.

THEOREM. *If* (T, o) *is reduced,* $\bigcup H^0(X_t, \Theta_t)$ *is, for* t *in a suitable neighborhood of* o, *the kernel of a vector bundle homomorphism* $u: E_1 \to E_2$ *where* E_1 *has dimension* $\dim H^0(X_0, \Theta_0)$.

If $\dim H^0(X_t, \Theta_t)$ is not independent of t near o, there would be a vector field on X_0 which is not extendible to any neighborhood of o in T. Griffiths has shown, however, that a formal extension implies an actual extension [3, p. 125]. Thus the condition given in Theorem 4.2 is necessary and sufficient.

COLUMBIA UNIVERSITY

REFERENCES

[1] A. DOUADY, *Le problème des modules pour les variétés analytiques complexes,* Seminaire Bourbaki 1964/5, Expose #277.

[2] H. GRAUERT, *Ein Theorem der analytischen Garbentheorie und die Modulräume komplexer Structuren,* Publ. Math. I.H.E.S. No. 5, 1960.

[3] P. A. GRIFFITHS, "The extension problem for compact submanifolds of complex manifolds: I", in Proc. Conf. on Complex Analysis, Minneapolis, 1964, Springer, New York, 1965.

[4] A. GROTHENDIECK, A general theory of fibre spaces with structure sheaf, University of Kansas Report #4, 2nd ed., 1958.

[5] ———, *Techniques de construction en géométrie analytique I–X,* Séminaire Cartan 1960/1 2nd ed., Paris, 1962.

[6] R. C. GUNNING and H. ROSSI, Analytic Functions of Several Complex Variables, Prentice-Hall, Englewood Cliffs, N.J., 1965.

[7] K. KODAIRA and D. C. SPENCER, *On deformations of complex analytic structures*: I, II, Ann. of Math. **67** (1958), 328–466.

[8] M. KURANISHI, *On the locally complete families of complex analytic structures,* Ann. of Math. **75** (1962), 536-577.

[9] ———, "New proof for the existence of locally complete families of

complex structures", in Proc. Conf. on Complex Analysis, Minneapolis, 1964, Springer, New York, 1965.
[10] ———, "A note on families of complex structures", (in this volume).
[11] D. C. SPENCER, "Some remarks on homological analysis and structures", in Proc. of Symposia in Pure Mathematics, Vol. III-Differential Geometry, American Mathematical Society, Providence, R. I., 1961.

(Received October 8, 1968)